T0235188

Lecture Notes in Computer Science　　10125

Commenced Publication in 1973
Founding and Former Series Editors:
Gerhard Goos, Juris Hartmanis, and Jan van Leeuwen

Editorial Board

More information about this series at http://www.springer.com/series/7412

César Beltrán-Castañón · Ingela Nyström
Fazel Famili (Eds.)

Progress in Pattern Recognition, Image Analysis, Computer Vision, and Applications

21st Iberoamerican Congress, CIARP 2016
Lima, Peru, November 8–11, 2016
Proceedings

 Springer

Editors
César Beltrán-Castañón
Pontificia Universidad Católica del Perú
Lima
Peru

Fazel Famili
University of Ottawa
Ottawa, ON
Canada

Ingela Nyström
Uppsala University
Uppsala
Sweden

ISSN 0302-9743   ISSN 1611-3349   (electronic)
Lecture Notes in Computer Science
ISBN 978-3-319-52276-0   ISBN 978-3-319-52277-7   (eBook)
DOI 10.1007/978-3-319-52277-7

Library of Congress Control Number: 2017930202

LNCS Sublibrary: SL6 – Image Processing, Computer Vision, Pattern Recognition, and Graphics

Printed on acid-free paper

This Springer imprint is published by Springer Nature
The registered company is Springer International Publishing AG
The registered company address is: Gewerbestrasse 11, 6330 Cham, Switzerland

Preface

The 21st Iberoamerican Congress on Pattern Recognition CIARP 2016 (Congreso IberoAmericano de Reconocimiento de Patrones), held during November 8–11, 2016, was the 21st edition of a yearly event organized by scientific associations of Iberoamerican countries in this field. In this special anniversary edition, as in previous years, the congress received contributions from many countries beyond Iberoamerica. The papers presented research results in the areas of pattern recognition, biometrics, image processing, computer vision, speech recognition, and remote sensing to name a few. The papers tackle theoretical as well as applied contributions in many fields related to the main topics of the conference. In this way, CIARP 2016 continued the tradition of an event that fosters scientific exchange, discussions, and cooperation among researchers.

CIARP 2016 received 131contributions authored by researchers from 29 countries, 11 of which are Iberoamerican countries. These contributions were reviewed in a double-blind process and 69 papers were accepted. Following tradition, CIARP 2016 was a single-track conference in which 35 papers were selected for oral presentation and 34 were presented in poster sessions. The type of presentation did not imply quality differences. The poster sessions were organized by topic to encourage discussions among authors and attendees.

The IAPR-CIARP Best Paper Award recognizes outstanding contributions and is aimed at acknowledging excellence and originality of both theoretical contributions to and practical applications in the field of pattern recognition and data mining. On the other hand, the CIARP Aurora Pons-Porrata Award is given to a living woman in recognition of her outstanding contribution to the field of pattern recognition or data mining.

Beside the presentation of the 69 selected contributions, four keynotes were given by Professors Yann LeCun (Director of AI Research, Facebook Founding Director of the NYU Center for Data Science), B.S. Manjunath (Director, Center for Bio-image Informatics, University of California), Xiaohui Liu (Design and Physical Sciences, Department of Computer Science, Brunel University London), and George Azzopardi (Intelligent Computer Systems, ICT Faculty University of Malta). CIARP 2016 was organized by the Peruvian Association for Pattern Recognition, including members from the Pontificia Universidad Católica del Perú, with the endorsement of the International Association for Pattern Recognition (IAPR) and the sponsorphip of the following national associations: Argentine Society for Pattern Recognition (SARP-SADIO), the Special Interest Group of the Brazilian Computer Society (SIGPR-SBC), the Chilean Association for Pattern Recognition (AChiRP), the Cuban Association for Pattern Recognition (ACRP), the Mexican Association for Computer Vision, Neural Computing and Robotics (MACVNR), the Spanish Association for

Pattern Recognition and Image Analysis (AERFAI), Uruguayan Association for Pattern Recognition (AURP), and the Portuguese Association for Pattern Recognition (APRP). We acknowledge the work of all members of the Organizing Committee and of the Program Committee for the rigorous work in the reviewing process.

November 2016

César Beltrán Castañón
Ingela Nyström
Fazel Famili

Organization

Scientific Committee

Ingela Nyström   Uppsala Universitet, Sweden
Fazel Famili   University of Ottawa, Canada
César Beltrán Castañón   Pontificia Universidad Católica del Perú, Peru

Steering Committee

Marta Mejail   SARP, Argentina
Helio Cortes Vieira Lopes   SIGPR-SBC, Brazil
Marcelo Mendoza   ACHiRP, Chile
Andrés Gago Alonso   ACRP, Cuba
Eduardo Bayro-Corrochano   MACVNR, Mexico
Cesar A. Beltrán Castañón   APeRP, Peru
Luis Filipi Barbosa de   APRP, Portugal
    Almeida Alexandre
Roberto Paredes Palacios   AERFAI, Spain
Álvaro Pardo   APRU, Uruguay

Local Committee: PUCP

Andrés Melgar Sasieta   Marco Soldevilla Cabezudo
Hugo Alatrista Salas   Eduardo Cortavitarte
Ivan Sipirán Mendoza   Victor Cárdenas Castañeda
Johan Baldeón Medrano   Manuel Solórzano C.
Layla Hirsh Martínez   Kevin Baba Yamakawa
Sofia Khlebnikov Nuñez   Renato Hermoza Aragonés
Rosario Medina Rodriguez   Emilio García Ríos
Ana Paula Galarreta   Edmundo Aparicio
María Elena Gonzáles   Christian Pérez
Arturo Oncevay Marcos   Kervy Rivas
Fernando Alva Manchego   Erasmo Montoya

Support Committee: PUCP

Patricia Harman
Araccelly Romero
Milca Bueno R.
Irma Palpan
Gloria Vargas Q.
Jesared Suarez
Iliana Castillo
Estrella Cuadrado

Program Committee

Daniel Acevedo	Universidad de Buenos Aires, Argentina
Cecilia Aguerrebere	Duke University, USA
Hugo Alatrista	Pontificia Universidad Católica del Perú, Peru
Enrique Alegre	University of Leon, Spain
Luis Alexandre	Universidade da Beira Interior, Portugal
Marco Alvarez	Rhode Island University, USA
Arnaldo Araujo	Universidade Federal de Minas Gerais, Brazil
Leticia Arco	Asociación Cubana de Reconocimiento de Patrones, Cuba
Akira Asano	Kansai University, Japan
George Azzopardi	University of Malta, Malta
Rafael Bello	Universidad Central de Las Villas, Cuba
Olga Bellon	Universidade Federal do Parana, Brazil
César Beltrán Castañón	Pontificia Universidad Católica del Perú, Peru
Rafael Berlanga	Universitat Jaume I, Spain
Paola Bermolen	Universidad de la República, Uruguay
Isabelle Bloch	ENST – CNRS, France
Jean-Francois Bonastre	Université d'Avignon et des Pays de Vaucluse, France
Gunilla Borgefors	Uppsala University, Sweden
Adrian Bors	University of York, UK
Henri Bouma	Toegepast Natuurwetenschappelijk Onderzoek, The Netherlands
Odemir-M. Bruno	Universidade de Sao Paulo, Brazil
Maria-Elena Buemi	Universidad de Buenos Aires, Argentina
José-Ramón Calvo De Lara	CENATAV, La Habana, Cuba
Guillermo Cámara Chavez	Universidade Federal de Ouro Preto, Brazil
Aldo Camargo	Pontificia Universidad Católica del Perú, Peru
Germán Capdehourat	Universidad de la República, Uruguay
Guillermo Carbajal	Universidad de la República, Uruguay
Miguel Carrasco	Universidad Adolfo Ibanez, Chile
Jesus-Ariel Carrasco-Ochoa	INAOE, Mexico
Benjamin Castañeda	Pontificia Universidad Católica del Perú, Peru

Maria-Jose Jimenez	Universidad de Sevilla, Spain
Javier Jo	Texas A&M University, College Station, USA
Pedro-Real Jurado	Universidad de Sevilla, Spain
Martin Kampel	Vienna University of Technology, Austria
Gisela Klette	Auckland University of Technology, New Zealand
Reinhard Klette	Auckland University of Technology, New Zealand
Vitaly Kober	CICESE, Mexico
Denis Laurendeau	Université Laval, Canada
Manuel Lazo-Cortés	INAOE, Puebla, Mexico
Federico Lecumberry	Universidad de la República, Uruguay
Yan LeCun	NYU Center for Data Science, USA
Xiaohui Liu	Brunel University, UK
Helio Lopes	PUC-Rio, Brazil
Filip Malmberg	Uppsala University, Sweden
B.S. Manjunath	University of California, USA
Rebeca Marfil	University of Malaga, Spain
Manuel-J. Marín-Jiménez	University of Cordoba, Spain
Ricardo Marroquim	Universidade Federal do Rio de Janeiro, Brazil
José-Fco. Martínez-Trinidad	INAOE, Puebla, Mexico
Nelson Mascarenhas	Universidade Federal de Sao Carlos, Brazil
José-E. Medina-Pagola	CENATAV, Cuba
Jesús Mena-Chalco	Federal University of ABC, Brazil
Marcelo Mendoza	Universidad Técnica Federico Santa María, Chile
David Menotti	Universidade Federal de Ouro Preto, Brazil
Domingo Mery	Pontificia Universidad Católica de Chile, Chile
Gustavo Meschino	Universidad Nacional de Mar de Plata, Argentina
Manuel Montes-y-Gómez	INAOE, Mexico
Claudio Moraga	Universität Dortmund, Germany
Eduardo Morales	INAOE, Mexico
Vadim Mottl	Russian Academy of Sciences, Russia
Pablo Musé	Universidad de la República, Uruguay
Pablo Negri	CONICET, Argentina
Heinrich Niemann	University of Erlangen-Nuremberg, Germany
Mark Nixon	University of Southampton, UK
Ingela Nyström	Uppsala University, Sweden
Lawrence O'Gorman	Alcatel-Lucent Bell Labs, USA
Kalman Palagyi	University of Szeged, Hungary
Alvaro Pardo	Universidad Católica del Uruguay, Uruguay
Constantinos Pattichia	University of Cyprus, Republic of Cyprus
Carmen Paz Suárez-Araujo	Universidad de Las Palmas de Gran Canaria, Spain
Glauco Pedrosa	Universidade de Sao Paulo, Brazil
Jian Pei	Simon Fraser University, Canada
Francisco-Jose Perales	Universitat de les Illes Balears, Spain
Alfredo Petrosino	Parthenope University of Naples, Italy
Hemerson Pistori	Universidade Católica Dom Bosco, Brazil

Bárbara Poblete   Universidad de Chile, Chile
Ignacio Ponzoni   Universidad Nacional del Sur, Argentina
Javier Preciozzi   Universidad de la República, Uruguay
Petia Radeva   Universitat de Barcelona, Spain
Ignacio Ramirez   Universidad de la Republica, Uruguay
Gregory Randall   Universidad de la República, Uruguay
M Raza Ali   Vision Research Division, Pakistan
Bernadete Ribeiro   University of Coimbra, Portugal
Antonio Rodriguez-Sanchez   University of Innsbruck, Austria
Paul Rosin   Cardiff University, UK
Jose Ruiz-Shulcloper   CENATAV, Cuba
Alessia Saggese   University of Salerno, Italy
Chen Sagiv   Sagiv Tech. Ltd., Israel
Cesar San Martin   Universidad de la Frontera, Chile
Lidia Sánchez González   Universidad de León, Spain
Antonio-José   Universitat Politècnica de València, Spain
  Sánchez-Salmerón
Alberto Sanfeliu   Universitat Politècnica de Catalunya, Spain
Gabriella Sanniti di Baja   Institute for High Performance Computing
                                and Networking, CNR, Italy
Mykola Sazhok   IRTC, Ukraine
William-Robson Schwartz   Federal University of Minas Gerais, Brazil
Giuseppe Serra   University of Florence, Italy
Ivan Sipirán   Pontificia Universidad Católica del Perú, Peru
Rafael Sotelo   Universidad de Montevideo, Uruguay
Axel Soto   Dalhousie University, Halifax, Canada
João-Manuel-R.S. Tavares   INEGI – University of Porto, Portugal
Mariano Tepper   Duke University, USA
Yvan Tupac   Universidad Católica San Pablo, Peru
Herwig Unger   FernUniversität in Hagen, Germany
Gustavo Vazquez   Universidad Catolica del Uruguay, Uruguay
Sergio Velastin   Kingston University, UK
Mario Vento   Università degli Studi di Salerno, Italy

Additional Reviewers

Joel Azzopardi   University of Malta, Malta
Rafael Baeta   Universidad Federal de Minas Gerais, Brazil
Ricardo Barata   University of Coimbra, Portugal
Paola Bermolen   Universidad de la República, Uruguay
André Bindilatti   Universidade Federal de Sao Carlos, Brazil
Pablo Cancela   Universidad de la República, Uruguay
Vincenzo Carletti   University of Salerno, Italy
Violeta Chang   Universidad de Chile, Chile
Henrique Costa   Universidade Federal do Parana, Brazil
Francesco Fontanella   Università degli Studi di Cassino, Italy

Contents

Direction-Based Segmentation of Retinal Blood Vessels 1
  M. Frucci, D. Riccio, G. Sanniti di Baja, and L. Serino

Identifying *Aedes aegypti* Mosquitoes by Sensors and One-Class Classifiers . . . 10
  Vinicius M.A. Souza

Partial Matching of Finger Vein Patterns Based on Point Sets Alignment
and Directional Information 19
  Maria Frucci, Daniel Riccio, Gabriella Sanniti di Baja, and Luca Serino

Highly Transparent and Secure Scheme for Concealing Text Within Audio . . . 27
  Diego Renza, Camilo Lemus, and Dora M. Ballesteros L.

Spatio-Colour Asplünd's Metric and Logarithmic Image Processing
for Colour Images (LIPC) 36
  Guillaume Noyel and Michel Jourlin

Two Compound Random Field Texture Models 44
  Michal Haindl and Vojtěch Havlíček

An Automatic Tortoise Specimen Recognition 52
  Matěj Sedláček, Michal Haindl, and Dominika Formanová

Autonomous Scanning of Structural Elements in Buildings. 60
  B. Quintana, S.A. Prieto, A. Adán, and A.S. Vázquez

Parallel Integer Motion Estimation for High Efficiency Video Coding
(HEVC) Using OpenCL. 68
  Augusto Gomez, Jhon Perea, and Maria Trujillo

Automatic Fruit and Vegetable Recognition Based on CENTRIST
and Color Representation. 76
  Jadisha Yarif Ramírez Cornejo and Helio Pedrini

Scale Sensitivity of Textural Features 84
  Michal Haindl and Pavel Vácha

Noise-Added Texture Analysis 93
  Tuan D. Pham

Breast Density Classification with Convolutional Neural Networks 101
  Pablo Fonseca, Benjamin Castañeda, Ricardo Valenzuela,
  and Jacques Wainer

Community Feature Selection for Anomaly Detection in Attributed Graphs . . .   109
Mario Alfonso Prado-Romero and Andrés Gago-Alonso

Lung Nodule Classification Based on Deep Convolutional Neural Networks. . .   117
Julio Cesar Mendoza Bobadilla and Helio Pedrini

A Hierarchical K-Nearest Neighbor Approach for Volume of Tissue
Activated Estimation   125
*I. De La Pava, J. Mejía, A. Álvarez-Meza, M. Álvarez, A. Orozco,
and O. Henao*

Discriminative Capacity and Phonetic Information of Bottleneck Features
in Speech.   134
Ana Montalvo and José Ramón Calvo

Boosting SpLSA for Text Classification.   142
Julio Hurtado, Marcelo Mendoza, and Ricardo Ñanculef

A Compact Representation of Multiscale Dissimilarity Data
by Prototype Selection.   150
*Yenisel Plasencia-Calaña, Yan Li, Robert P.W. Duin,
Mauricio Orozco-Alzate, Marco Loog, and Edel García-Reyes*

A Kernel-Based Approach for DBS Parameter Estimation   158
*V. Gómez-Orozco, J. Cuellar, Hernán F. García, A. Álvarez, M. Álvarez,
A. Orozco, and O. Henao*

Consensual Iris Segmentation Fusion.   167
Dailé Osorio Roig and Eduardo Garea Llano

Depth Estimation with Light Field and Photometric Stereo Data
Using Energy Minimization   175
Doris Antensteiner, Svorad Štolc, and Reinhold Huber-Mörk

Distributed and Parallel Algorithm for Computing Betweenness Centrality . . .   184
Mirlayne Campuzano-Alvarez and Adrian Fonseca-Bruzón

A New Parallel Training Algorithm for Optimum-Path
Forest-Based Learning.   192
Aldo Culquicondor, César Castelo-Fernández, and João Paulo Papa

Face Composite Sketch Recognition by BoVW-Based
Discriminative Representations   200
*Yenisel Plasencia-Calaña, Heydi Méndez-Vázquez,
and Rainer Larin Fonseca*

Efficient Sparse Approximation of Support Vector Machines Solving
a Kernel Lasso   208
    Marcelo Aliquintuy, Emanuele Frandi, Ricardo Ñanculef,
    and Johan A.K. Suykens

Metric Learning in the Dissimilarity Space to Improve Low-Resolution
Face Recognition   217
    Mairelys Hernández-Durán, Yenisel Plasencia-Calaña,
    and Heydi Méndez-Vázquez

Video Temporal Segmentation Based on Color Histograms
and Cross-Correlation   225
    Anderson Carlos Sousa e Santos and Helio Pedrini

Automatic Classification of Herbal Substances Enhanced with
an Entropy Criterion   233
    Victor Mendiola-Lau, Francisco José Silva Mata, Yoanna Martínez-Díaz,
    Isneri Talavera Bustamante, and Maria de Marsico

Extended LBP Operator to Characterize Event-Address
Representation Connectivity   241
    Pablo Negri

Definition and Composition of Motor Primitives Using Latent
Force Models and Hidden Markov Models   249
    Diego Agudelo-España, Mauricio A. Álvarez, and Álvaro A. Orozco

Similarity Measure for Cell Membrane Fusion Proteins Identification   257
    Daniela Megrian, Pablo S. Aguilar, and Federico Lecumberry

Abnormal Behavior Detection in Crowded Scenes Based on Optical
Flow Connected Components   266
    Oscar E. Rojas and Clesio Luis Tozzi

Identifying Colombian Bird Species from Audio Recordings   274
    Angie K. Reyes, Juan C. Caicedo, and Jorge E. Camargo

GMM Background Modeling Using Divergence-Based Weight Updating   282
    Juan D. Pulgarin-Giraldo, Andres Alvarez-Meza,
    David Insuasti-Ceballos, Thierry Bouwmans,
    and German Castellanos-Dominguez

Bayesian Optimization for Fitting 3D Morphable Models
of Brain Structures   291
    Hernán F. García, Mauricio A. Álvarez, and Álvaro A. Orozco

Star: A Contextual Description of Superpixels for Remote Sensing
Image Classification 300
    Tiago M.H.C. Santana, Alexei M.C. Machado, Arnaldo de A. Araújo,
    and Jefersson A. dos Santos

A Similarity Indicator for Differentiating Kinematic Performance
Between Qualified Tennis Players. 309
    J.D. Pulgarin-Giraldo, A.M. Alvarez-Meza, L.G. Melo-Betancourt,
    S. Ramos-Bermudez, and G. Castellanos-Dominguez

Subsampling the Concurrent AdaBoost Algorithm: An Efficient Approach
for Large Datasets. 318
    Héctor Allende-Cid, Diego Acuña, and Héctor Allende

Deep Learning Features for Wireless Capsule Endoscopy Analysis 326
    Santi Seguí, Michal Drozdzal, Guillem Pascual, Petia Radeva,
    Carolina Malagelada, Fernando Azpiroz, and Jordi Vitrià

Interactive Data Visualization Using Dimensionality Reduction
and Similarity-Based Representations. 334
    P. Rosero-Montalvo, P. Diaz, J.A. Salazar-Castro, D.F. Peña-Unigarro,
    A.J. Anaya-Isaza, J.C. Alvarado-Pérez, R. Therón,
    and D.H. Peluffo-Ordóñez

Multi-labeler Classification Using Kernel Representations and Mixture
of Classifiers 343
    D.E. Imbajoa-Ruiz, I.D. Gustin, M. Bolaños-Ledezma,
    A.F. Arciniegas-Mejía, F.A. Guasmayan-Guasmayan,
    M.J. Bravo-Montenegro, A.E. Castro-Ospina, and D.H. Peluffo-Ordóñez

Detection of Follicles in Ultrasound Videos of Bovine Ovaries. 352
    Alvaro Gómez, Guillermo Carbajal, Magdalena Fuentes,
    and Carolina Viñoles

Decision Level Fusion for Audio-Visual Speech Recognition
in Noisy Conditions 360
    Gonzalo D. Sad, Lucas D. Terissi, and Juan C. Gómez

Improving Nearest Neighbor Based Multi-target Prediction
Through Metric Learning. 368
    Hector Gonzalez, Carlos Morell, and Francesc J. Ferri

An Approximate Support Vector Machines Solver with Budget Control. 377
    Carles R. Riera and Oriol Pujol

Multi-biometric Template Protection on Smartphones: An Approach Based
on Binarized Statistical Features and Bloom Filters   385
   Martin Stokkenes, Raghavendra Ramachandra, Kiran B. Raja,
   Morten Sigaard, Marta Gomez-Barrero, and Christoph Busch

Computing Arithmetic Operations on Sequences of Handwritten Digits   393
   Andrés Pérez, Angélica Quevedo, and Juan C. Caicedo

Automatic Classification of Non-informative Frames in Colonoscopy
Videos Using Texture Analysis.   401
   Cristian Ballesteros, Maria Trujillo, Claudia Mazo, Deisy Chaves,
   and Jesus Hoyos

Efficient Training Over Long Short-Term Memory Networks for Wind
Speed Forecasting.   409
   Erick López, Carlos Valle, Héctor Allende, and Esteban Gil

Classifying Estimated Stereo Correspondences Based
on Delaunay Triangulation   417
   Cristina Bustos, Elizabeth Vargas, and Maria Trujillo

Data Fusion from Multiple Stations for Estimation of PM2.5 in Specific
Geographical Location.   426
   Miguel A. Becerra, Marcela Bedoya Sánchez, Jacobo García Carvajal,
   Jaime A. Guzmán Luna, Diego H. Peluffo-Ordóñez, and Catalina Tobón

Multivariate Functional Network Connectivity for Disorders
of Consciousness   434
   Jorge Rudas, Darwin Martínez, Athena Demertzi, Carol Di Perri,
   Lizette Heine, Luaba Tshibanda, Andrea Soddu, Steven Laureys,
   and Francisco Gómez

Non-parametric Source Reconstruction via Kernel Temporal Enhancement
for EEG Data   443
   C. Torres-Valencia, J. Hernandez-Muriel, W. Gonzalez-Vanegas,
   A. Alvarez-Meza, A. Orozco, and M. Alvarez

Tsallis Entropy Extraction for Mammographic Region Classification   451
   Rafaela Alcântara, Perfilino Ferreira Junior, and Aline Ramos

Edge Detection Robust to Intensity Inhomogeneity: A 7T MRI Case Study . . .   459
   Fábio A.M. Cappabianco, Lucas Santana Lellis, Paulo Miranda,
   Jaime S. Ide, and Lilianne R. Mujica-Parodi

Fine-Tuning Based Deep Convolutional Networks for Lepidopterous
Genus Recognition   467
   Juan A. Carvajal, Dennis G. Romero, and Angel D. Sappa

Selection of Statistically Representative Subset from a Large Data Set   476
Javier Tejada, Mikhail Alexandrov, Gabriella Skitalinskaya,
and Dmitry Stefanovskiy

Non-local Exposure Fusion   484
Cristian Ocampo-Blandon and Yann Gousseau

Analysis of the Geometry and Electric Properties of Brain Tissue
in Simulation Models for Deep Brain Stimulation   493
Hernán Darío Vargas Cardona, Álvaro A. Orozco,
and Mauricio A. Álvarez

Spatial Resolution Enhancement in Ultrasound Images from Multiple
Annotators Knowledge.   502
Julián Gil González, Mauricio A. Álvarez, and Álvaro A. Orozco

How Deep Can We Rely on Emotion Recognition   511
Ana Laranjeira, Xavier Frazão, André Pimentel, and Bernardete Ribeiro

Sparse Linear Models Applied to Power Quality Disturbance Classification . . .   521
Andrés F. López-Lopera, Mauricio A. Álvarez, and Álvaro Á. Orozco

Trading off Distance Metrics vs Accuracy in Incremental
Learning Algorithms   530
Noel Lopes and Bernardete Ribeiro

Author Index   539

Direction-Based Segmentation of Retinal Blood Vessels

M. Frucci[1], D. Riccio[1,2], G. Sanniti di Baja[1], and L. Serino[1(✉)]

[1] Institute for High Performance Computing and Networking,
CNR, Naples, Italy
{maria.frucci,gabriella.sannitidibaja,
luca.serino}@cnr.it, daniel.riccio@unina.it
[2] University of Naples Federico II, Naples, Italy

Abstract. An unsupervised method is introduced for retinal blood vessels segmentation. The direction map is built by assigning to each pixel a discrete direction out of twelve possible ones. Under- and over-segmented images are obtained by applying two different threshold values to the direction map. Almost all foreground pixels in the under-segmented image can be taken as vessel pixels. Missing vessel pixels in the under-segmented image are recovered by using the over-segmented image. The method has been tested on the DRIVE dataset [1] producing satisfactory results, and its performance has been compared to that of other unsupervised methods.

Keywords: Retinal image · Blood vessel segmentation · Direction map

1 Introduction

Automatic procedures to analyze retinal blood vessels are useful in ophthalmology to allow an early diagnosis of a number of diseases, such as diabetic retinopathy, arteriosclerosis, hypertension, cardiovascular diseases and stroke [2, 3]. The automatic analysis of the structure of retinal vessels is also useful in biometrics [4], since the structure is different for each individual and even for the left and the right eye of the same person. In the literature, segmentation techniques based on matched filters, e.g., [5], wavelet transform, e.g., [6, 7], line detectors, e.g., [8–12], and morphological image processing [8, 9, 13–16] are available. In this paper we present an unsupervised retinal blood vessels segmentation method based on directional information. Each pixel of the green channel G of a RGB retinal image is assigned a direction out of twelve possible discrete directions. Two different threshold values are then employed to roughly segment the so obtained direction map DM_G. The foreground in the under-segmented image G_u, obtained in correspondence with the higher threshold value, includes pixels that can be most possibly interpreted as vessel pixels. The foreground in the over-segmented image G_o, obtained in correspondence with the lower threshold value, includes several more pixels than G_u. Some pixels of G_o are detected as missing vessel pixels of G_u, and are added to G_u to improve the quality of the segmentation. The method has been tested on the DRIVE database [1] producing satisfactory results, and its performance has been compared to that of other unsupervised methods.

© Springer International Publishing AG 2017
C. Beltrán-Castañón et al. (Eds.): CIARP 2016, LNCS 10125, pp. 1–9, 2017.
DOI: 10.1007/978-3-319-52277-7_1

2 Building the Direction Map DM_G

We work with the green channel G of RGB color retinal images, as done by the majority of authors in the field, since G is the channel characterized by the highest contrast. Gray-levels in G are in the range [0,255], where lighter pixels are those with larger gray-level values. Vessel pixels are thin structures whose pixels are generally darker than their neighboring non-vessel pixels and are aligned along different directions.

The direction of any pixel p of G can be computed by taking into account the gray-levels of the pixels in an $n \times n$ window W centered on p, so as to build the Direction Map DM_G. Selecting the proper value for n is crucial to obtain a correct segmentation: W should be large enough to include both vessel and non-vessel pixels, even when centered on the most internal pixels of the thickest vessels, so as to have appreciable variations of gray-levels within the window; on the other hand, W should not be too large so as to avoid the inclusion of vessel pixels belonging to close vessels. By taking into account the average width and distance of vessels in the DRIVE database, we set $n = 7$. Since in an $n \times n$ window, $2 \times (n-1)$ discrete straight lines can be built, 12 directions with an angle of 15° between any two successive directions are obtained. See Fig. 1 top, where the twelve directions are shown in different colors.

Fig. 1. The twelve directions d_i, top, and the twelve directional templates T_d_i, bottom. (Color figure online)

To build DM_G we use the twelve 7×7 directional templates T_d_i, $i = 1,2,\ldots,12$, shown in Fig. 1 bottom. In the i-th template, the pixels aligned along any direction out of d_{i-1}, d_i, and d_{i+1} are set to 1, and all the remaining pixels are set to 0. Of course, for $i = 1$ ($i = 12$) d_{i-1}, d_i, d_{i+1} are respectively d_{12}, d_1, d_2 (d_{11}, d_{12}, d_1).

Given two arrays $DIFF$ and DM_G with the same size as G and initially empty, for each pixel p of G and for any directional template T_d_i, $i = 1,2,\ldots,12$, we compute:

- the arithmetic mean, $Ad_i(p)$, of gray-levels of pixels of G matching 1's in T_d_i,
- the arithmetic mean, $NAd_i(p)$, of gray-levels of pixels of G matching 0's in T_d_i,
- the difference $\Delta_i(p) = NAd_i(p) - Ad_i(p)$.

The direction d_i for which $\Delta_i(p)$ has the maximum value $M(p)$ is taken as direction of p; the score $M(p)$ and the index i are respectively stored in the homologous position of p in $DIFF$ and DM_G. For the green channel G of the image 05_test of the DRIVE database, shown in Fig. 2 left, the obtained DM_G is shown in Fig. 2 middle, where colors correspond to the directions as in Fig. 1 top.

Fig. 2. The image G, left; the direction map DM_G before, middle, and after, right, direction updating. (Color figure online)

Actually, some pixels in DM_G have been assigned a direction different from the one along which they appear to be aligned. Thus, an updating of DM_G is done for each of the twelve directions and by using an auxiliary array AUX with the same size as G and whose pixels are initially set to 255. For the i-th direction d_i, the pixels of DM_G with assigned direction d_i are set to 0 in AUX. Thus, AUX becomes a gray-level image with only two values, where the foreground consists of the pixels with direction d_i, and the background includes all the remaining pixels. The idea is to build the direction map of AUX, DM_{AUX}, by following the procedure described to compute DM_G and to compare the direction assigned to any pixel p in the two maps. Only pixels for which the same direction is obtained in DM_G and DM_{AUX} are confirmed as foreground pixels in DM_G. All other pixels of DM_G are assigned to the background. Actually, when transferring in AUX pixels with direction d_i, also the pixels that in DM_G where characterized by the directions d_{i-1} and d_{i+1} are set to 0 in AUX, while updating is done only for the pixels with direction d_i. The purpose is to avoid that only a few sparse pixels of AUX are foreground pixels, for which almost all directions would be equally possible, so biasing the correct updating of the directions. The updated DM_G is shown in Fig. 2 right, where white pixels are those assigned to the background.

3   The Segmentation Method

Segmentation can be achieved by assigning to the background any pixel p whose score $M(p)$ in $DIFF$ is smaller than a threshold θ, which is set based on the directional features of the processed image. To this aim, we compute the arithmetic mean m of the scores $M(p)$ of all pixels that are currently foreground pixels, i.e., pixels that in DM_G are different from zero, and set θ equal to a given percentage of m.

Actually, we use two different percentages of m, so as to achieve two different values for θ, $\theta\_u = u \times m$ and $\theta\_o = o \times m$ with $u > o$, leading to two different segmented images, G_u and G_o. When the higher threshold value is adopted, only the pixels with higher probability to be true vessel pixels are assigned to the foreground in G_u. The resulting image is generally under-segmented, since pixels belonging to slightly lighter vessels or to capillaries may not survive thresholding. With the lower threshold value, also pixels with lower probability to be vessel pixels are selected, so achieving an over-segmented image G_o. The number of pixels that are not true vessel pixels is much smaller in G_u than in G_o. For the running example, the under-segmented

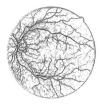  

Fig. 3. The image G_u, left, the image G_o, middle, and the image difference G_{o-u}, right.

image G_u and the over-segmented image G_o, obtained for $u = 1.8$ and $o = 0.9$, are respectively shown in Fig. 3 left and middle. The image difference G_{o-u} includes the pixels in the foreground of G_o but not in the foreground of G_u. See Fig. 3 right.

Linking and Cleaning
The image difference G_{o-u} is rather noisy. In fact, the low threshold value $\theta\_o = o \times m$ adopted to obtain G_o guarantees the detection of almost all the true vessel pixels, but also causes the wrong assignment to the foreground of a number of false vessel pixels. Thus, we consider as necessary some cleaning of G_{o-u}, based on removal of small size components. Before doing the size-based cleaning, we link with each other small close components that we interpret as parts of vessels resulting in distinct connected foreground components of G_{o-u} since some of their pixels were selected as foreground pixels when adopting both the lower and the higher threshold value.

For the linking process we consider components of G_{o-u} with size smaller than a given maximum value *max*, since only small size components risk to be removed during the successive cleaning task. On the other hand, we do not consider components with very little size, say with size smaller than a given minimum value *min*, since these components most possibly consist of false vessel pixels. Thus, we consider for the linking process every component of G_{o-u} whose size s is such that $min \leq s \leq max$. We have experimentally found that the better performance of the method is achieved in the average by setting $min = 16$ and $max = 150$. We use an iterated expansion process. At each iteration, the background neighbors of any pixel p in the components of G_{o-u} selected for the linking process are assigned to the components provided that they are foreground pixels in G_o, and share with p the same direction. The first requirement guarantees that linking pixels were at least tentatively selected as vessel pixels in the over-segmented image; the second requirement guarantees that linking regards exclusively components that actually are part of vessels aligned along given directions. The number of iterations depends on the maximal distance between two components to be connected. Since G_o includes many false vessel pixels, we limit the number of iterations to two, which means that we can link to each other only components with a maximal distance of at most four pixels. The expansion process is followed by an iterated topological removal process, aimed at assigning again to the background all pixels added by the expansion process except those that favored linking of components. When no more pixels are removed, the surviving added pixels are assigned to the foreground in G_{o-u}.

Size-based cleaning is performed, by removing from G_{o-u} all components with area smaller than a threshold σ. We have found that the best performance is obtained in the

Fig. 4. The image $G_{o\text{-}u}$ after linking and cleaning, left, expansion, middle, and shrinking, right. (Color figure online)

average for $\sigma = 64$. The result obtained after linking and cleaning can be seen in Fig. 4 left. Different colors denote the identity labels of the components of $G_{o\text{-}u}$.

Selection of Foreground Components Consisting of True Vessel Pixels

We use the main feature characterizing vessels, i.e., the fact that vessels have linear structure, to distinguish in $G_{o\text{-}u}$ the components to be maintained in the foreground from those to be assigned to the background. We perform a small number of iterations of an expansion process, set to three in this work, during which thin holes and concavities possibly present in any component of $G_{o\text{-}u}$ are filled in, without creating an unwanted merging of foreground components. See Fig. 4 middle showing the resulting $G_{o\text{-}u}$ after the three iterations. Then, we perform an equal number of iterations of topological shrinking, during which pixels added to any component by the expansion process are removed provided that they have at least a neighbor in any other component, including the background. The resulting image $G_{o\text{-}u}$ is shown in Fig. 4 right. It can be observed that while components originally characterized by linear structure are not remarkably modified by the expansion/shrinking process as regards their size, components with a more complex structure, erroneously assigned to the foreground, have at the end of the process a significantly larger size. Actually, besides the changes in size, we also take into account the changes in maximal width of the components of $G_{o\text{-}u}$. This last feature is easily measured as the maximal value in the distance transform of each component before and after the expansion/shrinking process. Let W_{in} and W_{exp} be the initial maximal width and the maximal width after the expansion/shrinking process, respectively. Moreover, let A_{in} and A_{exp} be the initial area and the area after the expansion/shrinking process. We assign to the background any component for which it results $A_{exp}/A_{in} \geq 0.40$, or it is $0.30 \leq A_{exp}/A_{in} < 0.40$ and $W_{exp}/W_{in} \geq 2$. All other components remain in the foreground.

All components of $G_{o\text{-}u}$ surviving the expansion/shrinking process are recognized as consisting of vessel pixels. The pixels that belonged to these components before the expansion/shrinking process are transferred into G_u. The currently obtained G_u can be seen in Fig. 5 left.

Improving Segmentation

To recover missed vessel pixels, we apply to the components of G_u a slightly modified version of the process adopted to link close components of $G_{o\text{-}u}$. As done when processing $G_{o\text{-}u}$, background neighbors of pixels in components of G_u are added to the components only if they also exist in G_o. Differently from the process applied to $G_{o\text{-}u}$, the number of iterations is not fixed a priori; moreover, we have some tolerance on the direction of the pixels to be added. In detail, the expansion process terminates when no

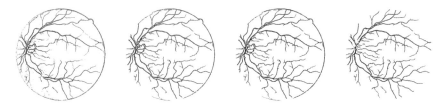

Fig. 5. The image G_u after adding the contribution of G_{o-u}, left; G_u after recovery of missed vessel pixels and removal of small components, middle left, G_u after foreground concavities and holes filling, middle right, and G_u after cleaning of the circular boundaries of retina and optical disc, right.

more pixels can be added; at each iteration, if d_i is the direction of a pixel p in a given component of G_u, any neighbor of p is parallelwise added to G_u, provided that its assigned direction is any out of d_{i-1}, d_i, d_{i+1}. Finally, differently from the process done on G_{o-u}, the expansion process is not followed by any iterated topological removal and we accept as vessel pixels all the recovered pixels. Finally, components of G_{o-u} that notwithstanding the recovery process are still characterized by small size are removed. We use the same threshold $\sigma = 64$ already adopted for size-based cleaning of G_{o-u}. The resulting image is shown in Fig. 5 middle left.

By taking into account the average width and distance of the vessels in the DRIVE database, we found adequate to fill foreground concavities up to 2-pixel wide. Since the concavities filling process is the same used in [12], for space limitation we do not describe again it here. Very small size holes (consisting of at most 16 pixels in this work) are also filled in. The resulting image G_u is shown in Fig. 5 middle right.

The last process is devoted to cleaning of the circular boundaries of retina and of optical disc. The masks available in the DRIVE database are used to extract only the circular part of G_u corresponding to the retina. However, some pixels in proximity of the boundary of the circular mask may have been erroneously interpreted as vessel pixels. Thus, we dilate the circular boundary of the mask by means of a structuring element of radius 20. Foreground pixels reached by the dilation process are marked as removable. Then, an iterated un-marking process is accomplished that removes the marker from any neighbor q of an un-marked vessel pixel p with direction d_i if the direction of q is compatible with the direction of p, i.e., the direction of q is any out of d_{i-1}, d_i, d_{i+1}. The number of iterations of the un-marking process is equal to 20, i.e., to the radius of the structuring element. Pixels that at the end of the un-marking process are still marked are assigned to the background.

To remove pixels erroneously detected as vessel pixels within the optical disc, we first identify in G a region of interest, ROI, that includes the optical disc. It can be noted that the optical disc includes pixels with remarkably lighter gray-levels, and its maximal diameter and position are predictable in the DRIVE database. Thus, we use a sliding window of size 121×121 which swipes the image in the rectangular area where the optical disc is expected to be located. The pixels in the sliding window with gray-level larger than 80% of the maximal gray-level in G are counted while the window swipes the image. The position of the window in correspondence of which the number of counted pixels has the highest value defines the ROI. We observe that some vessels exist in the optical disc and these are characterized by rather dark gray-level and almost horizontal direction. Thus, out of all pixels of the ROI that have been assigned to the foreground of G_u, we assign to the background those with gray-level larger than a suitable percentage, *perc*, of the arithmetic mean of the gray-levels within the ROI and with direction i which is not compatible with the horizontal direction. We have obtained the best performance by setting *perc* = 90% and by considering compatible with the horizontal direction d_i, characterized by $i = 7$, any direction out of d_5, d_6, d_7, d_8, and d_9. The final segmentation result is shown in Fig. 5 right.

4 Experimental Results and Concluding Remarks

The suggested segmentation method has been checked on the 40 images of the DRIVE database, by using as ground truth the manually segmented images included in the database. Actually, two manually segmented images are available for each of the 20 retinal images forming the test set, and one manual segmentation for the remaining 20 images forming the training set. Since our segmentation method is unsupervised, we have not done any difference between the test set and the training set.

For a qualitative evaluation of the method, see Fig. 6, where the results for the four images 09_test, 19_test, 27_training, and 32_training are shown.

To quantitatively evaluate the method, let TP and TN count the pixels correctly classified as vessel pixels and as non-vessel pixels, and FP and FN the pixels incorrectly classified as vessel pixels and as non-vessel pixels, and compute:

Accuracy = (TP + TN)/(TP + FN + TN + FP)
Sensitivity = TP/(TP + FN)
Specificity = TN/(TN + FP)

The average values of Accuracy, Sensitivity and Specificity have been computed for the 20 test images with respect to the first ground truth to compare the performance of our method with that of other 6 unsupervised segmentation methods in the literature,

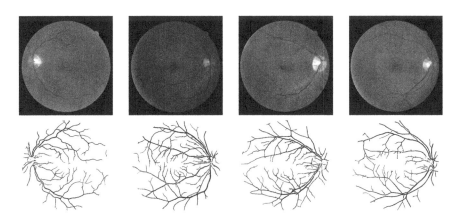

Fig. 6. Some examples, top, and the segmentation results, bottom.

Table 1. Performance comparisons.

	Accuracy	Specificity	Sensitivity
Mendonca and Campilho [8]	0.946	0.976	0.734
Al-Rawi et al. [17]	0.953		
Zhang et al. [18]	0.938	0.972	0.712
Zhao et al. [19]	0.954	0.982	0.742
Azzopardi et al. [11]	0.944	0.970	0.765
Frucci et al. [12]	0.955	0.985	0.640
Our method	0.956	0.985	0.660

[8, 11, 12, 17–19]. See Table 1. Our method is slightly better as regards Accuracy and Specificity, and inferior for Sensitivity, but generally produces results qualitatively more satisfactory.

References

1. Staal, J.J., Abramoff, M.D., Niemeijer, M., Viergever, M.A., van Ginneken, B.: Ridge based vessel segmentation in color images of the retina. IEEE Trans. Med. Imaging **23**, 501–509 (2004)
2. Teng, T., Lefley, M., Claremont, D.: Progress towards automated diabetic ocular screening: a review of image analysis and intelligent systems for diabetic retinopathy. Med. Biol. Eng. Comput. **40**, 2–13 (2002)
3. Haddouche, A., Adel, M., Rasigni, M., Conrath, J., Bourennane, S.: Detection of the foveal avascular zone on retinal angiograms using Markov random fields. Digit. Sig. Process. **20**, 149–154 (2010)
4. Jain, A.K., Ross, A., Prabhakar, S.: An introduction to biometric recognition. IEEE Trans. Circ. Syst. Video Technol. **14–1**, 4–20 (2004)
5. Li, Q., You, J., Zhang, D.: Vessel segmentation and width estimation in retinal images using multiscale production of matched filter responses. Expert Syst. Appl. **39**(9), 7600–7610 (2012)
6. Rangayyan, R., Zhu, X., Ayres, F., Ells, A.: Detection of the optic nerve head in fundus images of the retina with Gabor filters and phase portrait analysis. J. Digit. Imaging **23**, 438–453 (2010)
7. Wang, Y., Ji, G., Lin, P., Trucco, E.: Retinal vessel segmentation using multiwavelet kernels and multiscale hierarchical decomposition. Pattern Recogn. **46–8**, 2117–2133 (2013)
8. Mendonça, A.M., Campilho, A.: Segmentation of retinal blood vessels by combining the detection of centerlines and morphological reconstruction. IEEE Trans. Med. Imaging **25**(9), 1200–1213 (2006)
9. Fraz, M.M., Barman, S.A., Remagnino, P., Hoppe, A., Basit, A., Uyyanonvara, B., Rudnicka, A.R., Owen, C.G.: An approach to localize the retinal blood vessels using bit planes and centerline detection. Comput. Methods Programs Biomed. **108**(2), 600–616 (2012)
10. Ricci, E., Perfetti, P.: Retinal blood vessel segmentation using line operators and support vector classification. IEEE Trans. Med. Imaging **26**(10), 1357–1365 (2007)

11. Azzopardi, G., Strisciuglio, N., Vento, M., Petkov, N.: Trainable COSFIRE filters for vessel delineation with application to retinal images. Med. Image Anal. **19**(1), 46–57 (2015)
12. Frucci, M., Riccio, D., Sanniti di Baja, G., Serino, L.: Severe, segmenting vessels in retina images. Pattern Recogn. Lett. **82**, 162–169 (2016)
13. Sun, K., Chen, Z., Jiang, S., Wang, Y.: Morphological multiscale enhancement, fuzzy filter and watershed for vascular tree extraction in angiogram. J. Med. Syst. **35**(5), 811–824 (2010)
14. Imani, E., Javidi, M., Pourreza, H.R.: Improvement of retinal blood vessel detection using morphological component analysis. Comput. Methods Programs Biomed. **118**(3), 263–279 (2015)
15. Frucci, M., Riccio, D., di Baja, G.S., Serino, L.: Using contrast and directional information for retina vessels segmentation. In: Proceedings of the SITIS 2014, pp. 592–597. IEEE CS (2014)
16. Zhao, Y.Q., Wang, X.H., Wang, X.F., Shih, F.Y.: Retinal vessels segmentation based on level set and region growing. Pattern Recogn. **47**(7), 2437–2446 (2014)
17. Al-Rawi, M., Qutaishat, M., Arrar, M.: An improved matched filter for blood vessel detection of digital retinal images. Comput. Biol. Med. **37**, 262–267 (2007)
18. Zhang, B., Zhang, L., Zhang, L., Karray, F.: Retinal vessel extraction by matched filter with first-order derivative of Gaussian. Comput. Biol. Med. **40**(4), 438–445 (2010)
19. Zhao, Y., Rada, L., Chen, K., Harding, S.P., Zheng, Y.: Automated vessel segmentation using infinite perimeter active contour model with hybrid region information with application to retinal images. IEEE Trans. Med. Imaging **34**(9), 1797–1807 (2015)

Identifying *Aedes aegypti* Mosquitoes by Sensors and One-Class Classifiers

Vinicius M.A. Souza[⊠]

Instituto de Ciências Matemáticas e de Computação,
Universidade de São Paulo, São Carlos, SP, Brazil
vsouza@icmc.usp.br

Abstract. Yellow fever, zika, and dengue are some examples of arboviruses transmitted to the humans by the *Aedes aegypti* mosquitoes. The efforts to curb the transmission of these viral diseases are focused on the vector control. However, without the knowledge of the exact location of the insects with a reduced time delay, the use of techniques as chemical control becomes costly and inefficient. Recently, an optical sensor was proposed to gather real-time information about the spatio-temporal distributions of insects, supporting different vector control techniques. In field conditions, the assumption of knowledge of all classes of the problem, it is hard to be fulfilled. For this reason, we address the problem of insect classification by one-class classifiers, where the learning is performed only with positive examples (target class). In our experiments, we identify *Aedes aegypti* mosquitos with an AUC = 0.87.

Keywords: Optical sensor · Insect classification · One-class classifiers

1 Introduction

The *Aedes aegypti* mosquito is one of the most important vectors of arboviruses that affect human health, including yellow fever, chikungunya, zika, Japanese encephalitis, and dengue. The viruses are passed on to humans through the bites of an infective female *Aedes* mosquito, which mainly acquires the virus while feeding on the blood of an infected person.

In May 2015, the Pan American Health Organization issued an alert regarding the first confirmed Zika virus infections in Brazil. Since this identification, the virus has spread rapidly throughout the America. The illness is usually mild with symptoms lasting for several days to a week after being bitten by an infected mosquito. However, Zika virus infection during pregnancy can cause a serious birth defect called microcephaly, as well as other severe fetal brain defects [1].

Dengue is the most important vector-borne viral disease of humans and likely more important than malaria globally in terms of morbidity and economic impact [2]. Studies estimate that 3.6 billion people living in areas of risk, with 390 million

V.M.A. Souza—The author thank the financial support of FAPESP (Grants #2011/17698-5, and #2015/16004-0).

C. Beltrán-Castañón et al. (Eds.): CIARP 2016, LNCS 10125, pp. 10–18, 2017.
DOI: 10.1007/978-3-319-52277-7_2

dengue infections per year globally, of which 96 million manifests clinically [3,4]. According to the World Health Organization, only 9 countries had experienced severe dengue epidemics before 1970. The disease is now endemic in more than 100 countries. In Latin America, the incidence and severity of this disease have increased rapidly in recent years. In 2015, 2.35 million cases of dengue were reported in the Americas alone, of which 10,200 cases were diagnosed as severe dengue causing 1,181 deaths [5].

Currently, no licensed vaccine against dengue infection is available, and the most advanced vaccine candidate did not meet expectations in a large trial [6]. Thus, the efforts to curb the transmission of these viral diseases are focused on the vector control in order to reduce the population of *Aedes aegypti*. There are many methods to insect control, as biological, genetic technology, environmental management and chemical control. However, without the knowledge of the exact location of the insects with a reduced time delay, the use of these techniques becomes costly and inefficient.

Recently, a new optical sensor was proposed as a tool to gather information about the spatio-temporal distributions of insects and to control disease vectors by the use of this sensor combined with an electronic trap [7]. The sensor captures insect flight information using a source light and automatically classifies the insects according to their species using machine learning algorithms. This sensor can provide real-time population estimates of insect species, supporting the effective use of traditional strategies to vector control.

The previous efforts related to insect classification by optical sensors have focused on multiclass classifiers, such as Support Vector Machines, k-Nearest Neighbors, Random Forest, Deep Neural Network, among others [7–10]. In multiclass classification, we have n predefined classes composed by the set of class labels $Y = \{y_1, y_2, \ldots, y_n\}$, where the main goal of a classifier is to assign the most probable class label $y_i \subset Y$ for an unknown example $\overrightarrow{x}$, where $\overrightarrow{x} \in \mathbb{R}^d$ is a feature vector with d dimensions. This procedure can be problematic when the example does not belong to any of predefined classes.

For the effective use of the sensor in field conditions, we note that the assumption of knowledge of all classes made by multiclass classifiers, it is hard to be fulfilled. For example, it is estimated that only the insects of the order *Diptera*, has more than 240,000 different species, where about 120,000 are cataloged [11]. Thus, it is impossible to conduct a comprehensive data collection that covers all possible species to build a classification model with all possible species. In practice, this means that there is a high probability of the sensor to deal with unknown species. In this case, a multiclass classifier will assign an incorrect class label to this insect, due the lack of data from other possible species.

Given the need of identification of *Aedes aegypti* mosquitoes by sensors to support methods of vector control and the challenge to cope with unknown species, in this paper we address this classification problem using one-class classifiers [12]. In one-class classification, the learning is performed only with positive examples (target class) and none or few unlabeled examples from negative class.

We evaluated eight algorithms learned with only data from *Aedes aegypti* mosquitoes. The test was conducted with a dataset with five insect species collected by optical sensors. In our experimental evaluation, we conclude that the Parzen and SVDD are the most accurate algorithms for this application to the identification of *Aedes aegypti* mosquitoes.

The rest of the paper is organized as follows. Section 2 presents the optical sensor for insect classification. Section 3 describes the main concepts of one-class classification. Section 4 shows the results of our experimental evaluation. Finally, our conclusions are presented in Sect. 5.

2   Optical Sensor and Insect Data

The data evaluated in this paper was obtained from an optical sensor built with low-cost components to remotely capture information about flying insects. The sensor uses a light source, as a low-powered planar laser, that is pointed at an array of phototransistors as illustrated in Fig. 1-a). When a flying insect crosses the laser, its wings partially occlude the light, causing small variations in the light captured by the phototransistors. These variations are recorded as an audio signal, as the example presented in Fig. 1-b), given an *Aedes aegypti* crossing.

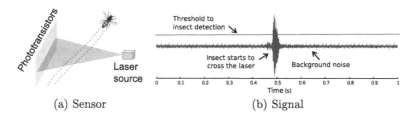

(a) Sensor   (b) Signal

Fig. 1. Illustration of the optical sensor to capture information about insects and an example of audio signal collected given the crossing of an *Aedes aegypoti* mosquito.

In general, the data consist of background noise with occasional "events", resulting in the brief moment that an insect flies across the sensor. Note that the signal generated by the passage of the insect has an amplitude that is significantly higher than the amplitude of the background noise. In this way, using a simple threshold it is a trivial task to identify signal sections in which there is an insect passage. In contrast, the correct classification of each passage according to the insect species that generated the event is a more elaborate task. Basically, this task consists in extracting discriminant features from the signals for each species and using these features with machine learning algorithms.

2.1   Data Collection

In our study, we use the stream insect dataset previously evaluated in [10]. In this dataset, the collection was performed during six consecutive days in laboratory

Table 1. Insect dataset distribution.

Species of insect	Examples	Distribution (%)
Musca domestica	917	17.22
Culex quinquefasciatus	1,285	24.13
Culex tarsalis	1,265	23.76
Drosophila melanogaster	954	17.91
Aedes aegypti	904	16.98

conditions in which the temperature varied slightly between 20°C and 22°C and humidity varied between 20% and 35%. This dataset has insect passage signals from two species of flies and three species of mosquitoes. The flies species are the *Drosophila melanogasler* and the *Musca domestica*. The mosquito species are the *Culex quinquefascialus, Culex tarsalis* and the *Aedes aegypti*. It is interesting to note that *Culex* are species visually similar to *Aedes* and predominant in the Latin America houses. Table 1 presents a general description of the dataset.

2.2 Feature Extraction

In this work, we use the Mel-Frequency Cepstral Coefficients (MFCC) as recommended in a previous evaluation with a wide variety of signal processing techniques for feature extraction [7]. MFCCs are popular features in various application domains, particularly speech and speaker recognition [13].

MFCCs are calculated by taking the magnitudes of frequency components using an acoustically-defined scale called *mel* [14]. This scale relates physical frequencies to the frequencies perceived by the human auditory system. Equation 1 shows the conversion from frequency (f) to mel-frequency (m). Next, we apply a Discrete Cosine Transform. The MFCC are the cepstrum coefficients obtained from this operation. Specifically, we consider the 40 first coefficients as features.

$$m = 2595 \times log_{10}(1 + \frac{f}{700}) \qquad (1)$$

3 One-Class Classification

Conventional multiclass classification algorithms aim to classify an unknown object into one of the several predefined categories. A problem arises when the unknown object does not belong to any of those categories. In one-class classification (OCC) [12,15], one of the classes (referred as target class) is well characterized by instances in the training data. For the other class (non-target), it has either no instances at all, very few of them, or they do not form a statistically representative sample of the negative concept.

In general, the problem of one-class classification is harder than the problem of conventional two-class or multiclass classification. For example, in binary classification problems, the decision boundary is supported from both sides by examples of both classes. Because in the case of one-class classification only one set of data is available, only one side of the boundary is supported. It is therefore hard to decide, on the basis of just one class, how strictly the boundary should fit around the data in each of the feature directions [15]. This task is often called *data domain description*.

This OCC problem is often solved by estimating the target density or by fitting a model to the data support vector classifier. Instead of using a hyperplane, to distinguish between two classes, a hypersphere around the target set is used. The volume of the hypersphere is minimized directly. This method is called support vector data description (svdd) [16]. In svdd, a spherically shaped decision boundary around a set of objects is constructed by a set of support vectors describing the sphere boundary.

Different methods for data domain description have been developed. In this work, we evaluated eight different algorithms from the *Data Description toolbox* (DDtools) [12,17]. Specifically, the following algorithms: *gausdd* (Gaussian target distribution), *svdd* (support vector data description), *parzendd* (Parzen density estimator data description), *kmeansdd* (k-means data description), *knndd* (k-nearest neighbor data description), *lpdd* (linear programming data description), *mstdd* (minimum spanning tree data description), and *mogdd* (mixture of Gaussians data description). Unfortunately, due to space constraints, it is not possible to describe the algorithms. We direct the interested readers to [12] and [17] for a detailed explanation. However, an intuition of the decision boundary considered for each algorithm is shown in Fig. 2, given an artificial data example.

4 Experimental Evaluation

In our experimental evaluation, the classifiers were learned only with data of *Aedes aegypti* (target class). More specifically, we have considered the data from the first 48 h of the data collection, which represents 347 examples. To test the classifiers, we consider the remaining 557 examples from the class *Aedes aegypti* that was not used to train the classifiers and the 4,421 examples from the other four species of insects, totaling 4,978 test examples.

Due to the imbalanced proportion of examples of target class compared to the non-target, a classifier that predicts the non-target class for all test examples achieves an accuracy around 90%. For this reason, we evaluate our results by the analysis of different performance measures, as Precision, Recall, F1-Score. Thus, given the rates of true positive (TP), true negative (TN), false positive (FP), and false negative (FN) observed in a confusion matrix built from the errors of a classifier, these measures are defined as follow:

$$Precision = \frac{TP}{TP+FP}, \quad Recall = \frac{TP}{TP+FN}, \quad F1 - Score = \frac{2 \times (Precision \times Recall)}{Precision + Recall}$$

In addition, we also consider the measure Area Under Curve (AUC). This measure is related to the observed area on the Receiver Operating Characteristic

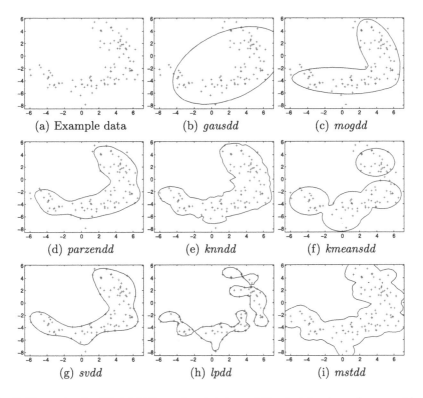

Fig. 2. Example of artificial bidimensional data and the decision boundary considered by each OCC algorithm evaluated.

curve (ROC curve). The ROC curve is a two-dimensional graphical representation which corresponds to false positive rate on the horizontal axis and the true positive rate on the vertical axis. Thus, in an ideal scenario, is expected a minimum value of false positives and a maximum value of true positives, which consequently leads to a value for AUC = 1.

The general results of the algorithms considering the five performance measures discussed are shown in Table 2. For each measure, the best result is highlighted. In this table, we also show the results achieved by a baseline which corresponds to a classifier that predicts the target class for all test examples.

We can see in Table 2 that the algorithm *parzendd* showed the best results for the measures Recall and AUC. On the other hand, the algorithm *svdd* showed the best results for the measures F1-Score and Accuracy. To better compare the results, the ROC curves achieved by the algorithms are shown in Fig. 3.

From the results showed in Table 2 and Fig. 3, we can note that both *parzendd* and *svdd* are very competitive, but the *svdd* showed results better balanced in terms of false positive and true positive rates. Although the *parzend* algorithm correctly identifies a higher number of *Aedes aegypti* mosquitoes, it also incorrectly identifies a higher number of insects from other species as *Aedes*. In Table 3 we shown more details about the errors of both algorithms.

Table 2. Results of one-class classifiers.

Algorithm	Precision	Recall	F1-Score	Accuracy	AUC
gausdd	0.45	0.76	0.57	86.96	0.82
mogdd	0.64	0.64	0.64	91.98	0.80
parzendd	0.41	**0.91**	0.56	84.35	**0.87**
knndd	0.43	0.87	0.57	85.42	0.86
kmeansdd	0.41	0.78	0.54	85.05	0.82
svdd	**0.78**	0.73	**0.75**	**94.62**	0.85
lpdd	0.32	0.85	0.46	77.90	0.81
mstdd	0.32	0.89	0.48	78.04	0.83
Baseline	0.10	1,00	0.18	10.98	0.50

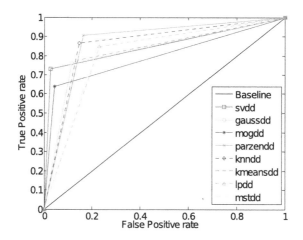

Fig. 3. The ROC curves of the OCC algorithms evaluated.

Table 3. Confusion matrices showed by the algorithms *parzendd* and *svdd*.

Actual	Predicted	
	Ae. aegypti	¬ *Ae. aegypti*
Ae. aegypti	505	52
¬ *Ae. aegypti*	727	3694

parzendd

Actual	Predicted	
	Ae. aegypti	¬ *Ae. aegypti*
Ae. aegypti	407	150
¬ *Ae. aegypti*	118	4303

svdd

5   Conclusion

In this paper, we showed an evaluation of one-class classifiers for the recognition of *Aedes aegypti* mosquitoes by optical sensors. *Aedes aegypti* is one of the most important vector of arboviruses as yellow fever, chikungunya, zika, and dengue. Thus, the recognition task is essential to support the efficient use of traditional methods to reduce the mosquitoes population, given the spatio-temporal informations provided by the sensors. From the results, we conclude that even with a reduced number of target examples for training the classifiers (347 examples) and the absence of non-target examples, we can learn accurate classifiers. Among the evaluated algorithms, *svdd* and *parzendd* showed the best results, with AUC = 0.85 and AUC = 0.87, respectively. In future works, we want to explore the combination of different OCC algorithms and feature sets, and in conditions with concept drifts and extreme latency to update the classification model [18, 19].

References

1. Plourde, A.R., Bloch, E.M.: A literature review of Zika virus. Emerg. Infect. Dis. **22**(7), 1185–1192 (2016)
2. Gubler, D.J.: The economic burden of dengue. Am. J. Trop. Med. Hyg. **86**(5), 743–744 (2012)
3. Beatty, M.E., Letson, G.W., Margolis, H.S.: Estimating the global burden of dengue. Am. J. Trop. Med. Hyg. **81**(5), 231 (2009)
4. Bhatt, S., Gething, P.W., Brady, O.J., Messina, J.P., Farlow, A.W., Moyes, C.L., Drake, J.M., Brownstein, J.S., Hoen, A.G., Sankoh, O.: The global distribution and burden of dengue. Nature **496**(7446), 504–507 (2013)
5. W.H.O.: Dengue and severe dengue. Technical report Fact Sheet 117, World Health Organization (2015)
6. Halstead, S.B.: Dengue vaccine development: a 75% solution? The Lancet **380**(9853), 1535–1536 (2012)
7. Silva, D.F., Souza, V.M.A., Ellis, D.P.W., Keogh, E.J., Batista, G.E.A.P.A.: Exploring low cost laser sensors to identify flying insect species. J. Intell. Robot. Syst. **80**(1), 313–330 (2015)
8. Qi, Y., Cinar, G.T., Souza, V.M.A., Batista, G.E.A.P.A., Wang, Y., Principe, J.C.: Effective insect recognition using a stacked autoencoder with maximum correntropy criterion. In: Proceedings of the International Joint Conference on Neural Networks, pp. 1–7 (2015)
9. Silva, D.F., Souza, V.M.A., Batista, G.E.A.P.A., Keogh, E., Ellis, D.P.W.: Applying machine learning and audio analysis techniques to insect recognition in intelligent traps. In: Proceedings of the International Conference on Machine Learning and Applications, pp. 99–10 (2013)
10. Souza, V.M.A., Silva, D.F., Batista, G.: Classification of data streams applied to insect recognition: initial results. In: Proceedings of the Brazilian Conference on Intelligent Systems, pp. 76–81 (2013)
11. Wiegmann, B., Yeates, D.K.: The Tree of Life Diptera (1996)
12. Tax, D.M.J.: One-class classification. Ph.D. thesis, TU Delft, Delft University of Technology (2001)
13. Zhen, B., Wu, X., Liu, Z., Chi, H.: On the importance of components of the MFCC in speech and speaker recognition. Acta Scietiarum Naturalium **37**(3), 371–378 (2001)
14. Stevens, S.S., Volkmann, J., Newman, E.B.: A scale for the measurement of the psychological magnitude pitch. J. Acoust. Soc. Am. **8**(3), 185–190 (1937)
15. Tax, D.M.J., Duin, R.P.: Uniform object generation for optimizing one-class classifiers. J. Mach. Learn. Res. **2**, 155–173 (2002)

16. Tax, D.M.J., Duin, R.P.W.: Support vector domain description. Pattern Recogn. Lett. **20**(11), 1191–1199 (1999)
17. Tax, D.: Ddtools, the data description toolbox for matlab version 2.1.2, June 2015
18. Souza, V.M.A., Silva, D.F., Gama, J., Batista, G.E.A.P.A.: Data stream classification guided by clustering on nonstationary environments and extreme verification latency. In: Proceedings of the SIAM International Conference on Data Mining, pp. 873–881 (2015)
19. Souza, V.M.A., Silva, D.F., Batista, G.E.A.P.A., Gama, J.: Classification of evolving data streams with totally delayed labels. In: Proceedings of the International Conference on Machine Learning & Applications, pp. 214–219 (2015)

Partial Matching of Finger Vein Patterns Based on Point Sets Alignment and Directional Information

Maria Frucci[1], Daniel Riccio[1,2(✉)], Gabriella Sanniti di Baja[1], and Luca Serino[1]

[1] Istituto Calcolo e Reti ad Alte Prestazioni, CNR, Naples, Italy
{maria.frucci,luca.serino}@cnr.it,
daniel.riccio@unina.it,
gabriella.sannitidibaja@girpr.org
[2] University of Naples, Federico II, Naples, Italy

Abstract. In recent years finger vein authentication has gained an increasing attention, since it has shown the potential of providing high accuracy as well as robustness to spoofing attacks. In this paper we presented a new finger verification approach, which does not need precise segmentation of regions of interest (ROIs), as it exploits a co-registration process between two vessel structures. We tested the verification performance on the MMCBNU_6000 finger vein dataset, showing that this approach outperforms state of the art techniques in terms of Equal Error Rate.

Keywords: Finger vein · Directional information · Partial matching

1 Introduction

In the last decade, finger vein authentication has emerged as one of the most promising biometric technologies. Finger blood vessels are subcutaneous structures, providing a vein pattern that is unique and distinctive for each person, so it can be considered as an effective biometric trait. Since blood vessels can be captured only in a living finger, the vein pattern shows two main desirable properties: (a) it is harder to copy/forge than other biometric traits such as fingerprint and face; (b) it is very stable, not being affected by environmental and skin conditions (humidity, skin disease, dirtiness, etc.). There are many further aspects making the finger vein pattern a good candidate for a new generation of biometric systems. First of all, each individual has unique finger vein patterns that remain unchanged despite ageing. Just like fingerprints, ten different vein patterns are associated to a person, one for each finger, so that if one finger is accidentally injured, other fingers can be used for personal authentication. However, finger vein patterns can be acquired contactless, and this makes them preferable to fingerprints, considering that they are comparable in terms of distinctiveness. Moreover, small size capturing devices are available on the market at a low price, that makes this biometric very convenient for a wide range of security applications (e.g. physical and remote access control, ATM login, etc.).

© Springer International Publishing AG 2017
C. Beltrán-Castañón et al. (Eds.): CIARP 2016, LNCS 10125, pp. 19–26, 2017.
DOI: 10.1007/978-3-319-52277-7_3

As for many other biometrics, still there are open issues affecting the authentication accuracy, so restraining a massive diffusion of this technology out of academic laboratories. Indeed, finger vein patterns are mainly affected by uneven illumination and finger posture changes that produce irregular shading and deformation of the vein patterns in the acquired image. The research in this field is mainly devoted to cope with these two major problems by designing either *ad hoc* preprocessing algorithms, or robust feature extraction/matching techniques. Generally, the most of the approaches from recent literature mainly act specifically on one of the four main stages of the finger vein authentication, that are: (i) image acquisition, (ii) preprocessing, (iii) vein feature extraction and (iv) feature matching.

The capturing device project infrared light onto the finger and scans the inner surface of the skin to read the vein pattern by means of an infrared camera. Many imaging sensors are available on the market, which may differ both in the quality and resolution of the acquired finger image. In general, they provide a 8 bit grayscale finger image, whose resolution may range from 240×180 pixels to 640×480 pixels.

The image preprocessing stage can be further subdivided in three main steps that are enhancement, normalization, and segmentation. Image enhancement techniques aim to suppress noisy pixels and to make the vein structure more prominent and easy to distinguish from the background. Problems related to uneven illumination and low local contrast, are also addressed in this stage. Some image enhancement methods are based on multi-threshold combination [1], directional information [2], Gabor filter [3], Curvelets transform [4] and, restoration algorithms [5].

Image enhancement is generally followed by a normalization stage, which is aimed to detect a ROI containing the most of distinctive vein features. The detection of the ROI is implemented by means of edge operators [6–8], while taking into account the finger structure (knuckles and phalanxes) and some basic optic knowledge. After the ROI has been detected and separated from the rest of the finger image, it undergoes to a segmentation stage, whose purpose is to extract finger veins (foreground) from a complicated background. Finger vein extraction represents a crucial point, as it is fundamental for vein pattern-based authentication methods, as demonstrated by the large number of techniques that have been proposed in literature to address this problem; line-tracking [9], threshold-based methods [10], mean curvature [11] and region growth-based feature [12] are just some examples.

The last stage into the finger vein recognition pipeline is represented by feature extraction and classification. All the approaches related to this stage can be grouped into three main categories: (i) methods characterizing geometric shape and topological structure of vein patterns [9, 13–16], (ii) approaches based on dimensionality reduction techniques and learning processes [17–20], (iii) methods exploiting local operators to generate histograms of short binary codes [21–24].

In this paper we propose a new technique for finger vein verification, namely partial matching of finger vein patterns based on point sets alignment and directional information (PM-FVP). This method introduces some novelties. First of all, the same directional information is used to both detect and match vessel structures. It has been previously tested on retina blood vessel segmentation [25], but it had not been experimented yet on finger veins. We propose a new enhancing operator that increases local contrast by exploiting grey level information in pixels neighborhood. We also introduce a vessel structures co-registration stage in the processing pipeline, so avoiding the need of segmenting ROIs. We define a new local directional based feature vector that outperforms the existing ones like Local Binary Pattern (LBP) [26] and Local Directional Code (LDC) [23], which can be considered as the state of the art. The proposed approach has been tested on finger vein images provided in the MMCBNU_6000 dataset [27], which includes both complete finger vein images and their associated ROIs. Experimental results show that PM-FVP outperforms other approaches in terms of verification accuracy.

2   The Proposed Approach

In this paper, we introduce a new finger vein verification technique (PM-FVP) that aims to solve the problem of partial matching between finger vein patterns without the need of detecting specific ROIs. Our approach implements four main steps: (i) image enhancement, (ii) image segmentation, (iii) finger vein patterns co-registration, and (iv) feature extraction and matching. By exploiting a rigid co-registration process, this method is able to align partial finger vein patterns and to perform matching according to local directional information extracted from the blood vessel structure.

Image Enhancement
Finger vein images are generally characterized by a very low contrast of blood vessels, which often jeopardizes results obtained by the image segmentation techniques. However, the detection of vein structure (foreground) from the rest of the image (background) represents a crucial point in finger vein authentication systems, since the better the extracted foreground, the higher the accuracy of the authentication process. In order to mitigate effects of noise and uneven illumination, PM-FVP implements a preprocessing stage with the purpose of enhancing local contrast before applying the segmentation algorithm. The input images in MMCBNU_6000 have a resolution of 640×480 pixels. We resize the input image to a fixed resolution of 128×96 pixels for reasons that will be explained in the following. From now on, we will consider as input images the ones obtained after size reduction. Subsequently, PM-FVP applies a contrast enhancement filter. The enhancing filter is a pixel-wise local operator. For each pixel p of the image, it considers two local windows W_1 and W_2 both centered in p, but with different sizes $n_1 \times n_1$ and $n_2 \times n_2$, respectively (n_1 and n_2 depend on the image resolution and in our case they have been set to 3 and 11, respectively). The maximum grey value max_1 (max_2) in W_1 (W_2) is computed. The new grey value assigned to p is computed according to the following equation:

$$p^{'} = p \cdot e^{\frac{max_1}{(1 + max_2)}} \tag{1}$$

Clearly $max_2 \geq max_1$ always holds, as W_1 is included in W_2, so that the exponent in (1) is always lower than one. Since the value of p' may be larger than 255, all grey values of the filtered image are mapped to the range [0, 255] by means of the min-max rule:

$$p'' = 255 \cdot \left(p' - m_1\right)/(m_2 - m_1) \tag{2}$$

where m_1 and m_2 are the minimum and maximum grey values of the filtered image, respectively.

Image Segmentation

The segmentation process exploits a technique presented in [25], where directional information associated to each pixel of the image is exploited to separate the foreground from the background. It is based on the assumption that blood vessels are generally darker than background, and they are thin and elongated structures with different directions. PM-FVP builds a direction map DM, where each pixel p of the image is assigned a discrete direction d_j, according to a directional mask, centered in p. The number of different discrete directions d_j depends on the size of the directional mask. In particular, for a mask of $n \times n$ pixels, $2 \times (n - 1)$ discrete directions are possible. In turn, the value of n should be selected by taking into account the thickness of the veins and of the background regions (valleys) separating them. If a too small mask is centered on the innermost pixels in a vein (valley), no significant grey-level changes can be detected along all directions, so that any direction can be assigned to the innermost pixels. On the contrary, if a too large mask is used, it can include vein pixels of different veins causing a non-correct direction assignment.

For MMCBNU_6000 images, a reasonably compromise between segmentation quality and computational cost, is to reduce the image size to 128×96 pixels and to use a directional mask m of size $n = 7$ generating twelve different directions represented by twelve templates t_j ($j = 1,...,12$). A graphical representation of all discrete directions generated by a 7×7 mask is given in Fig. 1.

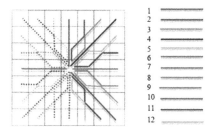

Fig. 1. Discrete directions generated by a 7×7 mask. Continue lines pass through the pixels along the direction in one of the two ways and dashed lines identify those in the opposite way.

Let be p a pixel of the image, for any direction d_j ($j = 1,2,...,12$), neighboring pixels that fall into the template t_j centered in p are partitioned into two sets indicated by S_1 and S_2. The set S_1 consists in the 7 pixels aligned along the direction d_j, while S_2 is formed by the 42 remaining pixels. Let be $Ad_j(p)$ and $NAd_j(p)$ the average values computed over S_1 and S_2, respectively. In the directional map DM, the pixel p is assigned the direction d_j, for which the score $s_j(p) = NAd_j(p) - Ad_j(p)$ is maximized. Doing so, a score map SM is built where $SM(p) = max_j(s_j(p))$. Starting from DM and SM, a preliminary segmentation is performed, which also undergoes to some refinement steps to obtain the final foreground. Details of these steps can be found in [25].

Fig. 2. The finger vein image used as running example (left), its enhancement (center), and the final segmentation (right).

In Fig. 2, from left to the right, a running example, with its enhancement and final segmentation is shown.

The Co-registration of Finger Vein Patterns

We formulate the problem of co-registering two finger vein patterns as a point set registration process, whose aim is to assign correspondences between two sets of points and to derive a transformation mapping one point set to the other. It is a challenging task due to many factors like the dimensionality of point sets and the presence of noise and outliers. In order to ensure that features extracted from different vessel structures are correctly matched, finger vein pattern must be previously align, so to guaranty a proper correspondence among homologous regions. To this aim, PM-FVP implements a co-registration process to align finger vein patterns. In other words, foreground pixels of vessel structures extracted from two different vein finger images may be considered as two different point sets. Since the dimensionality of point sets heavily affects the computational cost of point set alignment methods, a skeletonization operation is applied to the foreground extracted from a finger vein image. The skeletons of two different vessel structures are then considered as two point sets P_1 and P_2, which are inputted to the co-registration algorithm.

Co-registration between point sets is performed by means of the Coherent Point Drift (CPD) algorithm presented in [27]. The CPD technique reformulates the point sets registration problem as a probability density estimation. The first point set P_1 represents centroids of a Gaussian Mixture Model (GMM) that must be fit to the second point set P_2, by maximizing a likelihood. During the fitting process, topological structure is preserved by inducing centroids to move coherently as a group. CPD has been designed to cope with both rigid and non-rigid transformation, but in our specific case we only search for a rigid affine transform T mapping P_1 in P_2. After the co-registration process, the directional features are extracted from both aligned vessel structures according to a local descriptor that is detailed in the following Section. To show that our method is able to perform a partial matching, in Fig. 3 we provide the co-registration result (right) obtained by aligning the vessel structures extracted from the whole finger (left) and its ROI (middle).

Fig. 3. The vessel structure of the MMCBNU_6000 image 002_01_L_Fore (left), the vessel structure of the corresponding ROI (middle), and the result of co-registration (right).

Feature Extraction and Matching

The most of the existing approaches designed for finger vein matching work on ROIs, which are extracted from finger vein images, to guarantee that corresponding vessel structures properly overlap. In turn, PM-FVP does not work on ROIs, but directly aligns vessel structures before extracting directional features. Before applying the feature extraction process, PM-FVP computes the overlapping region between the two aligned vessel structures and limits next processing steps only to this area. The feature extraction process exploits the directional information that has been calculated to segment vessels, and that is provided by the corresponding directional maps DM_1 and DM_2, and score maps SM_1 and SM_2. In other words, PM-FVP just crops DM_1, DM_2, SM_1 and SM_2 to obtain DMc_1, DMc_2, SMc_1 and SMc_2 corresponding to the overlapping region. PM-FVP partitions DMc_k and SMc_k ($k = 1,2$) in non-overlapping sub-windows W_i with fixed dimensions $r \times r$, where r is proportional to the dimension n of the directional mask m that has been adopted to compute directional/score maps (in our case $r = 16$). For each sub-window W_i, PM-FVP computes a 12-bins histogram H, where each bin b_j corresponds to a discrete direction d_j, while the value of each bin is computed by averaging the scores $SM(p)$ of all pixels p in W_i, whose assigned direction is d_j. The feature vector V_k is then built by concatenating all histograms in a global feature vector $V_k = H_1 \oplus H_2 \oplus \ldots \oplus H_n$. After the feature extraction stage, two vessel structures are assigned with the corresponding feature vectors V_1 and V_2, which have the same length. Thus, matching between V_1 and V_2, can be easily performed by applying a distance measure. In our experiments we tested the cosine distance.

3 Experiments and Results

All the experiments have been performed on the MMCBNU_6000 finger vein dataset [28], using MATLAB (R2010a) on an Intel Core2 Duo U7300 @ 1.30 GHz with 4 GB of RAM. The MMCBNU_6000 dataset consists of 6000 finger vein grayscale images captured from 100 persons. It also provides 6000 segmented ROIs with a resolution of 64×128 pixels. For the sake of simplicity we will indicate the set of images of whole fingers as A and the set of corresponding ROIs as B.

The experiments consist of two different tests: (a) evaluating the verification accuracy on set A (Experiment I), (b) computing the verification accuracy while comparing set B to set A (Experiment II). We measured the verification accuracy in terms of Equal Error Rate (EER), Feature Dimensionality (FD) and Processing Time (PT). In both experiments, five finger vein images from one subject have been considered as the gallery set, while the remaining five images have been used as the probe set. This led to 3000 genuine matches and 1797000 imposter matches, respectively. The performance of PM-FVP has been compared with that of some most commonly used state of the art technique for experiment one. Performance results are reported in Table 1 and show the effectiveness of the proposed method.

Table 1. Performance in terms of EER, FD, PT.

Experiment	Methods	Equal error rate (%)	Feature dimensionality	Processing time (sec)
Experiment I	LDC [23]	3.15	7680	0.297
	LBP [26]	2.40	5760	0.221
	PM-FVP	0.80	768	0.512
Experiment II	PM-FVP	1.24	768	0.312

4 Conclusion

Finger vein authentication is emerging as one of the most promising biometric traits, as solves many of the limitations of existing biometrics like fingerprints, but offering comparable performances in terms of verification accuracy. In this paper we presented a new approach, namely PM-FVP, whose main contributions are: (i) designing a new local contrast enhancement filter, (ii) using new local directional features to both segmenting and matching finger vein pattern, (iii) co-registering vessel structures without needing precise segmentation of ROIs. Experimental results show that PM-FVP outperforms state of the art techniques in terms of verification accuracy. In future works we aim to combine both rigid and non rigid transformation to further improve the quality of the co-registration results, as we expect that this will further increase the verification performance.

References

1. Yu, C.-B., Zhang, D.-M., Li, H.-B.: Finger vein image enhancement based on multi-threshold fuzzy algorithm. In: 2nd International Congress on Image and Signal Processing, pp. 1–3. IEEE Press (2009)
2. Park, Y.H., Park, K.R.: Image quality enhancement using the direction and thickness of vein lines for finger-vein recognition. Int. J. Adv. Robot. Syst. **9**, 1–10 (2012)
3. Yang, J.F., Yang, J.L.: Multi-channel gabor filter design for finger vein image enhancement. In: 5th International Conference on Image and Graphics, pp. 87–91. IEEE Press (2009)
4. Zhang, Z., Ma, S., Han, X.: Multiscale feature extraction of finger-vein patterns based on curvelets and local interconnection structure neural network. In: 18th International Conference on Pattern Recognition, pp. 145–148. IEEE Press (2006)
5. Yang, J., Shi, Y.: Towards finger-vein image restoration and enhancement for finger-vein recognition. Inf. Sci. **268**, 33–52 (2014)
6. Yang, J., Li, X.: Efficient finger vein localization and recognition. In: 20th International Conference on Pattern Recognition, pp. 1148–1151. IEEE Press (2010)
7. Yang, J.F., Shi, Y.H.: Finger-vein ROI localization and vein ridge enhancement. Pattern Recogn. Lett. **33**, 1569–1579 (2012)
8. Lu, Y., Xie, S.J., Yoon, S., Yang, J.C., Park, D.S.: Robust finger vein ROI localization based on flexible segmentation. Sensor **13**(11), 14339–14366 (2013)

9. Miura, N., Nagasaka, A., Miyatake, T.: Feature extraction of finger-vein patterns based on repeated line tracking and its application to personal identification. Mach. Vis. Appl. **15**(4), 194–203 (2004)
10. Bakhtiar, A.R., Chai, W.S., Shahrel, A.S.: Finger vein recognition using local line binary pattern. Sensors **11**, 11357–11371 (2012)
11. Song, W., Kim, T., Kim, H.C., Choi, J.H., Kong, H.J., Lee, S.R.: A finger-vein verification system using mean curvature. PRL **32**(11), 1541–1547 (2011)
12. Huafeng, Q., Lan, Q., Chengbo, Y.: Region growth-based feature extraction method for finger vein recognition. Opt. Eng. **50**(2), 281–307 (2011)
13. Miura, N., Nagasaka, A., Miyatake, T.: Extraction of finger vein patterns using maximum curvature points in image profiles. IEICE Trans. Inf. Syst. **E90D**(8), 1185–1194 (2007)
14. Kumar, A., Zhou, Y.B.: Human identification using finger images. IEEE Trans. Image Process. **21**(4), 2228–2244 (2012)
15. Qin, H.F., Yu, C.B., Qin, L.: Region growth-based feature extraction method for fingervein recognition. Opt. Eng. **50**(5), 057208 (2011)
16. Liu, T., Xie, J.B., Yan, W., Li, P.Q., Lu, H.Z.: An algorithm for finger-vein segmentation based on modified repeated line tracking. Imaging Sci. J. **61**(6), 491–502 (2013)
17. Wu, J.D., Liu, C.T.: Finger-vein pattern identification using principal component analysis and the neural network technique. Expert Syst. Appl. **38**(5), 5423–5427 (2011)
18. Wu, J.D., Liu, C.T.: Finger-vein pattern identification using SVM and neural network technique. Expert Syst. Appl. **38**(11), 14284–14289 (2011)
19. Yang, G.P., Xi, X.M., Yin, Y.L.: Finger vein recognition based on $(2D)^2$ PCA and metric learning. J. Biomed. Biotechnol. **2012**, 1–9 (2012)
20. Liu, Z., Yin, Y.L., Wang, H., Song, S., Li, Q.: Finger vein recognition with manifold learning. J. Netw. Comput. Appl. **33**(3), 275–282 (2010)
21. Lee, E.C., Jung, H., Kim, D.: New finger biometric method using near infrared imaging. Sensors **11**(3), 2319–2333 (2011)
22. Rosdi, B.A., Shing, C.W., Suandi, S.A.: Finger vein recognition using local line binary pattern. Sensors **11**(12), 11357–11371 (2011)
23. Meng, X.J., Yang, G.P., Yin, Y.L., Xiao, R.Y.: Finger vein recognition based on local directional code. Sensors **12**(11), 14937–14952 (2012)
24. Dai, Y.G., Huang, B.N., Li, W.X., Xu, Z.Q.: A method for capturing the finger-vein image using nonuniform intensity infrared light. In: Proceedings of Congress on Image and Signal Processing, pp. 501–505 (2008)
25. Frucci, M., Riccio, D., di Baja, G.S., Serino, L.: SEVERE: SEgmenting VEssels in REtina images. Pattern Recogn. Lett. **82**, 162–169 (2016). doi:10.1016/j.patrec.2015.07.002
26. Lee, E.C., Lee, H.C., Park, K.R.: Finger vein recognition using minutia-based alignment and local binary pattern-based feature extraction. Int. J. Imaging Syst. Technol. **19**(3), 179–186 (2009)
27. Myronenko, A., Song, X.: Point set registration: coherent point drift. IEEE Trans. Pattern Anal. Mach. Intell. **32**(12), 2262–2275 (2010)
28. Lu, Y., Xie, S.J., Yoon, S., Wang, Z.H., Park, D.S.: A available database for the research of finger vein recognition. In: Proceedings of the 6th International Congress on Image and Signal Processing, pp. 410–415 (2013)

Highly Transparent and Secure Scheme for Concealing Text Within Audio

Diego Renza[✉], Camilo Lemus, and Dora M. Ballesteros L.

Universidad Militar Nueva Granada, Bogotá DC, Colombia
{diego.renza,camilo.lemus,dora.ballesteros}@unimilitar.edu.co

Abstract. This paper presents a highly transparent and secure scheme for concealing text within audio, based on Quantization Index Modulation and Orthogonal Variable Spreading Factor. The audio signal is decomposed through the Discrete Wavelet Transform and the approximation coefficients are selected to embed the text. Every character of the text is represented by a 256-bit orthogonal code, through mapping operations between the ASCII integer representation of the character and an external key. For improving the quality of recovered data, a repetition code is applied in the embedding process. Several tests were performed in order to measure the transparency of the output audio signal (i.e. stego signal) and the security of the recovered one. The main advantage of our proposal is the good trade-off among transparency, security and hiding capacity.

Keywords: Steganography · Quantization Index Modulation · Orthogonal Variable Spreading Factor · Discrete Wavelet Transform · Security

1 Introduction

Data hiding has been adopted as a way to securely embed and transport information within a digital media. There are three main parameters to measure the effectiveness of a data embedding system: transparency, hiding capacity and robustness. Transparency measures the level of imperceptibility of a hidden message within a media; hiding capacity refers to the amount of information that a scheme is able to successfully hide without adding perceptual/statistical distortion; and finally, robustness evaluates the resistance of the hidden data against attacks or modifications of the stego signal. Data hiding systems can be classified into watermarking and steganography; in watermarking the most important parameter is the robustness, while in steganography it is transparency [9].

Although there are a lot of works of image and video steganography, steganographic techniques have also been applied to audio signals [1]. Specifically, concealing text into audio can be used in applications of non-perceptual marking [7], authentication [10] and data transportation [6]. In general, current methods for audio steganography can be divided into three main categories depending

© Springer International Publishing AG 2017
C. Beltrán-Castañón et al. (Eds.): CIARP 2016, LNCS 10125, pp. 27–35, 2017.
DOI: 10.1007/978-3-319-52277-7_4

on the domain that is used in each one: temporal domain techniques, frequency domain techniques and transform domain techniques [5]. Many temporal domain techniques are based on bit modification methods, like the least significant bit (LSB); they are characterized by an easy implementation offering a high embedding capacity and transparency, but a weak security against intentional attacks [3]. In the second category, frequency domain techniques, the embedding process is applied in the frequency components of the signal, using, in most cases, techniques such as frequency masking [3], spread spectrum, shifted spectrum or Discrete Cosine Transform [4]. In the third category, the Discrete Wavelet Transform (DWT) has been used to embed the secret message into time-frequency coefficients of the audio signal. The most important advantage of the time-frequency domain over the other ones is a better relationship between transparency and hiding capacity [1].

Regarding hiding capacity, different methods have been focused on enhancing hiding capacity without missing transparency, Quantization Index Modulation (QIM) is one of them. This method is widely used nowadays in information hiding systems, because of its good performance in terms of transparency, quality of the recovered data and computational cost [2]. Nevertheless, the security level of the secret message can be a weakness for this kind of schemes. The application of spreading by means of orthogonal codes in the text data before embedding can improve security of the secret message [7].

According to the above, in this paper we propose a scheme for concealing text within audio, by using the QIM method in the wavelet domain to obtain transparent stego signals and by using Orthogonal Variable Spreading Factor (OVSF) codes to increase security of the secret message.

2   Proposed Audio Steganography Model

2.1   Embedding Module

The embedding module involves five main parts: *text processing and permutation, decomposition, repetition code, embedding,* and *reconstruction* (Fig. 1). The audio file and the secret text to be embedded are the inputs of this module.

Text Processing and Permutation: The input of this block is the secret text (S) and the outputs are the random vector (K_1) and the selected orthogonal codes (O_p). Firstly, a random vector K_1 is generated with the numbers of 1 to 256 in disorderly places. Secondly, the ASCII integer value of every character of S is obtained and kept in the vector I. For example, if $S = HOLA$, then ASCII(H) = 72, ASCII(O) = 79, ASCII(L) = 76, ASCII(A) = 65, and the result is $I = [72\ 79\ 76\ 65]$. In the third place, with the values (or elements) of I, locations of K_1 are sought with the following mapping function 1.

$$T_0 = K_1(I) \tag{1}$$

Suppose that $K_1(72) = 15$, $K_1(79) = 5$, $K_1(76) = 246$, $K_1(65) = 129$. Then, the result of this step is $T_o = [15\ 5\ 246\ 129]$. Fourthly, a 256-bit OVSF code is

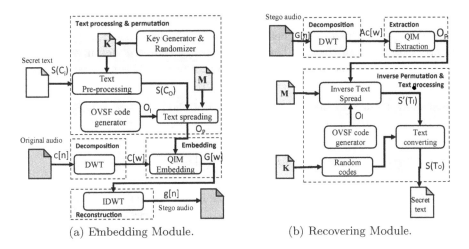

(a) Embedding Module.   (b) Recovering Module.

Fig. 1. Proposed audio steganography model.

constructed. The result is the matrix O_i which has 256 rows each one of 256 bits. Fifthly, with the numbers contained in T_o, the rows of the matrix O_i are selected. The output is the O_p matrix, whose number of rows equals to the number of characters in the secret text. Each row has 256 bits, according to Eq. 2.

$$O_p = \begin{bmatrix} O_i(T_o(1)) \\ O_i(T_o(2)) \\ \dots \\ O_i(T_o(N)) \end{bmatrix} = \begin{bmatrix} O_i(15) \\ O_i(5) \\ O_i(246) \\ O_i(129)) \end{bmatrix} \tag{2}$$

Where N is the total number of characters of the secret text. The right side matrix shows the example where $T_o = [15\ 5\ 246\ 129]$. In Summary, every character of the text is transformed to 256 bits, according to Table 1.

Table 1. Example of the result of the text processing and permutation block.

Text	I	$T_o = K_1(I)$	$O_p = O_i(T_o)$
H	72	15	$O_i(15)$
O	79	5	$O_i(5)$
L	76	246	$O_i(246)$
A	65	129	$O_i(129)$

Decomposition: The aim of this step is to obtain the wavelet coefficients of the audio signal, according to Eq. 3.

$$[A_c, D_c] = DWT(c) \tag{3}$$

The *DWT* function performs the wavelet decomposition of the signal c. The outputs are the approximation coefficients A_c and the detail coefficients D_c. Since the approximation coefficients have higher energy than the detail coefficients, they are selected to embed the binary word, O_p, repeated M times (see repetition code block).

Repetition Code: To increase the accuracy of the recovered text, the binary word O_p is repeated M times for adding a repetition code to itself.

$$M = \left\lfloor \frac{L}{256 \times N} \right\rfloor \tag{4}$$

Where L is the total number of approximation coefficients, N is the total number of characters of the secret text and $\lfloor . \rfloor$ is the integer part of data. With the current example, suppose that $L = 10240, N = 4$, then $M = 10$. It means, the secret text is repeated ten times within the audio signal. The result of this step is a binary data string of length L, named O_w, and organized as follows:

Finally, the *secret key* is constructed with the values of M and K_1.

Embedding: Once the text has been converted to $256 \times M$ bits per character, the bit string O_w is embedded into the approximation coefficients by the QIM technique. The process is to quantize every approximation coefficient, according to the bit to be hidden and the quantization step (Δ). With the current example, if there are 10240 approximation coefficients, and O_w has 10240 bits, every coefficient is quantized according to the bit value and Δ, as follows:

$$A'_c(i) = \begin{cases} \Delta \left\lfloor \frac{A_c(i)}{\Delta} \right\rfloor & if & O_w = 0 \\ \Delta \left\lceil \frac{A_c(i)}{\Delta} \right\rceil + \frac{\Delta}{2} & if & O_w = 1 \end{cases} \tag{5}$$

Where $A_c(i)$, $A'_c(i)$ are the approximation coefficient before and after the quantization process, respectively; $O_w(i)$ is the bit to be hidden; and i is the index position.

Reconstruction: The last stage of this module consists of reconstructing the audio signal from A'_c and the original D_c.

$$G = DWT^{-1}[A'_c, D_c] \tag{6}$$

After performing all the above steps, the stego audio G signal is generated and transmitted.

2.2   Recovering Module

In the extraction process, G and K_1 are the inputs of the recovering module, and the output is the recovered text. This process consists of applying three main steps: *decomposition, extraction* and *inversing permutation and text processing* (Fig. 1(b)).

Decomposition: The DWT is applied to the input signal G to obtain the approximation and the detail coefficients of the stego signal (A_{cs} and D_{cs}). It uses the same conditions of the wavelet base and number of decomposition levels of the embedding module.

Extraction: Inverse QIM is used for extracting the binary string from the approximation coefficients of the stego signal, through Eq. 7, where $O_{wr}(i)$ is the recovered bit.

$$O_{wr}(i) = \begin{cases} bit = 1 & if & \frac{\Delta}{4} < A_{cs}(i) - \Delta \left\lfloor \frac{A_{cs}(i)}{\Delta} \right\rfloor \leq \frac{3\Delta}{4} \\ bit = 0 & if & Otherwise \end{cases} \tag{7}$$

Inverse Permutation and Text Processing: Once the inverse QIM has been performed and the system knows all the recovered bits (one per approximation coefficient), this string is divided in 256-bit frames. The total number of characters of the text is obtained through the value of M contained into the *secret key*, as follows: $N = \left\lfloor \frac{L}{M \times 256} \right\rfloor$.

Where L is the total number of approximation coefficients, M is the number of redundancy, N is the number of characters, and $\lfloor \rfloor$ is the integer part of data. For example, if $L = 10240$ and $M = 10$, then $N = 4$.

The string O_{wr} is organized in a similar way of the string O_w (Fig. 2). Then, every $N \times 256$ bits, the recovered bits should be the same due to the repetition code. For example, with the current data, every 1024 bits (i.e. 4×256) should be the same.

Fig. 2. Distribution of the O_w vector.

The following step consists of comparing the first 256-bit string with every row of the matrix O_i, until a match is found. The number of the row is kept in the vector $T'_o(1)$. If none of the rows of O_i is equal to the bit string, the bits corresponding to the first character in the 2nd group are selected, and a comparison to the rows of O_i is made again. The above procedure is performed until a match between the OVSF code of the first character and a row of the matrix O_i is found. Then, the second 256-bit string is selected and the comparison process begins again. The length of T'_o is equal to the value of N. Subsequently, the values in T'_o vector are sought in the K_1 vector (obtained from the *secret key*) and the index of every match is kept in the vector I'. In other words, a vector I' that satisfies Eq. 8 is obtained:

$$K_1(I') = T'_0 \tag{8}$$

For example if $T_0' = [15\ 5\ 246\ 129]$ and $K_1(72) = 15$, $K_1(79) = 5$, $K_1(76) = 246$, $K_1(65) = 129$, then, the vector I' that satisfies Eq. 8 is $I' = [72\ 79\ 76\ 65]$. The last step consists of looking for the corresponding ASCII value of I', the result is S'. For the current example, the recovered text is $S' = HOLA$.

3   Experimental Results and Discussion

Experimental validation is focused on the measurement of transparency and security analysis of the proposed scheme. Since the proposed method can work with different audio signals regardless sampling frequency, it was tested with 20 music files downloaded from www.freesound.org database and two speech files recorded by the authors. Also the length of the text ranged from 1 to 12 characters. According to preliminary tests, the *Haar* base is selected to decompose the audio signal because of the requirement of perfect reconstruction. This parameter is fixed in both the embedding and recovering modules.

3.1   Transparency

Transparency is measured through the following parameters: PSNR (Peak Signal to Noise Ratio), and HER (Histogram Error Ratio). PSNR calculates the correlation between the maximum value of original signal and the noise added by embedding process; HER measures the normalized difference between two histograms. They are calculated as follows:

$$PSNR = 10Log_{10}(\frac{MAX^2}{MSE}) \tag{9}$$

Where MAX is the maximum value of the host signal (C), and MSE is the mean squared error between the stego (G) and the host signal. A higher value of PSNR means less distortion between two signals [2].

$$HER = \frac{\sum_{i=2}^{N}(Ch_i - Gh_i)^2}{\sum_{i=2}^{N}(Ch_i)^2} \tag{10}$$

Where, Ch and Gh are the histogram of the host and stego signals, respectively. The ideal value for HER is 0 [8]. Figure 3 shows the PSNR results for different Δ values. All approximation coefficients of the first decomposition level were modified. Figure 4 shows the comparison between a zoom of the original audio signal and the stego audio signal in time domain. According to the results, the PSNR is higher than 100 dB, which represents a very low distortion in the host signal and then high transparency. In HER case, values are in the order of 10^{-16}, which means that data distribution of the stego signal is very similar to the one of the host signal, and again, it works with high transparency.

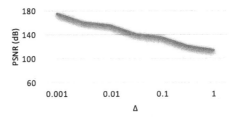

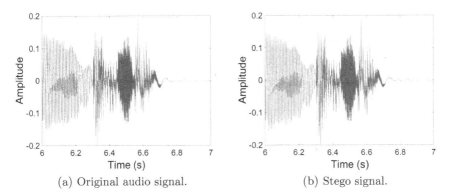

Fig. 3. PSNR average of the stego signal for all the tested signals when $\Delta = [0\ 1]$.

(a) Original audio signal.   (b) Stego signal.

Fig. 4. Comparison of a zoom of original audio signal and stego-signal.

3.2   Security

The most important contribution of this proposal is the security of the text through the use of OVSF codes and the vector K_1. It is only, if the user has both data, that she/he can reveal the secret content. Since the total number of rows of the matrix O_i (which has the orthogonal codes) is 256, the total number of mapping choices between the input character and its orthogonal code is 256. If the secret text has N characters, each one can be mapped with 1 of the 256 orthogonal codes and then, the total number of available choices is: $\#keys = 256^N$. Where $\#keys$ is the total number of available keys that an intruder must test to discover the secret text. For example, if $N = 4$, an intruder must test around 4.2×10^9 choices. The higher the value of N, the longer is the total number of available keys.

With the purpose to illustrate the security of the proposed scheme, Table 2 is presented. The original vector K_1 is slightly modified and used it to recover the text. The result is completely different to the original secret text.

Although a text is always recovered, there is not a way to know if the recovered text matches the original text, since an alphanumeric string will always be obtained and the original text could not have a logic or linguistic sequence.

Table 2. Test on extracting secret text with slightly modification of (K_1).

Audio type	Secret text	Text length (characters)	Recovered text
Music	UMNG2016	8	td Í " ˆ
Music	UMNG2016	8	SÆ'ÊRWi
Music	12Stego34	9	•(2Mõ%o!Q
Music	12Stego34	9	f>&\fi,{Ó
Music	12Stego34	9	5 – '0¶>$D\pm$
Music	M@CB%ok4?	10	)Áó×QJp 3/4 Ü
Music	<T"EST]>	10	B5><éK5
Music	r9e803bd?	10	Ÿ"–²Ê0nL,,
Music	M@CB%ok4?	10	çžT" 1_

4   Conclusion

A scheme of concealing text into audio (voice or music) has been presented in this paper. The strengths of the proposal are: high transparency of the stego signal and high security of the secret content. The first characteristic obeys the very low distortion introduced by the quantization process (i.e. by the QIM method) in the wavelet domain. A good selection of (Δ) gives stego signals with PSNR higher than 100 dB. The second characteristic is related to the mapping process between every character of the secret text and a 256-bit orthogonal code by the use of OVSF codes. Only the correct key (composed by K_1 and M) allows obtaining the correct text. On the other hand, a repetition code is used for providing a fortress against possible errors in recovering bits. Then, some errors are tolerated in the detection of the OVSF codes, because the secret text is embedded many times into the approximation coefficients of the host signal. Results of several simulations show perfect recovering of the secret text.

Acknowledgment. This work is supported by the "Universidad Militar Nueva Granada-Vicerrectoría de Investigaciones" (grant IMP-ING-2136 of 2016).

References

1. Ballesteros L, D.M., Moreno A, J.M.: Highly transparent steganography model of speech signals using efficient wavelet masking. Expert Syst. Appl. **39**(10), 9141–9149 (2012)
2. Basu, P.N., Bhowmik, T.: On embedding of text in audio. In: International Conference on Recent Trends in Information, Telecommunication and Computing, pp. 203–206. IEEE (2010)
3. Djebbar, F., Ayad, B., Hamam, H., Abed-Meraim, K.: A view on latest audio steganography techniques. In: International Conference on Innovations in Information Technology, pp. 409–414. IEEE (2011)

4. Famili, Z., Faez, K., Fadavi, A.: A new steganography based on χ^2 technic. In: Bayro-Corrochano, E., Eklundh, J.-O. (eds.) CIARP 2009. LNCS, vol. 5856, pp. 1062–1069. Springer, Heidelberg (2009). doi:10.1007/978-3-642-10268-4_124

5. Lin, Y., Abdulla, W.H.: Audio Watermark. Springer, Heidelberg (2015)

6. Qadir, M.A., Ahmad, I.: Digital text watermarking: secure content delivery and data hiding in digital documents. IEEE Aerosp. Electron. Syst. Mag. **21**(11), 18–21 (2006)

7. Renza, D., Ballesteros, D.M., Ortiz, H.D.: Text hiding in images based on QIM and OVSF. IEEE Lat. Am. Trans. **14**(3), 1206–1212 (2016)

8. Shahadi, H.I., Jidin, R., Way, W.H.: Lossless audio steganography based on lifting wavelet transform and dynamic stego key. Indian J. Sci. Technol. **7**(3), 323–334 (2014)

9. Zamani, M., Manaf, A., Ahmad, R.B., Jaryani, F., Taherdoost, H., Zeki, A.M.: A secure audio steganography approach. In: International Conference for Internet Technology and Secured Transactions, pp. 1–6. IEEE (2009)

10. Zmudzinski, S., Steinebach, M.: Perception-based audio authentication watermarking in the time-frequency domain. In: Katzenbeisser, S., Sadeghi, A.-R. (eds.) IH 2009. LNCS, vol. 5806, pp. 146–160. Springer, Heidelberg (2009). doi:10.1007/978-3-642-04431-1_11

Spatio-Colour Asplünd's Metric and Logarithmic Image Processing for Colour Images (LIPC)

Guillaume Noyel[1(✉)] and Michel Jourlin[1,2]

[1] International Prevention Research Institute, 95 cours Lafayette, 69006 Lyon, France
guillaume.noyel@i-pri.org
http://www.i-pri.org
[2] Lab. H. Curien, UMR CNRS 5516, 18 rue Pr. B. Lauras, 42000 St-Etienne, France

Abstract. Asplünd's metric, which is useful for pattern matching, consists in a double-sided probing, i.e. the over-graph and the sub-graph of a function are probed jointly. This paper extends the Asplünd's metric we previously defined for colour and multivariate images using a marginal approach (i.e. component by component) to the first spatio-colour Asplünd's metric based on the vectorial colour LIP model (LIPC). LIPC is a non-linear model with operations between colour images which are consistent with the human visual system. The defined colour metric is insensitive to lighting variations and a variant which is robust to noise is used for colour pattern matching.

Keywords: Asplünd's metric · Spatio-colour metric · Colour Logarithmic Image Processing · Double-sided probing · Colour pattern recognition

1 Introduction

The Asplünd's metric initially defined for binary shapes [1,4] has been extended to grey-scale images by Jourlin et al. [6,7] and to colour and multivariate images in the LIP framework by Noyel and Jourlin [13]. It consists in probing a function by two homothetic template functions, i.e. the probes which are computed by the LIP multiplication.

The Logarithmic Image Processing (LIP) model initially defined for grey level images by Jourlin and Pinoli [8,9] is perfectly suited for images acquired by transmitted light (i.e. when the observed object is located between the source and the sensor) and by reflected light because of its consistency with the Human Vision [3]. The necessity to analyse together the channels of the colour images (i.e. by a vectorial analysis) has led to the introduction of the Logarithmic Image Processing for Colour images (LIPC) by Jourlin et al. [5].

The LIP Asplünd's metric was defined in [13] in a marginal way (i.e. channel by channel). In this paper, our contribution is to extend this metric by using the spatio-colour properties [11,12] of the colour LIPC framework.

After some prerequisites about the colour LIPC model and about the marginal LIP Asplünd's metric, we will define a spatio-colour Asplünd's metric in the LIPC framework. Then we will perform spatio-colour pattern matching which is robust to noise. Examples will illustrate the definitions.

© Springer International Publishing AG 2017
C. Beltrán-Castañón et al. (Eds.): CIARP 2016, LNCS 10125, pp. 36–43, 2017.
DOI: 10.1007/978-3-319-52277-7_5

2   Prerequisites

2.1   LIPC Model

A colour image $\mathbf{f}$, defined on a domain $D \subset \mathbb{R}^N$, with values in $\mathcal{T}^3 = [0, M[^3$, $M \in \mathbb{R}$, is written:

$$\mathbf{f} : \begin{cases} D \to & \mathcal{T}^3 = [0, M[^3 \\ x \to \mathbf{f}(x) = (f_R(x), f_G(x), f_B(x)) \end{cases} \tag{1}$$

f_R, f_G, f_B are the red, green and blue channels (i.e. components) of $\mathbf{f}$, $\mathbf{f}(x)$ is a vector-pixel and x is the spatial coordinate of the vector-pixel. The real value M is equal to $2^8 = 256$ for 8 bits images. Given P the number of pixels, the matrix $\mathbf{F}$ of $E \to \mathcal{T}$, $E = 3 \times P$, associated to the image $\mathbf{f}$ is written:

$$\mathbf{F} = \begin{bmatrix} f_R(x_1) & f_R(x_2) & \dots & f_R(x_P) \\ f_G(x_1) & f_G(x_2) & \dots & f_G(x_P) \\ f_B(x_1) & f_B(x_2) & \dots & f_B(x_P) \end{bmatrix} \tag{2}$$

To make the comments easier, the word "image" designates both the matrix $\mathbf{F}$ and the image $\mathbf{f}$. The image space for 24-bits images $\mathbf{F}$ is written $\mathcal{I}^3$.

A colour image is a particular case of a multivariate image defined as $\mathbf{f}_\lambda$: $D \to \mathcal{T}^L$, where $L \in \mathbb{N}$ is the number of channels [11,12].

As for the grey-level LIP, the colour LIPC framework is based on colour transmittance [5]. It is valid for transmitted and reflected images [3]. It models the human perceptual system approach by taking into account: (i) the sensitivity of the human eye in the visible domain characterised by colour matching functions of Stiles and Burch [14] and (ii) the spectral distribution of light with the D65 illuminant [15].

In the LIPC framework, the transmittance of the sum of two images $\mathbf{T}_{\mathbf{F} \triangle_c \mathbf{G}}$ is equal to the product of their transmittances $\mathbf{T}_\mathbf{F}$ and $\mathbf{T}_\mathbf{G}$: $\mathbf{T}_{\mathbf{F} \triangle_c \mathbf{G}} = \mathbf{T}_\mathbf{F} * \mathbf{T}_\mathbf{G}$. The symbol of the LIPC addition is $\triangle_c$ and $*$ represents the element-wise multiplication [5]. The addition of two images $\mathbf{F}, \mathbf{G} \in \mathcal{I}^3$ is:

$$\mathbf{F} \triangle_c \mathbf{G} = \acute{\mathbf{K}}^{-1} \acute{\mathbf{U}} (\acute{\mathbf{U}}^{-1} \acute{\mathbf{K}} \mathbf{F} * \acute{\mathbf{U}}^{-1} \acute{\mathbf{K}} \mathbf{G}). \tag{3}$$

$\acute{\mathbf{K}}$ and $\acute{\mathbf{U}}$ are real matrices of size 3×3 corresponding to the LIPC mixing model[1]. From the LIPC addition, a multiplication by a scalar $\alpha \in \mathbb{R}$ has been defined:

$$\alpha \triangle_c \mathbf{F} = \acute{\mathbf{K}}^{-1} \acute{\mathbf{U}} (\acute{\mathbf{U}}^{-1} \acute{\mathbf{K}} \mathbf{F})^\alpha. \tag{5}$$

---

[1] With colour matching functions of Stiles and Burch (1959) and D65 illuminant [5], matrices $\acute{\mathbf{K}}$ and $\acute{\mathbf{U}}$ equal to:

$$\acute{\mathbf{U}} = \begin{bmatrix} 25.0440 & 53.1416 & 176.8144 \\ 21.3002 & 185.9744 & 47.7254 \\ 229.2474 & 19.9944 & 5.7583 \end{bmatrix} \acute{\mathbf{K}} = \begin{bmatrix} 0.6991 & 0.2109 & 0.0899 \\ 0.1947 & 0.8002 & 0.0049 \\ 0.0681 & 0.0002 & 0.9315 \end{bmatrix} \tag{4}$$

The space $(\mathcal{I}^3, \triangle_c, \triangle_c)$ is the positive cone of a vector space with robust mathematical properties.

Physical interpretation [5]: the LIPC addition corresponds to the superposition of two semi-transparent layers. A LIPC multiplication by a scalar $\alpha \in \,]0,1[$ brightens the result by suppressing layers, while a scalar $\alpha \in \,]1,+\infty[$ darkens the result by superimposing α times the image on itself.

2.2 Marginal Asplünd's Metric for Colour and Multivariate Images

In [13], an Asplünd's metric between colour images was defined with the LIP model by using a marginal approach (i.e. channel by channel) [11,12].

Definition 1. *The Asplünd's metric (with LIP multiplication) between two colour images* $\mathbf{f}$ *and* $\mathbf{g}$ *on a region* $Z \subset D$ *is*

$$d_{As,Z}^{\triangle}(\mathbf{f}, \mathbf{g}) = \ln(\lambda/\mu) \qquad (6)$$

with $\lambda = \inf\{k, \forall x \in Z, k \triangle g_R(x) \geq f_R(x), k \triangle g_G(x) \geq f_G(x), k \triangle g_B(x) \geq f_B(x)\}$ *and* $\mu = \sup\{k, \forall x \in Z, k \triangle g_R(x) \leq f_R(x), k \triangle g_G(x) \leq f_G(x), k \triangle g_B(x) \leq f_B(x)\}$.

In particular, by the property of the distance $d_{As,Z}^{\triangle}(\mathbf{f}, \mathbf{g}) = d_{As,Z}^{\triangle}(\mathbf{g}, \mathbf{f})$.

3   Asplünd's Metric Defined in the Logarithmic Image Processing Colour (LIPC) Framework

Given two colours $C_1 = (r_1, g_1, b_1)$, $C_2 = (r_2, g_2, b_2) \in \mathcal{T}^3$, as we are only looking for lower and upper bounds, a marginal order [2] is used: $C_1 \geq C_2 \Leftrightarrow \{r_1 \geq r_2$ and $g_1 \geq g_2$ and $b_1 \geq b_2\}$.

Definition 2. *Given two colours* C_1, $C_2 \in \mathcal{T}^3$, *their Asplünd's distance (with LIPC multiplication) is equal to:*

$$d_{As}^{\triangle_c}(C_1, C_2) = \ln(\mu/\lambda) \qquad (7)$$

$\lambda = \inf_k\{k \triangle_c C_2 \geq C_1\}$ *and* $\mu = \sup_k\{k \triangle_c C_2 \leq C_1\}$.

Strictly speaking, $d_{As}^{\triangle_c}$ is a metric if the colours C_n are replaced by their equivalence classes $\tilde{C}_n = \{C \in \mathcal{T}^3 / \exists \alpha \in \mathbb{R}^+, \ \alpha \triangle_c C = C_n\}$.

Comment: in Eq. 7 contrary to the Asplünd's distance (with LIP multiplication) defined in [13] (Eq. 6), we have $\lambda \leq \mu$ because, by definition of the colour LIPC model the scales are inverted as compared to the grey LIP model [5].

Colour metrics (with LIPC multiplication) between two colour images $\mathbf{f}$ and $\mathbf{g}$ may be defined as the sum (d_1 metric) or the supremum (d_∞) of $d_{As}^{\triangle_c}(C_1, C_2)$ on the region of interest $Z \subset D$ of cardinal $\#Z$

$$\begin{aligned} d_{1,Z}^{\triangle_c}(\mathbf{f}, \mathbf{g}) &= \tfrac{1}{\#Z} \sum_{x \in Z} d_{As}^{\triangle_c}(\mathbf{f}(x), \mathbf{g}(x)) \\ d_{\infty,Z}^{\triangle_c}(\mathbf{f}, \mathbf{g}) &= \sup_{x \in Z} d_{As}^{\triangle_c}(\mathbf{f}(x), \mathbf{g}(x)) \end{aligned} \qquad (8)$$

The Asplünd's metric can be extended to colour functions.

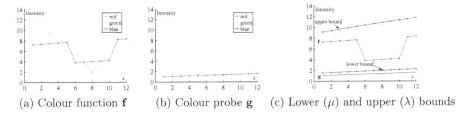

(a) Colour function $\mathbf{f}$ (b) Colour probe $\mathbf{g}$ (c) Lower (μ) and upper (λ) bounds

Fig. 1. Computation of the Asplünd's distance between two colour functions $d_{As,D}^{\triangle_c}(\mathbf{f}, \mathbf{g}) = 0.43$. Each colour channel is represented by a line of the same colour. (Colour figure online)

Definition 3. *The colour Asplünd's metric (with LIPC multiplication) between two colour images $\mathbf{f}$ and $\mathbf{g}$ on a region $Z \subset D$ is*

$$d_{As,Z}^{\triangle_c}(\mathbf{f}, \mathbf{g}) = \ln(\mu/\lambda) \tag{9}$$

$\lambda = \inf_k \{\forall x \in Z, k \triangle_c \mathbf{g}(x) \geq \mathbf{f}(x)\}$ *and* $\mu = \sup_k \{\forall x \in Z, k \triangle_c \mathbf{g}(x) \leq \mathbf{f}(x)\}$.

In Fig. 1, the Asplünd's metric has been computed between the colour probe $\mathbf{g}$ and the colour function $\mathbf{f}$ on their definition domain D.

Comment: the lower (resp. upper) bound $\mu \triangle_c \mathbf{g}$ (resp. $\lambda \triangle_c \mathbf{g}$) may not be equal to any point of the function $\mathbf{f}$ but strictly less (or greater) than the function. Indeed, one can demonstrate that the following assertion is verified: "given $\mathbf{C}_0 \in T^3, \forall C \in T^3, \nexists \lambda \in \mathbb{R}^+/\lambda \triangle_c C_0 = C$".

The metric $d_{As,Z}^{\triangle_c}$ can be adapted to local processing with a colour template image (i.e. the probe) $\mathbf{t}$ defined on a spatial support $D_t \subset D$. For each point $x \in D$, the distance $d_{As,D_t}^{\triangle_c}(\mathbf{f}_{|D_t(x)}, \mathbf{t})$ is computed on the neighbourhood $D_t(x)$ centred in x where $\mathbf{f}_{|D_t(x)}$ is the restriction of $\mathbf{f}$ to $D_t(x)$.

Definition 4. *Given a colour image $\mathbf{f}$ defined on D with values in T^3, $(T^3)^D$, a colour probe $\mathbf{t}$ defined on D_t with values in T^3, $(T^3)^{D_t}$, and $D_t(x)$ the neighbourhood D_t centred in $x \in D$, the map of Asplünd's distances (with $\triangle_c$) is:*

$$As_{\mathbf{t}}^{\triangle_c}\mathbf{f} : \begin{cases} (T^3)^D \times (T^3)^{D_t} \rightarrow & (\mathbb{R}^+)^D \\ (\mathbf{f}, \mathbf{t}) \rightarrow As_{\mathbf{t}}^{\triangle_c}\mathbf{f}(x) = d_{As,D_t}^{\triangle_c}(\mathbf{f}_{|D_t(x)}, \mathbf{t}) \end{cases} \tag{10}$$

In Fig. 2, the map of Asplünd's distances is computed between a colour function and a colour probe. The minima of the map corresponds to the location of a pattern which is similar to the probe.

Asplünd's distance is sensitive to noise because the probe lays on regional extrema that may be caused by noise (Fig. 1). In [7,13], definitions of Asplünd's distance with a tolerance on the extrema have been introduced. In this paper, we extend this definition for colour images with LIPC model.

To reduce the sensitivity of Asplünd's distance to the noise, the "Measure metric" or "M-metric" has been defined in the context of "Measure Theory".

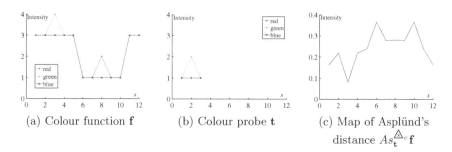

(a) Colour function **f**   (b) Colour probe **t**   (c) Map of Asplünd's distance $As_t^{\triangle_c}\mathbf{f}$

Fig. 2. (c) Map of the Asplünd's distances $As_t^{\triangle_c}\mathbf{f}$ between a colour function and a probe. (a) and (b) Each colour channel is represented by a line of the same colour. (Colour figure online)

The image being digitized, the number of pixels of D is finite and the "measure" of a subset of D is linked to the cardinal of this subset, e.g. the percentage P of its elements with respect to D. We are looking for a subset D' of D, such that $\mathbf{f}_{|D'}$ and $\mathbf{g}_{|D'}$ are neighbours for Asplünd's metric and the complementary set $D \setminus D'$ of D' into D is of small size when compared to D. This last condition is written as: $P(D \setminus D') = \frac{\#(D \setminus D')}{\#D} \leq p$, where p is an acceptable percentage and $\#D$ is the number of elements in D.

Given ϵ a small positive real number, the neighbourhood of function $\mathbf{f}$ is

$$N_{P,d_{As},\epsilon,p}(\mathbf{f}) = \left\{ \mathbf{g} \setminus \exists D' \subset D, d_{As,D'}^{\triangle_c}(\mathbf{f}_{|D'}, \mathbf{g}_{|D'}) < \epsilon \text{ and } \frac{\#(D \setminus D')}{\#D} \leq p \right\} \quad (11)$$

The closest points of the probe to the function are discarded as in [5,13].

Definition 5. *Given two constant vector-pixels* $\mathbf{c}_\mu, \mathbf{c}_\lambda \in \mathcal{T}^3$, *a percentage* p *of points to be discarded. The colour Asplünd's metric (with LIPC multiplication) with tolerance between two colour images* $\mathbf{f}$ *and* $\mathbf{g}$ *on a region* $Z \subset D$ *is*

$$d_{As,Z,p}^{\triangle_c}(\mathbf{f}, \mathbf{g}) = \ln(\mu'/\lambda') \quad (12)$$

$\lambda' = \inf_k \{\forall x \in Z, k \triangle_c \mathbf{g}(x) \geq \mathbf{f}(x) - \mathbf{c}_\lambda\}$ *and* $\mu' = \sup_k \{\forall x \in Z, k \triangle_c \mathbf{g}(x) \leq \mathbf{f}(x) + \mathbf{c}_\mu\}.\mathbf{c}_\mu$ *and* $\mathbf{c}_\lambda$ *are increased such as a percentage* p *of points is discarded.*

In Fig. 3, a tolerance of $p = 20\%$ is used to discard two points. The Asplünd's distance decreases from 0.43 to 0.21.

A map of Asplünd's distances (with $\triangle_c$) can now be defined.

Definition 6. *Given a colour image* $\mathbf{f}$ *of* $(\mathcal{T}^3)^D$, *a colour probe* $\mathbf{t}$ *of* $(\mathcal{T}^3)^{D_t}$ *and a tolerance* $p \in [0,1]$, *the map of Asplünd's distances with a tolerance is:*

$$As_{t,p}^{\triangle_c}\mathbf{f} : \begin{cases} (\mathcal{T}^3)^D \times (\mathcal{T}^3)^{D_t} \to & (\mathbb{R}^+)^D \\ (\mathbf{f}, \mathbf{t}) & \to As_{t,p}^{\triangle_c}\mathbf{f}(x) = d_{As,D_t,p}^{\triangle_c}(\mathbf{f}_{|D_t(x)}, \mathbf{t}) \end{cases} \quad (13)$$

$D_t(x)$ *is the neighbourhood* D_t *centred in* $x \in D$.

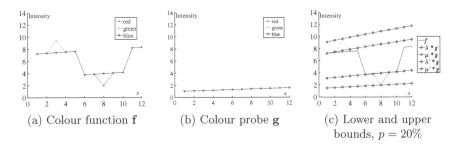

(a) Colour function **f**

(b) Colour probe **g**

(c) Lower and upper
bounds, $p = 20\%$

Fig. 3. Colour Asplünd's distance with a tolerance of $p = 20\%$. (μ, λ) are the scalars multiplying the probe without tolerance. (μ', λ') are the scalars multiplying the probe with tolerance. $d_{As,D}^{\triangle c}(\mathbf{f}, \mathbf{g}) = 0.43$ and $d_{As,D,p=20\%}^{\triangle c}(\mathbf{f}, \mathbf{g}) = 0.21$ (Colour figure online)

4   Examples

In Fig. 4, we look for the bricks of a wall, similar to a colour probe. A blue brick has been added to the wall. In the image without noise **f**, the regional minima of the map $As_{\mathbf{t}}^{\triangle c}\mathbf{f}$ (dark points in Fig. 4b) correspond to the centre of the bricks similar to the probe (according to the Asplünd's distance). The white

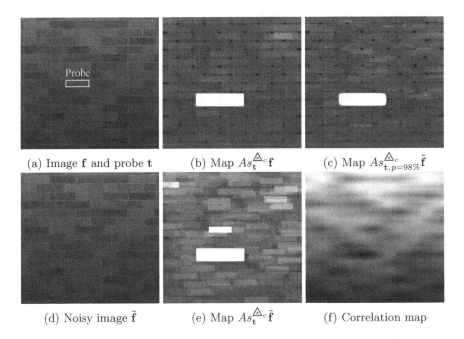

(a) Image **f** and probe **t**

(b) Map $As_{\mathbf{t}}^{\triangle c}\mathbf{f}$

(c) Map $As_{\mathbf{t},p=98\%}^{\triangle c}\tilde{\mathbf{f}}$

(d) Noisy image $\tilde{\mathbf{f}}$

(e) Map $As_{\mathbf{t}}^{\triangle c}\tilde{\mathbf{f}}$

(f) Correlation map

Fig. 4. Maps of Asplünd's distances without tolerance $As_{\mathbf{t}}^{\triangle c}\tilde{\mathbf{f}}$ and with $As_{\mathbf{t},p}^{\triangle c}\tilde{\mathbf{f}}$. $\tilde{\mathbf{f}}$ image with a white noise ($\sigma^2 = 2.6$, spatial density 1%). (f) Correlation map. (Colour figure online)

rectangle corresponds to the maxima of the distances between the blue brick and the probe. Therefore, the distance is sensitive to colour (i.e. the hue). In the image with noise $\tilde{\mathbf{f}}$, the map without tolerance $As_{\mathbf{t}}^{\triangle_c}\tilde{\mathbf{f}}$ is more sensitive to noise (Fig. 4e) than the map with tolerance $As_{\mathbf{t},p}^{\triangle_c}\tilde{\mathbf{f}}$ (Fig. 4c). Indeed, the minima are preserved into the map with tolerance (Fig. 4c) compared to the map without (Fig. 4e). The minima can be extracted using mathematical morphology [10,16]. Importantly, all the maps of Asplünd's distances are insensitive to the vertical lighting drift. Moreover, a correlation map is useless to find the location of the bricks (Fig. 4f).

In Fig. 5, two images of the same scene, a bright image $\mathbf{f}$ and a dark image $\tilde{\mathbf{f}}$, are acquired with two different exposure times. The probe $\mathbf{t}$ is extracted in the bright image and used to compute the map of Asplünd's distance $As_{\mathbf{t}}^{\triangle_c}\tilde{\mathbf{f}}$ in the darker image. By finding the minima of the map, all the balls are detected and their contours are added to the image in Fig. 5(b). One can notice that the Asplünd's distance is very robust to the lighting variations.

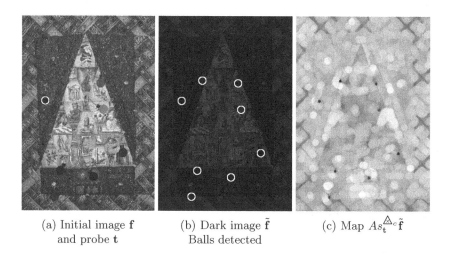

| (a) Initial image $\mathbf{f}$ and probe $\mathbf{t}$ | (b) Dark image $\tilde{\mathbf{f}}$ Balls detected | (c) Map $As_{\mathbf{t}}^{\triangle_c}\tilde{\mathbf{f}}$ |

Fig. 5. Detection of coloured balls on a dark image $\tilde{\mathbf{f}}$ with a probe $\mathbf{t}$ extracted in the bright image $\mathbf{f}$. (a) The border of the probe $\mathbf{t}$ is coloured in white. (Colour figure online)

5   Conclusion and Perspectives

A new spatio-colour Asplünd's distance based on colour LIPC model has been defined. It is a true colour (i.e. vectorial) metric based on a colour model consistent with the human visual system. It is also consistent with the previous properties given in [7,13]. An extension of this metric robust to noise has been presented and illustrated on pattern recognition examples. This double-sided probing distance is efficient for colour pattern matching and performs better than traditional correlation methods. In future work, we will evaluate in details the properties of this colour distance on practical applications (e.g. in medical, remote sensing or industrial images). We will compare it to the marginal colour Asplünd's distance and we will study the links between Asplünd's probing and mathematical morphology.

References

1. Asplünd, E.: Comparison between plane symmetric convex bodies and parallelograms. Math. Scand. **8**, 171–180 (1960)
2. Barnett, V.: The ordering of multivariate data. J. R. Stat. Soc. Ser. A (Gen.) **139**(3), 318–355 (1976)
3. Brailean, J., Sullivan, B., Chen, C., Giger, M.: Evaluating the EM algorithm for image processing using a human visual fidelity criterion. In: 1991 International Conference on Acoustics, Speech, and Signal Processing, ICASSP 1991, vol. 4, pp. 2957–2960, April 1991
4. Grünbaum, B.: Measures of symmetry for convex sets. In: Proceedings of Symposia in Pure Mathematics, vol. 7, pp. 233–270 (1963)
5. Jourlin, M., Breugnot, J., Itthirad, F., Bouabdellah, M., Closs, B.: Chapter 2 - Logarithmic image processing for color images. In: Hawkes, P.W. (ed.) Advances in Imaging and Electron Physics, vol. 168, pp. 65–107. Elsevier, Amsterdam (2011)
6. Jourlin, M., Carré, M., Breugnot, J., Bouabdellah, M.: Chapter 7 - Logarithmic image processing: additive contrast, multiplicative contrast, and associated metrics. In: Hawkes, P.W. (ed.) Advances in Imaging and Electron Physics, vol. 171, pp. 357–406. Elsevier, Amsterdam (2012)
7. Jourlin, M., Couka, E., Abdallah, B., Corvo, J., Breugnot, J.: Asplünd's metric defined in the logarithmic image processing (LIP) framework: a new way to perform double-sided image probing for non-linear grayscale pattern matching. Pattern Recogn. **47**(9), 2908–2924 (2014)
8. Jourlin, M., Pinoli, J.: A model for logarithmic image processing. J. Microsc. **149**(1), 21–35 (1988)
9. Jourlin, M., Pinoli, J.: Logarithmic image processing: the mathematical and physical framework for the representation and processing of transmitted images. In: Hawkes, P.W. (ed.) Advances in Imaging and Electron Physics, vol. 115, pp. 129–196. Elsevier, Amsterdam (2001)
10. Matheron, G.: Eléments pour une théorie des milieux poreux. Masson, Paris (1967)
11. Noyel, G., Angulo, J., Jeulin, D.: Morphological segmentation of hyperspectral images. Image Anal. Stereol. **26**(3), 101–109 (2007)
12. Noyel, G., Angulo, J., Jeulin, D., Balvay, D., Cuenod, C.A.: Multivariate mathematical morphology for DCE-MRI image analysis in angiogenesis studies. Image Anal. Stereol. **34**(1), 1–25 (2014)
13. Noyel, G., Jourlin, M.: Asplünd's metric defined in the logarithmic image processing (LIP) framework for colour and multivariate images. In: 2015 IEEE International Conference on Image Processing (ICIP), pp. 3921–3925, September 2015
14. Stiles, W.S., Burch, J.M.: N.P.L. Colour-matching investigation: final report (1958). Optica Acta: Int. J. Opt. **6**(1), 1–26 (1959). doi:10.1080/713826267
15. Schanda, J.: Colorimetry: Understanding the CIE System. Wiley, New York (2007)
16. Serra, J., Cressie, N.: Image Analysis and Mathematical Morphology, vol. 1. Academic Press, London (1982)

Two Compound Random Field Texture Models

Michal Haindl$^{(\boxtimes)}$ and Vojtěch Havlíček

The Institute of Information Theory and Automation
of the Czech Academy of Sciences, Prague, Czech Republic
{haindl,havlicek}@utia.cz

Abstract. Two novel models for texture representation using parametric compound random field models are introduced. These models consist of a set of several sub-models each having different characteristics along with an underlying structure model which controls transitions between them. The structure model is a two-dimensional probabilistic mixture model either of the Bernoulli or Gaussian mixture type. Local textures are modeled using the fully multispectral three-dimensional causal autoregressive models. Both presented compound random field models allow to reproduce, compress, edit, and enlarge a given measured color, multispectral, or bidirectional texture function (BTF) texture so that ideally both measured and synthetic textures are visually indiscernible.

Keywords: Texture · Texture synthesis · Compound random field model · CAR model · Two-dimensional Bernoulli mixture · Two-dimensional Gaussian mixture · Bidirectional texture function

1 Introduction

Physically correct and visually convincing virtual models require object surfaces covered with realistic nature-like surface material textures to present realism in virtual scenes. The primary purpose of any synthetic texture approach is to reproduce and enlarge a given measured material texture so that ideally both natural and synthetic texture will be visually indiscernible. The appearance of real materials dramatically changes with illumination and viewing variations and its most advanced current texture representation is the seven-dimensional Bidirectional Texture Function (BTF) [9]. Unfortunately, measured texture data are nearly always too limited to reliable estimate these complex seven-dimensional models, thus their modeling requires some simplifying factorization [9], such as the presented compound random field models which serve as the the three-dimensional factor model in the complex overall BTF material model [9].

Compound random field models (CRF) consist of several sub-models each having different characteristics along with an underlying structure model which controls transitions between these sub models [11]. Compound Markov filed models were successfully applied to image restoration [2,3,11,12], segmentation [13], and modeling [6,7,10]. However, these models always require demanding numerical

© Springer International Publishing AG 2017
C. Beltrán-Castañón et al. (Eds.): CIARP 2016, LNCS 10125, pp. 44–51, 2017.
DOI: 10.1007/978-3-319-52277-7_6

solutions with all their well known drawbacks. Our exceptional CMRF [6] model allows analytical synthesis at the cost of a slightly compromised compression rate.

We propose two textural models - CRF$^{BM-3CAR}$, CRF$^{GM-3CAR}$, based on complex spatial probabilistic mixture models. These control field models are either probabilistic Bernoulli of Gaussian mixture models.

2   Compound Random Field Texture Models

Let us denote a multiindex $r = (r_1, r_2)$, $r \in I$, where I is a discrete 2-dimensional rectangular lattice and r_1 is the row and r_2 the column index, respectively. $X_r \in \{1, 2, \ldots, K\}$ is a random variable with natural number value (a positive integer), Y_r is multispectral pixel at location r and $Y_{r,j} \in \mathcal{R}$ is its j-th spectral plane component. Both random fields (X, Y) are indexed on the same lattice I. Let us assume that each multispectral or BTF observed texture $\tilde{Y}$ (composed of d spectral planes) can be modelled by a compound random field model, where the principal random field X controls switching to a regional local model $Y = \bigcup_{i=1}^{K} {}^i Y$. Single K regional sub-models ${}^i Y$ are defined on their corresponding lattice subsets ${}^i I$, ${}^i I \cap {}^j I = \emptyset$ $\forall i \neq j$ and they are of the same RF type. They differ only in their contextual support sets ${}^i I_r$ and corresponding parameters sets ${}^i \theta$. The CRF model has posterior probability $P(X, Y \mid \tilde{Y}) = P(Y \mid X, \tilde{Y}) P(X \mid \tilde{Y})$ and the corresponding optimal MAP solution is: $(\hat{X}, \hat{Y}) = \arg\max_{X \in \Omega_X, Y \in \Omega_Y} P(Y \mid X, \tilde{Y}) P(X \mid \tilde{Y})$, where Ω_X, Ω_Y are corresponding configuration spaces for random fields (X, Y). To avoid an iterative MCMC MAP solution, we propose the following two step approximation [6]:

$$(\check{X}) = \arg \max_{X \in \Omega_X} P(X \mid \tilde{Y}), \tag{1}$$

$$(\check{Y}) = \arg \max_{Y \in \Omega_Y} P(Y \mid \check{X}, \tilde{Y}). \tag{2}$$

This approximation significantly simplifies the CRF$^{BM-3CAR}$, CRF$^{GM-3CAR}$ estimation because it allows us to take advantage of a straightforward analytical estimation of the regional RF models ${}^i Y$ in (2).

2.1   Region Switching Model

The control RF $(P(X \mid \tilde{Y}))$ is supposed to be represented by two-dimensional Bernoulli or Gaussian distribution mixture model, respectively. The mixture distribution $P(Y_{\{r\}})$ has the form:

$$P(Y_{\{r\}}) = \sum_{m \in M} P(Y_{\{r\}} \mid m)\, p(m) = \sum_{m \in M} \prod_{s \in I_r} p_s(Y_s \mid m)\, p(m) \tag{3}$$

where $Y_{\{r\}} \in \mathcal{K}^\eta$, $\mathcal{M} = \{1, 2, \ldots, M\}$, $I_r \subset I$ is an index set, $\eta = cardinality\{I_r\}$, and $p(m)$ are probability weights $\sum_{m \in \mathcal{M}} p(m) = 1$. The maximum-likelihood parameter estimates $p(m)$ (probability weights),

μ_{ms}, σ_{ms} (Gaussian mixture component means and standard deviation), $\theta_{m,s}$ (Bernoulli mixture component parameters) are computed using the EM algorithm [1,4] $p_s^{(t+1)}(. \,|\, m)$ and

$$q^{(t)}(m \,|\, X_{\{r\}}) = \frac{p^{(t)}(m)\, P^{(t)}(X_{\{r\}} \,|\, m)}{\sum_{j \in \mathcal{M}} p^{(t)}(j) P^{(t)}(X_{\{r\}} \,|\, j)}, \tag{4}$$

$$p^{(t+1)}(m) = \frac{1}{|\mathcal{S}|} \sum_{X_{\{r\}} \in \mathcal{S}} q^{(t)}(m \,|\, X_{\{r\}}). \tag{5}$$

Bernoulli Distribution Mixture Model. We assume that control field pixel $X_r \in \mathcal{K}$ where $\mathcal{K}$ is the index set of K distinguished sub-models. The distribution $P(X_{\{r\}})$ is assumed to be multivariable Bernoulli mixture (BM) and the control field is further decomposed into separate binary bit planes of binary variables $\xi \in \mathcal{B}$, $\mathcal{B} = \{0,1\}$ which are separately modeled and can be learned from much smaller training texture than a multi-level discrete mixture model. We suppose that a bit factor of a control field can be fully characterised by a marginal probability distribution of binary levels on pixels within the scope of a window centered around the location r and specified by the index set $I_r \subset I$, i.e. $X_{\{r\}} \in \mathcal{B}^\eta$ and $P(X_{\{r\}})$ is the corresponding marginal distribution of $P(X \,|\, \tilde{Y})$. The component distributions $P(\cdot \,|\, m)$ are factorisable, and multivariable Bernoulli:

$$P(X_{\{r\}} \,|\, m) = \prod_{s \in I_r} \theta_{m,s}^{X_s} (1 - \theta_{m,s})^{1 - X_s} \qquad X_s \in X_{\{r\}}. \tag{6}$$

The mixture model parameters (6) include component weights $p(m)$ and the univariate discrete distributions of binary levels. They are simply defined by one parameter $\theta_{m,s}$ as a vector of probabilities:

$$p_s(\cdot \,|\, m) = (\theta_{m,s}, 1 - \theta_{m,s}). \tag{7}$$

The EM solution is (4), (5) and

$$p_s^{(t+1)}(\xi \,|\, m) = \frac{1}{|\mathcal{S}|\, p^{(t+1)}(m)} \sum_{X_{\{r\}} \in \mathcal{S}} \delta(\xi, X_s)\, q^{(t)}(m \,|\, X_{\{r\}}), \quad \xi \in \mathcal{B}. \tag{8}$$

The total number of mixture (3), (7) parameters is thus $M(1+\eta)$ – confined to the appropriate norming conditions. The advantage of the multivariable Bernoulli model (7) is a simple switch-over to any marginal distribution by deleting superfluous terms in the products $P(X_{\{r\}} \,|\, m)$.

Gaussian Mixture Model. The discrete control field can be alternatively modeled by a continuous RF if we map single indices into continuous random variables with uniformly separated mean values and small variance. The synthesis results are subsequently inversely mapped back into a corresponding synthetic discrete control field. We assume the joint probability distribution $P(X_{\{r\}})$,

$X_{\{r\}} \in \mathcal{K}^\eta$ in the form of a normal mixture and the mixture components are defined as products of univariate Gaussian densities

$$P(X_{\{r\}} \mid \mu_m, \sigma_m) = \prod_{s \in I_{\{r\}}} p_s(X_s \mid \mu_{ms}, \sigma_{ms}), \tag{9}$$

$$p_s(X_s \mid \mu_{ms}, \sigma_{ms}) = \frac{1}{\sqrt{2\pi}\sigma_{ms}} \exp\left\{ -\frac{(X_s - \mu_{ms})^2}{2\sigma_{ms}^2} \right\},$$

i. e., the components are multivariate Gaussian densities with diagonal covariance matrices. The maximum-likelihood estimates of the parameters $p(m), \mu_{ms}, \sigma_{ms}$ can be computed by EM algorithm [1,4]. Anew we use a data set $\mathcal{S}$ obtained by pixel-wise shifting the observation window within the original texture image $\mathcal{S} = \{X_{\{r\}}^{(1)}, \dots, X_{\{r\}}^{(K)}\}, \ X_{\{r\}}^{(k)} \subset X$. The corresponding log-likelihood function is maximized by the EM algorithm $(m \in \mathcal{M}, n \in \mathcal{N}, X_{\{r\}} \in \mathcal{S})$ and the iterations are $(4), (5)$ and

$$\mu_{m,n}^{(t+1)} = \frac{1}{\sum_{X_{\{r\}} \in \mathcal{S}} q^{(t)}(m \mid X_{\{r\}})} \sum_{X_{\{r\}} \in \mathcal{S}} X_n \, q(m \mid X_{\{r\}}), \tag{10}$$

$$(\sigma_{m,n}^{(t+1)})^2 = -(\mu_{m,n}^{(t+1)})^2 + \frac{\sum_{X_{\{r\}} \in \mathcal{S}} X_n^2 \, q^{(t)}(m \mid X_{\{r\}})}{\sum_{X_{\{r\}} \in \mathcal{S}} q(m \mid X_{\{r\}})}. \tag{11}$$

Control Field Synthesis. We can assume at a given position r of the contextual neighbourhood I_r to have some part of the pixel-wise synthesised control field $X_{\{r\}}$ already specified. If $X_{\{\rho\}}$ is a sub-vector of all of $X_{\{r\}}$ pixels previously specified within this window and $I_\rho \subset I_r$ the corresponding index subset, then the statistical properties of the remaining unspecified variables are fully described by the corresponding conditional distribution:

$$p_{n \mid \rho}(X_n \mid X_{\{\rho\}}) = \sum_{m=1}^{M} W_m(X_{\{\rho\}}) \, p_n(X_n \mid m), \tag{12}$$

where $W_m(X_{\{\{\rho\}\}})$ are the a posteriori component weights corresponding to the given sub-vector $X_{\{\rho\}}$:

$$W_m(X_{\{\rho\}}) = \frac{p(m)P_\rho(X_{\{\rho\}} \mid m)}{\sum_{j=1}^{M} p(j)P_\rho(X_{\{\rho\}} \mid j)}, \tag{13}$$

$$P_\rho(X_{\{\rho\}} \mid m) = \prod_{n \in \rho} p_n(X_n \mid m).$$

X_n can be randomly generated by the conditional distribution $p_{n \mid \rho}(X_n \mid X_{\{\rho\}})$ where by Eq. (12) can be applied to all the unspecified variables $n = \eta - \text{card}\{\rho\}$ given a fixed position of the control field. Each newly generated X_n is used to upgrade the conditional weights $W_m(X_{\{\rho\}})$.

2.2 Local Markov Models

Local i-th texture region (not necessarily continuous) is represented by the adaptive 3D causal auto-regressive random (3DCAR) field model [5,8]. This model can be analytically estimated as well as easily synthesised. The model can be defined in the following matrix equation (i-th model index is further omitted to simplify notation):

$$Y_r = \gamma Z_r + \epsilon_r, \tag{14}$$

where $Z_r = [Y_{r-s}^T : \forall s \in I_r]^T$ is the $\eta d \times 1$ data vector with multiindices r, s, t, $\gamma = [A_1, \ldots, A_\eta]$ is the $d \times d\,\eta$ unknown parameter matrix with parametric sub-matrices A_s. The model functional contextual neighbour index shift set is denoted I_r and $\eta = cardinality(I_r)$. All CAR model statistics can be efficiently estimated analytically [8]. Given the known 3DCAR process history $Y^{(t-1)} = \{Y_{t-1}, Y_{t-2}, \ldots, Y_1, Z_t, Z_{t-1}, \ldots, Z_1\}$ the parameter estimation $\hat{\gamma}$ can be accomplished using fast, numerically robust and recursive statistics [8]:

$$\hat{\gamma}_{t-1}^T = V_{zz(t-1)}^{-1} V_{zy(t-1)}, \qquad \dot{V}_{t-1} = \tilde{V}_{t-1} + V_0,$$

$$\tilde{V}_{t-1} = \begin{pmatrix} \sum_{u=1}^{t-1} Y_u Y_u^T & \sum_{u=1}^{t-1} Y_u Z_u^T \\ \sum_{u=1}^{t-1} Z_u Y_u^T & \sum_{u=1}^{t-1} Z_u Z_u^T \end{pmatrix} = \begin{pmatrix} \tilde{V}_{yy(t-1)} & \tilde{V}_{zy(t-1)}^T \\ \tilde{V}_{zy(t-1)} & \tilde{V}_{zz(t-1)} \end{pmatrix},$$

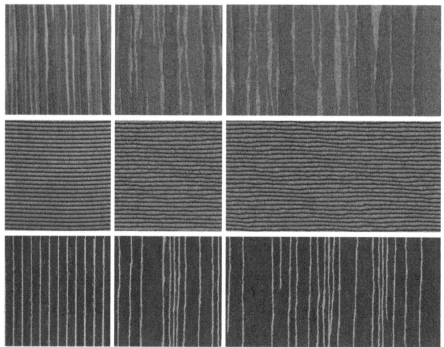

Fig. 1. Cloth and frosted planks (left column) synthesis and enlargement (right column) using the $CRF^{BM-3CAR}$ model.

where V_0 is a positive definite matrix (see [8]). Although, an optimal causal functional contextual neighbourhood I_r can be solved analytically by a straightforward generalisation of the Bayesian estimate in [8], we use faster approximation which does not need to evaluate statistics for all possible I_r configurations. This approximation is based on spatial correlations. Starting from the causal part

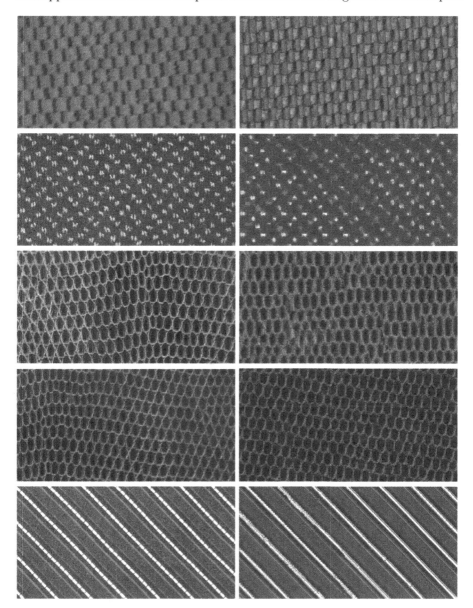

Fig. 2. Measured three textile and two cobra skin textures (left column) and their synthesis using the $CRF^{GM-3CAR}$ model.

of a hierarchical non-causal neighbourhood, neighbours locations corresponding to spatial correlations larger than a specified threshold (>0.6) are selected. The i-th model pixel-wise synthesis is simple direct application of (14) for all 3DCAR models. 3DCAR models provide better spectral modelling quality than the alternative spectrally decorrelated 2D models for motley textures at the cost of small increase of number of parameters to be stored.

3   Experiments

Both presented compound random field models ($CRF^{BM-3CAR}$, $CRF^{GM-3CAR}$) are well suited for near-regular textures such as textile materials which are notoriously difficult for Markov random field type of textural models [6,9]. The dimension of the estimated control field model distribution is not too high ($\eta \approx 10^1 - 10^2$) and the number of the training data vectors is relatively large ($|\mathcal{S}| \approx 10^4 - 10^5$). Nevertheless the window should always be kept reasonably small and the sample size as large as possible. In our experiments we have used a regular left-to-right and top-to-down shifting of the generating window. Figure 1 illustrates the $CRF^{BM-3CAR}$ model applied to a frosted planks and two textile textures synthesis and enlargement, while Fig. 2 shows three textile materials and two skin samples synthesized using the $CRF^{GM-3CAR}$ model.

4   Conclusion

Both presented CRF ($CRF^{GM-3CAR}, CRF^{BM-3CAR}$) methods show good visual performance on selected real-world measured materials. The appearance of such materials should consist of several types of relatively small regions with fine-granular inner structure such as fabric, skin, or wood. The models offer large data compression ratio (only tens of parameters per BTF) easy simulation and fast seamless synthesis of any required texture size. The methods can be easily generalised for colour or BTF texture editing by estimating some local models from different target materials. The model does not compromise spectral correlation thus it can reliably model motley textures. A drawback of the presented CRF models is the need to have sufficiently large learning data for both mixture sub-models.

Acknowledgements. This research was supported by the Czech Science Foundation project GAČR 14-10911S.

References

1. Dempster, A., Laird, N., Rubin, D.: Maximum likelihood from incomplete data via the EM algorithm. J. Roy. Stat. Soc., B **39**(1), 1–38 (1977)
2. Figueiredo, M., Leitao, J.: Unsupervised image restoration and edge location using compound Gauss - Markov random fields and the MDL principle. IEEE Trans. Image Process. **6**(8), 1089–1102 (1997)

3. Geman, S., Geman, D.: Stochastic relaxation, Gibbs distributions and Bayesian restoration of images. IEEE Trans. Pattern Anal. Mach. Intell. **6**(11), 721–741 (1984)
4. Grim, J., Haindl, M.: Texture modelling by discrete distribution mixtures. Comput. Stat. Data Anal. **41**(3–4), 603–615 (2003)
5. Haindl, M., Havlíček, V.: A multiscale colour texture model. In: Kasturi, R., Laurendeau, D., Suen, C. (eds.) Proceedings of the 16th International Conference on Pattern Recognition, pp. 255–258. IEEE Computer Society, Los Alamitos, August 2002. http://dx.doi.org/10.1109/ICPR.2002.1044676
6. Haindl, M., Havlíček, V.: A compound MRF texture model. In: Proceedings of the 20th International Conference on Pattern Recognition, ICPR 2010, pp. 1792–1795. IEEE Computer Society CPS, Los Alamitos, August 2010. http://doi.ieeecomputersociety.org/10.1109/ICPR.2010.442
7. Haindl, M., Remeš, V., Havlíček, V.: Potts compound Markovian texture model. In: Proceedings of the 21st International Conference on Pattern Recognition, ICPR 2012, pp. 29–32. IEEE Computer Society CPS, Los Alamitos, November 2012
8. Haindl, M.: Visual data recognition and modeling based on local Markovian models. In: Florack, L., Duits, R., Jongbloed, G., Lieshout, M.C., Davies, L. (eds.) Mathematical Methods for Signal and Image Analysis and Representation, Computational Imaging and Vision, vol. 41, chap. 14, pp. 241–259. Springer, Heidelberg (2012). http://dx.doi.org/10.1007/978-1-4471-2353-8_14
9. Haindl, M., Filip, J.: Visual Texture. Advances in Computer Vision and Pattern Recognition. Springer-Verlag, London (2013)
10. Haindl, M., Havlíček, V.: A plausible texture enlargement and editing compound Markovian model. In: Salerno, E., Çetin, A.E., Salvetti, O. (eds.) MUSCLE 2011. LNCS, vol. 7252, pp. 138–148. Springer, Heidelberg (2012). doi:10.1007/978-3-642-32436-9_12. http://www.springerlink.com/content/047124j43073m202/
11. Jeng, F.C., Woods, J.W.: Compound Gauss-Markov random fields for image estimation. IEEE Trans. Sig. Process. **39**(3), 683–697 (1991)
12. Molina, R., Mateos, J., Katsaggelos, A., Vega, M.: Bayesian multichannel image restoration using compound Gauss-Markov random fields. IEEE Trans. Image Proc. **12**(12), 1642–1654 (2003)
13. Wu, J., Chung, A.C.S.: A segmentation model using compound Markov random fields based on a boundary model. IEEE Trans. Image Process. **16**(1), 241–252 (2007)

An Automatic Tortoise Specimen Recognition

Matěj Sedláček[1], Michal Haindl[1(✉)], and Dominika Formanová[2]

[1] The Institute of Information Theory and Automation
of the Czech Academy of Sciences, Prague, Czech Republic
{sedlacek,haindl}@utia.cz
[2] Czech Environmental Inspectorate, Prague, Czech Republic
formanova_dominika@cizp.cz

Abstract. The spur-thighed tortoise (*Testudo graeca*) is listed among endangered species on the CITES list and the need to keep track of its specimens calls for a noninvasive, reliable and fast method that would recognize individual tortoises one from another. We present an automatic system for the recognition of tortoise specimen based on variable-quality digital photographs of their plastrons using an image classification approach and our proposed discriminative features. The plastron image database, on which the recognition system was tested, consists of 276 low-quality images with a variable scene set-up and of 982 moderate-quality images with a fixed scene set-up. The achieved overall success rates of automatically identifying a tortoise in the database were 43,0% for the low-quality images and 60,7% for the moderate-quality images. The results show that the automatic tortoise recognition based on the plastron images is feasible and suggests further improvements for a real application use.

Keywords: Tortoise recognition · *Testudo graeca*

1 Introduction

The spur-thighed tortoise population in Mediterranean decreases as their natural environment dwindles and they are simultaneously collected for the illegal pet trade [6]. As other endangered species, the spur-thighed tortoise is listed on the EC Council regulation No. 338/97, Annex A and on the Appendix II of the Convention on International Trade in Endangered Species of Wild Fauna and Flora (CITES II), which classifies it as being directly threatened with extinction in the wild. It may become so unless the trade in its specimens is subject to a strict regulation. Several tortoise species are thus bred in captivity and only captive born animals can be sold and kept as pets. European countries try to curb the illegal pet trade by identifying and monitoring tortoises in captivity so that the illegally collected and imported specimens could be intercepted, returned to the wild and the poachers prosecuted. Therefore all spur-thighed tortoises held in captivity must be registered and their owners cannot trade them unless they obtain exemption from the prohibition of trade. We propose an automatic system for the recognition of individual tortoises from low and unconstrained quality

© Springer International Publishing AG 2017
C. Beltrán-Castañón et al. (Eds.): CIARP 2016, LNCS 10125, pp. 52–59, 2017.
DOI: 10.1007/978-3-319-52277-7_7

plastron photographs required for their official registration. To our knowledge this is the first attempt of the automatic tortoise specimen recognition.

2   Discriminative Tortoise Features

The primary problem for a tortoise recognition is that there are not know any time-steady discriminative visual features for such longevous animal [3,4]. However, without such time-invariant features we cannot hope to uniquely identify a tortoise specimen anytime during its lifetime, if the possible time span between an acquired registration photograph and the corresponding living animal validation can be several decades.

Features such as the angles between plastron seams, areas of plastron scutes and color or texture of the plastron are all highly variable during a tortoise life. Based on the work of biologists, we hypothesize, that the discriminative features might be the normalized lengths of the segments of the central seam on the plastron [2,5,7]. Using marginal seams, as proposed in [7], is not possible in our case as the sides of the plastron are not always well visible due to uneven lightning or a rotated position of the plastron.

We choose as the discriminative features the lengths of the segments on the central seam that are bounded by connections of two consecutive side seams, one on the left side and the other one on the right side of the central seam, respectively (see Fig. 1-left). These features $l_i = \frac{L_i}{L}$ and $r_i = \frac{R_i}{L}$ $i = 1, \ldots, 6$ are normalized by the total length $L = \sum_{k=1}^{6} L_k = \sum_{k=1}^{6} R_k$ of the central seam to eliminate plastron size changes due to the growth of the tortoise and variable resolution of the tortoise database images.

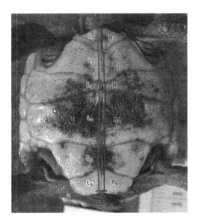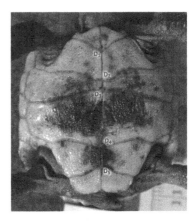

Fig. 1. The individual left and right central seam segments are marked in red and in blue (left). The differences between connections of the side seams on the left and right of the central seam are marked in red (right). (Color figure online)

Additional selected normalized features $d_i = \frac{D_i}{L}$ $i = 1, \ldots, 5$ are the differences between connection nodes of the lateral left and right seam branches for the five inner junctions on the central seam (see Fig. 1-right).

The complete feature vector for a given i-th plastron image has seventeen features

$$\mathbf{f}^i = [l_1^i, \ldots, l_6^i, r_1^i, \ldots, r_6^i, d_1^i, \ldots, d_5^i]^T \in \mathbb{R}^{17}. \tag{1}$$

If the plastron is clean, then the central seams are well visible on all tortoise images in our database. Our feature hypothesis was successfully verified on manually-measured proposed features. This proves that the chosen features are sufficiently discriminative provided we are able to automatically localize them with sufficient precision.

3   Automatic Detection of Tortoise Features

The positions of all seam junctions on the central seam have to be detected in order to measure the lengths of individual central seam segments. This junctions localization has to be done with the highest possible precision because it directly influences the final tortoise recognition accuracy.

3.1   Plastron Localization

The first step is the plastron localization. The position and the orientation of the plastron is detected using the generalized Hough transform [1] applied to the general plastron template shown in Fig. 2.

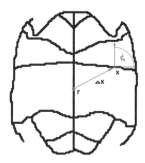

Fig. 2. Plastron template for the generalized Hough transformation.

3.2   Central Seam Junctions Localization

The central seam junctions are approximately localized one by one using a graph algorithm for finding the positions of the side seams and the central seam. As can be seen on the close-ups of a plastron image and its corresponding edge image in Fig. 3, the plastron seams create almost continuous long edge curves, which is not the case for other parts of the plastron.

To locate left and right side seams, the edges of the plastron area were evaluated one by one from the left, resp. the right, end side of the plastron, respectively, towards the central seam. We assign a value to each edge pixel, which is

 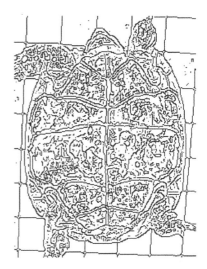

Fig. 3. The side seams as well as the central seam of the plastron image (left) form almost continuous curves of edges in the corresponding edge image (right).

the minimum from the sums of the distances to the edge pixels to the left, resp. to the right, and the already assigned values to those pixels. Therefore, each pixel is assigned the shortest "path" to the left, reps. to the right, side of the plastron. The distance between two edges is calculated as

$$d(e_1, e_2) = |e_{1_y} - e_{2_y}| + (|e_{1_x} - e_{2_x}| + 1)^2. \tag{2}$$

Since the difference in absolute value of the x coordinates in (2) is squared, the edge values are smaller where there is a close succession of vertices along the x axis direction (i.e. the direction of the side seams). The local minima of the edge values near the position of the central seam give the position of the seam junctions. Similarly, the precise position of the central seam can be found using this method in the direction of the central seam (along y axis). Given the positions of the central seam junctions, the sought features can be calculated.

4   Tortoise Classification

The classification whether two tortoise images belong to a same specimen or not is done using the kNN classifier based on the absolute differences vector of the measured features from the two images (the classification feature vectors) $\Delta \mathbf{f}^{ij} = |\mathbf{f}^i - \mathbf{f}^j|$, $i \neq j$. The feature vectors set is divided into two classes. The class 1 with feature (difference) vectors of identical animal plastrons and the class 2 with plastron feature vectors from distinct animals. It was known beforehand which plastron images in the database belong to which specific tortoise.

5   Experimental Results

The plastron image database, on which the suitability of the features and the successfulness of the automatic recognition was tested, consists of 276 varied and mostly low-quality plastron images with a variable scene set-up (CEI dataset), see Fig. 4, and of 982 moderate-quality plastron images with a constant scene set-up (Bender's dataset), see Fig. 5. The Bender's dataset [2] contains also several series of the same animals recorder in different age steps (up to 16 images).

Fig. 4. Four plastron examples illustrating the CEI dataset diversity.

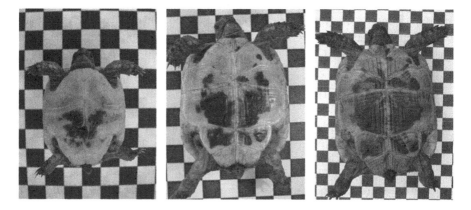

Fig. 5. Examples of plastron images from the Bender's dataset. Photographs of a tortoise at the age of 6 days (left), 272 days (middle) and 3 years 59 days (right).

5.1 Feature Discriminability Validation

The discriminability hypothesis of our selected tortoise features (1) is verified on the manually-measured features on the complete tortoise database. A set of 5204 class 1 feature vectors and a set of 5203 class 2 feature vectors are calculated and both sets are equally divided into a training and test sub-sets. The kNN classification is done for two different $k = 5$ and $k = 30$ values, which were estimated experimentally as giving the best results. These results are presented in Table 1-manual where single columns denotes, true positive (class 1, class 2), false negative (FN), and false positive (FP) values, respectively. High true positive values in Table 1-manual suggest acceptable discriminability of the selected features. These results are obviously relative to available tortoise data, where only limited aging measurements for single individuals are available.

Table 1. The classification results for the manually or automatically measured features. The settings and the results for the smallest achieved false negative (FN) error are highlighted.

k	Class 1	Class 2	FN	FP
	Manually-measured			
5	**99,42**	**88.54**	**0,58**	**11,46**
30	99,33	84,57	0,67	15,43
	Automatically-measured			
5	63,32	89,05	36,68	10,95
30	**67,79**	**87,88**	**32,21**	**12,12**

5.2 Classification of Automatically Measured Features

The overall success rates of the automatic localization of the junctions composed of the plastron detection and the localization of the individual branch junctions was obtained by counting all images, where none of these methods failed (see Table 2).

Table 2. Overall results of the automatic junction localization.

Dataset	All localization methods combined	
	# of successes	success rate %
CEI	175 of 276	**63,41**
Bender	880 of 982	**89,61**

The results of the success/error rate for the hold-out rate estimation (disjunct training and test sets) based on the automatically measured features are given in Table 1 - automatic. The test set was composed of 4182 class 1 feature vectors and 4183 class 2 feature vectors, which are calculated from the automatically measured features on all plastron images from the database where the automatic feature extraction was successful. The training set is composed of 5204 class 1 and 5203 class 2 manually-measured feature vectors. The classification was not carried out separately for the CEI dataset and the Bender's dataset as the CEI dataset contained only few time-lagged photographs of the same tortoises and therefore the results of the class 1 feature vectors classification would not be reliable. The overall success rates of the proposed tortoise recognition (see Table 3) were obtained by multiplying the success rates of the automatic approximate junction localization (Table 2) with the success rates of the feature classification based on the automatically measured features (Table 1-automatic).

Table 3. Overall success rates of the automatic tortoise recognition for the CEI and Bender's data-sets.

Dataset	Class 1	Class 2
CEI	42,99	55,72
Bender	60,75	78,75

The results deterioration (18–23%) between Bender's and CEI databases is due to uneven and mostly lower quality of the CEI images. Comparison between the best achievable recognition results (Table 1-manual) suggests a need for further research to improve the salient points detection accuracy especially for the identical animal detection class.

6   Conclusion

We have proposed a set of discriminative features for the individual tortoise identification. These features compose of the normalized lengths of segments of the central seam on the plastron. Using a classification of these manually-measured features, it was experimentally verified that these chosen features are stable relative to our limited tortoise aging test set. We were able to reach correct individual animal detection with 0.99 probability for any time-lagged plastron image of a specific tortoise, which is essential for the task of inspection, i.e., to check whether a tortoise has a time-lagged image in the database thus it is a registered animal. Similarly, two images of different tortoises were classified correctly as not belonging to a same tortoise with 0.88 probability. Thus these features can be used for individual tortoise recognition. Fully automatic tortoise recognition has lower recognition rate for both classes due to imprecise salient points localization, which is the topic for our further research.

Acknowledgements. This research was supported by the Czech Science Foundation project GAČR 14-10911S. The CEI database is courtesy of the Czech Environmental Inspectorate and the Bender's database is courtesy of Dr. Caroline Bender.

References

1. Ballard, D.H.: Generalizing the hough transform to detect arbitrary shapes. Pattern Recogn. **13**(2), 111–122 (1981)
2. Bender, C., Henle, K.: Individuelle fotografische identifizierung von land-schildkröten-arten (testudinidae) des anhangs a der europaisch artenschutzverord-nung. Salamandra **37**(4), 193–204 (2001)
3. Carretero, M., Znari, M., Harris, D., Mace, J.: Morphological divergence among populations of testudo graeca from west-central Morocco. Anim. Biol. **55**(3), 259–279 (2005)
4. Fritz, U., Hundsdorfer, A., Siroky, P., Auer, M., Kami, H., Lehmann, J., Mazanaeva, L., Turkozan, O., Wink, M.: Phenotypic plasticity leads to incongruence between morphology-based taxonomy and genetic differentiation in western palaearctic tortoises (testudo graeca complex; testudines, testudinidae). Amphibia-Reptilia **28**(1), 97–121 (2007)
5. Mosimann, J.: Variation and relative growth in the plastral scutes of the turtle kinosternon integrum leconte (1956)
6. Pérez, I., Giménez, A., Sánchez-Zapata, J., Anadón, J., Martinez, M., Esteve, M.: Non-commercial collection of spur-thighed tortoises (testudo graeca graeca): a cultural problem in southeast Spain. Biol. Conserv. **118**(2), 175–181 (2004)
7. Tichỳ, L., Kintrová, K.: Specimen identification from time-series photographs using plastron morphometry in testudo graeca ibera. J. Zool. **281**(3), 210–217 (2010)

Autonomous Scanning of Structural Elements in Buildings

B. Quintana$^{(\boxtimes)}$, S.A. Prieto, A. Adán, and A.S. Vázquez

3D Visual Computing and Robotics Lab, Universidad de Castilla-La Mancha,
Ciudad Real, Spain
{Blanca.Quintana,Samuel.Prieto,Antonio.Adan,
Andress.Vazquez}@uclm.es

Abstract. This paper presents a method for automatic scanning of structural elements in indoors. Whereas most of the next-best-scan (NBS) based methods do not separate clutter and useful data, we present a scanning strategy in which potential structural components (SE) of the building are recognized as a new scan is carried out. This makes our method more efficient and less time consuming compared with the rest. Besides, our approach gives a response to essential issues in the scanning world, such as the data discrimination, the hypotheses about the workspace and the complexity of the scanned scene. The method has been tested in indoors under occlusion and clutter yielding promising results. Additionally, a comparison with three techniques close to ours is included in the paper.

Keywords: 3D sensing · 3D data processing · Next best view · 3D modeling

1 Introduction: Key Points and Contributions

Nowadays automatic scanning of buildings is a challenging and very active research in which there are still underlying questions (scan's objective, hypothesis and the scene complexity) that are rarely debated and that determine the validity of the method.

In the majority of the approaches the scanning strategy does not depend on the final objective [1] (preliminary version of our approach). Thus, the objective is frequently to scan everything which lies inside [2] or outside [3] the building. Data redundancy and cluttering are therefore ignored. Those methods are inefficient because a great part of the gathered 3D data can be irrelevant for the final goal. In contrast with these methods, our proposal aims to capture the data belonging to structural elements of the scene (i.e. ground, walls and ceiling) to further generate a realistic 3D CAD model of the building. Consequently, we can highly reduce the volume of data and simplify the algorithmia of future processes.

Hypotheses in the scanning process determine the soundness and versatility of a proposal. Some NBS algorithms a priori assume bounding boxes or convex-hulls that contain the scene [2]. Others take manually a set of preliminary sparse scans of the environment and later tackle the NBS algorithm [3].

© Springer International Publishing AG 2017
C. Beltrán-Castañón et al. (Eds.): CIARP 2016, LNCS 10125, pp. 60–67, 2017.
DOI: 10.1007/978-3-319-52277-7_8

An important issue related with this idea, ignored in most of the papers, is the updating of the workspace with a new scan. This makes the earlier methods less reliable and credible in real scenarios composed of several irregular adjacent rooms. In this matter, we define a dynamic workspace that contains the accumulated point cloud and that it is updated with new scans. Therefore, the boundaries of our workspace are not hypothesized and fixed but they are updated as a new scan is added.

The complexity of the scene is determined essentially by the shape of the room, the occlusion properties and the number of rooms to be scanned. Depending on the complexity of the geometry of the sensed area, we find simple [4] or complex 3D data processing [3, 5]. As regard interiors, most of the works deal with scenes composed of a corridor and several rectangular rooms connected to it [5–7]. However, indoors composed of concave rooms are rarely addressed in the field of 3D reconstruction. An exception can be found in the work by Jun et al. [8]. Besides, not all the approaches are able to overcome occlusion and clutter problems. [4, 9] and only a few of them consider obstacles or clutter in the scene [6, 7].

Our proposal is able to deal with concave rooms connected by doors in a building story. Due to this type of scenes our approach addresses the occlusion issue from the beginning. Figure 1 shows a prototype of the scene in which we are testing our automatic scanning method.

Fig. 1. Prototype of the scenario in our work: story with several non-rectangular inhabited rooms connected by doors.

2   A Brief Overview of the Method

Although the NBS procedure is the main objective of the paper, this is just a part of a more complex system composed of a mobile robot that carries a 3D laser scanner.

We assume that the scenario is composed of several rooms and that the system carries out the complete scanning of the current room before passing to an adjacent one. Thus, when the room scanning process ends, the robot places under the doorframe that separates adjacent rooms and launches the first scan of a new room. To detect openings (in this case, doors) we follow the algorithm described in [10].

The automatic scanning of the current room is viewed as a cyclic process which begins with the data acquisition from a 3D laser scanner with a field of view of

360° × 100°, and ends with the output of the NBS algorithm, that is, the coordinates of the next scan position in the world coordinate system. The main stages of a room scanning process are shown in Fig. 2. The stages are: (1) raw point cloud preprocessing and alignment, (2) RoI (Region of Interest) definition, (3) wall identification, (4) space labeling and NBS decision. The next sections are devoted to briefly explaining these stages.

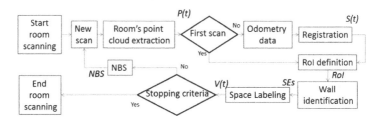

Fig. 2. Outline of the scanning process.

3 Structural Elements Recognition

3.1 Finding the Region of Interest

Our automatic scanning approach is a cyclic process in which each single point cloud coming from a new position of the scanner is registered into the accumulated point cloud of the scene. First, a coarse registration is carried out through the robot localization sensors. Then, the earlier registration is refined by applying a 6D (x, y, z, roll, pitch, yaw) ICP (Iterative Closest Point) technique [11]. We will denote $S(t)$ as the accumulated point cloud at time t.

In this framework, the region of interest (RoI) is defined as the region which establishes the boundaries of the current space at time t. Thus, the RoI can be implemented as the prism that contains $S(t)$.

In practice, the RoI is obtained through the top projection of $S(t)$. The projection of the points is quantized over a horizontal grid that finally we convert into a binary image. See Fig. 3(a) for a better understanding. A polygonal contour is calculated in this image with the help of Hough and Harris' algorithms. On one hand, the segments that compose this contour lead us to obtain, first, the planes that fit to the data points and, second, the vertical parallelograms nearest the points. Note that, these contours themselves determine the polygons at the top and bottom of the RoI, which represents the ceiling and floor of the scene (right column of Fig. 3(a)). Figure 3(b) illustrates the RoI updating for three consecutive scans. The polygons that represent the ceiling and the floor are not shown for a better vision.

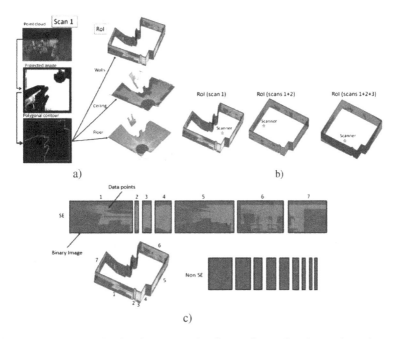

Fig. 3. (a) Steps to obtain the first RoI. The figure shows the data points (in magenta) superimposed to the faces of the RoI (in blue). (b) RoI evolution for the first three scans. (c) Structural elements classified in the first RoI. (Color figure online)

3.2   Structural Elements Classification

After having obtained the RoI, we identify which of its faces are structural elements (SE) and which are not. Note that, apart from all RoI's sides, we know the data points which are near to them. Each face and its associated data points generate a binary image in which we can infer whether the face is considered as SE or not. Of course, top and bottom polygons (which correspond to ceiling and floor) are a priori assumed structural elements, so that the SE classification is uniquely accomplished for walls.

The decision function that classifies the faces of the RoI as SE or non-SE has been implemented by means of a binary Support Vector Machine (SVM) classifier.

Let us assume I be a binary image generated from a polygon of the RoI in which a white (magenta in Fig. 3(c)) pixel means a data point. Let also assume d_1d_2 be the size of I (d_1 and d_2 are whatever image dimensions in pixels) and n be the number of white pixels contained in I. We consider a feature vector $F(\alpha, \delta, \varepsilon)$ that contains the following information:

– Occupancy percentage (α). This means the occupancy of the hypothetic wall. In Eq. (1), d_1 and d_2 are the image dimensions.

$$\alpha = \frac{n}{d_1 d_2} \tag{1}$$

– Clusters' compactness (β). We cluster the data points in I and calculate the density per cluster. The cluster's density is defined as the as the ratio of the number of data points (n_i) to the area of its corresponding bounding box ($d_{1i}d_{2i}$). Equation (2) provides the mean clusters density or the clusters' compactness (where k is the number of clusters).

$$\delta = \frac{1}{k}\sum_i^k \frac{n_i}{d_{1i}d_{2i}} \tag{2}$$

– Data dispersion (ε). This is calculated through the area occupied by the bounding boxes of the clusters.

$$\varepsilon = \frac{\sum_i d_{1i}d_{2i}}{d_1 d_2} \tag{3}$$

Figure 3(c) shows a set of images I corresponding to the faces of the RoI for the first scan and the classification result. A deeper discussion and explanation of this method can be done in a more extended publication.

4   NBS

The NBS is computed in a discretized 3D space, that is, in a voxel-space V, where a voxel is a small cube. We define five labels in the space V, divided into two classes (occupied and non-occupied):

– Occupied voxels are: *Clutter* (the voxel contains points that do not belong to SEs) and *Structure* (the voxel contains points belonging to SEs).
– Non-occupied voxels are: *Empty* (the voxel has been seen from earlier scanner positions but does not contain data), *Occluded-clutter* (the voxel has not been sensed because it lies between an occupied voxel and a SE) and *Occluded-structure* (the voxel has not been sensed because it is behind an occupied voxel and lies in SEs).

To calculate the NBS we estimate the amount of *Occluded-structure* voxels that would turn into *Structure* voxels from a set of next valid positions (NVP) of the scanner. Since a mobile robot carries the scanner, a NVP is defined under robot path-planning requirements, that is, there must be at least one secure path from the current to the next position of the robot. A secure path entails that the robot moves through *Empty* voxels and the distance from such voxels to any occupied voxel along the path is higher than certain security distance (in our case 20 cm). From each NVP and by means of a ray-tracing algorithm, we calculate the number of conversions from *Occluded-structure* to *Structure* voxels. A ranked list of NVP is then established taking into account the amount of new *Structure* voxels. The NBS corresponds to the first NVP of the list. Figure 4 shows the labeled space V for two consecutive scans. *Structure* voxels are in blue, *Occluded-structure* in green and *Clutter* (*Occluded-clutter* is not shown for a better visualization). Note the increment of *Structure* voxels from the next best position of the scanner.

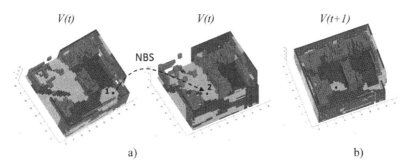

Fig. 4. (a) Current voxel-space V(t) with the current position (1) and the NBS position (2) in black (b) Next voxel-space V(t + 1). (Color figure online)

5 Test and Experimental Comparison

In this section we present an experimental comparison of our scanning approach with the ones of Stachniss and Burgard [12], Blaer and Allen [3] and Potthast Sukhatme [2], which can be considered related proposals. The comparison has been done in a scene composed of 5 adjacent concave-shape rooms, with clutter and occlusion (See Fig. 1(a)). This complex scenario has been created in Blender and its add-on Blensor [13]. This tool simulates real scanning with commercial 3D laser scanners similar to ours (Riegl VZ-400).

It is worth mentioning that, in order to make possible the experimental comparison, we needed to make some adaptations on those methods. Since the original version of those methods do not detect doors and besides impose a fixed-size occupancy grid for the scenario, we added our door detection algorithm to their codes and also updated the size of their voxel spaces with the beginning of a new scan. A brief report of the results follows.

Despite the adaptations, methods [2, 12] were not able to scan completely rooms #4 and #5. This is mainly owing to the fact that these approaches do not deal with concave-shape regions. Our method completed the scanning process taking one or two scan less than the rest per room. A total of 22 scans were taken for the whole scenario. Apart from ours, approach [3] was also able to complete the scanning process after taking 25 samples.

The computational cost was measured in terms of processing time. On this point, our algorithm spent less time than the others. Our total time for scanning and processing was 14868 s. We reduced 78,7%, 83,7% and 33,6% the time spent by [2, 3, 12] respectively. These reductions would have been much impressive without code adaptations.

Our percentage of the total structural surface sensed is also higher than the rest. We reached 88,57% compared with 87,26% for [4]. Note that Blaer et al.'s approach does not recognize structural elements. For rooms #1, #2 and #3, [2, 12] achieved 87,0% and 84,3% respectively. In summary, our method takes less scans and achieves higher percentages.

With regard to the size of the processed data points, we obtain smaller point clouds in most of the rooms. Note that we uniquely process SE points, whereas the rest of the approaches deal with the whole point cloud. We processed a total of 23.576.419 points. The mean reduction percentages in the number of processed points with respect to the other approaches were 7,2% [12], 23% [3] and 14% [2].

Concerning the path length, our approach yielded better results in rooms #1, #2, and #4. For rooms #3 and #5, the results were better than in [2, 12], and similar to the ones of [3]. We cover 30% less distance than [12] (43, 34 m versus 62, 28 m in rooms #1, #2 and #3), 25% less than [2] (32.71 m versus 43, 82 m in rooms #1 and #3) and 2,8% more than [3] (88, 26 versus 85, 82 m).

Figure 5 illustrates some details of the scanning process in room #2 and the whole point cloud of structural elements superimposed to the CAD model generated.

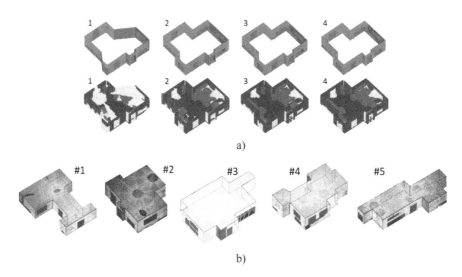

Fig. 5. (a) Evolution of RoI and Voxel-space of room #2 for four scans. (b) Data points assigned to structural elements in five different rooms and the corresponding CAD models superimposed to them.

6   Conclusions

The primary goal in our work is to accumulate data belonging to structural elements of a building to obtain a precise 3D model. The main contributions of our proposal are:

- A new NBS algorithm addressed to capture structural elements that highly reduces the volume of data and alleviates the algorithmic complexity in further processes.
- A dynamic RoI which allows effectively dealing with more complex scenarios composed of several non-rectangular spaces with occlusion and clutter.

Many aspects need to be improved in future developments. Our method works very well for flat structural elements but neither for curve shapes nor for several ceilings/floors within the same room. Therefore, new approaches for more complex scenes, including several stories, must be addressed in the future.

Acknowledgements. This work was supported by the Spanish Economy and Competitiveness Ministry [DPI2013-43344-R project] and by Castilla-La Mancha Government [PEII-2014-017-P project].

References

1. Adán, A., Quintana, B., Vázquez, A., Olivares, A., Parra, E., Prieto, S.: Towards the automatic scanning of indoors with robots. Sensors **15**(5), 11551–11574 (2015)
2. Potthast, C., Sukhatme, G.S.: A probabilistic framework for next best view estimation in a cluttered environment. J. Vis. Commun. Image Represent. **25**(1), 148–164 (2014)
3. Blaer, P.S., Allen, P.K.: Data acquisition and view planning for 3-D modeling tasks. In: International Conference on Intelligent Robots and Systems, pp. 417–422 (2007)
4. Charrow, B., Kahn, G., Patil, S., Liu, S.: Information-theoretic planning with trajectory optimization for dense 3D mapping. In: Proceedings of Robotics: Science and Systems (2015)
5. Borrmann, D., Nüchter, A., Đakulović, M., Maurovic, I., Petrović, I., Osmankovic, D., Velagić, J.: A mobile robot based system for fully automated thermal 3D mapping. In: Advanced Engineering Informatics (2014)
6. Biswas, J., Veloso, M.: Depth camera based indoor mobile robot localization and navigation. In: IEEE International Conference on Robotics and Automation (2012)
7. Strand, M., Dillrnann, R.: Using an attributed 2D-grid for next-best-view planning on 3D environment data for an autonomous robot. In: Proceedings of the IEEE International Conference on Information and Automation (2008)
8. Jun, C., Youn, J., Choi, J., Doh, N.L.: Convex Cut: a realtime pseudo-structure extraction algorithm for 3D point cloud data. In: IEEE/RSJ International Conference on Intelligent Robots and Systems (2015)
9. Thrun, S., Hähnel, D., Montemerlo, M., Ferguson, D., Triebel, R., Burgard, W.: A system for volumetric robotic mapping of underground mines. In: IEEE International Conference on Robotics and Automation (2003)
10. Quintana, B., Prieto, S.A., Adan, A., Bosché, F.: Doors detection with 3D laser. Scanners for autonomous navigation in multi-room scenes. In: IPIN: Indoor Positioning and Indoor Navigation Conference (in press)
11. Besl, P., McKay, N.: A method for registration of 3-D shapes. IEEE Trans. Pattern Anal. Mach. Intell. **14**(2), 239–256 (1992)
12. Stachniss, C., Burgard, W.: Exploring unknown environments with mobile robots using coverage maps. In: Proceedings of the International Conference on Artificial Intelligence, pp. 1127–1132 (2003)
13. Gschwandtner, M., Kwitt, R., Uhl, A., Pree, W.: BlenSor: blender sensor simulation toolbox. In: Bebis, G., et al. (eds.) ISVC 2011. LNCS, vol. 6939, pp. 199–208. Springer, Heidelberg (2011). doi:10.1007/978-3-642-24031-7_20

Parallel Integer Motion Estimation for High Efficiency Video Coding (HEVC) Using OpenCL

Augusto Gomez[✉], Jhon Perea, and Maria Trujillo

Multimedia and Computer Vision Group, Universidad Del Valle, Cali, Colombia
{augusto.gomez,jhon.edinson.perea,maria.trujillo}@correounivalle.edu.co

Abstract. High Efficiency Video Coding is able to reduce the bit-rate up to 50% compared to H.264/AVC, using increasingly complex computational processes for motion estimation. In this paper, some motion estimation operations are parallelised using Open Computing Language in a Graphics Processing Unit. The parallelisation strategy is three-fold: calculation of distortion measurement using 4×4 blocks, accumulation of distortion measure values for different block sizes and calculation of local minima. Moreover, we use 3D-arrays to store the distortion measure values and the motion vectors. Two 3D-arrays are used for transferring data from GPU to CPU to continue the encoding process. The proposed parallelisation is able to reduce the execution time, on average 52.5%, compared to the HEVC Test Model. Additionally, there is a negligible impact on the compression efficiency, as an increment in the BD-BR, on average 2.044%, and a reduction in the BD-PSNR, on average 0.062%.

Keywords: GPU · HEVC · Motion estimation · OpenCL · Parallel programming

1 Introduction

High Efficiency Video Coding (HEVC) achieves increase coding efficiency, while preserving video quality with lower bit-rate compared to H.264/AVC. However, it also increases dramatically the encoding computational complexity [15]. Motion estimation is the most time consuming task in a video encoder. In HEVC, the motion estimation requires about 77%–81% of the total encoding time due to distortion measure calculations and flexibility in block sizes [8]. Moreover, motion estimation configurations, such as Advanced Motion Vector Prediction (AMVP) [19], generate data dependency between neighbour blocks that make difficulties in using hardware with Single Instruction Multiple Data (SIMD) architecture [9]. The HEVC specifications include the following parallel processing options: Wavefront Parallel Processing (WPP) [5], Slices [11] and Tiles [3]. Moreover, there are approaches on parallel motion estimation presented in the literature, for instance *Wang* et al. in [17].

Graphics Processor Units (GPUs) have been used for reducing the encoding time, especially for motion estimation in HEVC – for instance in [12]. *Luo* et al. presented in [10] a quaternion of (L, T, R, B) for indexing a look-up table with

C. Beltrán-Castañón et al. (Eds.): CIARP 2016, LNCS 10125, pp. 68–75, 2017.
DOI: 10.1007/978-3-319-52277-7_9

PU sizes from 4×4 to 32×32; the performance evaluation was implemented on HM10.0 [6]. *Wang* et al. proposed in [16] a two-fold approach: an initial motion estimation is obtained on GPU and then, it is refined on CPU. The authors considered 256 symmetric partitions of a Coding Tree Unit (CTU) using an initial partition of 4×4 and accumulating block sizes to reach a 64×64 block; the implementation was done on x265 video encoder [13]. *Jiang* et al. introduced in [7] an approach focused on reducing the overhead, which is based on dividing a CTU into 256 blocks of 4×4, and the implementation was done on HM12 [6]. They reported results, using the PeopleOnStreet video sequence, of a time reduction of 34.64% with QP equal to 32 and 40.83% with QP equal to 42 compared to the HM12.

In this paper, we propose a parallel strategy for integer motion estimation with negligible impact on the coding efficiency, using OpenCL as programming framework [4]. The proposed approach has been divided into three main steps: distortion measure calculations, distortion measure accumulations and rate distortion minima estimations, as the foundations of estimating motion vectors. Two 3D-arrays are built for storing distortion measure values and estimated motion vectors. In this way, data is transferred to CPU. Experimental results shown an average of 52.5% reduction of the execution time compared to the HEVC Test Model [6], with a slight increment of the BD-BR, of 2.044% on average, and a slight reduction of the BD-PSNR, of 0.062% on average, using the Bjøntegaard Delta [1] as a metric to compare compression efficiency.

2   Parallel Integer Motion Estimation

The goal of the Parallel Integer Motion Estimation (PIME) is to reduce the computational time required for calculating Motion Vectors (MV)s during the encoding process. HEVC supports CTU with maximum size of 64×64. A CTU is divided into Coding Units (CU) of different sizes. A total of 593 CUs are obtained from a CTU for inter-frame prediction. Unlike the HEVC standard, PIME does not create the tree structure top-down for calculating MVs. Instead, a CTU is divided into 4×4 size blocks and distortion measures are calculated for creating the tree structure bottom-up.

The integer motion estimation is performed in the GPU and MVs are transferred to the CPU for continuing the encoding process. For each CTU, luma data and the search area are transferred to the GPU global memory. The workflow of PIME is presented in Fig. 1. Using a block matching algorithm: Firstly, distortion measures are calculated. Secondly, distortion measures are accumulated. Finally, minima of rate distortions are estimated.

2.1   Distortion Measure Calculations

The calculation of the distortion measure is computed using OpenCL, it uses work-items for running a single instruction. A work-item executes the calculation of a distortion measure in a 4×4 block. Also, work-items are arranged in a

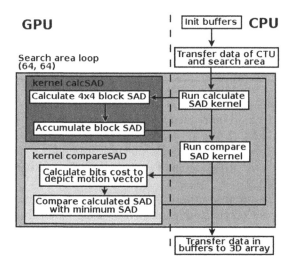

Fig. 1. Workflow of Parallel Integer Motion Estimation (PIME)

work-group for accessing a local memory and taking advantage of transferring speed. A work-group contains 16×16 work-items − a total of 256 work-items. Obtained values are stored into three temporal buffers: horizontal, vertical and asymmetric (AMP).

2.2 Distortion Measure Accumulations

Using the temporal buffers, distortion measure values are recursively accumulated horizontally and vertically to estimate the distortion in larger blocks, starting from 4×4 blocks to 64×64 block, including asymmetric partitions using parallel reduction [2]. A work-item adds a pair of corresponding horizontal or vertical distortion measures to obtain the distortion measure for each block. Temporal buffers, in local memory, store recursively accumulations and the buffer, in global memory, stores distortion measures of 593 blocks, defined in the HEVC specifications.

Table 1 shows positions in the global memory buffer of each stored CU.

Table 1. Index used for storing distortion measure values in the GPU global buffer (U = Up, D= Down, L = Left, R = Right)

Index	PU	Index	PU	Index	PU	Index	PU	Index	PU
0–127	8×4	336–351	$4 \times 16R$	516–519	$32 \times 8D$	544–559	16×16	580	$16 \times 64L$
128–255	4×8	352–367	$12 \times 16L$	520–523	$32 \times 24U$	560–567	32×16	581	$16 \times 64R$
256–271	$16 \times 4U$	368–383	$12 \times 16R$	524–527	$32 \times 24D$	568–575	16×32	582	$48 \times 64L$
272–287	$16 \times 4D$	384–447	8×8	528–531	$8 \times 32L$	576	$64 \times 16U$	583	$48 \times 64R$
288–303	$16 \times 12U$	448–479	16×8	532–535	$8 \times 32R$	577	$64 \times 16D$	584–587	32×32
304–319	$16 \times 12D$	480–511	8×16	536–539	$24 \times 32L$	578	$64 \times 48U$	588–589	64×32
320–335	$4 \times 16L$	512–515	$32 \times 8U$	540–543	$24 \times 32R$	579	$64 \times 48D$	590–591	32×64

2.3 Estimating Minima of Rate-Distortion

A global minima buffer, in the global memory, is used for storing the distortion measures corresponding to the obtained minima calculated using the rate distortion measure, over the search area. Rate distortion is a widely used measure, defined in [14]:

$$J_{\mathrm{MV}} = SAD(MV) + \lambda \, R(MV). \tag{1}$$

where J_{MV} is the cost function, $SAD(MV)$ is a distortion function, λ is the Lagrangian multiplier and $R(MV)$ symbolised the bits required to code the MV.

Two additional buffers are created to keep the $x-$ and the $y-$ coordinates of obtained minima. In this case, the work-items are independent, instead of being organised in a work-group. A work-item compares the current J_{MV} to the obtained local minimum at the moment. Once the block matching is completed, the distortion measure minima are stored in the global minima buffer. Figure 2 show the architecture of PIME.

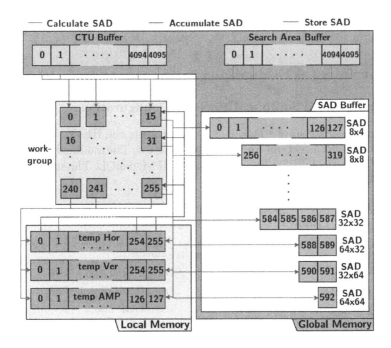

Fig. 2. Parallel integer motion estimation architecture

2.4 GPU and CPU Communication

The PIME process is performed in the GPU device while the encoding process is performed in the CPU (not concurrently). We introduce the use of 3D-arrays for data communication between GPU and CPU. In this way, the OpenCL buffers are mapped into two 3D-arrays in CPU for storing distortion measures and MVs. A 3D-array is built as follows: the first dimension corresponds to the reference frame lists (eRefPicList), the second dimension points to the index for the reference frame (iRefIdxPred) and the third dimension points to the indexes for accessing the CU (CUindex) in Table 1. A 3D-array is illustrated in Fig. 3.

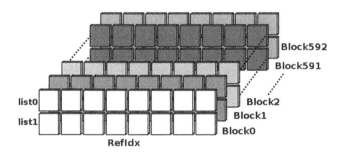

Fig. 3. The 3D-array used for storing

3   Experimental Evaluation

3.1   Experimental Setup and Encoder Configuration

Performance evaluation of PIME is conducted using the HEVC Test Model [6] version 16.4, with the default profile Random Access. The modified source code is available at https://github.com/gomezpirry/HM-OpenCL. The used parameter configurations are in Table 2.

Table 2. Parameters configuration of the HM

Encoder setting/parameter					
MaxCUSize:	64	HadamardME:	True	Search:	Full
MaxPartitionDepth:	4	Fast decision:	True	RateControl:	false
IntraPeriod:	32	AMP:	True	GOP size:	2
SearchRange:	64	Bi-SearchRange:	4	QP:	22, 27, 32, 37

The hardware platform used in these experiments is composed of a CPU Intel Core i7–4500U processor with 1.8 GHz clock base frequency and 8 GB DDR3 of system memory. The GPU used is a NVidia GeForce 840 M, with 384 cores, 2 GB DDR3 of memory and running at 1032 MHz. The encoder was compiled using GCC 4.8.4 and OpenCL 1.2, executed on Ubuntu 14.04 64 bits.

The compression efficiency is judged with the Bjøntegaard Delta Bit-Rate (BD-BR) and the Bjøntegaard Delta PSNR (BD-PSRN). Also, the speed-up using the proposed model is measured using the delta of the execution time (ΔTR %):

$$\Delta TR(\%) = \frac{T_{\mathrm{HM}} - T_{\mathrm{HM-PIME}}}{T_{\mathrm{HM}}} \times 100. \tag{2}$$

where T_{HM} is the execution time required by the HM, and $T_{\mathrm{HM-PIME}}$ is the execution time required by the proposed approach.

The selected test video sequences for presenting evaluation results include one video class A (2560×1600p) and three videos class B (1920×1080p) [18]. All video sequences have a bit depth of 8 and 4:2:0 chroma sub-sampling. For evaluation purposes, 150 frames were encoded from each video sequence.

3.2 Experimental Results

The results in Table 3 show that the proposed parallel strategy is able to reduce the execution time of 52.5% on average at the cost of less than 2.044% coding efficiency. Moreover, there is a negligible impact on the frame reconstruction quality that is reflected in a reduction on the PSNR of 0.062%.

Table 3. Average values over QPs of the BD-BR, the BD-PSRN and the ΔTR

Class	Video sequence	BD-BR(%)	BD-PSRN(%)	ΔTR(%)
A	PeopleOnStreet	2.483	−0.089	51.2
B	PedestrianArea	2.805	−0.084	52.9
B	ParkScene	2.096	−0.056	53.0
B	Kimono	0.793	−0.020	53.0
Mean values		2.044	−0.062	52.5

Jiang et al. in [7] reported a time reduction on average of 37.74% using the PeopleOnStreet video sequence. Table 3 shows on average 2.48% BD-BR, −0.089% BD-PSRN and 51.2% ΔTR, using the proposed approach with the same video sequence.

Figure 4 contains the Rate-Distortion curves calculated using the four QP values. Differences between curves are almost imperceptible.

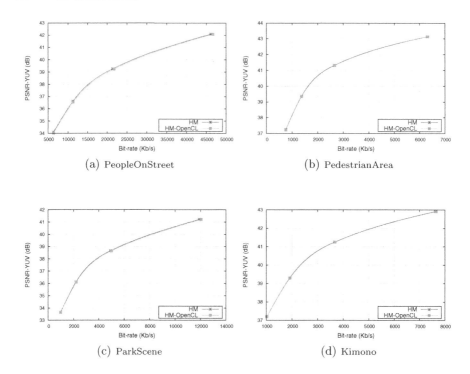

(a) PeopleOnStreet

(b) PedestrianArea

(c) ParkScene

(d) Kimono

Fig. 4. Rate-Distortion curves of the PeopleOnStreet (top-left), the PedestrianArea (top-right), the ParkScene (bottom-left) and the Kimono (bottom-right) sequences calculated using different QPs

4 Conclusions

In this paper, a parallelisation strategy is presented for reducing the computational time required for calculating motion vectors. The proposed strategy is based on OpenCL and achieves higher transfer speed by arranging work-items into a work-group during distortion measure calculations. Additionally, we introduced the use of 3D-arrays for data communication between GPU and CPU. The experimental tests have shown a significant reduction in the execution time using the proposed strategy whilst the compression efficiency suffers from a slight reduction. The parallel hardware may break data flow during the parallel processing and it may be the cause of loss of compression efficiency.

As a future work, the proposed approach will be adjusted for up-scaling in order to evaluate more CTUs at the same time. In this case, the use of a GPU with more processing units is required for mapping each 4 × 4 block into one processing unit (each CTU needs 256 processing units). For the communication, a dimension in the 3D-array is added in order to store a motion vector per CTU. Moreover, it will be explored the use of fast search algorithms and embedded hardware for calculating CTUs in a concurrent way.

References

1. Bjøntegaard, G.: Calculation of average PSNR differences between RD-curves. Technical report, ITU-T Video Coding Experts Group (VCEG) (VCEG-M33 2001) (2001)
2. Catanzaro, B.: OpenCL optimization case study: simple reductions (2010). http://developer.amd.com/resources/documentation-articles/articles-whitepapers/opencl-optimization-case-study-simple-reductions/. Accessed Mar 2015
3. Fuldseth, A., Horowitz, M., Xu, S., Segall, A., Zhou, M.: Tiles. Technical report JCTVC-F335, March 2011
4. Group, K: Open Computing Language (OpenCL). https://www.khronos.org/opencl/
5. Henry, F., Pateux, S.: Wavefront parallel processing. Technical report JCTVC-E196, March 2011
6. JCT-VC: HEVC Test Model (HM). https://hevc.hhi.fraunhofer.de/
7. Jiang, X., Song, T., Shimamoto, T.L.W.: High efficiency video coding (HEVC) motion estimation parallel algorithms on GPU. In: IEEE International Conference on Consumer Electronics, pp. 115–116, May 2014
8. Kim, S., Park, C., Chun, H., Kim, J.: A novel fast and low-complexity motion estimation for UHD HEVC. In: Picture Coding Symposium (PCS), pp. 105–108, December 2013
9. Lawson, H.: Parallel Processing in Industrial Real-Time Applications. Prentice Hall Series in Innovative Technology. Prentice Hall, New York (1992)
10. Luo, F., Ma, S., Ma, J., Qi, H., Su, L., Gao, W.: Multiple layer parallel motion estimation on GPU for High Efficiency Video Coding (HEVC). In: IEEE International Symposium on Circuits and Systems (ISCAS), pp. 1122–1125, May 2015
11. Misra, K., Zhao, J., Segall, A.: Entropy slices for parallel entropy coding. Technical report JCTVC-B111, July 2010
12. Monteiro, E., Maule, M., Sampaio, F., Diniz, C., Zatt, B., Bampi, S.: Real-time block matching motion estimation onto GPGPU. In: IEEE International Conference on Image Processing (ICIP), pp. 1693–1696, October 2012
13. MulticoreWare: x265 HEVC Encoder. http://x265.org/
14. Sullivan, G., Wiegand, T.: Rate-distortion optimization for video compression. IEEE Signal Process. Mag. **15**, 74–90 (1998)
15. Sullivan, G., Ohm, J., Han, W.J., Wiegand, T.: Overview of the high efficiency video coding (HEVC) standard. IEEE Trans. Circuits Syst. Video Technol. **22**, 1649–1668 (2012)
16. Wang, F., Zhou, D., Goto, S.: OpenCL based high-quality HEVC motion estimation on GPU. In: IEEE International Conference on Image Processing (ICIP), pp. 1263–1267, October 2014
17. Wang, X., Song, L., Chen, M., Yang, J.: Paralleling variable block size motion estimation of HEVC on CPU plus GPU Platform. In: IEEE International Conference on Multimedia and Expo Workshops (ICMEW), pp. 1–5, July 2013
18. xiph.org: Xiph.org Video Test Media [derf's collection]. https://media.xiph.org/video/derf/. Accessed Mar 2015
19. Zhao, L., Guo, X., Lei, S., Ma, S., Zhao, D.: Simplified AMVP for high efficiency video coding. In: IEEE Visual Communications and Image Processing (VCIP), pp. 1–4, November 2012

Automatic Fruit and Vegetable Recognition Based on CENTRIST and Color Representation

Jadisha Yarif Ramírez Cornejo and Helio Pedrini[✉]

Institute of Computing, University of Campinas, Campinas, SP 13083-852, Brazil
helio@ic.unicamp.br

Abstract. Automatic fruit and vegetable recognition from images is still a very challenging task. In this work, we describe and analyze an efficient and accurate fruit and vegetable recognition system based on fusing two visual descriptors: Census Transform Histogram (CENTRIST) and Hue-Saturation (HS) Histogram representation. Initially, background subtraction is applied to the fruit and vegetable images. CENTRIST and HS-Histogram are extracted, as well as Color CENTRIST features for comparison purpose. Then, the feature vector is reduced through Principal Component Analysis (PCA) and Linear Discriminant Analysis (LDA). In the recognition process, K-nearest neighbor (K-NN) and Support Vector Machine (SVM) classifiers are employed. Experiments conducted on a benchmark demonstrate that combining CENTRIST and HS-Histogram representation reached high and competitive recognition accuracy rates compared to other similar works in the literature.

Keywords: Fruit and vegetable recognition · CENTRIST · Color features

1 Introduction

The development of effective recognition systems has many practical applications, however, it is still a challenging task. One of the main problems remains in determining whether or not the selected data contains representative and discriminative information for the classes under consideration. A challenge is to achieve comparable accuracy rates to human recognition.

Fruit and vegetable recognition methods can be applied to crop monitoring, supermarket checkout systems, educational purposes, among others [1,5,9]. Accurate and efficient identification of different species of fruits and vegetables is essential in supermarkets, which must invest time and money in training qualified cashiers. Possible solutions to this problem include the use of price tags or barcodes on each item of the products. When a price change is required, barcodes have the advantage of not having a fixed price encoded in the tag, such that a scanner at the checkout line is employed to identify the price of each product from a computer system. However, some strategies demand manpower and material resources, or impose limited customer choices for selecting, packaging and weighting their products.

© Springer International Publishing AG 2017
C. Beltrán-Castañón et al. (Eds.): CIARP 2016, LNCS 10125, pp. 76–83, 2017.
DOI: 10.1007/978-3-319-52277-7_10

Several approaches have been proposed in the literature for fruit and vegetable recognition based on color, texture and shape features individually. Some previous works based on only one feature have not met certain application requirements, such as recognition accuracy rates. More recently, feature fusion using texture, color and shape have been applied to fruit and vegetable recognition, achieving superior results [3,4].

A novel and effective automatic fruit and vegetable recognition methodology, based on fusing CENTRIST features and a Hue-Saturation Histogram representation, is proposed and analyzed in this work. Initially, background subtraction is performed on the images. Then, a set of features is extracted and fused. Finally, fruit and vegetable species are recognized.

The remainder of the paper is organized as follows. Section 2 briefly describes the Census Transform Histogram (CENTRIST) and Color CENTRIST. Section 3 presents the methodology proposed in this work, describing the preprocessing, the feature extraction, the feature reduction and classification stage. Section 4 describes and discusses the experiments and results. Section 5 concludes the paper with final remarks.

2   Background

Concepts related to the topic under investigation are briefly described in this section.

2.1   CENTRIST

CENsus TRansform hISTogram (CENTRIST) [8] was originally proposed for recognizing topological places and scene categories. This visual descriptor provides an accurate and stable performance on various scene images. Based on Census Transform (CT), this operator is a holistic representation, which models distribution of local structures. It is characterized for providing a high generalization for categorization, suppressing detailed textural information.

The descriptor compares the intensity value of a central pixel with its eight neighboring pixels. If the central pixel intensity value is higher than or equal to one of its neighboring pixels, bit 1 is set to the corresponding location, otherwise bit 0 is set. The resulting 8 bits can be concatenated in any order, converting the 8-bit binary representation to a base-10 number, being the Census Transform value (CT value) of the current central pixel. After computing a CT value for each pixel, a histogram descriptor of CT values is constructed with 256 bins, called CENTRIST.

2.2   Color CENTRIST

The Color CENTRIST [2] was introduced to embed color information into CENTRIST framework. It has proven to improve recognition accuracy rates for color scene categorization. This operator describes color information by working over the hue-saturation-value (HSV) color space, where the three channels are normalized into the range of 0 to 255, requiring 8 bits to represent each channel. In order to propose a comparable representation to CT values of CENTRIST, the Color CENTRIST devises a framework for representing color information by 8 bits and different quantization for the three channels.

To represent hue, saturation and value components, 1, 2 and 5 bits are allocated, respectively. The three channels are uniformly quantized. The hue axis is divided into $2^1 = 2$ ranges, i.e., $[0, 127]$ and $[128, 255]$. The saturation axis is divided into $2^2 = 4$ ranges, i.e., $[0, 63]$, $[64, 127]$, $[128, 191]$, $[192, 255]$. The value axis is divided into $2^5 = 32$ ranges, i.e., $[0, 7]$, $[8, 15]$, ..., $[240, 247]$, and $[248, 255]$. The hue component for a pixel is transformed into a color index (base 10) by $i_h = \lceil \frac{h \times 2^1}{256} \rceil$, then the resulting value is converted into a binary number through 1 bit. In a similar way, the color indices for saturation and value components are represented by $i_s = \lceil \frac{s \times 2^2}{256} \rceil$ and $i_v = \lceil \frac{v \times 2^5}{256} \rceil$, then transformed into binary numbers through 2 and 5 bits, respectively.

This operator gives the highest priority to value component. Hence, the color indices (base 2) of a pixel are concatenated as $(i_v i_s i_h)$. Then, the binary stream is converted into a base-10 system. Thereby, the CENTRIST operator is applied over the resulting color indices (base 10), which generates a histogram descriptor of CT values of length 256.

3   Methodology

The proposed methodology for fruit and vegetable recognition is composed of four main stages: (i) preprocessing, (ii) feature extraction, (iii) feature reduction, and (iv) classification. These steps are described as follows.

3.1   Preprocessing

The image preprocessing stage is crucial for the recognition task, whose main purpose is to perform background subtraction [1]. Background subtraction is basically used for segmenting out objects of interest in a scene. This preprocessing procedure consists of the following six steps: (i) conversion of input image into the HSV color space [7]; (ii) performing Sobel operator over saturation component; (iii) filling small holes using closing morphological operator with a disk structuring element; (iv) removing noise using median filtering; (v) finding a rectangular mask to find an interest region; and (vi) cropping the interest region. Figure 1 illustrates the stages in some samples of images.

3.2   Feature Extraction

Three types of features – CENTRIST [8], Color CENTRIST [2] and Hue-Saturation Histogram – were extracted and some of them fused for fruit and vegetable recognition.

(a)   (b)   (c)

Fig. 1. (a) Original images from the Supermarket Produce data set [6]; (b) interest region detection from (a); (c) cropped interest regions from (b).

The CENTRIST [8] features were extracted from the entire images, forming a vector of length 256. This feature vector was normalized by dividing each intensity level r by the total pixels n of the image, as follows

$$p(r) = \frac{h(r)}{n} \tag{1}$$

where $h(r)$ defines the occurrence frequency of each intensity level in the image, and $p(r)$ gives the probability of occurrence intensity. CENTRIST is able to capture local structures of the image. Figure 2 shows a sample of a census transformed fruit image.

(a)   (b)

Fig. 2. (a) Cropped image from the supermarket produce data set [6]; (b) census transformed image from (a).

In a similar way to CENTRIST feature extraction, we applied the Color CENTRIST [2] operator over the entire image. The resulting feature vector was normalized, forming a vector of 256. Figure 3 illustrates the process.

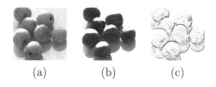

(a)   (b)   (c)

Fig. 3. (a) Cropped image from the supermarket produce data set; (b) HSV channels from (a) transformed into color indices; (c) census transformed image from (b).

After transforming all images from the RGB to the HSV color space, we performed several experiments over the different HSV channels and decided to generate the Hue-Saturation Histogram. From both hue and saturation channels, we computed two histograms, resulting in a feature vector of length 512 ($= 2 \times 256$). We also performed histogram normalization (Eq. 1).

3.3   Feature Reduction and Classification

Two techniques for feature reduction, Principal Component Analysis (PCA) and Linear Discriminant Analysis (LDA), were sequentially used. Thereby, we applied this procedure individually for each feature set. We then used K-Nearest Neighbors (K-NN) and Support Vector Machine (SVM) classifiers for comparing the recognition rates.

4   Experimental Results

The proposed methodology was tested on the available Supermarket Produce data set proposed in [6]. This data set contains 2633 images of 15 different fruit and vegetable categories: Agata Potato (201), Asterix Potato (182), Cashew (210), Diamond Peach (211), Fuji Apple (212), Granny-Smith Apple (155), Honeydew Melon (145), Kiwi (171), Nectarine (247), Onion (75), Plum (264), Taiti Lime (106), Watermelon (192), and Williams Pear (159).

The Supermarket Produce data set comprises illumination deviations (Fig. 4), pose differences, variability in the number of elements and different maturity degrees (Fig. 5). Furthermore, this data set contains elements packed into plastic bags, creating reflections and shadows, as well as other occlusions, such as hands and cropping elements (Fig. 6).

Initially, we randomly choose 80% of samples of each class for the training set and the remaining 20% for the testing set. We set 20 different randomized data collections to conduct experiments. Thus, we performed experiments using CENTRIST, Color CENTRIST, HS-Histogram, and the fusion of CENTRIST

Fig. 4. Different illumination deviations for Asterix Potato category.

Fig. 5. Distinct maturity degrees for Williams Pear category.

(a)   (b)   (c)

Fig. 6. Examples of occlusions. (a) cashew's plastic repository; (b) hand presence; (c) cropping Watermelon element.

with HS-Histogram, through six methods: K-NN, PCA+K-NN, PCA+LDA+K-NN, SVM, PCA+SVM and PCA+LDA+SVM. The results are shown in Table 1, whose values represent the fruit and vegetable recognition accuracy rates by running 20 experiments with different data collections.

From our experiments, we can notice that the PCA+LDA approach achieves better recognition accuracy than just using PCA. Moreover, we can observe that the fruit and vegetable recognition using the fusion of CENTRIST and HS-Histogram representation is much higher when compared to the other texture and color representations, especially when is followed by SVM classifier. From Table 1, it is possible to see that the CENTRIST representation works much

Table 1. Average accuracy, in percentage, using CENTRIST, Color CENTRIST, HS-Histogram, and the fusion of CENTRIST with HS-Histogram, for the Supermarket Produce [6] data set.

Recognition method	CENTRIST (%)	Color CENTRIST (%)	HS-Histogram (%)	CENTRIST+ HS-Histogram(%)
K-NN	79.58	69.90	88.12	92.40
PCA + K-NN	79.58	71.09	89.51	92.56
PCA + LDA + K-NN	85.96	76.04	90.50	96.60
SVM	86.18	74.85	90.50	**97.23**
PCA + SVM	89.15	78.02	91.29	94.47
PCA + LDA + SVM	**90.64**	**80.20**	**91.89**	95.96

better than the Color CENTRIST representation, although the latter considers further information, as color features. We can see that the accuracy using HS-Histogram is slightly better than using a CENTRIST representation.

Table 2 shows the confusion matrix using the fusion of CENTRIST and HS-Histogram features for SVM classifier. It is possible to observe that accuracy rates up to 90% are achieved for each class. Furthermore, the results show that the proposed approach is able to proper recognize the variety of fruit and vegetable species, for instance, Fuji Apple and Granny-Smith Apple, and Diamond Peach and Nectarine.

Table 2. Confusion matrix using fusion of CENTRIST and HS-Histogram representation with SVM classifier for the Supermarket Produce data set.

	Agata Potato	Asterix Potato	Cashew	Diamond Peach	Fuji Apple	Granny-Smith Apple	Honeydew Melon	Kiwi	Nectarine	Onion	Orange	Plum	Taiti Lime	Watermelon	Williams Pear	Sensitivity (%)
Agata Potato	40	0	0	0	0	0	0	0	0	1	0	0	0	0	0	97.56
Asterix Potato	0	37	0	1	0	0	0	0	0	0	0	0	0	0	0	100.00
Cashew	0	0	42	0	1	0	0	0	0	0	0	0	0	0	0	97.67
Diamond Peach	0	0	0	40	0	0	0	0	1	0	0	2	0	0	0	93.02
Fuji Apple	0	0	0	1	42	0	0	0	0	0	0	0	0	0	0	97.67
Granny-Smith Apple	0	0	0	0	0	32	0	0	0	0	0	0	0	0	0	100.00
Honewdew Melon	0	0	0	0	0	0	29	0	0	0	0	0	1	0	0	96.67
Kiwi	3	0	0	0	0	1	0	32	0	0	0	0	0	0	0	91.43
Nectarine	0	0	1	0	0	0	0	0	49	0	0	0	0	0	0	98.00
Onion	0	0	0	0	0	0	0	0	0	16	0	0	0	0	0	100.00
Orange	0	0	0	0	0	0	1	0	0	0	19	0	0	1	0	90.48
Plum	0	0	0	0	0	0	0	0	0	0	0	53	0	0	0	100.00
Taiti Lime	0	0	0	0	0	0	1	0	0	0	0	0	31	0	0	96.88
Watermelon	0	0	0	0	0	0	0	0	0	0	0	0	0	22	0	100.00
Williams Pear	0	0	0	0	0	0	0	0	0	0	0	0	0	0	7	100.00
Precision (%)	93.02	100.00	97.67	97.56	97.67	100.00	93.55	100.00	98.00	94.12	100.00	100.00	91.18	95.65	100.00	**97.23**

Table 3. Comparison of average accuracy rates (%) for fruit and vegetable recognition.

Approach	Strategy	Accuracy (%)
Proposed	CENTRIST and HS-Histogram representation + SVM	**97.23**
Rocha et al. [6]	Bagging Ensemble of Linear Discriminant Analysis	97.00
Dubey et al. [3]	Global Color Histogram and Local Binary Patterns + SVM	93.84
Arivazhagan et al. [1]	Minimum distance criterion	86.00

We compared our proposed method to others available in the literature. Table 3 summarizes the comparison results on the Supermarket Produce data set. We can see that the proposed approach, i.e., the fusion of CENTRIST and HS-Histogram representation with SVM, obtains superior results. Table 3 is sorted in descending order by accuracy rate.

5   Conclusions

In this work, experimental results have shown that the fusion of CENTRIST features with HS-Histogram can achieve high accuracy rates for fruit and vegetable recognition. Our method has also demonstrated to be robust due the presence of different occlusions, for instance, shadows, plastic bags, cashier's hand.

Despite the fact that the PCA and LDA approach did not provide a higher accuracy rate when combining CENTRIST and HS-Histogram, this approach allowed to reduce and select discriminative features, as well as increasing recognition accuracy significantly when features were used individually. Furthermore, computing CENTRIST and HS-Histogram features does not demand high computational resources. The generation of the combined feature vector is a very fast and efficient process, allowing fruit and vegetable recognition in real time.

Acknowledgements. The authors are thankful to FAPESP (grant #2011/22749-8) and CNPq (grant #307113/2012-4) for their financial support.

References

1. Arivazhagan, S., Shebiah, R.N., Nidhyanandhan, S.S., Ganesan, L.: Fruit recognition using color and texture features. J. Emerg. Trends Comput. Inf. Sci. **1**(2), 90–94 (2010)
2. Chu, W.T., Chen, C.H., Hsu, H.N.: Color CENTRIST: embedding color information in scene categorization. J. Vis. Commun. Image Representation **25**(5), 840–854 (2014)
3. Dubey, S.R., Jalal, A.S.: Fruit and vegetable recognition by fusing colour and texture features of the image using machine learning. Int. J. Appl. Pattern Recogn. **2**(2), 160–181 (2015)
4. Gao, G., Zhao, S., Zhang, C., Yu, X., Li, Z.: Study on fruit recognition methods based on compressed sensing. J. Comput. Theor. Nanosci. **12**(9), 2937–2942 (2015)
5. Leemans, V., Destain, M.F.: A real-time grading method of apples based on features extracted from defects. J. Food Eng. **61**(1), 83–89 (2004)
6. Rocha, A., Hauagge, D.C., Wainer, J., Goldenstein, S.: Automatic fruit and vegetable classification from images. Comput. Electron. Agric. **70**(1), 96–104 (2010)
7. Sural, S., Qian, G., Pramanik, S.: Segmentation and histogram generation using the hsv color space for image retrieval. In: International Conference on Image Processing, vol. 2, pp. 589–592. IEEE (2002)
8. Wu, J., Rehg, J.M.: CENTRIST: a visual descriptor for scene categorization. IEEE Trans. Pattern Anal. Mach. Intell. **33**(8), 1489–1501 (2011)
9. Zawbaa, H.M., Abbass, M., Hazman, M., Hassenian, A.E.: Automatic fruit image recognition system based on shape and color features. In: Hassanien, A.E., Tolba, M.F., Taher Azar, A. (eds.) AMLTA 2014. CCIS, vol. 488, pp. 278–290. Springer, Heidelberg (2014). doi:10.1007/978-3-319-13461-1_27

Scale Sensitivity of Textural Features

Michal Haindl[(⊠)] and Pavel Vácha

The Institute of Information Theory and Automation of the Czech Academy
of Sciences, Prague, Czech Republic
{haindl,vacha}@utia.cz

Abstract. Prevailing surface material recognition methods are based
on textural features but most of these features are very sensitive to scale
variations and the recognition accuracy significantly declines with scale
incompatibility between visual material measurements used for learning
and unknown materials to be recognized. This effect of mutual incompatibility between training and testing visual material measurements scale
on the recognition accuracy is investigated for leading textural features
and verified on a wood database, which contains veneers from sixty-six
varied European and exotic wood species. The results show that the
presented textural features, which are illumination invariants extracted
from a generative multispectral Markovian texture representation, outperform the most common alternatives, such as Local Binary Patterns,
Gabor features, or histogram-based approaches.

Keywords: Textural features · Texture scale recognition sensitivity ·
Surface material recognition · Markovian illumination invariant features

1 Introduction

Visual scene understanding is based on shapes and materials. While the shape
is stable visual attribute, the surface material appearance vastly changes under
variable observation conditions [3], which significantly and negatively affect its
recognition as well as its realistic synthesis. Reliable computer-based interpretation of visual information which would approach human cognitive capabilities is very challenging and impossible without significant improvement of the
corresponding sophisticated visual information models capable to handle huge
variations of possible observation conditions. The appropriate paradigm for such
a surface reflectance function models is a multidimensional visual texture. Generative visual texture models are useful not only for modelling physically correct
virtual objects material surfaces in virtual or augmented reality environments
or restoring images but also for contextual recognition applications such as segmentation, classification or image retrieval.

2 Textural Features

Numerous textural features have been published with miscellaneous recognition successfulness. Only the leading and commonly used textural features are

© Springer International Publishing AG 2017
C. Beltrán-Castañón et al. (Eds.): CIARP 2016, LNCS 10125, pp. 84–92, 2017.
DOI: 10.1007/978-3-319-52277-7_11

selected for this texture scale sensitivity study. These are the two-dimensional causal auto-regressive (2DCAR), local binary patterns (LBP), Gabor, and colour histogram features, respectively.

2.1 2DCAR Illumination Invariant Features

The texture is factorised into K levels of the Gaussian down-sampled pyramid and subsequently each pyramid level is modelled by a wide-sense Markovian type of model - the Causal Auto-regressive Random (CAR) model. The model parameters are estimated and illumination invariants are computed from them. Finally, the illumination invariants from all the pyramid levels are concatenated into one feature vector. Let us assume that each multispectral (colour) texture is composed of C spectral planes (usually $C = 3$), $Y_r = [Y_{r,1}, \ldots, Y_{r,C}]^T$ is the multispectral pixel at location r. The multiindex $r = (r_1, r_2)$ is composed of row index r_1 and column index r_2. The spectral planes are mutually decorrelated by the Karhunen-Loeve transformation and subsequently modelled using either a three-dimensional model or a set of C two-dimensional models. The two-dimensional models assumes that the j-th spectral plane of pixel at position r can be modelled as:

$$Y_{r,j} = \gamma_j Z_{r,j} + \epsilon_r, \qquad Z_{r,j} = [Y_{r-s,j} : \forall s \in I_r]^T$$

where $Z_{r,j}$ is the $\eta \times 1$ data vector, $\gamma_j = [a_1, \ldots, a_\eta]$ is the $1 \times \eta$ unknown parameter vector. Some selected contextual neighbour index shift set is denoted I_r and $\eta = cardinality(I_r)$. The texture is analysed in a chosen direction, where multiindex t changes according to the movement on the image lattice I. Given the known CAR process history $Y^{(t-1)} = \{Y_{t-1}, Y_{t-2}, \ldots, Y_1, Z_t, Z_{t-1}, \ldots, Z_1\}$, $\hat{\gamma}$ can be estimated using fast, numerically robust recursive statistics [2]:

$$V_{t-1} = \begin{pmatrix} \sum_{u=1}^{t-1} Y_u Y_u^T & \sum_{u=1}^{t-1} Y_u Z_u^T \\ \sum_{u=1}^{t-1} Z_u Y_u^T & \sum_{u=1}^{t-1} Z_u Z_u^T \end{pmatrix} + V_0 = \begin{pmatrix} V_{yy(t-1)} & V_{zy(t-1)}^T \\ V_{zy(t-1)} & V_{zz(t-1)} \end{pmatrix},$$

$$\lambda_{t-1} = V_{yy(t-1)} - V_{zy(t-1)}^T V_{zz(t-1)}^{-1} V_{zy(t-1)},$$

where the positive definite matrix V_0 represents prior knowledge. The following features are proved to be colour invariant [10,11]:

$$a_s, \ \forall s \in I_r, \qquad\qquad \alpha_1 = 1 + Z_t^T V_{zz(t)}^{-1} Z_t,$$

$$\alpha_2 = \sqrt{\sum_{\forall r \in I} (Y_r - \hat{\gamma}_t Z_r)^T \lambda_t^{-1} (Y_r - \hat{\gamma}_t Z_r)},$$

$$\alpha_3 = \sqrt{\sum_{\forall r \in I} (Y_r - \mu)^T \lambda_t^{-1} (Y_r - \mu)} \},$$

$$\beta_1 = \ln\left(\frac{\psi(r)^C}{\psi(t)^C} |\lambda_t| |\lambda_r|^{-1}\right), \qquad \beta_8 = \left(\frac{\psi(r)^C}{\psi(t)^C} |\lambda_t| |\lambda_r|^{-1}\right)^{\frac{1}{2C}},$$

$$\beta_2 = \ln\left(\frac{\psi(r)^C}{\psi(t)^C} |V_{zz(t)}| |V_{zz(r)}|^{-1}\right), \ \beta_9 = \left(\frac{\psi(r)^C}{\psi(t)^C} |V_{zz(t)}| |V_{zz(r)}|^{-1}\right)^{\frac{1}{2C\eta}},$$

$$\beta_3 = \ln\left(|V_{zz(t)}| |\lambda_t|^{-\eta}\right), \qquad \beta_{10} = \left(|V_{zz(t)}| |\lambda_t|^{-\eta}\right)^{\frac{1}{2C}},$$

$$\beta_4 = \ln\left(|V_{zz(t)}| |V_{yy(t)}|^{-\eta}\right), \qquad \beta_{11} = \left(|V_{zz(t)}| |V_{yy(t)}|^{-\eta}\right)^{\frac{1}{2C}},$$

$$\beta_5 = tr\left\{V_{yy(t)} \lambda_t^{-1}\right\}, \qquad \beta_{12} = \sqrt{|V_{yy(t)}| |\lambda_t|^{-1}},$$

where μ is the mean value of Y_r. The dissimilarity between two feature vectors of two textures is computed using fuzzy contrast [8] in its symmetrical form FC_3.

Table 1. Classification accuracy for 2DCAR features and variable scale inconsistency between test and learning texture scales computed separately on all RGB spectral channels.

Scale	Test											
Train	50	55	60	65	70	75	80	85	90	95	100	∅
50	100	100	100	100	99.24	96.97	95.45	88.64	79.55	68.94	64.39	90.29
55	100	100	100	100	100	99.24	99.24	99.24	92.42	87.12	78.79	96.01
60	100	100	100	100	100	100	100	99.24	98.48	94.70	89.39	98.35
65	100	100	100	100	100	100	100	100	100	99.24	96.97	99.66
70	100	100	100	100	100	100	100	100	100	100	100	100
75	99.24	100	100	100	100	100	100	100	100	100	100	99.93
80	94.70	99.24	100	100	100	100	100	100	100	100	100	99.45
85	84.09	95.45	100	100	100	100	100	100	100	100	100	98.14
90	75.00	90.15	96.97	100	100	100	100	100	100	100	100	96.56
95	66.67	82.58	93.18	97.73	99.24	100	100	100	100	100	100	94.49
100	61.36	72.73	88.64	94.70	98.48	99.24	100	100	100	100	100	92.29

2.2 Local Binary Patterns

Local Binary Patterns [6] are histograms of texture micro patterns. For each pixel, a circular neighbourhood around the pixel is sampled, P is the number of samples and R is the radius of circle. The sampled point values are thresholded by the central pixel value and the pattern number is formed:

$$LBP_{P,R} = \sum_{s=0}^{P-1} \text{sgn}\,(Y_s - Y_c)\,2^s, \tag{1}$$

where sgn is the sign function, Y_s is the grey value of the sampled pixel, and Y_c is the grey value of the central pixel. Subsequently, the histogram of patterns is computed. Because of the thresholding, the features are invariant to any monotonic grey-scale change. The multiresolution analysis is done by growing of the circular neighbourhood size. All LBP histograms were normalised to have unit L_1 norm. The similarity between LBP feature vectors is measured by means of Kullback-Leibler divergence as the authors suggested. We have tested features $LBP_{8,1+8,3}$, which are combination of features with radii 1 and 3. They were computed either on gray images or on each spectral plane of color image and concatenated. We also tested uniform version $LBP_{16,2}^u$, but their results were inferior.

2.3 Gabor Features

The Gabor filters [1,7] can be considered as orientation and scale tunable edge and line (bar) detectors and statistics of Gabor filter responses in a given region

Table 2. Classification accuracy for LBP features on gray-scale textures.

Scale	Test											
Train	50	55	60	65	70	75	80	85	90	95	100	∅
50	100	100	99.24	93.94	75.76	63.64	46.21	35.61	26.52	20.45	19.70	61.91
55	100	100	100	99.24	93.94	79.55	65.15	45.45	36.36	28.79	26.52	70.45
60	99.24	100	100	100	99.24	95.45	81.82	65.91	46.21	38.64	38.64	78.65
65	86.36	99.24	100	100	100	100	96.21	83.33	66.67	50.00	46.97	84.44
70	71.97	91.67	100	100	100	100	100	98.48	85.61	71.97	63.64	89.39
75	48.48	71.21	93.94	100	100	100	100	100	98.48	87.12	83.33	89.33
80	39.39	51.52	72.73	96.97	100	100	100	100	100	99.24	93.94	86.71
85	27.27	38.64	53.03	74.24	96.97	100	100	100	100	100	95.45	80.51
90	22.73	28.79	40.15	57.58	77.27	98.48	100	100	100	100	96.97	74.72
95	18.18	21.97	28.79	40.91	59.85	84.09	98.48	100	100	100	96.21	68.04
100	16.67	20.45	28.03	37.12	52.27	76.52	89.39	93.94	96.21	96.21	100	64.26

are used to characterise the underlying texture information. A two dimensional Gabor function $g(r) : \mathbb{R}^2 \to \mathbb{C}$ can be specified as

$$g(r) = \frac{1}{2\pi\sigma_{r_1}\sigma_{r_2}} \exp\left[-\frac{1}{2}\left(\frac{r_1^2}{\sigma_{r_1}^2} + \frac{r_2^2}{\sigma_{r_2}^2}\right) + 2\pi i V r_1\right],$$

where $\sigma_{r_1}, \sigma_{r_2}, V$ are filter parameters. The convolution of a texture image and Gabor filter extracts edges of given frequency and orientation range. The whole filter set was obtained by four dilatations and six rotations of the function $g(r)$. The filter set is designed so that Fourier transformations of filters cover most of image spectrum, see [5] for details. Finally, given a single spectral image with values $Y_{r,j}$, $r \in I$, $j = 1$, its Gabor wavelet transform is defined as

$$W_{k\phi,j}(r_1, r_2) = \int\limits_{u_1, u_2 \in \mathbb{R}} Y_{r,j} \, g_{k\phi}^*(r_1 - u_1, r_2 - u_2) \, du_1 \, du_2,$$

where $(\cdot)^*$ indicates the complex conjugate, ϕ and k are orientation and scale of the filter. The Gabor features [5] are defined as the mean μ_j and the standard deviation σ_j of the magnitude of filter responses W. The Gabor features of colour images have been computed either on grey images or on each spectral plane separately and concatenated to form a feature vector. The distance between two textures T, S is measured as the sum:

$$L1_{STD}(T, S) = \sum_{i=0}^{p} \left| \frac{f_i^{(T)} - f_i^{(S)}}{\sigma(f_i)} \right|, \tag{2}$$

where $\sigma(f_i)$ is standard deviations of a feature f_i computed over all database, and p is the size of the feature vector. Another extension of the Gabor filters to colour textures [4] is based on adding a chromatic antagonism, while the spatial antagonism is modelled by Gabor filters themselves. Opponent Gabor features consists of the monochrome part of features: $\eta_{i,m,n} = \sqrt{\sum_r W_{i,m,n}^2(r)}$, where $W_{i,m,n}$ is the response to Gabor filter of orientation m and scale n, i is i-th spectral band of the colour texture T. The opponent part of features is:

$$\psi_{i,j,m,m',n} = \sqrt{\sum_r \left(\frac{W_{i,m,n}(r)}{\eta_{i,m,n}} - \frac{W_{j,m',n}(r)}{\eta_{j,m',n}} \right)^2},$$

for all i,j with $i \neq j$ and $|m - m'| \leq 1$. (Opponent features could be also expressed as correlation between spectral planes responses.) The distance between textures T, S using the Opponent Gabor features is measured as sum

$$L2_{STD}(T,S) = \sqrt{\sum_{i=0}^{p} \left(\frac{f_i^{(T)} - f_i^{(S)}}{\sigma(f_i)} \right)^2}, \tag{3}$$

where $\sigma(f_i)$ is standard deviations of feature f_i computed over all database, and p is the size of feature vector.

2.4 Histogram Based Features

The simplest features used in this study are based on histograms of colours or intensity values. Although, these features cannot be considered as proper textural features, because they are not able to describe spatial relations which are the key texture properties, their advantage is robustness to various geometrical transformations, fast, and easy implementation. The cumulative histogram proposed in [9] is defined as the distribution function of an image histogram. The i-th bin H_i is computed as $H_i = \sum_{\ell \leq i} h_\ell$, where h_ℓ is the ℓ-th bin of ordinary histogram. The distance between two cumulative histograms is computed in L_1 metric. The cumulative histogram is more robust than the ordinary histogram, because a small intensity change characterised by a one-bin shift in the ordinary histogram, have only negligible effect on the cumulative histogram.

3 Experiments

The scale sensitivity of the selected textural features was tested on the wood database, which contains veneers from varied European and exotic wood species, each with two sample images only. The original images of 66 wood species were acquired by a colour scanner device. These images were scaled down to $95\%, 90\%, 85\%, \ldots, 50\%$ of their original size. Finally, regions with the same resolution were cropped out, see examples in Fig. 1. As a consequence, image of scale 50% covers doubles size of the original texture image, but with half of details than scale 100%. In the experiment, the training set was composed of images of a single selected scale and the classification accuracy was tested for all scales, separately. To avoid training and testing on the same images the scaled

Table 3. Classification accuracy for colour histogram and opponent Gabor features.

Scale	Test											
Train	50	55	60	65	70	75	80	85	90	95	100	∅
Colour histogram features												
50	100	100	98.48	96.21	93.94	91.67	87.12	82.58	79.55	76.52	65.91	88.36
55	100	100	100	98.48	96.97	95.45	92.42	88.64	85.61	81.82	70.45	91.80
60	98.48	100	100	100	99.24	99.24	96.21	93.18	87.12	86.36	74.24	94.01
65	96.21	99.24	100	100	100	100	99.24	97.73	92.42	90.91	78.79	95.87
70	93.18	98.48	100	100	100	100	100	97.73	96.21	94.70	82.58	96.63
75	90.15	93.94	98.48	100	100	100	100	100	97.73	95.45	84.85	96.42
80	84.09	92.42	93.18	99.24	100	100	100	100	100	96.97	85.61	95.59
85	82.58	87.12	90.15	93.94	99.24	100	100	100	100	100	87.12	94.56
90	78.79	82.58	85.61	90.91	95.45	98.48	99.24	100	100	100	86.36	92.49
95	74.24	78.03	82.58	85.61	92.42	95.45	98.48	99.24	100	100	87.12	90.29
100	65.91	73.48	76.52	78.79	78.79	83.33	85.61	87.12	86.36	86.36	100	82.02
Opponent gabor features												
50	100	100	100	98.48	96.21	92.42	87.12	82.58	75.00	67.42	62.12	87.40
55	100	100	100	100	100	99.24	96.97	93.18	87.12	80.30	70.45	93.39
60	100	100	100	100	100	100	98.48	98.48	96.97	93.18	85.61	97.52
65	100	100	100	100	100	100	100	98.48	98.48	96.21	90.15	98.48
70	100	100	100	100	100	100	100	100	100	96.97	95.45	99.31
75	99.24	100	100	100	100	100	100	100	100	99.24	96.97	99.59
80	95.45	99.24	100	100	100	100	100	100	100	100	97.73	99.31
85	87.88	96.21	99.24	100	100	100	100	100	100	100	98.48	98.35
90	81.06	90.91	98.48	100	100	100	100	100	100	99.24	97.73	97.04
95	72.73	85.61	93.18	97.73	99.24	100	100	100	100	100	98.48	95.18
100	62.12	75.00	87.88	93.94	98.48	98.48	99.24	99.24	99.24	99.24	100	92.08

images were split to upper half and lower half with 812 × 1034 pixel resolution. The final result were computed as an average of classification accuracy tested on lower and upper halfs (training was performed on the other half). The computed feature vectors were compared using the suggested distances and classified with the Nearest Neighbour (1-NN) classifier. Scale sensitivity of textural features is visualised in the graph on Fig. 2, which is created from two experiment results. The left part is based on training on scale 100% and testing on downscaled images 100%, 95%, ..., 50%, while the right part (scaling factor 1–2) uses training on scale 50% and testing on 55%, 60%, ..., 100%, which simulates upscaled images which avoids heavy interpolation. Results on Tables 1, 2, 3 and Fig. 2 illustrate the best and the most robust scale invariant performance of the 2DCAR features, which benefits from multiscale approach, followed by opponent

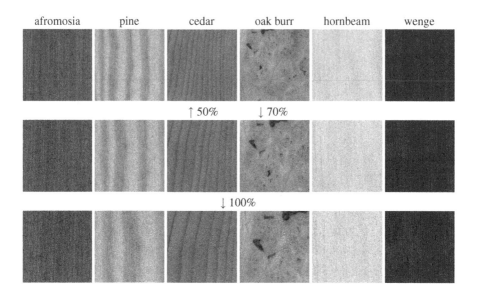

Fig. 1. The afromosia, pine, cedar, oak burr, hornbeam, and wenge veneer textures, respectively. The rows correspond to different resolution (50% top, 70% middle, 100% bottom) setups.

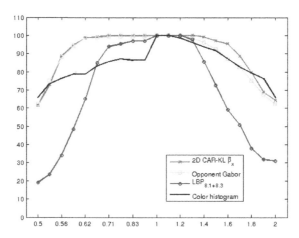

Fig. 2. Classification accuracy (y: [%]) for test x: scaling factor(0.5–2), training at scale factor 1, and average over 66 classes.

Gabor features. Histogram feature are the least sensitive to extreme scale variation, but they simultaneously they lack sufficient discriminability for recognition at similar scales. LBP are very sensitive to scale changes, because they have small support. Multiscale approach could probably improve their performance, but it will increase the already large number of LBP features.

4   Conclusion

The presented results indicate that Markovian illumination invariant texture features (2DCAR), based on Markovian descriptive model, are the most robust textural features for texture classification when learning and classified textures differ in scale. The 2DCAR features outperformed tested textural features, i.e., the LBP, Gabor or histogram texture features, respectively. Their additional advantage is their fast and numerically robust estimation. Additionally, our colour Markovian textural features were successfully applied in recognition of wood veneers using a smart phone camera. The method's correct recognition accuracy improvements are between 20% and 40%, compared to the Local Binary Patterns (LBP) features and up to 8% compared to the opponent Gabor features which is the second best alternative from all tested textural features.

It is worth to note that the presented results apply for recognition with bounded scale variation. If the expected scale variation would go to extremes, the fully scale invariant textural features should be considered. On the other hand, fully invariant features usually losses some discriminability, therefore each application need to carefully balance invariance to expected variability and discriminability. In the future, the presented results on wood recognition will be extended to a larger and more general texture database.

Acknowledgements. This research was supported by the Czech Science Foundation project GAČR 14-10911S.

References

1. Bovik, A.: Analysis of multichannel narrow-band filters for image texture segmentation. IEEE Trans. Sig. Process. **39**(9), 2025–2043 (1991)
2. Haindl, M.: Visual data recognition and modeling based on local Markovian models. In: Florack, L., Duits, R., Jongbloed, G., Lieshout, M.C., Davies, L. (eds.) Mathematical Methods for Signal and Image Analysis and Representation, Computational Imaging and Vision, vol. 41, pp. 241–259. Springer, London (2012). chap. 14
3. Haindl, M., Filip, J.: Visual Texture. Advances in Computer Vision and Pattern Recognition. Springer-Verlag, London (2013)
4. Jain, A.K., Healey, G.: A multiscale representation including opponent color features for texture recognition. IEEE Trans. Image Process. **7**(1), 124–128 (1998)
5. Manjunath, B.S., Ma, W.Y.: Texture features for browsing and retrieval of image data. IEEE Trans. Pattern Anal. Mach. Intell. **18**(8), 837–842 (1996)
6. Ojala, T., Pietikäinen, M., Mäenpää, T.: Multiresolution gray-scale and rotation invariant texture classification with local binary patterns. IEEE Trans. Pattern Anal. Mach. Intell. **24**(7), 971–987 (2002)
7. Randen, T., Husøy, J.H.: Filtering for texture classification: a comparative study. IEEE Trans. Pattern Anal. Mach. Intell. **21**(4), 291–310 (1999)
8. Santini, S., Jain, R.: Similarity measures. IEEE Trans. Pattern Anal. Mach. Intell. **21**(9), 871–883 (1999)
9. Stricker, M.A., Orengo, M.: Similarity of color images. In: SPIE, vol. 2420, pp. 381–392 (1995)

10. Vácha, P., Haindl, M.: Natural material recognition with illumination invariant textural features. In: Proceedings of the 20th International Conference on Pattern Recognition, ICPR 2010, pp. 858–861. IEEE Computer Society CPS, Los Alamitos (August 2010)
11. Vacha, P., Haindl, M.: Image retrieval measures based on illumination invariant textural MRF features. In: Proceedings of the 6th ACM International Conference on Image and Video Retrieval, CIVR 2007, pp. 448–454. NY, USA. ACM, New York (2007)

Noise-Added Texture Analysis

Tuan D. Pham[✉]

Department of Biomedical Engineering, Linkoping University,
58183 Linkoping, Sweden
tuan.pham@liu.se

Abstract. Noise is unwanted signal that causes a major problem for the task of image classification and retrieval. However, this paper reports that adding noise to texture at certain levels can improve classification performance without training data. The proposed method was tested with images of different texture categories degraded with various noise types: Gaussian (additive), salt-and-pepper (impulsive), and speckle (multiplicative). Experimental results suggest that the inclusion of noise can be useful for extracting texture features for image retrieval.

Keywords: Texture · Noise · Geostatistics · Spectral distortion · Image retrieval

1 Introduction

Texture analysis is an active area of research in image processing. The main reason for the importance of texture analysis is that its applications are pervasive in many fields of study, such as biometric identification [1], remote sensing [2], science and technology [3], biology [4], medicine [5], and visual arts [6]. However, the concept of texture in images is inherently imprecise, making it difficult and limited for the mathematical formulation of its complex property. Furthermore, noise is well known as a major factor that hinders the performance of extracting effective texture features for image classification, motivating many efforts to develop techniques for removing or robustly working with noise in images [7–10].

In general, noise in signals is undesirable for image analysis, but there are circumstances when some certain amount of noise can be useful, for example, to prevent discretization artifacts in color banding or posterization of an image [11]. The preserve of film grain noise can also help enhance the subjective perception of sharpness in images, known as acutance in photography, although it degrades the signal-to-noise ratio [12]. The intentional inclusion of noise in processing digital audio, image, and video data is called dither [13,14].

Another useful aspect of noise for texture analysis is reported in this paper. The motivation is mainly based on the concept of geostatistics that employs random models to characterize spatial attributes of natural phenomena [15]. In statistics, numerical methods are developed to deal with variables or random

© Springer International Publishing AG 2017
C. Beltrán-Castañón et al. (Eds.): CIARP 2016, LNCS 10125, pp. 93–100, 2017.
DOI: 10.1007/978-3-319-52277-7_12

variables. However, many variables of natural phenomena that vary in space and/or time may not be all random. Some variables may be totally deterministic, and some may take place somewhere between randomness and. Thus, the concept of a regionalized variable as a variable distributed in space is introduced to capture a behavior that is characterized with both of randomness and determinism [15]. In this regard, a regionalized variable is deterministic within a spatial domain, but beyond that it behaves as a random variable. The semi-variogram of geostatistics is formulated to quantify the behavior of a regionalized variable [16]. In this sense, adding noise at some certain levels into a texture space can give rise to discriminative power in the quantification of the random part of a regionalized variable of the image intensity by means of the semi-variogram. Due to the limitation of the semi-variogram for expressing the spatial continuity of a regionalized variable with experimental data, a spectral distortion measure is adopted to measure the dissimilarity of regionalized variables for texture retrieval.

2   Regionalized Variables in an Image

Since the intensities of pixels are variables distributed in space, they can be modeled as regionalized variables, each of which is considered as a single realization of a random function. In one sense, this regionalized variable is not related to its neighboring variables (pixels). In other sense, this variable has a spatial structure, depending on the distance separating the pixels. Thus, with the combination of the random and structured properties of a regionalized variable in a single function, the spatial variability of an image can be described on the basis of the spatial structure of these variables [17].

Without loss of generality, let $Z(x)$ be a regionalized variable, which is a realization of a random function Z, x and h be a spatial location and a lag distance in the sampling space, respectively. The variogram of the random function is defined as [16]

$$2\gamma(h) = Var[Z(x) - Z(x+h)], \tag{1}$$

where $\gamma(h)$ is the semi-variogram of the random function. This definition of the variogram, $2\gamma(h)$, or semi-variogram, $\gamma(h)$, assumes that the random function changes within the space, but $\gamma(h)$ is independent of spatial location and depends only on the distance of the pair of the considered variates. To simplify technical jargon, the semi-variogram is now referred to as the variogram, unless mathematical expression requires a precise definition.

Based on Eq. (1), the variogram is equivalent to

$$\gamma(h) = \frac{1}{2}E\left[\{Z(x) - Z(x+h)\}^2\right] \tag{2}$$

Let $Z(x_i)$, $i = 1, 2, \ldots, n$, be a sampling of size n, the unbiased estimator for the variogram, which is called the experimental variogram, of the random function is expressed as

$$\gamma(h) = \frac{1}{2N(h)} \sum_{i=1}^{N(h)} [Z(x_i) - Z(x_i + h)]^2, \tag{3}$$

where $N(h)$ is the number of pairs of variables separated by distance h.

The formulation of Eq. (3) is based on the assumption that the spatial auto-correlation structure is isotropic. This means that the semi-variogram depends only on the magnitude of the lag (h). When the spatial autocorrelation pattern is not the same in different directions in the sampling space, an anisotropic semi-variogram should be used to accommodate these differences. There are two types of anisotropy of the semi-variogram: geometric and zonal [18]. The geometric anisotropy occurs when the range of the semi-variogram varies with different directions. The zonal anisotropy takes place when both the range and sill of the semi-variogram change with different directions. The sill, which is the upper bound of the semi-variogram, represents the variance of the random field, whereas the range, at which the semi-variogram reaches the sill, indicates the distance at which data are no longer autocorrelated. These two parameters of the semi-variogram expressed by the spherical (theoretical) function [16], $\gamma^T(h)$, are shown in Fig. 1.

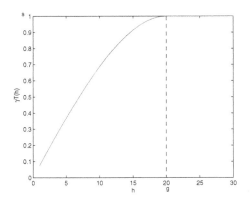

Fig. 1. Theoretical semi-variogram using spherical model with $s = 1$ (sill) and $g = 20$ (range)

With two coordinates for a 2D case, where $\mathbf{h} = (h_1, h_2)$, then a model for the geometric anisotropic semi-variogram can be defined as [18]

$$\gamma(r) = \gamma\left(\sqrt{\mathbf{h}^T \mathbf{B} \mathbf{h}}\right), \tag{4}$$

where $\mathbf{B} = \mathbf{Q}^T \mathbf{\Lambda} \mathbf{Q}$, and $\mathbf{Q}$ is the transformation matrix:

$$\mathbf{Q} = \begin{bmatrix} \cos\theta & \sin\theta \\ -\sin\theta & \cos\theta \end{bmatrix}, \tag{5}$$

where θ is the rotation angle, and Λ is the diagonal matrix of eigenvalues:

$$\Lambda = \begin{bmatrix} \lambda_1 & 0 \\ 0 & \lambda_2 \end{bmatrix}. \tag{6}$$

A model for the zonal anisotropic semi-variogram is defined as [18]

$$\gamma(\mathbf{h}) = \gamma_1(\mathbf{h}) + \gamma_2(\mathbf{h}), \tag{7}$$

where $\gamma_1(\mathbf{h})$ is the isotropic semi-variogram in one direction whose sill is much larger than the sill produced in the other direction, and $\gamma_2(\mathbf{h})$ is the geometric anisotropic semi-variogram.

3   Measuring Dissimilarity of Regionalized Variables

Given an experimental variance at lag h, $\gamma(h)$ has recently been proposed to be approximated as a linear combination of the past p variances [19]

$$\tilde{\gamma}(h) = -\sum_{i=1}^{p} a_i \gamma(h - i) \tag{8}$$

where a_i, $i = 1, \ldots, p$ are the linear predictive coding (LPC) coefficients [20], and to be optimally determined as follows.

The error between $\tilde{\gamma}(h)$ and $\gamma(h)$ is given by

$$e(h) = \gamma(h) + \sum_{i=1}^{p} a_i \gamma(h - i) \tag{9}$$

By minimizing the sum of squared errors, the pole parameters $\{a_i\}$ of the LPC model can be determined as follows.

$$\mathbf{a} = -\mathbf{R}^{-1}\,\mathbf{r} \tag{10}$$

where $\mathbf{a}$ is a $p \times 1$ vector of the LPC coefficients, $\mathbf{R}$ is a $p \times p$ autocorrelation matrix, and $\mathbf{r}$ is a $p \times 1$ autocorrelation vector whose elements are defined as

$$r_i = \sum_{h=0}^{N} \gamma(h)\gamma(h + i), \ i = 1, \ldots, p. \tag{11}$$

Let $S(\omega)$ and $S'(\omega)$ be the spectral density functions of the semi-variograms $\gamma(h)$ and $\gamma'(h)$, respectively, where ω is a normalized frequency ranging from $-\pi$ to π. The spectral density $S(\omega)$ is defined as [20]

$$S(\omega) = \frac{\sigma^2}{|A|^2}, \tag{12}$$

where $\sigma^2 = \mathbf{a}^T\mathbf{R}\mathbf{a}$, and $A = 1 + a_1 e^{-i\omega} + \cdots + a_p e^{-ip\omega}$.

The log-likelihood-ratio (LLR) distortion measure between $S(\omega)$ and $S'(\omega)$, denoted as $D_{LLR}(S, S')$, is defined as [21]

$$D_{LLR}(S, S') = \log \frac{\mathbf{a}'^T\mathbf{R}\mathbf{a}'}{\mathbf{a}^T\mathbf{R}\mathbf{a}}, \tag{13}$$

where $\mathbf{a}'$ is the vector of the LPC coefficients of S'.

4   Experiments

The proposed method was tested using four subsets of the Brodatz database [22] to represent four manually-labeled types of texture [23]: (1) fine-periodic, (2) fine-aperiodic, (3) coarse-periodic, and (4) coarse-aperiodic. Each subset consists of 90 images of 215×215 pixels, which are produced by dividing each of the 9 corresponding original Brodatz images into 9 non-overlapping smaller samples. The Brodatz indices of the fine-periodic texture are: D3, D6, D14, D17, D21, D34, D36, D38, D49, and D52. The image indices of the fine-aperiodic texture are: D4, D9, D16, D19, D24, D26, D28, D29, D32, and D39. For the coarse-periodic texture, the ten images are: D1, D8, D10, D11, D18, D20, D22, D25, D35, and D47. For the coarse-aperiodic texture, the ten images are: D2, D5, D7, D12, D13, D15, D23, D27, D30, and D31.

The isotropic model of the semi-variogram was used to compute the semi-variances for 30 lags of all the images, because the ranges of the images are almost the same, the use of the isotropic model has been found suitable for extracting image features, and the computation of the anisotropic variogram is very time-consuming, thus, it is not possible for large images.

The images were degraded with white Gaussian noise (additive noise), speckle noise (multiplicative noise), and salt & pepper noise (impulsive noise). Figure 2 shows a sub-image of D30 representing a coarse-aperiodic texture degraded with white Gaussian noise of zero mean and 0.1 variance, speckle noise with zero mean and 0.1 variance, and salt & pepper noise with 0.1 noise density.

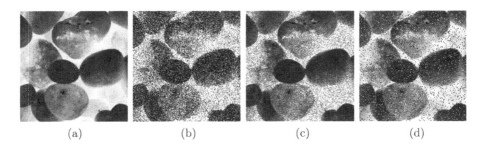

    (a)   (b)   (c)   (d)

Fig. 2. Brodatz coarse-aperiodic D30 (a), and corresponding degraded images with white Gaussian noise with zero mean and 0.1 variance (b), speckle noise with zero mean and 0.1 variance (c), and salt & pepper noise with 0.1 noise density (d).

For the matching of pattern similarity between two images, D_{LLR} was used with $p = 20$, to compute the spectral distortion of the semi-variograms of the two corresponding images. The retrieval was carried out for each of the 360 $(10 \times 9 \times 4)$ degraded images by searching for 8 images of the same texture among the most k similar images, where $k = 21$ in this case. Figure 3 shows the plots of the retrieval rates of the four Brodatz image subsets degraded with three noise models and a variety of noise levels.

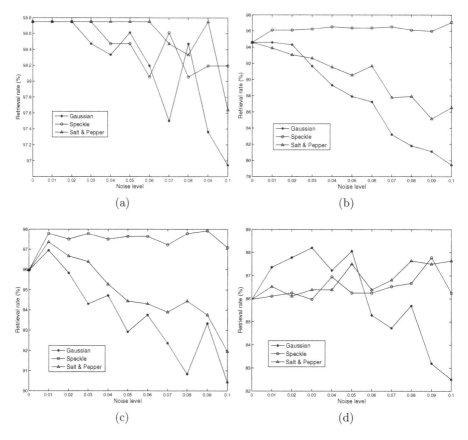

Fig. 3. Retrieval rates of four Brodatz image subsets degraded with various noise models for four texture categories: fine-periodic (a), fine-aperiodic (b), coarse-periodic (c), and coarse-aperiodic (d); where the zero noise level indicates the corresponding retrieval rate of undegraded images.

For the fine-periodic texture (Fig. 3(a)), the retrieval rates of images degraded with salt & pepper noise maintain the same rate in comparison with the undegraded images up to the noise level of 0.06, whereas 0.2 and 0.3 for Gaussian noise and speckle noise, respectively. For the fine-aperiodic texture, Fig. 3(b) shows the consistently high performance of the images added with speckle noise, whereas the retrieval rate of the images degraded with Gaussian noise with 0.01 variance is higher than that of the undegraded images, and the rate using the inclusion of salt & pepper noise at 0.01 density level is slightly lower than the retrieving performance without noise. For the coarse-periodic texture shown in Fig. 3(c), the retrieval rates are higher for images degraded at the noise level of 0.01 using all the three noise models than the retrieval rate for undegraded images, where speckle noise gives the highest result. The same results also apply to the coarse-aperiodic (Fig. 3(d)), where Gaussian noise has the best performance. Keeping

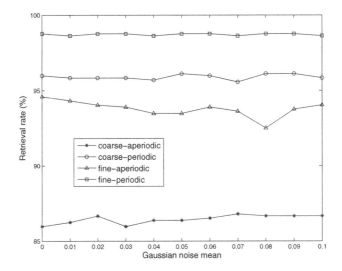

Fig. 4. Retrieval rates of four Brodatz subsets degraded with Gaussian noise with 0.01 variance and variable means, where the zero Gaussian noise mean indicates the corresponding retrieval rate of undegraded images.

the Gaussian noise variance constant at 0.01, while varying the Gaussian noise mean from 0.01 to 0.1, with the exception of the fine-aperiodic texture, Fig. 4 shows the high performance of the retrieval task by adding Gaussian noise to the images of the other three types of texture.

5   Conclusion

Texture analysis using the spectral distortion measure of semi-variograms can be enhanced by adding noise at certain levels to the images. The set of 40 Brodatz images were selected in this study because they fairly represent the 4 types of texture. The addition of different types of noise to other types of texture is worth pursuing further investigation. Furthermore, in this paper, only empirical evidence was provided to support the argument of the useful application of noise to texture, theoretical verification of the power of geostatistics in finding structures in random-biased texture is therefore important to follow.

References

1. Batool, N., Chellappa, R.: Fast detection of facial wrinkles based on Gabor features using image morphology and geometric constraints. Pattern Recogn. **48**, 642–658 (2015)
2. Petrou, Z., et al.: Discrimination of vegetation height categories with passive satellite sensor imagery using texture analysis. IEEE. J. Sel. Top. Appl. Earth Obs. Remote Sens. **8**, 1442–1455 (2015)

3. Zheng, C., Sun, D.W., Zheng, L.: Recent applications of image texture for evaluation of food qualities-a review. Trends Food Sci. Technol. **17**, 113–128 (2006)
4. Todorovic, V.: A visual aid for cellular imaging analysis. Nat. Methods **12**, 175 (2015)
5. Hatt, M., et al.: FDG PET uptake characterization through texture analysis: investigating the complementary nature of heterogeneity and functional tumor volume in a multi-cancer site patient cohort. J. Nucl. Med. **56**, 38–44 (2015)
6. Shamir, L.: What makes a Pollock Pollock: a machine vision approach. Int. J. Arts Technol. **8**, 1–10 (2015)
7. Lin, C.H., Tsai, J.S., Chiu, C.T.: Switching bilateral filter with a texture/noise detector for universal noise removal. IEEE Trans. Image Process. **19**, 2307–2320 (2010)
8. Song, T., et al.: Noise-robust texture description using local contrast patterns via global measures. IEEE Signal Process. Lett. **21**, 93–96 (2014)
9. Zhu, Z., et al.: An adaptive hybrid pattern for noise-robust texture analysis. Pattern Recogn. **48**, 2592–2608 (2015)
10. Pham, T.D.: Estimating parameters of optimal average and adaptive Wiener filters for image restoration with sequential Gaussian simulation. IEEE Signal Process. Lett. **22**, 1950–1954 (2015)
11. Langford, M.: The Darkroom Handbook. Dorling Kindersley, New York (1981)
12. Oh, B.T., Lei, S.M., Kuo, C.C.J.: Advanced film grain noise extraction and synthesis for high-definition video coding. IEEE Trans. Circ. Syst. Video Technol. **19**, 1717–1729 (2009)
13. Roberts, L.G.: Picture coding using pseudo-random noise. IEEE Trans. Inf. Theory **8**, 145–154 (1962)
14. Vanderkooy, J., Lipshitz, S.P.: Dither in digital audio. J. Audio Eng. Soc. **35**, 966–975 (1987)
15. Matheron, G.: Principles of geostatistics. Econ. Geol. **58**, 1246–1266 (1963)
16. Olea, R.A.: Geostatistics for Engineers and Earth Scientists. Kluwer Academic Publishers, Boston (1999)
17. Starck, J.L., Murtagh, F.D., Bijaoui, A.: Image Processing and Data Analysis: The Multiscale Approach. Cambridge University Press, New York (1998)
18. Wackernagel, H.: Multivariate Geostatistics: An Introduction with Applications, 3rd edn. Springer, Berlin (2003)
19. Pham, T.D.: The semi-variogram and spectral distortion measures for image texture retrieval. IEEE Trans. Image Process. **25**, 1556–1565 (2016)
20. Rabiner, L., Juang, B.H.: Fundamentals of Speech Recognition. Prentice Hall, New Jersey (1993)
21. Nocerino, N., et al.: Comparative study of several distortion measures for speech recognition. In: Proceedings of the IEEE International Conference on Acoustics, Speech and Signal Processing, vol. 10, pp. 25–28 (1985)
22. Brodatz, P.: Textures: A Photographic Album for Artists and Designers. Dover, New York (1966)
23. Arivazhagan, S., Nidhyanandhan, S.S., Shebiah, R.N.: Texture categorization using statistical and spectral features. In: Proceedings of the International Conference on Computing, Communication and Networking, 8 p. (2008)

Breast Density Classification with Convolutional Neural Networks

Pablo Fonseca[1]([⊠]), Benjamin Castañeda[2], Ricardo Valenzuela[3],
and Jacques Wainer[1]

[1] RECOD Lab, Institute of Computing, University of Campinas, Campinas, Brazil
pablo.arroyo@ic.unicamp.br
[2] Laboratorio de Imágenes Médicas, Pontificia Universidad Católica del Perú,
Lima, Peru
[3] Imaging Lab, Eldorado Research Institute, Campinas, Brazil

Abstract. Breast Density Classification is a problem in Medical Imaging domain that aims to assign an American College of Radiology's BIRADS category (I-IV) to a mammogram as an indication of tissue density. This is performed by radiologists in an qualitative way, and thus subject to variations from one physician to the other. In machine learning terms it is a 4-ordered-classes classification task with highly unbalance training data, as classes are not equally distributed among populations, even with variations among ethnicities. Deep Learning techniques in general became the state-of-the-art for many imaging classification tasks, however, dependent on the availability of large datasets. This is not often the case for Medical Imaging, and thus we explore Transfer Learning and Dataset Augmentationn. Results show a very high squared weighted kappa score of 0.81 (0.95 C.I. [0.77,0.85]) which is high in comparison to the 8 medical doctors that participated in the dataset labeling 0.82 (0.95 CI [0.77, 0.87]).

1 Introduction

Breast cancer is a major health treat as it accounts for the 13.7% of cancer deaths in women according to the World Cancer Report [1]. Moreover, it is the second most common type of cancer worldwide and recent statistics show that one in every ten women will develop it at some point of their lives. However, it is important to notice that when detected at an early stage, the prognosis is good, opening the door to Computer Aided Diagnosis Systems that target the prevention of this disease. Medical research towards the prevention of breast cancer has shown that breast parenchymal density is a strong indicator of cancer risk [2]. Specifically, the risk of developing breast cancer is increased only in 5% related to mutations in the genetic biomarkers BRCA 1 and 2; this risk, on the other hand, is increased to 30% for breast densities higher than 50% [3,4]. Because

B. Castañeda—Supported by grant DGI 2012-0141 from the Pontificia Universidad Católica del Perú.

of this, the breast density can be seen as very valuable information in order to perform preventive tasks and assess cancer risk. However, this behavior varies from one ethnicity to the other, even with different breast density distributions across populations. A comparative study of our dataset used in this research with other populations can be found in Casado et al. [5].

1.1 Breast Density

According to Otsuka et al. [6], mammographic density refers to the proportion of radiodense fibroglandular tissue relative to the area or volume of the breast. In order to assess breast composition, there are both qualitative and quantitative methods. One of the best known qualitative methods is the Breast Imaging Reporting and Data System (BI-RADS), the target of this research [7]. Among quantitative methods, there is one developed by Boyd, the quantitative ACR and other computer-assisted methods. Some previous work of automatic breast composition classification include Oliver et al. [8] where several methods are tested on the MIAS database using the BIRADS standard and Oliver et al. [9] where a method that included segmentation, extraction of morphological and texture features and bayesian classifier combination. Also there are commercial tools such as Volpara (TM) [10] and Quantra (TM) [11].

1.2 American College of Radiology: BIRADS Categories

The American College of Radiologists (ACR) developed four qualitative categories for breast density which are presented below along with the meaning of each composition category and sample mammograms in the study. The main goal of this research is to classify mammograms into the BIRADS categories using convolutional neural networks based techniques. There are two approaches that were explored: Random Filter Convolutional networks as feature extractors coupled with a Linear SVM and the well known Krizhevsky Deep Convolutional Network.

These four categories are qualitatively judged by radiologists according to their density. When it comes to assessing breast density in mammograms some challenges might arise even for experienced radiologists such as reported in the comparative study of inter- and intra-observer agreement among radiologists [12].

1.3 Radiologist Agreement and Performance Evaluation

Rater agreement need to be measured in a way that it is consistent with the actual performance of radiologists as well as automated algorithms. Accuracy would not be very informative as kappa indexes for this kind of task. Cohen's kappa can target agreement on both categorical and ordinal variables. In Rendondo et al. [12] the performance of radiologists is evaluated for a stratified sample of 100 mammogram in the BIRADS categories which include two separated target measuares: assesment and breast density. For breast density, the study found high interobserver and intraobserver agreement ($\kappa = 0.73$, with a 95% confidence interval 0.72–0.74 and $\kappa = 0.82$, with a 95% confidence interval 0.80–0.84, respectively) where the squared weighted kappa is used. In that study, there were 21 radiologists with average experience in reading mammograms of 12 years (range 4–22). Even if according to the Landis and Koch's strenght of agreement [13] the intraobserver kappa value shows an almost perfect agreement, the value 0.82 can show the subjectivity of the BIRADs measure, even for trained professionals.

2   The Breast Density Dataset

In this paper, we work with the dataset presented by Casado et al. [5], here we briefly survey its idiosyncrasies, but for more details please refer to that paper. The mammograms were obtained from two medical centers in Lima, Peru. Some of those images are shown in Fig. 1. All subjects were women who underwent routine breast cancer screening. The age of the subjects ranged from 31 to 86 years (mean age of 56.7 years, standard deviation of 9.5 years). A total of 1060 subjects were included in the sample population. The mammograms were collected in a craniocaudal view using two different systems. The first one was a Selenia Dimensions (Hologic, Bedford, MA) which produced digital mammograms with a pixel pitch of $100\,\mu m$, The second system was a Mammomat 3000 (Siemens Medical, Iselin, NJ) in combination with a CR 35 digitizer (Agfa Healthcare, Mortsel, Belgium) that allowed producing digital images with a depth of 16 bits and a pixel pitch of $50\,\mu m$. Approximately 16% and 84% of the mammograms used in this study were acquired with the Selenia Dimensions and Mammomat 3000 systems, respectively.

The mammograms were blindly classified by eight radiologists with varying degrees of experience assessing mammograms between 5 and 25 years. The mode of the breast density classification by the eight radiologists was considered as ground truth for this study. These medical doctors are named from A to H in no particular order. To serve the purposes of this research the Region of Interest (ROI) was manually selected to include only the breast region. The cropped

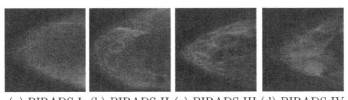

(a) BIRADS I (b) BIRADS II (c) BIRADS III (d) BIRADS IV

Fig. 1. Sample Mammograms in the study (BIRADS I-IV). Here 4 mammograms are shown along with their BIRADS density classification: the mammogram in figure (a) is the less dense and the mammogram in figure (d) is the most dense. Images were resized to an 1:1 ratio for feature extraction.

Table 1. Distribution of densities across dataset

BIRADS	I	II	III	IV
Total dataset	32.64%	45.19%	19.62%	2.55%
(1060 images)	346 images	479 images	208 images	27 images
Training subset	32.66%	45.16%	19.62%	2.55%
(744 images)	243 images	336 images	146 images	19 images
Test subset	32.59%	45.25%	19.62%	2.53%
(316 images)	103 images	143 images	62 images	8 images

images were also resized to a fixed size of 200×200 pixels, it must be noticed that radiologists had the full resolution DICOM mammogram. The semantics of the problem shows that finding regions of high radio dense tissue is meaningful only with respect of the total area of the breast. Initial experiments show that a better behavior of CNNs was found when the resize was performed. The general machine learning setup was to do a stratified division of the dataset in training and test sets as shown in Table 1.

3   Methods

We explore two methods based on convolutional neural networks that might deal with small datasets as ours. First, the HT-L3 architecture in Fig. 2 in Sect. 3.1 and previously studied for Breast Density Classification in [14] and initially proposed by [15]. In this first case, the filters are generated randomly, so a major focus is quantifying how stable are under different initializations. The second method in Sect. 3.2 is the well known Krizhevsky network [16] also identified as AlexNet. The network was trained on the ImageNet dataset an fine tuned for Breast Density Classification.

3.1   HT-L3 Visual Features with Random Filters

The HT-L3 family of features described by Cox and Pinto [17] and Pinto et al. [15] can be seen as a parameterizable image description function that was inspired on the primates visual system. It performs three consecutive layers of filtering, activation, pooling and normalization leading to a high dimensional representation of the image which can be used to feed a classifier such as the Support Vector Machines (SVM) with linear kernel.

Random noise filters were used in the filtering stage, although they were normalized to have zero mean and unit standard deviation no further training was performed on them. It is the aim of some of the experiments carried on to characterize the stability of such setup. Grid search was used to find the most suitable architectures. This means that several candidate architectures of the HT-L3 family are screened in order to chose the top performing ones. More

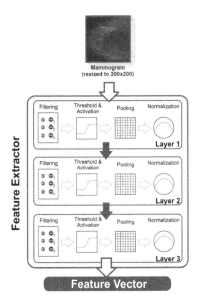

Fig. 2. HT-L3 Convolutional Neural Network: Here the linear and non-linear operations are shown in the order in which they are performed. Each layer (1–3) has the same operations, and the output of one layer is the input for the next one. However, each layer will produce more deep multiband images.

details on the implementation of this architecture for Breast Density Classification were reported in [14], a previous work of our group. From these results we chose the top three performing architectures which are presented in Table 2.

Table 2. Top performing architectures: here the top 3 performing architectures are shown alongside they parameters. S. L1, 2, 3 defines the filter size of the layer, # L1, 2, 3 defines the number of filters per layer, exponent ans α are fixed to 2.

	S. L1	S. L2	S. L3	# L1	# L2	# L3	Exp.	α
Arch1	5×5	5×5	5×5	128	128	64	2	2
Arch2	5×5	7×7	7×7	128	32	128	2	2
Arch3	5×5	5×5	7×7	32	32	64	2	2

3.2 Deep Learning Trained with the Image.net and Fine-Tuned with the Breast Density Dataset

In their 2012 paper, Krizhevsky et al. [16] presented a convolutional neural network architecture that got the best results on the Image.net dataset, a natural images dataset. Using the Caffe framework for Deep Learning [18], a convolutional network as described by [16] trained with the Image.net 1k classes subset was fine-tuned with the breast density dataset.

Changes to the net included the last layer being changed for a 4-way softmax in order to perform BIRADS classifications. The same sets of training and testing in the evaluating the HT-L3 features were used. Data aumengtation was used generating random crops of the mammogram.

4   Results

The HT-L3 features with random filters worked well and they are quite stable on multiple runs (we initialized 30 times with different seeds each architecture), which implied generating random filters with different seeds can indicate that a great deal of the discriminative power of these features lay on the architecture, leaving a classifier such as the SVM with linear kernel the task of finding separations across different classes, given the feature representations.

On the other side, in the Krizhevsky network the learning is performed with back-propagation and stochastic gradient descent, which means that the filters are actually learned as well as the weights of the fully connected layers that produce the final classification. As the size of the dataset is small for Deep Learning standards (1k images). On Table 3 we report the behavior of four automatic classification approaches, the first one is the Krizhevsky network trained with Image.net database and then fine-tuned with our breast density database and the three top performing HT-L3 architectures.

Table 3. Agreement with the golden standard on the test set

	κ-squared
Radiologist A	0.72
Radiologist B	0.88
Radiologist C	0.89
Radiologist D	0.74
Radiologist E	0.89
Radiologist F	0.88
Radiologist G	0.80
Radiologist H	0.88
Radiologists (95% CI)	[0.77, 0.87]
Krizhevsky NET (95% CI)	[0.77, 0.85]
HTL3 Arch1 (95% CI - 30 runs)	[0.75, 0.77]
HTL3 Arch2 (95% CI - 30 runs)	[0.72, 0.74]
HTL3 Arch3 (95% CI - 30 runs)	[0.71, 0.73]

5   Conclusions

Deep learning techniques for image classification aim an end-to-end learning procedure, specially in contrast to hand-engineered features as a previous approach presented in [19]. Convolutional neural networks, on the other hand, use a series of well-defined operations however parameterizable that both obtain a representation of the image and then perform the classification. Two techniques were explored, a three layer convolutional network with random filters used to produce high dimensional features later used to train a Linear SVM classifier and the Krizhevsky Network trained with the Image.net database and then fine tuned with our Breast Density Dataset. The main difference between these two approaches is that in the first one, no learning was performed by the convolutional neural network and thus the filter were randomly generated and the classification was actually performed by a linear SVM.

We found satisfactory results as measure by the kappa score, as a measure of agreement between the golden standard and the proposed methods. We would like to stress the point that even if results show a very high squared-weighted kappa score of 0.81 (0.95 C.I. [0.77,0.85]) of our best performing approach which is high in comparison to the 8 medical doctors that participated in the dataset labeling 0.82 (0.95 CI [0.77, 0.87]), radiologist-like behavior might not be fully well measured with this agreement metric.

We also been successfully in proving the stability of classification when random filters are used, this might indicate that a great deal of the discriminative power of these features lay on the model assumptions, at least for the problem targeted here. However, this HT-L3 based approaches did not performed better than the Krizhevsky Network.

Future work should include exploration of a better metric for assessing radiologist-like performance and further Deep Learning Architectures as well as Open Mammogram Databases.

References

1. Boyle, P., Levin, B., et al.: World Cancer Report 2008. IARC Press, International Agency for Research on Cancer (2008)
2. Wolfe, J.N.: Risk for breast cancer development determined by mammographic parenchymal pattern. Cancer **37**(5), 2486–2492 (1976)
3. Boyd, N., Martin, L., Yaffe, M., Minkin, S.: Mammographic density and breast cancer risk: current understanding and future prospects. Breast Cancer Res. **13**(6), 1 (2011)
4. Ursin, G., Qureshi, S.A.: Mammographic density-a useful biomarker for breast cancer risk in epidemiologic studies. Nor. Epidemiol. **19**(1), 59–68 (2009)
5. Casado, F.L., Manrique, S., Guerrero, J., Pinto, J., Ferrer, J., Castañeda, B.: Characterization of Breast Density in Women from Lima, Peru (2015)
6. Otsuka, M., Harkness, E.F., Chen, X., Moschidis, E., Bydder, M., Gadde, S., Lim, Y.Y., Maxwell, A.J., Evans, G.D., Howell, A., Stavrinos, P., Wilson, M., Astley, S.M.: Local Mammographic Density as a Predictor of Breast Cancer (2015)

7. D'Orsi, C.J.: Breast Imaging Reporting and Data System (BI-RADS). American College of Radiology (1998)
8. Oliver, A., Freixenet, J., Martí, R., Zwiggelaar, R.: A comparison of breast tissue classification techniques. In: Larsen, R., Nielsen, M., Sporring, J. (eds.) MICCAI 2006. LNCS, vol. 4191, pp. 872–879. Springer, Heidelberg (2006). doi:10.1007/11866763_107
9. Oliver, A., Freixenet, J., Marti, R., Pont, J., Perez, E., Denton, E., Zwiggelaar, R.: A novel breast tissue density classification methodology. IEEE Trans. Inf. Technol. Biomed. **12**(1), 55–65 (2008)
10. Volpara Solutions (2015). http://volparasolutions.com
11. Hologic (2015). www.hologic.com/
12. Redondo, A., Comas, M., Maciá, F., Ferrer, F., Murta-Nascimento, C., Maristany, M.T., Molins, E., Sala, M., Castells, X.: Inter-and intraradiologist variability in the BI-RADS assessment and breast density categories for screening mammograms. Br. J. Radiol. **85**(1019), 1465–1470 (2012). PMID: 22993385
13. Landis, J.R., Koch, G.G.: The measurement of observer agreement for categorical data. Biometrics **33**(1), 159–174 (1977)
14. Fonseca, P., Mendoza, J., Wainer, J., Ferrer, J., Pinto, J., Guerrero, J., Castañeda, B.: Automatic Breast Density Classification using a Convolutional Neural Network Architecture Search Procedure (2015)
15. Pinto, N., Doukhan, D., DiCarlo, J.J., Cox, D.D.: A high-throughput screening approach to discovering good forms of biologically inspired visual representation. PLoS Comput. Biol. **5**(11), e1000579 (2009)
16. Krizhevsky, A., Sutskever, I., Hinton, G.E.: Imagenet classification with deep convolutional neural networks. In: Advances in Neural Information Processing Systems, pp. 1097–1105 (2012)
17. Cox, D., Pinto, N.: Beyond simple features: a large-scale feature search approach to unconstrained face recognition. In: 2011 IEEE International Conference on Automatic Face Gesture Recognition and Workshops (FG 2011), pp. 8–15 (2011)
18. Jia, Y., Shelhamer, E., Donahue, J., Karayev, S., Long, J., Girshick, R., Guadarrama, S., Darrell, T.: Caffe: Convolutional Architecture for Fast Feature Embedding. arXiv preprint arXiv:1408.5093 (2014)
19. Angulo, A., Ferrer, J., Pinto, J., Lavarello, R., Guerrero, J., Castañeda, B.: Experimental Assessment of an Automatic Breast Density Classification Algorithm Based on Principal Component Analysis Applied to Histogram Data (2015)

Community Feature Selection for Anomaly Detection in Attributed Graphs

Mario Alfonso Prado-Romero$^{(\boxtimes)}$ and Andrés Gago-Alonso

Advanced Technologies Application Center (CENATAV),
7a ♯ 21406, Rpto. Siboney, Playa, CP 12200 Havana, Cuba
{mprado,agago}@cenatav.co.cu

Abstract. Anomaly detection on attributed graphs can be used to detect telecommunication fraud, money laundering, intrusions in computer networks, atypical gene associations, or people with strange behavior in social networks. In many of these application domains, the number of attributes of each instance is high and the curse of dimensionality negatively affects the accuracy of anomaly detection algorithms. Many of these networks have a community structure, where the elements in each community are more related among them than with the elements outside. In this paper, an adaptive method to detect anomalies using the most relevant attributes for each community is proposed. Furthermore, a comparison among our proposal and other state-of-the-art algorithms is provided.

Keywords: Anomaly detection · Feature selection · Attributed graphs

1 Introduction

Many phenomena, from our world, like neural networks, bank transactions, social networks, or genes in our DNA can be modeled as networks of interconnected elements. In these networks, each element has a set of features, and it also has relationships with other elements. Anomaly detection refers to the problem of finding patterns in data that do not conform to expected behavior [1]. Anomaly detection on the previously mentioned networks can be used to detect intrusions in computer networks, money laundering, identity thief in telecommunications, or strange gene associations, among other applications.

Traditional anomaly detection techniques only analyze the information about the elements [1], or the information regarding their relationship [2]. However, in many application domains like social networks, online shopping or bank transactions, both types of information can be found. Techniques for detecting anomalies in attributed graphs can deal with this heterogeneous data. Most of these techniques take advantage of the community structure present in the graphs, to analyze each element in a context relevant to it detecting subtle anomalies like products with a higher price than its co-purchased products.

© Springer International Publishing AG 2017
C. Beltrán-Castañón et al. (Eds.): CIARP 2016, LNCS 10125, pp. 109–116, 2017.
DOI: 10.1007/978-3-319-52277-7_14

In many application domains the number of features describing an element can be very high, thus identify relevant patterns in the data become very difficult, this is known as the curse of dimensionality [3]. To overcome this fact, it is important that anomaly detection algorithms identify the most meaningful features to be used in the detection process. In this paper, a method for improving anomaly detection in attributed graphs, using an unsupervised feature selection algorithm to select the most relevant features for each community of elements, is proposed. The advantages of this method are shown on the Amazon co-purchase network of Disney products[1] [4].

In the next sections, the existing approaches to anomaly detection are analyzed (Sect. 2), some basic concepts are introduced (Sect. 3), our method is presented (Sect. 4) and experimental results on a real data set are analyzed (Sect. 5). Conclusions are presented in Sect. 6.

2   Related Work

The three major approaches for anomaly detection discussed in this section are the vector based approach, graph based one, and hybrid one. Furthermore, the advantages and disadvantages of each approach, and how it tackles with the curse of dimensionality, are discussed.

The commonly reported anomaly detection techniques are designed to deal with vector valued data [1]. Some of these are distance-based algorithms [5], density-based algorithms [6,7] and algorithms to find clustered anomalies [8–10], but none of them avoid the curse of dimensionality. Some recent techniques rank objects based on the selection of a relevant subset of its attributes to tackle the curse of dimensionality [11,12]. Nonetheless, none of them takes into consideration relationships among the elements, ignoring part of the information in the data set, thus these algorithms cannot identify the most significant features for each community of elements.

Relationships among elements give valuable information about the structure of a network, due to this, many algorithms to detect anomalous nodes in graphs, using their relationships, has been proposed [2]. Commonly used techniques include the analysis of the structural characteristics of network elements finding deviations from normal behavior [13], and searching for infrequent structures in the network [14,15]. None of these algorithms uses the attributes of the elements, thus they are not affected by the curse of dimensionality, but they present low accuracy in heterogeneous data sets because they ignore part of the existing information.

Anomaly detection in attributed graphs where both graph and vector data are analyzed has not received much attention. The algorithm described in [16] combines community detection and anomaly detection in a single process, finding elements deviated from its community behavior. This algorithm uses the full attribute space; thus, its results are affected by the curse of dimensionality.

---

[1] http://www.ipd.kit.edu/~muellere/GOutRank/.

The technique described in [4] proposes an outlier ranking function capable of use only a subset from the node attributes. In a first step, this technique uses a state-of-the-art graph clustering algorithm considering subsets of the node attributes [17–19] and in a second step detects elements whose behavior deviates from the one of its group. This technique does not take into consideration that, in some application domains, the relationships among elements give a context to analyze them, instead of directly indicate than an element is anomalous. Thus, the performance of this algorithm and the quality of its detection, in this application domains, is affected.

3   Basic Concepts

It is important to introduce some fundamental concepts before presenting our proposal.

Definition 1 (Attributed Graph). *An attributed graph $G = <V, E, a>$ is a tuple where:*

(i)   The set $V = \{v_1, v_2, \ldots, v_n\}$ contains the vertices of the graph.
(ii) The set $E = \{(v, u)|v, u \in V\}$ contains the edges of the graph.
(iii) The function $a : V \to \mathbb{R}^d$ assigns an attribute vector of size d to each vertex from G.

The vertices of the graph are the elements of the network, and the edges the relationships among them.

Definition 2 (Disjoint Clustering). *A disjoint clustering $C = \{C_1, C_2, \ldots, C_k\}$ of elements from G is a set where:*

(i) $\forall C_i \in C, C_i \subset V$
(ii) For each $C_i, C_j \in C, i \neq j, C_i \cap C_j = \emptyset$
(iii) $\cup_{i=1}^{k} C_i = V$.

In this work, the disjoint clusterings are referred just as clusterings.

Definition 3 (Outlier Ranking). *An outlier ranking from a graph G is a set $R = \{(v, r)|v \in V, r \in [0, 1]\}$ of tuples, each one containing a vertex from G and its outlierness score.*

4   Improving Anomaly Detection Using Community Features

Many real networks are structured in communities. A community is a group of elements more connected among them, than with external elements. Usually, the elements of a same community share similar features. We propose to perform feature selection per community, selecting those features that better represent an element in its context. The elements outside the community of an element

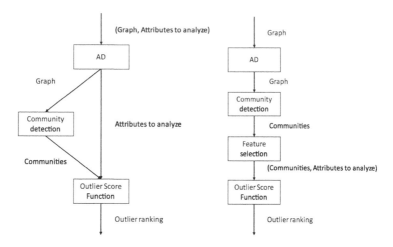

Fig. 1. Components interaction in the base algorithm (left) and in the improved algorithm (right)

are not relevant for it, and could affect the quality of the feature selection, for this reason they are ignored.

This idea was applied to our Glance algorithm, designed for anomaly detection in attributed graphs where the edges behave as contextual attributes. The algorithm originally received as parameters the attributed graph and the attributes to be considered by the anomaly score. In a first stage, the algorithm uses Louvain community detection method [20] to find communities of elements in the graph using connections among elements. In a second stage, it iterates over each community and uses an outlierness score function to determine the outlierness degree of each element. This function receives, as a parameter, the features to be used for the score.

The changes made to the original algorithm include removing the need of external parameters set by the user. Also, in the second stage, the algorithm selects the most relevant features for each community using a Laplacian Score [21] and then applies an outlierness score to each element of the community. The Laplacian Score ranks as more representatives those features, with large variance, that are similar in near elements. Using this feature selection technique, the anomaly detection algorithm becomes completely unsupervised. In Fig. 1, a comparison among the components interaction in the base algorithm and in the improved one can be observed.

The Glance algorithm with community feature selection can be observed in more detail in Algorithm 1. The algorithm receives an attributed graph G and returns an outlierness score of the vertices from G. In the second line, The Louvain community detection algorithm is used to find relevant groups of elements in G. In lines 3 to 10 the algorithm iterates over each community of G. In line 4 the more descriptive features for the community are selected. In lines 6 to 9, the outlierness score for each element in C_i is calculated using Glance score function. This function defines the outlierness score of an element as the percent of elements in its same community that have a difference with it greater than the mean difference among the community members. Finally a ranking R of the vertices from G is returned in line 11.

Algorithm 1. Glance Algorithm with Community Feature Selection

---

Input: G // *Attributed Graph*
Output: R // *An anomaly ranking of the vertices from G*

1 $R \leftarrow \varnothing$
2 $C \leftarrow \text{Clustering}(G)$
3 **foreach** $C_i \in C$ **do**
4      $A \leftarrow \text{FeatureSelection}(C_i)$
5      $P_{C_i} \leftarrow$ mean values of attributes from A in C_i
6   **foreach** $v_j \in C_i$ **do**
7         $R_{v_j} \leftarrow$ a dictionary containing for each attribute a_l from v_j the number of elements u satisfying $|a_l(v_j) - a_l(u)| > a_l(P_{C_i})$
8         $R \leftarrow R \cup \{(v_j, max(a_l \in R_{v_j}))\}$
9   **end**
10 **end**
11 **return** R

It is important to notice that community feature selection splits the number of elements to analyze in groups. This fact is useful with algorithms $O(V^2)$ or higher because it reduces the number of operations required to process the data. Thus the performance of the algorithm is increased in real networks where the number of communities is usually high. Also, this property can be useful to process the data in parallel. In the next section, the performance of the algorithm on a real network is analyzed.

5 Experimental Results

In this section, a comparison between Glance with community feature selection and the base algorithm is performed. Furthermore, our proposal is compared with other state-of-the-art algorithms.

The comparison was performed on the amazon co-purchase network of Disney products (124 nodes with 334 edges). This database was used as benchmark in [4], and the authors provided a labeled outlier ground truth. The ground truth was built from a user experiment where outliers were labeled by high school students. In the experiment, the products where clustered using a modularity technique, and then the students were asked to find outliers in each cluster. The experiment used the edges only to give context to the elements, and the attributes of the vertices were used to identify anomalies in each cluster. Thus, this database and its labeled anomaly ground truth represents a problem of anomaly detection in attributed graph where the edges are contextual attributes.

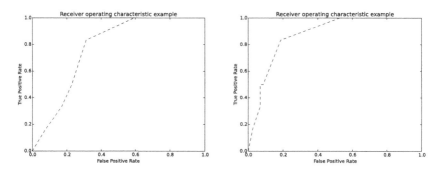

Fig. 2. ROC of the base Glance algorithm (left) and the one using community feature selection (right)

In Fig. 2, a comparison between the base Glance algorithm and the one improved with community feature selection is displayed. The former has an AUC (Area Under the ROC Curve) value of 77.4% and a runtime of 150 ms, assuming that the user selected all features as relevant. Thus the results were affected by the curse of dimensionality. The later has an AUC value of 87.43% with a runtime of 93 ms, and can be observed that the ROC curve rises faster than in the base algorithm. Using community feature selection, an improve of more than 10% was achieved and also the performance of the algorithm was improved due to a reduction in the dimensions of the data to be processed.

Table 1. AUC results for all algorithms on the Amazon database (Disney DVD selection).

Used data	Paradigm	Algorithm	AUC[%]	Runtime[ms]
Attribute data only	Full space outlier analysis	LOF [6]	56.85	41
	Subspace outlier analysis	SOF [11]	65.88	825
		RPLOF [12]	62.50	7
Graph structure only	Graph clustering	SCAN [22]	52.68	4
Attributes and graph data	Full space outlier analysis	CODA [16]	50.56	2596
	Subspace outlier analysis	GOutRank [4]	86.86	26648
	Contextual edges	Glance	77.40	150
		Glance + CFS	87.43	93

In Table 1, it is displayed the AUC measure and the runtime for different approaches to anomaly detection. In the approaches considering just the attribute vectors, only those performing subspace analysis can overcome the curse of dimensionality. Thus, these techniques have better accuracy compared with the ones performing full subspace analysis. Nonetheless, none of them can detect the complex outliers present in the data, because they ignore the relationship among elements. The approach considering only the graph structure has poor results in this database. This is mainly because, in this problem, edges are contextual attributes and they do not directly determine the outlierness of an element. The only approach able to detect the complex outliers in this database is the one that considers both attribute data and graph structure. Although the CODA algorithm belongs to this approach, it is greatly affected by the curse of dimensionality. The GOutRank algorithm obtains good results in this database, but ignores the contextual nature of the edges, affecting the quality of its results. Also, it is the most time consuming algorithm used in this comparison. The Glance algorithm using all the features achieves better detection than other algorithms but is affected by the curse of dimensionality. The Glance algorithm with community feature selection has the best AUC value and also has better runtimes than the other algorithms for anomaly detection in attributed graph. This results are an example of the potential of community feature selection to improve anomaly detection in attributed graphs.

6 Conclusions

In this paper, an adaptive method to improve anomaly detection in attributed graphs using community feature selection was proposed. The method was used to improve an algorithm to detect anomalies in attributed graphs where the edges behave as contextual attributes, and an improve in the result of more than a 10% was achieved. Also, the improved algorithm was compared with others from the state of the art and it achieved better results.

There are many open challenges in the field of anomaly detection in attributed graphs, we will focus on some of them in future work. The first one is to integrate feature selection and anomaly detection in a single process for avoiding redundant calculations and improve the performance of the algorithms. Finally, the algorithm could be parallelized to improve its performance, because it selects features and detects anomalies in disjoint communities.

References

1. Chandola, V., Banerjee, A., Kumar, V.: Anomaly detection: a survey. ACM Comput. Surv. (CSUR) **41**, 15 (2009)
2. Akoglu, L., Tong, H., Koutra, D.: Graph based anomaly detection and description: a survey. Data Min. Knowl. Disc. **29**, 626–688 (2015)
3. Beyer, K., Goldstein, J., Ramakrishnan, R., Shaft, U.: When is "nearest neighbor" meaningful? In: Beeri, C., Buneman, P. (eds.) ICDT 1999. LNCS, vol. 1540, pp. 217–235. Springer, Heidelberg (1999). doi:10.1007/3-540-49257-7_15
4. Müller, E., Sánchez, P.I., Mülle, Y., Böhm, K.: Ranking outlier nodes in subspaces of attributed graphs. In: 2013 IEEE 29th International Conference on Data Engineering Data Engineering Workshops (ICDEW), pp. 216–222 (2013)

5. Knorr, E.M.: Outliers and Data Mining: Finding Exceptions in Data. The University of British Columbia, Vancouver (2002)
6. Breunig, M.M., Kriegel, H.-P., Ng, R.T., Sander, J.: LOF: identifying density-based local outliers. ACM Sig. Rec. **29**(2), 93–104 (2000)
7. Papadimitriou, S., Kitagawa, H., Gibbons, P.B., Faloutsos, C.: LOCI: fast outlier detection using the local correlation integral. In: ICDE, pp. 315–326 (2003)
8. Xiong, Y., Zhu, Y., Yu, P.S., Pei, J.: Towards cohesive anomaly mining. In: AAAI (2013)
9. Liu, F.T., Ting, K.M., Zhou, Z.-H.: Isolation-based anomaly detection. ACM Trans. Knowl. Discov. Data (TKDD) **6**(1), 3 (2012)
10. Liu, F.T., Ting, K.M., Zhou, Z.-H.: On detecting clustered anomalies using SCi-Forest. In: Balcázar, J.L., Bonchi, F., Gionis, A., Sebag, M. (eds.) ECML PKDD 2010. LNCS (LNAI), vol. 6322, pp. 274–290. Springer, Heidelberg (2010). doi:10.1007/978-3-642-15883-4_18
11. Aggarwal, C.C., Yu, P.S.: Outlier detection for high dimensional data. ACM Sigmod Rec. **30**(2), 37–46 (2001)
12. Lazarevic, A., Kumar, V.: Feature bagging for outlier detection. In: Proceedings of the Eleventh ACM SIGKDD International Conference on Knowledge Discovery in Data Mining, pp. 157–166 (2005)
13. Akoglu, L., McGlohon, M., Faloutsos, C.: OddBall: spotting anomalies in weighted graphs. In: Zaki, M.J., Yu, J.X., Ravindran, B., Pudi, V. (eds.) PAKDD 2010. LNCS (LNAI), vol. 6119, pp. 410–421. Springer, Heidelberg (2010). doi:10.1007/978-3-642-13672-6_40
14. Eberle, W., Holder, L.: Discovering structural anomalies in graph-based data. In: Data Mining Workshops 2007. ICDM Workshops 2007, pp. 393–398 (2007)
15. Noble, C.C., Cook, D.J.: Graph-based anomaly detection. In: Proceedings of the Ninth ACM SIGKDD International Conference on Knowledge Discovery and Data Mining, pp. 631–636 (2003)
16. Gao, J., Liang, F., Fan, W., Wang, C., Sun, Y., Han, J.: On community outliers and their efficient detection in information networks. In: Proceedings of the 16th ACM SIGKDD International Conference on Knowledge Discovery and Data Mining, pp. 813–822 (2010)
17. Moser, F., Colak, R., Rafiey, A., Ester, M.: Mining cohesive patterns from graphs with feature vectors. In: SDM, vol. 9, pp. 593–604 (2009)
18. Günnemann, S., Farber, I., Boden, B., Seidl, T.: Subspace clustering meets dense subgraph mining: a synthesis of two paradigms. In: Data Mining (ICDM) 10th International Conference on Data Mining, pp. 845–850 (2010)
19. Akoglu, L., Tong, H., Meeder, B., Faloutsos, C.: PICS: parameter-free identification of cohesive subgroups in large attributed graphs. In: SDM, pp. 439–450 (2012)
20. Blondel, V.D., Guillaume, J.-L., Lambiotte, R., Lefebvre, E.: Fast unfolding of communities in large networks. J. Stat. Mech. Theory Exp. **2008**(10), P10008 (2008)
21. He, X., Cai, D., Niyogi, P.: Laplacian Score for feature selection. In: Advances in Neural Information Processing Systems, pp. 507–514 (2005)
22. Xu, X., Yuruk, N., Feng, Z., Schweiger, T.A.: Scan: a structural clustering algorithm for networks. In: Proceedings of the 13th ACM SIGKDD International Conference on Knowledge Discovery and Data Mining, pp. 824–833 (2007)

Lung Nodule Classification Based on Deep Convolutional Neural Networks

Julio Cesar Mendoza Bobadilla and Helio Pedrini[✉]

Institute of Computing, University of Campinas,
Campinas, SP 13083-852, Brazil
helio@ic.unicamp.br

Abstract. Lung nodule classification is one of the main topics on computer-aided diagnosis (CAD) systems for detecting nodules. Although convolutional neural networks (CNN) have been demonstrated to perform well on many tasks, there are few explorations of their use for classifying lung nodules. In this work, we present a method for classifying lung nodules based on CNNs. Training is performed by balancing the mini-batches on each stochastic gradient descent (SGD) iteration to address the lack of nodule samples compared to background samples. We show that our method outperforms a base feature-engineering method using the same techniques for other stages of lung nodule detection, and show that CNNs obtain competitive results when compared to state-of-the-art methods evaluated on Japanese Society of Radiological Technology (JSRT) dataset [13].

Keywords: Lung cancer · Chest imaging · Image classification · Convolutional Neural Networks

1 Introduction

Lung cancer affects men and women around the world and can be characterized by abnormal cell growth in the lung tissues. A chest radiography (CXR) or a computerized tomography (CT) is typically one of the first diagnostic steps performed by radiologists [2]. Lung nodules are small masses of tissue, which appear as white shadows on a CXR image or CT scan. Benign nodules have approximately 5–30 millimeters, whereas larger nodules are more likely to be malignant [1].

The detection of lung nodules on an imaging test is a challenging task due to low contrast of the lesions, possible overlap with ribs and large pulmonary vessels, as well as their irregular size, shape and opacity [8]. Pulmonary radiologists usually detect lung nodules by considering brightness and shape of circular objects within the thoracic cavity. More conclusive diagnosis of lung cancer can be conducted through histological examination of the tissue [7].

Advances in imaging modalities and Computer-Aided Diagnosis (CAD) systems have helped radiologists identify lung nodules more quickly and accurately.

© Springer International Publishing AG 2017
C. Beltrán-Castañón et al. (Eds.): CIARP 2016, LNCS 10125, pp. 117–124, 2017.
DOI: 10.1007/978-3-319-52277-7_15

Potential locations of nodules can be automatically detected by considering bright masses of expected size, shape and texture of a lung nodule. Techniques available for detecting lung nodules generally involve three main stages: lung segmentation, candidate detection, and nodule classification.

The main contribution of this work is the evaluation of Convolutional Neural Networks (CNN) for lung nodule classification. Our CNN is trained through stochastic gradient descent (SGD) algorithm. The input of the CNN is a set of candidate regions. The amount of false positive candidates is large when compared to the number of true positives, leading to an unbalanced classification problem. We address this issue by balancing the mini-batches on each SGD iteration. The proposed method is evaluated on the Japanese Society of Radiological Technology (JSRT) dataset [13]. Our method outperforms the results obtained with a feature-engineering approach [6] used as baseline, which uses the same algorithms for previous stages to nodule classification, and obtains competitive results when compared to other methods of the literature.

The remainder of the paper is organized as follows. Section 2 briefly describes background information on previous work. Section 3 presents the methodology proposed in this work. Section 4 describes and discusses our experimental results. Section 5 concludes the paper with final remarks and directions for future work.

2   Background

The lung nodule detection typically involves three sequential steps. The first one consists in segmenting the lung area to prevent the detection of structures similar to nodules located outside the lung using methods such as active contour models, active shape models, regression, among others. The second step is the detection of nodule candidates through appropriate filtering strategies, such as convergence index-based filters, Laplacian of Gaussian filters, pixel classification, among others. The third stage is the lung nodule classification. Most of the available methods follow the classical pattern recognition pipeline: feature extraction, feature selection, and classification. On the other hand, deep learning approaches have been rarely used for lung nodule classification.

2.1   Lung Nodule Classification

A method proposed by Coppini et al. [3] uses a multi-scale neural network architecture to distinguish true nodules from background patterns. A set of Gabor filters and a Laplacian of Gaussian filter are used to extract features from the input regions and to feed a 3-layer neural network. Then, an optimal linear combiner determines the nodule probability for each candidate.

Most approaches segment nodule regions to capture features from particular regions of the nodules. Schilham et al. [11] use multi-scale techniques to segment nodule candidates. Then, they convolve the input regions with a set of Gaussian filters to extract statistics from the inner and band regions of the nodule. A two-step classification is performed using approximate k-nearest neighbor algorithm.

Some methods perform transformations on the input regions. Wei et al. [15] transform the candidate regions using Adaptive Ring filter, Iris filter, and Sobel filter. They segment the nodule candidates using active contour models, and extract geometric features, contrast features, first and second order statistics from the inner and outer regions on the original and transformed candidates. The best features are selected using the forward stepwise selection algorithm. Classification is performed by using a statistical method based on the Mahalanobis distance measure.

Shiraishi et al. [14] proposed a method that describes nodules using geometric, intensity, background and edge gradient features from the original and density corrected, nodule enhanced, and contralateral subtracted images of each suspicious nodule. Classification is performed using artificial neural networks.

Hardie et al. [6] proposed an adaptive distance-based thresholding approach for nodule segmentation. Then, a set of geometric, gradient and intensity features are extracted from the local enhanced, normalized, nodule-mask and original versions of each nodule candidate. A Fisher Linear Discriminant is used for classification. Similar to [15], Hardie et al. use a sequential forward selection algorithm to find a proper subset of features.

A method proposed by Chen and Suzuki [2] uses a regression model to suppress rib-like structures on the images. Nodule candidates are segmented using dilatation and erosion followed by a watershed-based segmentation. Morphological and gray-level-based features are extracted from each nodule candidate in the original and rib-suppressed images. An SVM classifier with Gaussian kernel is used to classify each nodule.

2.2 Deep Learning

Several knowledge domains have benefited from deep learning research [9,12], such as image and video classification, audio and speech recognition, bioinformatics, natural language processing, and data mining. Different deep learning architectures have been developed to address various tasks, including CNNs, belief networks, Boltzmann machines, and recurrent neural networks.

Methods based on deep learning can be applied to supervised and unsupervised learning problems. In supervised tasks, the algorithms avoid the process of constructing feature descriptors by transforming the data into compact intermediate representations and deriving layered structures [4]. In unsupervised tasks, the algorithms explore unlabeled data, which are typically more available than labeled data.

3   Proposed Method

Our lung nodule classification method performs the candidate analysis directly on the pixels of the images using a CNN, instead of extracting features and using classifiers. The initial stages of lung segmentation and candidate detection follow the approach developed by Hardie et al. [6].

The classification starts by preprocessing and augmenting the candidate regions obtained in the segmentation process. A CNN is then trained with backpropagation using the augmented dataset. Finally, we reduce overlapping detections through an adjacent candidate rejection rule.

3.1 Preprocessing and Data Augmentation

We applied a z-score normalization to the pixel intensities over the dataset to have zero mean and unit variance. Candidate detection stage produces approximately 150 negative samples for each positive one. To balance the dataset and prevent overfitting, we augment the data available by increasing the amount of positive samples and applying the following transformations to all candidates, including negative samples: (i) horizontal/vertical random shift with a factor of each dimension of the region of interest, selected over a uniform distribution between 0 and 0.1; (ii) random zoom with a scale factor selected over a logarithmic distribution of 1.0 and 1.2; (iii) horizontal flip: yes or no with a probability of 0.5; (iv) random rotation with an angle between $-5°$ and $5°$.

3.2 Architecture

Inspired by the work developed by Graham [5], we explored a family of CNNs known as DeepCNets. DeepCNets are composed of alternating convolutional and max-pooling layers with a linearly increasing number of filters. We evaluated architectures with 3 to 6 convolutional layers, achieving better results with architectures with 6 convolutional layers. Figure 1 depicts our architecture.

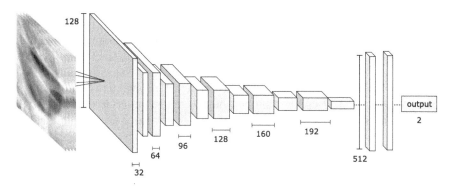

Fig. 1. Our CNN architecture. Nodule candidates are fed to the network composed of six alternating convolution and max-pooling layers, and two fully connected layers.

The network has 6 convolutional layers, 6 max-pooling layers, and 2 fully connected layers. Filters used on convolutional layers have a 3×3 receptive field. Max-pooling is performed over windows of 2×2, with stride 2. We use Leaky Rectifier Linear Units with alpha 0.333 for all convolutional and fully connected layers. We use dropout regularization with a drop ratio 0.25 on convolutional layers and 0.5 on fully connected layers. The output of the last layer is connected to a softmax layer that produces a distribution over the two classes.

3.3   Learning

For the learning stage, we use the Stochastic Gradient Descent (SGD) with Nesterov momentum. We initialized the weights on each layer with the orthogonal random matrix initialization procedure proposed by Saxe et al. [10]. The learning rate is decreased when the error on the test set stops improving. The learning rate is decreased three times. We experimented with weight decay regularization, however, it did not improve our results. Therefore, the final model does not use weight decay regularization.

Motivated by the work proposed by Yan et al. [16], we balance the amount of positives and negatives samples for each batch used to feed the CNN in the training step. For positives, we select half of the batch size samples randomly. For negatives, we select half of the batch size samples iteratively. Positive and negative samples are perturbed before they feed the CNN on each SGD iteration.

3.4   Adjacent Rejection

The trained CNN is used to measure the probability of a candidate being a nodule. Since more than one candidate can be related to a positive sample, we use the same adjacency rejection rule explained in [6]. This procedure reduces the number of final detections by preserving the candidates that have the maximum detection probability on their radial neighborhood of 22 mm.

4   Experimental Results

This section describes the dataset used in our experiments, the results obtained with our method, as well as a comparison of the results to other methods available in the literature.

4.1   Dataset

We use the Japanese Society of Radiological Technology (JSRT) database [13] composed of 154 nodule and 93 non-nodule chest radiographs. The images have 2048×2048 pixels in size and 4096 gray levels. Nodules are classified as malignant or benign and graded as to their visual subtlety: extremely subtle, very subtle, subtle, relative obvious and obvious. The images were annotated by three radiologists and validated using CT images to provide groundtruth information.

We validate our method with 10-fold cross-validation. We use sensitivity and false positive per image (FPPI) measures to compare our method to others in the literature. Sensitivity and FPPI are meaningful on lung nodule detection, in the sense that methods with high sensitivity tend to produce a large amount of false positives and, consequently, demanding unavailing attention of the physicians in the analysis of samples with low probability of being a nodule.

4.2   Results

Our CNN was trained with SGD optimization with Nesterov momentum of 0.9, a batch size of 32 samples, learning rate of 0.1 on 55 epochs, decreasing the learning rate by 10 on 15, 25 and 40 epochs. Our model takes approximately 10 h to be trained on a Tesla K40c GPU device.

Due to the variability in the validation, labeling and scoring procedures, it is not simple to make a precise comparison of the methods that use the JSRT database. In our experiments, we use the validation, labeling and scoring procedures as described in [6]. Our CNN was trained by excluding 14 images containing nodules on the opaque regions of the lung. In Fig. 2(a), we compare the performance of our method to the one proposed by Hardie et al. [6]. A comparison with other methods available in the literature is provided in Fig. 2(b). Since some methods report their results with the entire database, we adjust the sensitivity values by considering excluded cases as missed.

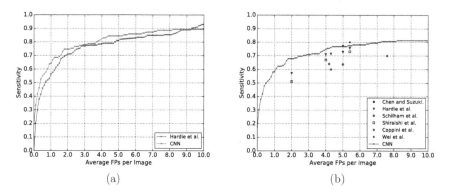

(a)   (b)

Fig. 2. (a) Comparison of the CAD performance between our method and the one proposed by Hardie et al. [6] on the JRST database. Opaque cases are excluded; (b) Comparison of the CAD performance on the JRST database. Sensitivity values are adjusted by considering opaque cases as missed.

Table 1 shows the sensitivity values obtained with our CNN method and state-of-the-art approaches at the same FPPI values. Chen and Suzuki [2] reported a slightly higher sensitivity. Their systems use dual-energy radiographs for rib-suppression on the training stage. The method proposed in [15] extracts a large set of features from the images at different scales. The methods described in [3,11,15] employ the entire JSRT database.

Figure 3 shows some samples of true positives and false positives detected with the proposed method. The samples are shown with respect to their probability values.

Table 1. CAD system performance comparison.

Average FPPI	Method	Reported sensitivity (%)	CNN sensitivity (%)
5.0	Chen and Suzuki [2][a,c]	**85**	84.5
5.0	Hardie et al. [6][a]	80.1	**84.5**
2.0	Schilham et al. [11]	51	**67.8**
4.0	Schilham et al. [11]	71	**74.6**
5.0	Shiraishi et al. [14][a,b]	70.1	**84.5**
4.3	Coppini et al. [3]	60	**76.1**
5.4	Wei et al. [15]	**80**	77.7

[a]Results reported in this row exclude opaque cases.
[b]Results based on 924 chest radiographs that include the JRST cases.
[c]A private database was used for training and the JSRT for testing.

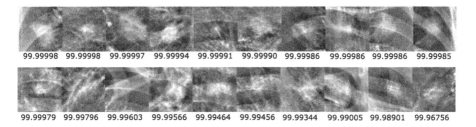

| 99.99998 | 99.99998 | 99.99997 | 99.99994 | 99.99991 | 99.99990 | 99.99986 | 99.99986 | 99.99986 | 99.99985 |

| 99.99979 | 99.99796 | 99.99603 | 99.99566 | 99.99464 | 99.99456 | 99.99344 | 99.99005 | 99.98901 | 99.96756 |

Fig. 3. Top detection results with their respective probability values (sorted in descending order). The first row shows the true positive samples, whereas the second row shows the false positive samples.

5 Conclusions

We proposed a method based on CNNs to classify lung nodules in chest radiographs. Our method outperforms the baseline work without the need of extracting several features from the images.

Experimental results show that CNNs can operate effectively on lung nodule classification through data augmentation and dropout regularization. By balancing the number of positive and negative samples on the batches, it was possible to train a CNN with only 140 positive samples.

Acknowledgements. The authors are thankful to FAPESP (grant #2011/22749-8) and CNPq (grant #307113/2012-4) for their financial support.

References

1. Brant, W.E., Helms, C.A.: Fundamentals of Diagnostic Radiology. Lippincott Williams & Wilkins, Philadelphia (2012)
2. Chen, S., Suzuki, K.: Computerized detection of lung nodules by means of "virtual dual-energy" radiography. IEEE Trans. Biomed. Eng. **60**(2), 369–378 (2013)
3. Coppini, G., Diciotti, S., Falchini, M., Villari, N., Valli, G.: Neural networks for computer-aided diagnosis: detection of lung nodules in chest radiograms. IEEE Trans. Inf. Technol. Biomed. **7**(4), 344–357 (2003)
4. Deng, L., Yu, D.: Deep learning: methods and applications. Found. Trends Sig. Process. **7**(3–4), 197–387 (2014)
5. Graham, B.: Spatially-Sparse Convolutional Neural Networks. arXiv preprint arXiv:1409.6070 (2014)
6. Hardie, R.C., Rogers, S.K., Wilson, T., Rogers, A.: Performance analysis of a new computer aided detection system for identifying lung nodules on chest radiographs. Med. Image Anal. **12**(3), 240–258 (2008)
7. Hauser, S., Josephson, S.A.: Harrison's Neurology in Clinical Medicine. McGraw Hill Professional, New York (2013)
8. Katsuragawa, S., Doi, K.: Computer-aided diagnosis in chest radiography. Comput. Med. Imaging Graph. **31**(4), 212–223 (2007)
9. LeCun, Y., Bengio, Y., Hinton, G.: Deep learning. Nature **521**(7553), 436–444 (2015)
10. Saxe, A.M., McClelland, J.L., Ganguli, S.: Exact Solutions to the Nonlinear Dynamics of Learning in Deep Linear Neural Networks. arXiv preprint arXiv:1312.6120 (2013)
11. Schilham, A.M., van Ginneken, B., Loog, M.: A computer-aided diagnosis system for detection of lung nodules in chest radiographs with an evaluation on a public database. Med. Image Anal. **10**(2), 247–258 (2006)
12. Schmidhuber, J.: Deep learning in neural networks: an overview. Neural Netw. **61**, 85–117 (2015)
13. Shiraishi, J., Katsuragawa, S., Ikezoe, J., Matsumoto, T., Kobayashi, T., Komatsu, K.I., Matsui, M., Fujita, H., Kodera, Y., Doi, K.: Development of a digital image database for chest radiographs with and without a lung nodule: receiver operating characteristic analysis of radiologists' detection of pulmonary nodules. Am. J. Roentgenol. **174**(1), 71–74 (2000)
14. Shiraishi, J., Li, Q., Suzuki, K., Engelmann, R., Doi, K.: Computer-aided diagnostic scheme for the detection of lung nodules on chest radiographs: localized search method based on anatomical classification. Med. Phys. **33**(7), 2642–2653 (2006)
15. Wei, J., Hagihara, Y., Shimizu, A., Kobatake, H.: Optimal image feature set for detecting lung nodules on chest X-ray images. In: Lemke, H.U., Inamura, K., Doi, K., Vannier, M.W., Farman, A.G., Reiber, J.H.C. (eds.) Computer Assisted Radiology and Surgery, pp. 706–711. Springer, Heidelberg (2002)
16. Yan, Y., Chen, M., Shyu, M.L., Chen, S.C.: Deep learning for imbalanced multimedia data classification. In: IEEE International Symposium on Multimedia, pp. 483–488, December 2015

A Hierarchical K-Nearest Neighbor Approach for Volume of Tissue Activated Estimation

I. De La Pava$^{(\boxtimes)}$, J. Mejía, A. Álvarez-Meza, M. Álvarez, A. Orozco, and O. Henao

Faculty of Engineerings, Automatic Research Group,
Universidad Tecnológica de Pereira, Pereira, Colombia
ide@utp.edu.co

Abstract. Deep brain stimulation (DBS) is a surgical technique used to treat movement disorders. The volume of tissue activated (VTA) is a concept that partly explains the effects of DBS. Its visualization as part of anatomically accurate reconstructions of the brain structures surrounding the DBS electrode has been shown to have important clinical applications. However, the computation time required to estimate the VTA with traditional methods makes it unsuitable for practical applications. In this study, we develop a hierarchical K-nearest neighbor approach (HKNN) for VTA computation to address that hurdle. Our method reduces the time to estimate the VTA by four orders of magnitude, to hundredths of a second. In addition, it keeps the error with respect to the standard method for VTA estimation in the same range of that obtained with alternative machine learning approaches, such as artificial neural networks, without the limitations entailed by them.

Keywords: k-nearest neighbors · Parkinson's disease · Volume of tissue activated · Deep brain stimulation

1 Introduction

Deep brain stimulation (DBS) is a treatment for movement disorders, such as Parkinson's disease, dystonia, or essential tremor. It usually consists of the implantation of a stimulator in the infraclavicular region connected to an electrode lead that is placed in a target structure in the basal ganglia (the subthalamic nucleus (STN) or the thalamus). The stimulator delivers electric pulses of a specific frequency, amplitude, and pulse-width to the target via the electrode, which results in symptom improvement [5]. Despite its effectiveness, the exact mechanisms of action of DBS remain elusive, and most of the current understanding of these mechanisms has come from computer simulations [7]. The volume of tissue activated (VTA), the spatial spread of direct neural activation in response to the DBS electric pulses, is a popular concept to partly explain the effects of DBS [2]. The visualization of the VTA as part of anatomically accurate reconstructions of the brain structures surrounding the implanted electrode allows

© Springer International Publishing AG 2017
C. Beltrán-Castañón et al. (Eds.): CIARP 2016, LNCS 10125, pp. 125–133, 2017.
DOI: 10.1007/978-3-319-52277-7_16

the clinical specialist to observe the areas that are responding directly to the delivered stimulus. This has been proven particularly useful for the post-surgical adjustment of the stimulation parameters, one of the most challenging aspects of the treatment. A faster adjustment of the DBS parameters saves the patient both time and discomfort. However, the computational burden for computing a patient-specific VTA is still tedious [4].

The gold standard for VTA estimation involves the computation of the brain tissue response to the electrical stimulation through a field of multi-compartment axon models [3]. Unfortunately, it is computationally intensive to be used as part of a system that allows the clinical specialist to visualize the VTA generated by a specific combination of stimulation parameters. This has led to machine learning methods that minimize the use of such models by taking advantage of the relation between the location of the axons in space and the electrical stimulation. Authors in [4] proposed the use of two artificial neural networks (ANNs) with the DBS stimulation parameters as inputs and the elliptic profiles defined by the active axons as outputs. Once trained, the neural networks can estimate the VTA for any combination of stimulation parameters. However, the elliptic profiles assumed generate small deviations from the actual contours of activation and this method only works under the assumption of isotropic tissue conductivity. Recently, we presented an alternative method that considers the problem of determining the VTA from a field of axons as equivalent to a binary classification task [6]. So, we use a Gaussian process classifier (GPC) to determine whether an axon at a given position in space is active due to DBS. This approach works independently of the assumed tissue conductivity conditions and cuts down to a tenth the time for VTA estimation. Nonetheless, this reduced computation time, in the range of minutes, is still too high for practical applications.

In this work, we explore a hierarchical K-nearest neighbor approach (HKNN) to accelerate the estimation of the VTA [1,8]. In this sense, HKNN is a data-driven approximation of an expected squared error functional towards a local weighted VTA averaging. The weights are computed as a softmax gating function applied over a set of features estimated from the DBS stimulation parameter space. Our aim is to reduce the computation time needed to estimate the VTA generated by the electrical stimulation delivered by the DBS electrode, for a specific configuration of stimulation parameters, while trying to reproduce the results that would be obtained with the standard method. Our results show that HKNN outperforms the state of the art approaches in terms of computational cost, achieving fast estimations and low errors with respect to the gold standard simulation. The remainder is as follows: Sect. 2 describes the materials and methods. Sections 3 and 4 describe the experimental set-up and the obtained results, respectively. Finally, the conclusions are outlined in Sect. 5.

2   Materials and Methods

Let $X \in \mathcal{X}$ and $Y \in \mathcal{Y}$ be a pair of random variables representing the DBS stimulation parameters and the VTA spaces $\mathcal{X}$ and $\mathcal{Y}$, respectively. Here, we introduce a data-driven approach to estimate a new VTA $y \in Y$ from a given DBS stimulation sample $x \in X$ through the minimization of the following expected square error functional:

$$f^* = \arg\min_f \boldsymbol{E}_X \left\{ \boldsymbol{E}_{Y|X} \left\{ (Y - f(g(X)))^2 | g(X) \right\} \right\}, \tag{1}$$

where $g : \mathcal{X} \to \mathcal{H}$ is a mapping function from the DBS stimulation parameter space $\mathcal{X}$ to a given feature representation space $\mathcal{H}$ coding relevant patterns, and $f : \mathcal{H} \to \mathcal{Y}$ is a regression function from $\mathcal{H}$ to the VTA space $\mathcal{Y}$. Then, a pointwise solution of Eq. (1) is carried out, yielding:

$$f(g(x)) = \boldsymbol{E}_{Y|X} \left\{ Y | g(X) = g(x) \right\}. \tag{2}$$

Now, let $\boldsymbol{X} \in \mathbb{R}^{N \times Q}$ and $\boldsymbol{Y} \in \{0,1\}^{N \times P}$ be a couple of sample matrices holding Q stimulation parameters, P axons, and N samples ($\mathcal{X} \subset \mathbb{R}^{N \times Q}$, $\mathcal{Y} \subset \{0,1\}^{N \times P}$). Namely, each row vector $\boldsymbol{x}_i \in \mathbb{R}^Q$ ($i \in \{1, \dots, N\}$) in $\boldsymbol{X}$ holds the stimulation parameters employed to compute the i-th VTA $\boldsymbol{y}_i \in \{0,1\}^P$, which is stored through axon concatenation. Note that $y_{ip} = 1$ ($p \in \{1, \dots, P\}$) if the p-th axon is activated by the DBS device, otherwise, $y_{ip} = 0$. So, to estimate f in Eq. (2) from $\boldsymbol{X}$ and $\boldsymbol{Y}$, a weighted average approximation is computed as:

$$\hat{f}(g(\boldsymbol{x})) = \sum_{\boldsymbol{y} \in \Omega_y} w_k \boldsymbol{y}_k, \forall \boldsymbol{y}_k \in \Omega_y \tag{3}$$

where Ω_y is a set containing the K nearest neighbors of $\boldsymbol{y}$ in $\boldsymbol{Y}$ and $w_k \in \mathbb{R}^+$ ($k \in \{1, \dots, K\}$). In particular, a hierarchical neighborhood is computed by considering both the feature representation space $\mathcal{H}$ and the DBS contacts as:

$$\Omega_y = \{\boldsymbol{y}_k : \delta\left(||\boldsymbol{c}||_1 - ||\boldsymbol{c}_k||_1\right) = 1, \mathrm{d}\left(g(\boldsymbol{x}), g(\boldsymbol{x}_k)\right) \leq \mathrm{d}\left(g(\boldsymbol{x}), g(\boldsymbol{x}_K)\right)\}, \tag{4}$$

where $\boldsymbol{c} \in \{1, 0, \text{-}1\}^C$ stores the configuration of the C DBS contacts (1, 0 and -1 represent anodic activation, inactivation or cathodic activation, respectively), $\delta(\cdot, \cdot)$ stands for the delta function that selects the contact configuration state (one or two active contacts) in a hierarchical process, and $\boldsymbol{x}_K$ is the K-th neighbor of $\boldsymbol{x}$ in $\boldsymbol{X}$ according to the Euclidean distance operator $\mathrm{d}(\cdot, \cdot)$. The specific form of $g(\cdot)$ is given in Sect. 3. Moreover, to assess the relative importance of $\boldsymbol{y}_k$, a softmax gating function is used for estimating w_k:

$$w_k = \frac{\exp\left(-\mathrm{d}\left(g(\boldsymbol{x}), g(\boldsymbol{x}_k)\right)\right)}{\sum_{\boldsymbol{x}_j \in \Omega_x} \exp\left(-\mathrm{d}\left(g(\boldsymbol{x}), g(\boldsymbol{x}_j)\right)\right)}, \tag{5}$$

where Ω_x is the set containing the corresponding elements of Ω_y in the DBS stimulation parameter space. Finally, a thresholding procedure ($\zeta \in [0, 1]$) is applied to estimate the new VTA $\hat{\boldsymbol{y}} \in \{0,1\}^P$ as follows:

$$\hat{y}_p = \begin{cases} 1 & \hat{f}(g(\boldsymbol{x})) > \zeta \\ 0 & \text{otherwise} \end{cases}. \tag{6}$$

3 Experimental Set-Up

We built a database of the VTAs generated by 1000 randomly selected combinations of realistic stimulation parameters, relevant in the context of VTA estimation, for the Medtronic ACTIVA-RC stimulator [9]. Thereby, the DBS stimulation parameter space is limited as follows: four contacts $c = \{c0, c1, c2, c3\}$, where a maximum of two simultaneously active contacts are analyzed, e.g., monopolar and bipolar configurations, amplitude values $a \in [0.5, 5.5]$ [V], and pulse-width values $w \in [60, 450]$ [μs]. The electric potentials are generated for 200 monopolar (one active contact) and 800 bipolar (two active contacts) stimulation configurations as in [6]. An isotropic and homogenous brain tissue medium, and an encapsulation layer of 0.5 [mm] are assumed. This procedure is repeated for three different encapsulation layer conductivities: 0.680 [Sm^{-1}], 0.128 [Sm^{-1}], 0.066 [Sm^{-1}], to represent low ($\sim$ 500 [Ω]), medium ($\sim$ 1000 [Ω]), and high ($\sim$ 1500 [Ω]) impedance conditions [4], yielding a total of 3000 electric potential distributions. Afterwards, the electric potentials are interpolated onto each section of a field of 4144 multi-compartment axon models of diameter 5.7 [μm], which are oriented perpendicularly to the axis of the electrode (see Fig. 1(a)), and positioned into a grid of width 9 [mm] and height 27.75 [mm]. Furthermore, the axons share a space of 0.25 [mm] between them in both the vertical and the horizontal directions. The response of the axons to the stimulating potentials defines each VTA: axons that fire an action potential per stimulation pulse are considered active and their positions in space shape the VTA [3]. Figure 1(b) depicts one of such shapes, a VTA for a monopolar configuration, and its spatial interaction with a representation of the STN. The use of multi-compartment axon models coupled to a stimulating electric field to estimate the VTA, as described above, is what we refer to as the gold standard for VTA estimation. All the axonal response simulations are implemented in NEURON 7.3.

Moreover, the g function in HKNN is defined as follows $g(\boldsymbol{x}) = [m_x \boldsymbol{c}, m_x abs(\boldsymbol{c})]$, where $abs(\cdot)$ is the element-wise absolute value operator, and $m_x \in \mathbb{N}$ is an approximation of the number of activated axons in $\boldsymbol{y}$ given $\boldsymbol{x}$. Namely, $m_x = \varrho(\boldsymbol{x}|a, w, \boldsymbol{c})$, where $\varrho(\cdot|\cdot)$ is a polynomial function. For concrete testing, the polynomial order of ϱ is fixed as six aiming to code nonlinear data relations. Unless otherwise is stated, the number of nearest neighbors in Eq. (4) is set to $K = 3$, and the threshold ζ in Eq. (6) is set to 0.6 (see Fig. 3 for details).

Two benchmarks are considered for validating the introduced HKNN approach. First, a classifier based on Gaussian processes (GPC) is used to estimate the VTA [6]. To this end, a random sample of 500 axons is taken from the total axonal population. Next, a multi-compartment simulation is executed to determine which of the sampled neural fibers are active during the stimulation. The information provided by the multicompartment simulation is converted to a set of labeled data. These data are used to train the GPC. The classifier is trained using a general purpose kernel (squared exponential covariance function with automatic relevance determination), and using Laplace's approximation to the posterior. The VTA is estimated by predicting which of the 4144 axonal fibers would be activated by the applied stimulus. Second, an artificial neural networks approach (ANNs) is tested as described in [4]. Hence, the ANN method allows modeling the spread of activation for both monopolar and bipolar stimulation. Particularly, two feed-forward networks are trained to estimate the activation towards an ellipse parametrization. The former describes the size of each activation region and the latter determines their placement along the electrode shaft. The ANNs are trained using one hidden layer with 20 elements, a sigmoid transfer function, and a linear output layer. As quantitative measures both the computational time and the prediction error are considered. With respect to the prediction error, each estimated VTA is compared with the VTA obtained with the gold standard approach as follows $\varepsilon = (FP + FN)/m$, where FP and FN are the false positives and false negatives with respect to the reference dataset (the VTAs from the database), and m is the reference number of active axons [4]. The errors are computed using a cross-validation scheme.

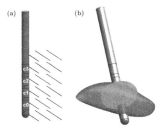

Fig. 1. (a) DBS electrode and VTA representations (Medtronic, model 3389). **(b)** 3D depiction of a VTA and its spatial interaction with the STN.

4   Results and Discussion

Figure 2 shows the VTA generated by a stimulation of 2.5 $[V]$, 330 $[\mu s]$, with contacts $c3$ and $c4$ configured as an anode and a cathode, respectively. The region of activation is formed by the axons that fired an action potential because of the applied stimulus (green dots). The axons that do not respond to the stimulation are depicted in black. Figure 2(a) shows the extent of activation obtained with the gold standard for VTA estimation. Figure 2(b) to 2(d) show the VTAs predicted by the HKNN approximation, the ANNs, and the GPC. Besides, their corresponding errors with respect to the gold standard are shown. In this regard, the errors for the methods studied arise from different factors: The GPC is based on the assumption that axons are independent from one another while ignoring the clear spatial relationship between active axons. The ANNs take into account the spatial relationships among axons, but in modeling it as ellipses they neglect the fine details of the distribution of active axons. Now, our approach assumes smooth variations in the VTA as a result of small variations in the DBS stimulation parameters space (provided that the activation status and

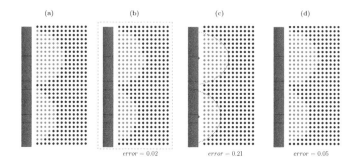

Fig. 2. VTA stimulation for 2.5 [V], 330 [μs], and $c2$ and $c3$ configured as an anode and a cathode. (**a**) Gold standard, (**B**) HKNN, (**c**) ANNs, and (**d**) GPC. (Color figure online)

polarity of the electrode contacts is held constant). So, it requires the presence of a set of VTAs similar to the desired VTA, thus, a database coding representative VTA shapes and sizes must be available for a successful estimation of the extent of neural activation. Figure 3 shows the error surfaces obtained with the HKNN algorithm for a parametric sweep over the number of nearest neighbors K and the binarization threshold ζ, for the (**a**) monopolar and (**b**) bipolar cases. The lowest errors occur when $K < 5$ and $0.4 \leq \zeta \leq 0.6$. That such a small number of neighbors suffices to estimate a new VTA points to the aforementioned smooth variations in the VTA in response to small changes in the stimulation parameters.

Despite these shortcomings, our results show that the proposed HKNN reproduces closely the results of the gold standard. Figure 4 shows the error distributions produced by the different approaches, discriminated by the number of active contacts and by the impedance condition. The top row corresponds to the errors for the monopolar case and the bottom row to the errors for the bipolar case. For all the methods studied, the error tends to grow with the number of active contacts, because of the higher complexity of the shape of the VTA, and with the impedance of the encapsulation tissue. A higher impedance will result in smaller VTAs, that is, a smaller number of active axons, and because

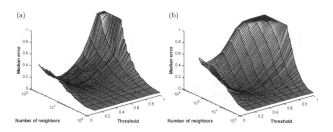

Fig. 3. Median error surfaces obtained with the proposed method for a parametric sweep over K and ζ for (**a**) monopolar case, (**b**) bipolar case.

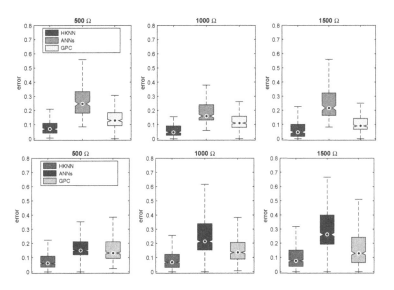

Fig. 4. Distributions of the error between the methods studied and the gold standard for VTA estimation, discriminated by the impedance condition and by the number of active contacts. Top row, monopolar case. Bottom row, bipolar case.

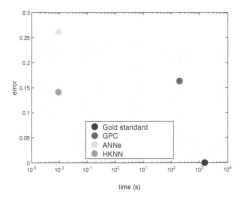

Fig. 5. Average errors versus the evaluation times for VTA estimation (models running in a Dell optiplex 990 with an Intel Core i7-2600 processor, and 8 GB RAM).

of the definition of the error measure, any false positive or false negative will be penalized more heavily. In addition, HKNN presents the lowest median error regardless of the number of active contacts or the impedance condition.

Figure 5 shows the average error versus the average evaluation time (computational load) for all the approaches tested across all DBS stimulation conditions considered. As seen, our method reduces the computation time to hundreds of a second, a figure that is well within the range of times needed for practical applications. This reduction also represents a decrease of four orders of magnitude in the VTA estimation time, compared with the gold standard. Hence, the introduced HKNN is faster than the GPC, and its evaluation time is of the same order than that of the ANNs. However, the HKNN outperforms the ANNs in terms of the prediction error.

5   Conclusions

In this study we develop a novel approach, termed HKNN, to speed up the computation of the VTA during DBS. The HKNN estimates a target VTA, given a set of stimulation parameters, from a database of precomputed VTAs. Our approach outperforms the state of the art approach in terms of computational cost, achieving fast estimation (fraction of a second) and low errors with respect to the gold standard simulation. In particular, it outperforms a GPC-based VTA estimation method [6], in terms of its computation time, while matching its prediction errors. In addition, HKNN matched the computation time of an ANN-based method [4], outperforming it in terms of the prediction error. Our method is heavily dependent on the size of the database used to interpolate the VTAs, and can produce larger errors when the VTA is comprised of a small number of active axons. We will tackle this issues exploring most robust machine learning techniques such as support vector machines (SVMs). Testing our approach for VTAs obtained under anisotropic conditions will also be a future line of work.

Acknowledgments. This study was developed under grant supported by the project 111065740687 funded by Colciencias.

References

1. Aci, M., İnan, C., Avci, M.: A hybrid classification method of k nearest neighbor, bayesian methods and genetic algorithm. Expert Syst. Appl. **37**(7), 5061–5067 (2010)
2. Butson, C.R., Cooper, E.A.: Patient-specific analysis of the volume of tissue activated during deep brain stimulation. Neuroimage **34**(2), 661–670 (2007)
3. Butson, C.R., McIntyre, C.C.: Tissue and electrode capacitance reduce neural activation volumes during deep brain stimulation. Clin. Neurophysiol. **116**(10), 2490–2500 (2005)
4. Chaturvedi, A., Luján, J.L., McIntyre, C.C.: Artificial neural network based characterization of the volume of tissue activated during deep brain stimulation. J. Neural Eng. **10**(5), 056023 (2013)
5. Da Cunha, C., Boschen, S., et al.: Toward sophisticated basal ganglia neuromodulation: review on basal ganglia deep brain stimulation. Neurosci. Biobehav. Rev. **58**, 186–210 (2015)
6. De La Pava, I., Gómez, V., Álvarez, M.A., Henao, Ó.A., Daza-Santacoloma, G., Orozco, Á.A.: A Gaussian process emulator for estimating the volume of tissue activated during deep brain stimulation. In: Paredes, R., Cardoso, J.S., Pardo, X.M. (eds.) IbPRIA 2015. LNCS, vol. 9117, pp. 691–699. Springer, Heidelberg (2015). doi:10.1007/978-3-319-19390-8_77

7. McIntyre, C.C., Hahn, P.J.: Network perspectives on the mechanisms of deep brain stimulation. Neurobiol. Dis. **38**(3), 329–337 (2010)
8. Valencia-Aguirre, J., Álvarez-Mesa, A., Daza-Santacoloma, G., Castellanos-Domínguez, G.: Automatic choice of the number of nearest neighbors in locally linear embedding. In: Bayro-Corrochano, E., Eklundh, J.-O. (eds.) CIARP 2009. LNCS, vol. 5856, pp. 77–84. Springer, Heidelberg (2009). doi:10.1007/978-3-642-10268-4_9
9. Ward, C., Heath, S., Janovsky, V., et al.: Care of the movement disorder patient with a deep brain stimulator. J. Neurosci. Nurs. **43**(2), 116–118 (2011)

Discriminative Capacity and Phonetic Information of Bottleneck Features in Speech

Ana Montalvo$^{(\boxtimes)}$ and José Ramón Calvo

Advanced Technologies Application Center, 7th A Street, #21406 Havana, Cuba
{amontalvo,jcalvo}@cenatav.co.cu
http://www.cenatav.co.cu

Abstract. The impressive gain in performance obtained using deep neural network (DNN) for automatic speech recognition have motivated their application to other speech related tasks such as speaker recognition and language recognition, but there is still uncertainty about what is deep training strategy extracting, from the acoustic data, to make it such a powerful learning tool. This paper compares the discriminative capacity and the phonetic information conveyed by the feature-space maximum likelihood linear regression (fMLLR) before and after passing through a DNN trained to discriminate between tri-phone tied states. The proposed experimentation reflected the superiority of DNN bottleneck features regarding its information content.

Keywords: Entropy · Phonetic information · Bottleneck features · Deep neural network

1 Introduction

Five years ago, most speech recognition systems used hidden Markov models (HMMs) to deal with the temporal variability of speech and gaussian mixture models (GMMs) to determine how well each state of each HMM fits a sequence of feature vectors that represents the acoustic input.

In recent years, discriminative hierarchical models such as deep neural networks (DNNs) [1] became feasible and significantly reduced the word error rate.

When neural nets were first used, they were trained discriminatively. It was only recently that researchers showed that significant gains could be achieved by adding an initial stage of generative pretraining. The successes achieved using pretraining led to a resurgence of interest in DNNs for acoustic modeling.

There are mainly two different ways to incorporate deep learning techniques in speech recognition related tasks [1].

In the first configuration which is called DNN-HMM or simply hybrid, a DNN is used to compute the posterior probabilities of context-dependent HMM states based on observed feature vectors, then a Viterbi decoding is performed with these posteriors. The second configuration, which is called tandem, uses the DNN to perform a nonlinear discriminative feature transformation, which can

© Springer International Publishing AG 2017
C. Beltrán-Castañón et al. (Eds.): CIARP 2016, LNCS 10125, pp. 134–141, 2017.
DOI: 10.1007/978-3-319-52277-7_17

be regarded as a bridge between low-level acoustic input and high-level phonetic information [2], hence exploiting the output from different layers of the DNN may lead to improved utterance representation.

Specifically in the bottleneck (BN) tandem approach proposed by [3], a DNN in which one of the internal layers has a small number of hidden units (relative to the size of the other layers) is trained. Whereas the linear output of this bottleneck layer is taken as output instead of the posteriors.

The impressive gains in performance obtained using BN features for automatic speech recognition have motivated their application to other speech related tasks such as speaker recognition and language recognition with their particular backends.

But, why are they so successful? What is deep training strategy extracting from the acoustic data? Which is the most appropriate setting to achieve success?

In the present work we assess the discriminative capacity and the phonetic information of bottleneck features applying GMMs and entropy based measures. As far as the authors know, this type of study has not been reported before. Taking into account these aspects would allow to have a priori information on the behavior of the BN features, which in turn could be very useful to adjust the layout of the DNN, in pursuit of maximizing information, without performing the entirely recognition process. This research is aimed to be applied in obtaining new representations for identifying languages on noisy and short duration signals.

We first briefly describe the BN framework in Sect. 2. In the following sections, we discuss about BN features discriminative capacity and phonetic information (3 and 4 respectively). Section 5 describes the corpus, the toolkit and the performed experiments. Section 6 presents some results and discussion, and Sect. 7 will conclude this work.

2   Bottleneck Features

BN features are conventionally generated from a neural network in which one of the hidden layers has a very small number of hidden units relative to the other layers. This layer of small size is described as bottleneck layer and can be viewed as an approach of nonlinear dimensionality reduction, since it constricts the neural network so that the information embedded in the input features will be forced into a low dimensional representation.

BN features have become a very powerful way to use learned representations from data, and a lot of different features have been tested in order to feed the neural network with the optimal acoustic representation toward the particular classification problem [4].

There is a complex relationship between acoustic and BN features, certain correlation can therefore exist between the two. BN features provide a different view of the same speech signals, complementary information characterizing the acoustic realization are also implicitly learnt by BN features.

In our experiments the BN features are extracted from a DNN consisting of stacked Restricted Boltzman Machines [1] pre-trained in an unsupervised manner [5].

The DNN configuration is $n \times 40\text{-}1024\text{-}1024\text{-}1024\text{-}42\text{-}1024\text{-}d_{softmax}$, where the input is a concatenation of the current frame with the preceding and following $(n-1)/2$ neighboring frames. In our case $n = 9$ considers 9 staked adjacent frames of the 40-dimensional acoustic feature for each time instance. $d_{softmax}$ is the number of units in the output layer, in practice, $d_{softmax}$ is set to 2171 according to english tri-phone tied states present in our database.

The outputs of the bottleneck layer yield a compact representation of both acoustic and phonetic information for each frame independently.

As input of the DNN we used standard Mel-Frequency Cesptral Coefficients (MFCC) followed by a feature space transformation using feature-space maximum likelihood linear regression (fMLLR). This feature has shown to be critical dealing with speaker variability [6].

3   Discriminative Capacity of the BN Features

A DNN trained to discriminate between tri-phone tied states must attenuate the information related to other sources such as the channel, speaker, gender and sessions information, so that the final linear classifier (softmax layer) can effectively discriminate between tri-phone classes. BN features extracted from such a classifier should perform well as phonotactic feature for language recognition, and be more robust to the variabilities above mentioned [7].

In order to compare the phoneme discriminative capacity of the BN features with their predecessors fMLLR, we modeled both feature distributions using Gaussian Mixture Models. For each distribution (BN features and fMLLR) a Gaussian Mixture Model-Universal Background Model (GMM-UBM) was trained and MAP adapted on phonetically segmented and annotated corpora, to build a GMMs representing each phoneme. Then a phoneme classification all across the test set was performed and identification rate (IR) was used to measure the phoneme discriminative capacity of each representation.

4   Phonetic Information Conveyed by the BN Features

In this paper we propose to measure the phonetic information conveyed by the BN features and fMLLR, using two entropy based metrics.

For comprehensiveness of exposition, we offer a brief outline of these methods and a recall of the definition of the entropy measure.

If Y and X are discrete random variables, Eqs. 1 and 2 gives the entropy of Y before and after observing X:

$$H(Y) = -\sum_{y \in Y} p(y) log_2(p(y)), \tag{1}$$

$$H(Y|X) = -\sum_{x \in X} p(x) \sum_{y \in Y} p(y|x) log_2(p(y|x)). \tag{2}$$

Information Gain (IG)

The information found commonly in this two variables is defined as the mutual information or IG:

$$IG = H(Y) - H(Y|X) \tag{3}$$

If entropy $H(Y)$ is regarded as a measure of uncertainty about a random variable Y, then $H(Y|X)$ is a measure of what X does not say about Y. IG can be read then as the amount of uncertainty in Y, minus the amount of uncertainty that remains in Y after X is known, in other words IG is the amount of uncertainty in Y which is removed by knowing X.

This corroborates the intuitive meaning of IG as the amount of information (that is, reduction in uncertainty) that knowing either variable provides about the other.

Symmetrical uncertainty (SU)

Unfortunately, IG is biased in favor of features with more values, that is, attributes with greater numbers of values will appear to gain more information than those with fewer values even if they are actually no more informative. SU compensates for information gains bias toward attributes with more values and normalizes its value to the range $[0, 1]$.

$$SU = \frac{IG}{H(Y) + H(X)} \tag{4}$$

5   Experimental Setup

We performed two experiments on TIMIT. The first experiment compares BN and fMLLR features regarding the discriminative power of each of the 48 GMMs (one per phoneme). The second experiment uses the same GMM-UBM, this time to estimate global (Eq. 1) and conditional entropy (Eq. 2), and to evaluate the performance of entropy based measures.

5.1   Data Corpus

The TIMIT corpus consists of 4288 sentences (approximately 3.5 hours) spoken by 630 speakers of 8 major dialects of American English [8]. The training set contains 3,696 sentences from 462 speakers. The development set contains 400 sentences from 50 speakers and the test set 192 sentences from 24 speakers. We defined our test set as the merge of the TIMIT's development and test sets.

When training the DNNs, we use 90 % of the training set's sentences (1,012,340 frames) as training data and the remaining 10 % (112,483 frames) as a validation set.

5.2   The DNN Toolkit

The DNN toolkit selected for our experiments was PDNN [9], together with Kaldi[1] [10] one of the most popular toolkits for constructing ASR systems.

PDNN is written in Python, and uses the Theano library [11] for all its computations. The initial model to force align the data and generate a label for each feature vector, was built using Kaldi (TIMIT "s5" recipe).

We used the *BNF Tandem* PDNN's recipe [9] to train the DNN. The resulting BN features are mean and variance normalized.

5.3   Experiments

The 13-dimensional MFCC features are spliced in time taking a context of ± 4 frames, followed by de-correlation and dimensionality reduction to 40 using linear discriminant analysis. The resulting features are further de-correlated using maximum likelihood linear transform [12]. This is followed by speaker normalization using fMLLR, also known as constrained MLLR [13].

To asses the discriminative capacity of both kind of features (BN and fMLLR), different GMM-UBMs were trained varying the number of gaussian components, all of them with shared diagonal covariance. 48 GMMs were obtained with MAP mean adaptation for all the GMM-UBM previously obtained. The relevance factor used was $\tau = 10$. The consecutive frames of each phoneme are grouped together and evaluated against the 48 models regarding the identification rate (IR).

To deeply understand the intrinsic information of the features and to evaluate how informative can the fMLLR become after passing through a DNN, two entropy based measures were used: Information Gain (IG) and Symmetrical Uncertainty (SU).

6   Results

The first experiments show that BN features are better than fMLLR discriminating phonemes, when modeled with GMM (Table 1).

Table 1. Analysis of the features discriminative capacity.

GMM dimension	BN features IR (%)	fMLLR IR (%)
4 components	**66.8**	53.2
8 components	66.4	54.0
16 components	66.7	56.4
32 components	64.9	**57**
64 components	62.8	56.8

---

[1] https://kaldi.svn.sourceforge.net/svnroot/kaldi/trunk.

It is worth noting how the best IR obtained with fMLLR is far from the worst obtained with BN features: BN features need less gaussian components to fully express its complexity. Using a GMM of 4 components yields the best discriminative capacity.

The fact that modeling BN features with gaussians allows the evaluation of phoneme discriminative capacity, could be useful to set the optimal DNN configuration without carrying out all the acoustic modeling and decoding part of the speech recognition process.

Analyzing the behavior of IG shown in Fig. 1, BN features outperforms fMLLR for all the gaussian models. For 4-components GMM is achieved the best performance and the IG distance between both features is wider.

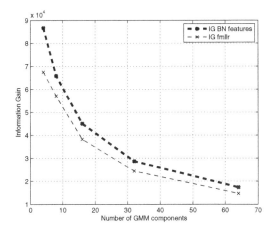

Fig. 1. Phonetic information: IG behavior for BN and fMLLR features.

Based on the previously computed entropy and conditional entropy, and looking at Fig. 2, one can conclude that the increase of the IG for BN features it is due to the intrinsic information carried by the features themselves, represented by the entropy.

In the other hand, the conditional entropy showed that both features are closely related with their respective phonetic models, so having small conditional entropy values, which is in complete correspondence with SU behavior.

The SU behaved pretty similar for all gaussian models (Table 2), showing a slightly higher performance in BN features.

SU values are close to one for both representations, indicating that the knowledge of features effectively predicts the labels. They are also quiet similar for both features because SU was defined to play a role in favor of variables with fewer values, and with such smalls values of conditional entropy this barely comes up.

In other words, small values of conditional entropy leave out the IG's bias weakness, and makes US less illustrative for the concerned phenomenon.

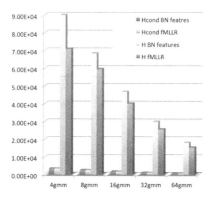

Fig. 2. Entropy contribution to IG.

Table 2. Symmetrical uncertainty.

Feature	4 components	8 components	16 components	32 components	64 components
BN	0.9632	0.9641	0.9642	0.9640	0.9631
fMLLR	0.9525	0.9586	0.9584	0.9585	0.9572

7   Conclusions

The main conclusion of this research is that BN features have more discriminative capacity than fMLLR and this in turn is a consequence of higher values of phonetic information and entropy.

These results show that the information content of BN features, is closely linked to its discriminative power. Then to calculate the entropy is a very easy way to evaluate features performance, avoiding conducting the recognition process as a whole, looking for an optimal network configuration.

As the best IR was obtained for a 4 components GMMs, we can conclude that BN features need less components to fully express its complexity than fMLLR.

It was also interesting to confirm that, the close dependence between features and phonetic labels, implies that the IG comes determined by the uncertainty contained in the features alone.

Future experiments will be conducted moving the BN layer, to asses its impact over language identification and to observ IG and US behavior.

References

1. Hinton, G., Deng, L., Yu, D., Dahl, G.E., Mohamed, A.R., Jaitly, N., Senior, A., Vanhoucke, V., Nguyen, P., Sainath, T.N., et al.: Deep neural networks for acoustic modeling in speech recognition: the shared views of four research groups. IEEE Sig. Process. Mag. **29**(6), 82–97 (2012)

2. Song, Y., Cui, R., Hong, X., Mcloughlin, I., Shi, J., Dai, L.: Improved language identification using deep bottleneck network. Proc. ICASSP **2015**, 1695–1699 (2015)
3. Gr'ezl, F., Karafiát, M., Kontar, S., Cernocký, J.: Probabilistic and bottle-neck features for LVCSR of meetings. In: IEEE International Conference on Acoustics, Speech and Signal Processing, pp. 757–760 (2007)
4. Rath, S., Povey, D., Vesely, K., Cernocky, J.: Improved feature processing for deep neural networks. In: Proceedings of Interspeech, pp. 109–113 (2013)
5. Albornoz, E.M., Sánchez-Gutiérrez, M., Martinez-Licona, F., Rufiner, H.L., Goddard, J.: Spoken emotion recognition using deep learning. In: Bayro-Corrochano, E., Hancock, E. (eds.) CIARP 2014. LNCS, vol. 8827, pp. 104–111. Springer, Heidelberg (2014). doi:10.1007/978-3-319-12568-8_13
6. Yu, D., Deng, L.: Automatic Speech Recognition a Deep Learning Approach. Springer, London (2015)
7. Jiang, B., Song, Y., Wei, S., Liu, J.-H., McLoughlin, I., et al.: Deep bottleneck features for spoken language identification. PLoS ONE **9**(7), e100795 (2014). doi:10.1371/journal.pone.0100795
8. Lopes, C., Perdigao, F.: Phone recognition on the TIMIT database. Speech Technol. **1**, 285–302 (2011)
9. Miao, Y.: Kaldi+PDNN: Building DNN-based ASR systems with Kaldi and PDNN. Computing Research Repository, abs/1401.6984 (2014)
10. Povey, D., Ghoshal, A., Boulianne, G., Burget, L., Glembek, O., Goel, N., Hannemann, M., Motlicek, P., Qian, Y., Schwarz, P., Silovsky, J., Stemmer, G., Vesely, K.: The Kaldi speech recognition toolkit. In: IEEE Workshop on Automatic Speech Recognition and Understanding. Signal Processing Society (2011)
11. Bergstra, J., Breuleux, O., Bastien, F., Lamblin, P., Pascanu, R., Desjardins, G., Turian, J., Warde-Farley, D., Bengio, Y.: Theano: a CPU and GPU math expression compiler. In: Proceedings of the Python for Scientific Computing Conference SciPy (2010)
12. Gopinath, R.: Maximum likelihood modeling with Gaussian distributions for classification. Proc. IEEE ICASSP **2**, 661–664 (1998)
13. Gales, M.J.F.: Maximum likelihood linear transformations for HMM-based speech recognition. Comput. Speech Lang. **12**(2), 75–98 (1998)

Boosting SpLSA for Text Classification

Julio Hurtado[1], Marcelo Mendoza[2(⊠)], and Ricardo Ñanculef[2]

[1] Pontificia Universidad Católica de Chile, Santiago, Chile
`julio.hurtado@puc.cl`
[2] Universidad Técnica Federico Santa María, Valparaíso, Chile
`{marcelo.mendoza,ricardo.nanculef}@usm.cl`

Abstract. Text classification is a challenge in document labeling tasks such as spam filtering and sentiment analysis. Due to the descriptive richness of generative approaches such as probabilistic Latent Semantic Analysis (pLSA), documents are often modeled using these kind of strategies. Recently, a supervised extension of pLSA (spLSA [10]) has been proposed for human action recognition in the context of computer vision. In this paper we propose to extend spLSA to be used in text classification. We do this by introducing two extensions in spLSA: (a) Regularized spLSA, and (b) Label uncertainty in spLSA. We evaluate the proposal in spam filtering and sentiment analysis classification tasks. Experimental results show that spLSA outperforms pLSA in both tasks. In addition, our extensions favor fast convergence suggesting that the use of spLSA may reduce training time while achieving the same accuracy as more expensive methods such as sLDA or SVM.

Keywords: Probabilistic Latent Semantic Analysis · Model Regularization · Label uncertainty · Text classification

1 Introduction

Given the large amount of data held in text format, modeling topics in text has been of growing interest in the last decade. Generative strategies based on probabilistic Latent Semantic Analysis (pLSA [6]) provide a solid theoretical base and a flexible framework for this task that allow for the modeling of various kind of corpora. The idea driving these models consists in introducing a set of latent variables that allow one to discover relationships between terms. These kind of strategies are known as topic models.

Topic models are fundamentally divided into two broad approaches: techniques derived from pLSA, which introduce latent variables without assuming distributional priors, and techniques based on Latent Dirichlet Allocation (LDA [2]) which assumes Dirichlet priors on topics and vocabularies.

From a machine learning perspective, topic models such as pLSA, correspond to unsupervised learning systems capable to discover structures in data without supervision. At the other end we have supervised learning systems, which are able to exploit annotated examples to predict future annotations. Classification techniques like the Multinomial Naive Bayes model [7] and the Support Vector

© Springer International Publishing AG 2017
C. Beltrán-Castañón et al. (Eds.): CIARP 2016, LNCS 10125, pp. 142–149, 2017.
DOI: 10.1007/978-3-319-52277-7_18

Machine (SVM [4]) are well-known examples of supervised systems with many applications in text mining, including e.g. spam filtering [1] and sentiment detection [8]. Unfortunately, though often very accurate, these models either lack the descriptive richness or the efficiency of text-oriented methods such as pLSA.

Recently, a supervised version of pLSA (spLSA [10]) has been proposed to tackle the problem of human action recognition in computer vision. By adding labels into the generative process, spLSA was endowed with more discriminative power for classification tasks when annotations are available. Surprisingly, spLSA has not yet been evaluated in text classification despite the fact that pLSA was originally designed for text modeling. To the best of our knowledge, in this paper we present the first evaluation of spLSA in text classification tasks. In addition, we introduce two extensions to spLSA: (a) Regularized spLSA, a variant that introduces label co-variance minimization into the Expectation-Maximization (EM) model fitting algorithm, and (b) Label uncertainty, an extension that allow us to deal with noisy labels, a common problem in human and machine annotated corpora [11].

2   Related Work

Due to its simplicity, the Multinomial Naive Bayes classifier (MNB [7]) is one of the most used methods for text classification. MNB assumes (class) conditional independence among terms, which reduces the complexity of the model in terms of the number of parameters and makes its estimation from data more efficient and reliable. However, in sparse data sets its performance tends to decrease. LDA-based methods introduce smoothing over the data set using Dirichlet priors [2], favoring classification tasks over sparse data sets. The supervised extension of LDA (sLDA [3]) shows good results in classification tasks but introduces difficulties in parameter tuning. In fact, due to the use of Dirichlet priors, the algorithm needs to tune distributional hyper-parameters. The lack of clear procedures for tuning is a drawback of this approach.

Discriminative approaches such as the Support Vector Machine and Logistic Regression [4] are also used for these tasks. In general, these approaches outperform generative approaches in terms of classification accuracy. However, they are less used due to difficulties associated with vocabulary characterization. In practice, generative approaches offer advantages regarding corpus descriptiveness, favoring tasks such as content analysis and term indexing, which are key tasks in information extraction and document processing.

3   Supervised pLSA and Our Extensions

3.1   Variables and Assumptions

An observation in a labeled corpus $\mathcal{D}$ is the realization of three observed random variables: A document ($\mathbf{d}$), a bag-of-words ($\mathbf{w}$) which describes the content of $\mathbf{d}$, and a label ($\mathbf{y}$) which indicates the membership of $\mathbf{d}$ to a given set of categories

Λ. The probability of observing a given realization of these variables $\langle \mathbf{d} = d, \mathbf{w} = w, \mathbf{y} = y \rangle$ is denoted by $P(d, w, y)$.

A document d is defined as a sequence of words selected from a vocabulary V. As documents share words, several semantic relationships among documents arise. By modeling hidden relationships between document and terms through latent variables, a latent semantic representation of the corpus is built.

While labels are typically noisy, words and documents convey information. By introducing latent factors, labels can be stressed by words and documents, allowing to model label uncertainty, discarding rare, unexpected labels and bearing out only data supported labels.

3.2   Supervised pLSA (spLSA)

The idea of a topic model is that documents are represented by mixtures of memberships over a latent space. Basically, spLSA considers a label as another random variable. Then, a generative approach where every observed variable depends on latent factors is used:

$$P_z(d, w, y) = \sum_z P(w|z, y, d) \cdot P(z, y, d) = \sum_z P(w|z) \cdot P(y, d|z) \cdot P(z)$$
$$= \sum_z P(w|z) \cdot P(y|z) \cdot P(d|z) \cdot P(z).$$

The last expression corresponds to a generative process of $\mathcal{D}$ from the model.

3.3   Model Fitting

Model parameters (θ) can be determined by maximizing the log-likelihood function of the generative process:

$$\mathcal{L}_\theta = \sum_{y \in \Lambda} \sum_{d \in \mathcal{D}} \sum_{w \in \mathcal{V}} n(d, w, y) \log P(d, w, y),$$

where $n(d, w, y)$ denotes the number of occurrences of (d, w, y) in $\mathcal{D}$. The Expectation Maximization algorithm (EM) is the standard procedure for parameter estimation in latent variable models. The E-step of the algorithm estimates $\hat{P}(z|d, w, y)$ from the generative model. Then, the M-step estimates the generative model components ($\hat{P}(w|z)$, $\hat{P}(y|z)$, $\hat{P}(d|z)$, and $\hat{P}(z)$) by likelihood marginalization.

3.4   Label Inference on New Documents

Document labeling using spLSA is straightforward. Starting from a uni-gram language model for a new document d with n_d terms, we have

$$P(d|z) \propto \prod_{i=1}^{n_d} P(w_i|z),$$

and then, by applying $P(y, d) = \sum_z P(y, d, z) = \sum_z P(y|z) \cdot P(d|z) \cdot P(z)$,

$$P_z(y = l, d) \propto \sum_{z \in \mathcal{Z}} \left[\prod_{i=1}^{n_d} P(w_i|z) \cdot P(y = l|z) \cdot P(z) \right].$$

Finally, d is labeled with $y = l$ if $P(y = l|d) \propto P(y = l, d)$ is greater than any other $P(y = \Lambda \backslash l|d)$.

3.5   A First Extension: SpLSA Regularization

Since the objective function in spLSA is not convex, the EM-based iterative method will tend to converge to local optima. As observed from the introduction of pLSA [6], many of these local optima are plagued by over-fitting. Therefore, it is advisable to endow the algorithm with a regularization technique. In this paper, we propose the incorporate additive Tikhonov-based regularizers [9] in the M-step of the model fitting phase. Concretely, we introduce regularization for latent factor co-variance minimization.

Our idea is to penalize solutions where the confusion matrix $(\hat{P}(y \mid z))$ is far from the identity $\mathcal{I}$. We do this by minimizing the co-variance between $P(y = l|z)$ and $P(y = l|Z \backslash z)$, forcing a match between each latent factor with only one label. Note that this procedure is equivalent to the diagonalization of the confusion matrix. The latent factor co-variance minimization modifies the likelihood function as follows:

$$\mathcal{L} = \mathcal{L}_\theta - \sum_{l \in \Lambda} \mathrm{Cov}\left(\hat{P}(y = l|z), \hat{P}(y = l|z^{'} \in Z \backslash z) \right).$$

The regularized maximum likelihood function can be maximized by modifying the M-step, where label probabilities conditioned on latent factors are estimated as in Sect. 3.3 with an additional step:

$$\hat{P}(y = l|z) = \hat{P}(y = l|z) \cdot \left[1 - \sum_{z^{'} \in Z \backslash z} \hat{P}(y = l|z^{'}) \right].$$

As a consequence, the EM-algorithm will penalize $\hat{P}(y = l|z)$ estimates for labels highly correlated to other latent variables. Note that $\hat{P}(y = l|z)$ is maximized when $(1 - \sum_{z^{'} \in Z \backslash z} \hat{P}(y = l|z^{'})) = 1$, which it is precisely what we are looking for.

Finally, we note that the co-variance regularizer may be applied to term probabilities or document probabilities, in a similar fashion. All these variants will be evaluated in our experiments.

3.6   A Second Extension: Modeling Label Uncertainty

Label uncertainty is common in text mining applications where annotations are obtained via crowd-sourcing or distant supervision [11]. To handle label uncertainty in spLSA, we modify the label estimation procedure by introducing the possibility to flip the label available in the data set. Formally, let $b \in [0, 1]$ the flipping probability parameter. To consider both possible labels (binary case), we introduce a convex combination conditioned on b in the M-step for $\hat{P}_b(y|z)$ as follows:

$$\hat{P}_b(y|z) = \frac{1}{Q_z} \cdot [(1 - b) \cdot \sum_{d \in \mathcal{D}} \sum_{w \in \mathcal{V}} n(d, w, y) \cdot \hat{P}(z|d, w, y = l)$$
$$+ b \cdot \sum_{d \in \mathcal{D}} \sum_{w \in \mathcal{V}} n(d, w, y) \cdot \hat{P}(z|d, w, y = \Lambda \backslash l)],$$

where l is the label provided by the data set. Note that b controls the level of uncertainty by blending $\hat{P}(y = l|z)$ and $\hat{P}(y = \Lambda \backslash l|z)$ into a unified model. When $b = 0$ the equation is equivalent to $\hat{P}(y = l|z)$ and when $b = 1$ is equivalent to $\hat{P}(y = \Lambda \backslash l|z)$. Thus, values between 0 and 1 will increase/decrease the confidence on the label provided by the data set. When $b = 0.5$ we give to the label a maximum degree of uncertainty, being both options (l or $\Lambda \backslash l$) equally probable.

Our method can be extended to the multi-class scenario, by applying a one-versus-the-rest strategy for $\hat{P}_b(y|z)$ estimation. In this case, it is enough to assume that all the labels that belong to $\Lambda \backslash l$ are equally probable.

4   Experiments

To evaluate our proposal, we performed experiments on two text labeled data sets. The first was designed for email spam filtering tasks [1]. We use this data set to illustrate the impact of regularization on text classification. The second data set contains tweets labeled as positive or negative regarding emotion polarity. It comprises a number of tweets labeled using distant supervision (it uses emoticons to automatically label each tweet) and thus it can be regarded as a data set with noisy labels. We use this data set to assess our strategy for handling label uncertainty.

4.1   Data Sets

Below, we provide a brief description of the data sets used in our experiments.

– **Email spam data set** [1]: This data set comprises a 700-email subset for training and a 260-email subset for testing. Both the training and testing subsets contain 50% spam messages and 50% non-spam messages. The data set comprises a vocabulary of 2.500 words. The data set is available in: http://csmining.org/index.php/ling-spam-datasets.html

– **Twitter distant supervision data set** [5]: It includes 1.600.000 tweets written in English, labeled as positive or negative regarding emotion polarity, inferring labels from emoticons. A set of 359 manually labeled tweets is provided for validation purposes (177 as negative and 182 as positive). The vocabulary is compounded by 267.013 terms. The data set is available in: http://cs.stanford.edu/people/alecmgo/trainingandtestdata.zip

4.2   Results

We start by exploring the performance of spLSA and co-variance regularization in text classification. We compare our proposals with pLSA [6], sLDA [3], Multinomial Naive Bayes (MNB [7]) and Support Vector Machines (SVMs [4]). For SVM, we used a C-SVM formulation provided by Liblinear [4], with a tuning over C based on five-fold cross validation ($C = 0.1$ in both data sets). For sLDA, we used default values for hyper-parameters[1]. We tested three variants of regularized spLSA, obtained by applying co-variance regularization on terms ($\hat{P}(w|z)$), documents ($\hat{P}(d|z)$), and labels ($\hat{P}(y|z)$). As both data sets rely on binary classes, we used two latent factors for the experiments (i.e. $T = 2$ for sLDA and pLSA).

We assess all the methods in terms of test accuracy and running time. In the first data set, experiments were conducted using a tolerance value of $1E-08$ for early stopping to guarantee local convergence (likelihood at single precision is enough for this experiment due to the small size of the data set). Training and testing performance are summarized in Table 1.

As Table 1 shows, the four variants of spLSA get good results in text classification. A slight difference in favor of SVM is observed regarding testing accuracy. As expected, training accuracies are better than testing accuracies. Note that the training times of spLSA are less than those incurred by pLSA, suggesting that spLSA helps to obtain fast convergence. In fact, the fastest solution is achieved using label regularization. Note also that spLSA and sLDA find the same solution, but spLSA is faster by one order of magnitude. In addition, this solution is better in terms of accuracy than the one achieved using MNB. In summary, results on this data set indicate that the use of regularization on spLSA reduces the number of iterations required for convergence without compromising classification accuracy.

Table 1. Classification accuracy performance on the email spam data set.

Method	Variant	Training acc.	Testing acc.	Time[s]
MNB	-	0.975	0.946	0.24
SVM	-	0.975	0.973	0.47
sLDA	-	0.975	0.965	3.12
pLSA	-	0.958	0.946	0.97
spLSA	-	0.975	0.965	0.24
spLSA	Reg. $\hat{P}(w|z)$	0.975	0.965	0.24
spLSA	Reg. $\hat{P}(d|z)$	0.975	0.965	0.24
spLSA	Reg. $\hat{P}(y|z)$	0.975	0.965	0.19

---

[1] $\alpha = \frac{50}{T}$, and $\beta = \frac{200}{W}$, T is the number of topics and W the vocabulary size.

Table 2. Classification accuracy performance on the Twitter distant supervision data set. Bold fonts indicate the best performance result.

Method	Variant	Training acc.	Testing acc.	Time[s]
MNB	-	0.797	0.777	312
SVM	-	0.817	0.816	624
sLDA	-	0.797	0.739	1245
pLSA	-	0.561	0.565	799
spLSA	-	0.817	0.777	326
spLSA	Reg. $\hat{P}(w|z)$	0.817	0.777	323
spLSA	Reg. $\hat{P}(d|z)$	0.817	0.777	331
spLSA	Reg. $\hat{P}(y|z)$	0.817	0.793	81

In the second data set, we used a tolerance value of $1E-16$ for early stopping (likelihood at double precision due to the size of the data set). Training and testing performance results are summarized in Table 2. These results show that spLSA outperforms pLSA by more than 20% accuracy points in its four variants, which, in turn, achieve a similar level of accuracy. As expected, the SVM is slightly more accurate at the expense of a greater computational cost ($\sim 2\times$ or more). Label regularization favors a significantly faster convergence, reducing training time to 81 secs with a testing accuracy better than the ones achieved by sLDA and MNB.

Finally, we analyze the impact of label uncertainty in the Twitter distant supervision data set, varying the value of the flipping probability parameter b in the range 0 to 0.6, with increments of 0.01. We use the label regularization variant for fast convergence. These results are shown in Fig. 1. The figure shows,

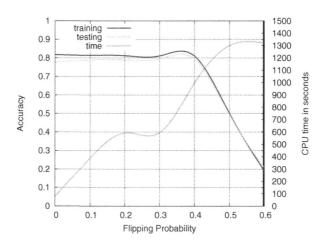

Fig. 1. Training and testing accuracy for different levels of label uncertainty.

the accuracy achieves values around 80% in both data partitions, with a small difference in favor of the training part of the data. This difference is reduced when b increases, until accuracy achieves its maximum for b in the range 0.3–0.4. As expected, the performance dramatically decreases when b tends to 0.5. In addition, we can observe high computational costs for high values of b. The best performance is achieved around $b = 0.35$, where our proposal reaches the same performance achieved by SVM.

5   Conclusions

Two extensions of spLSA for text classification has been proposed. Experimental results show that the both methods are feasible, achieving very competitive results in terms of accuracy. Two main findings arise. First, label regularization in spLSA allows to obtain faster convergence and thus lower training times. Second, handling label uncertainty in spLSA allows improvements in terms of test accuracy but increases computational costs.

References

1. Androutsopoulos, I., Koutsias, J., Chandrinos, K., Paliouras, G., Spyropoulos, C.: An evaluation of naive bayesian anti-spam filtering. In: 11th ECML, Workshop on Machine Learning in the New Information Age, pp. 9–17 (2000)
2. Blei, D., Ng, A., Jordan, M.: Latent dirichlet allocation. J. Mach. Learn. Res. (JMLR) **3**(4–5), 993–1022 (2003)
3. Blei, D., McAuliffe, J.: Supervised topic models. In: Proceedings of the Twenty-First Annual Conference on Neural Information Processing Systems (NIPS), Vancouver, British Columbia, Canada, pp. 121-128 (2007)
4. Fan, R., Chang, K., Hsieh, C., Wang, X., Lin, C.: LIBLINEAR: A library for large linear classification. J. Mach. Learn. Res. (JMLR) **9**, 1871–1874 (2008)
5. Go, A., Bhayani, R., Huang, L.: Twitter sentiment classification using distant supervision. CS224N Project report, Stanford University (2009)
6. Hofmann, T.: Unsupervised learning by probabilistic latent semantic analysis. Mach. Learn. **42**(2), 177–196 (2001)
7. McCallum, A., Nigam, K.: A comparison of event models for naive bayes text classification. In: Proceedings of the International Conference on Machine Learning (ICML), Madison, WI, USA, pp. 41–48 (1998)
8. Pang, B., Lee, L.: Opinion mining and sentiment analysis. Found. Trends in Inf. Retrieval **2**(1–2), 1–135 (2008)
9. Tikhonov, A., Arsenin, V.: Solutions of Ill-posed Problems. Winston & Sons, Great Falls (1977)
10. Wang, T., Liu, C.: Human action recognition using supervised pLSA. Int. J. Signal Process. Image Process. Pattern Recognit. **6**(4), 403–414 (2013)
11. Frenay, B., Verleysen, M.: Classification in the presence of label noise: a survey. IEEE Trans. Neural Netw. Learn. Syst. **25**(5), 845–869 (2014)

A Compact Representation of Multiscale Dissimilarity Data by Prototype Selection

Yenisel Plasencia-Calaña[1,2(✉)], Yan Li[2], Robert P.W. Duin[2],
Mauricio Orozco-Alzate[3], Marco Loog[2], and Edel García-Reyes[1]

[1] Advanced Technologies Application Center, 7ma A ♯ 21406, Playa, Havana, Cuba
{yplasencia,egarcia}@cenatav.co.cu
[2] Faculty of Electrical Engineering, Mathematics and Computer Sciences,
Delft University of Technology, Delft, The Netherlands
yanli.grass@gmail.com, r.duin@ieee.org, m.loog@tudelft.nl
[3] Departamento de Informática y Computación,
Universidad Nacional de Colombia - Sede Manizales,
Kilómetro 7 vía al Aeropuerto, Campus La Nubia – Bloque Q, Piso 2, Manizales,
Colombia
morozcoa@unal.edu.co

Abstract. Multiscale information provides an opportunity to improve the outcomes of data analysis processes. However, if the multiscale information is not properly summarized in a compact representation, this may lead to problems related to high dimensional data. In addition, in some situations, it is convenient to define dissimilarities directly for the multiscale data obtaining in this way a multiscale dissimilarity representation. When these dissimilarities are specifically designed for the problem, it is even possible that they do not fulfill metric requirements. Therefore, standard statistical analysis techniques may not be easily applicable. We propose a new method to combine non-metric multiscale dissimilarities in a compact representation which is used for classification. The method is based on the extended multiscale dissimilarity space and prototype selection, which allows us to handle the potentially non-metric nature of the dissimilarities and exploit the multiscale information at the same time. This is achieved in such a way that the most informative examples per scale are selected. Experimental results show that the approach is promising since it finds a better trade-off in accuracy and efficiency than its counterpart approaches.

Keywords: Extended multiscale dissimilarity space · Multiscale data · Prototype selection · Genetic algorithms

1 Introduction

The term multiscale refers to data represented at different scales of resolution. If the multiscale data is provided by an expert in the form of dissimilarities, the

---

Parts of the work described have been published as part of the first author's Ph.D. Thesis [1].

Yan Li is now with Lely Technologies, Maassluis, The Netherlands.

© Springer International Publishing AG 2017
C. Beltrán-Castañón et al. (Eds.): CIARP 2016, LNCS 10125, pp. 150–157, 2017.
DOI: 10.1007/978-3-319-52277-7_19

following question remains open: how to properly use the information contained in multiscale dissimilarities for classification? In the literature on supervised pattern recognition for multiscale similarities, we can find two main approaches: scale selection [2], and scale combining [3,4]. Scale selection has been tackled, for example, by Multiple Kernel Learning (MKL) [2], which is similar to the problem of selecting the best kernels for a given problem. For scale combining, all the different scales may be combined in the form of similarity or kernel matrices using MKL as well. However, note that, for potentially non-metric data, it is not possible to use kernel methods in a straightforward manner. One option is creating kernels from the dissimilarities, and after an eigen-analysis, correcting eigenvalues of the non-metric matrix by applying a strategy as spectrum clipping etc [5], which leads to some information loss.

There are other approaches that deal with a dissimilarity matrix which is potentially non-metric; for instance, the k Nearest Neighbour classifiers (k-NN) directly applied to the matrix, and the classifiers in the Dissimilarity Space (DS) [6,7]. In both cases, classifiers can be trained on the individual scales and combination may be performed by standard classifier combiners. Another possibility to combine the scales is by computing a weighted average of the dissimilarities [4]. The disadvantage of this approach is its high computational cost since, for an incoming test object, the dissimilarities with all the prototypes that span the DS for all the scales must be computed. Another approach, to which little attention has been paid so far, is constructing an Extended Multiscale Dissimilarity Space (EMDS) from the dissimilarity matrices [8]. Despite the fact that the first results presented in [8] using the EMDS were discouraging, we consider that a smarter selection of the set of prototypes can lead to better results.

In this paper we propose the use of a selection criterion in the EMDS optimized by a simple Genetic Algorithm [9] to perform such a smart selection. This criterion preserves the most important information from all the scales using the most informative prototypes. In our approach, a smart compromise is obtained between scale selection and scale combination, avoiding also expensive methods such as classifier combination.

The remaining part of the paper is organized as follows. Section 2 introduces the EMDS, presents the related work on prototype selection and the description of the proposed strategy to reduce the EMDS. Section 3 presents the data, experimental setup, results and analysis. Conclusions are drawn in Sect. 4.

2 Proposed Method

The DS was proposed by Pekalska and Duin [6] as an alternative to represent dissimilarity data. The DS is an adequate option to handle measures that are non-Euclidean or even non-metric. All the statistical pattern recognition procedures suitable for Euclidean spaces can be applied to the DS. Let X be the space of objects into consideration which may not be a feature vector space but a non-standard one such as a graph space. A set of prototypes $R = \{r_1, r_2, \ldots, r_l\} \in X$, also called representation set, is used for the creation of the DS. A training set

$T = \{x_1, x_2, \ldots, x_n\} \in X$ is represented in the DS by the dissimilarities of objects in T with objects in R. In general, for a representation set of l proto-types, and a suitable dissimilarity measure for the problem $d : X \times X \to \mathbb{R}_0^+$, we obtain a dissimilarity matrix $D(T, R)$; the mapping to a DS is represented as $\phi_R^d : X \to \mathbb{R}^l$. The representation of an object x in the DS is the vector of its dissimilarities with the prototypes: $\phi_R^d(x) = [d(x, r_1)\ d(x, r_2)\ \ldots\ d(x, r_l)]$.

The extended space representation is created from the individual represen-tations in a DS for each scale. For a multiscale problem with M scales, denot-ing $D_m = D_m(T, R)$ the dissimilarity matrix computed for scale m, we have $D_1, D_2, \ldots, D_M$, normalized dissimilarity matrices. The representation of train-ing objects in the EMDS is created by the concatenation of the individual dissim-ilarity matrices for each scale: $[D_1\ D_2\ \ldots\ D_M]$. The embedding of any object is obtained by the mapping $\Theta_R^d : X \to \mathbb{R}^{lM}$, which returns the vector of the dissimilarities with the prototypes from all the scales:

$$\Theta_R^d(x) = [\phi_{R_1}^d(x^1)\ \phi_{R_2}^d(x^2) \ldots \phi_{R_M}^d(x^M)], \tag{1}$$

where $R_m = \{r_1^m, r_2^m, \ldots, r_l^m\} \in X_m, m = 1\ldots M$, is the representation set in scale m and X_m is the space of objects for scale m; $x^m \in X_m, m = 1\ldots M$, are the representations of x under the different scales.

The main problem with the EMDS is its high dimensionality. It is a cause of overfitting and the "curse of dimensionality" phenomenon. Another problem is the increase of the computational costs involved in classification. In order to be able to use the multiscale information avoiding these problems, a prototype selection must be performed to create a reduced EMDS. As the EMDS presents different conditions compared to a standard DS, it is not possible to use most of the prototype selection procedures such as the KCentres or ModeSeek [7] unless they are applied on a single scale as in [8]. These methods require a direct com-parison of the prototypes being analyzed, but for the EMDS these prototypes are not directly comparable since they belong to different scales. Another good method, the Forward Selection (FS) [7], is not adequate for the EMDS due to the high dimensionality of this space and the method being quadratic in complex-ity with respect to this dimensionality. However, we found that GAs are more adequate for dealing with the EMDS, therefore we focus on this optimization strategy to select the prototypes in the EMDS.

We consider that GAs are specially suitable for prototype selection in dis-similarity representations, since, similar or nearby objects carry a similar infor-mational value and they can be chosen indistinctively as prototypes. Therefore, in our prototype selection problem, a thorough search is not needed, and there may be many suboptimal solutions that are sufficiently good and very close to the optimal one. Due to this, a GA can find good solutions for the prototype selection problem very fast, in contrast to other applications where GAs may converge slowly. The GA for prototype selection in a DS was proposed in [9], where it showed a good performance in standard DSs of moderate dimension-ality. However, its performance for very high dimensional spaces such as the extended ones has not been studied.

Our criterion focuses on selecting the prototypes in the EMDS taking into account the information provided by all the scales simultaneously, and not by a single scale as in previous approaches. The proposed criterion counts the matching labels of the prototypes and their nearest objects in each scale. For a given cardinality, the winning set of prototypes is the one that maximizes this number, note that selected prototypes for one scale are not necessarily selected for other scales. Only the combination (prototypes; scale) with significant contribution to the maximization is selected. This criterion can be formulated as follows: $J = \sum_{x_i \in S} ML(x_i)$, where:

$$ML(x_i) = \begin{cases} 1, \lambda_S(x_i) = \lambda_S(x_k) \\ 0, \lambda_S(x_i) \neq \lambda_S(x_k) \end{cases}, x_k = \operatorname*{argmin}_{x_j \in S \setminus \{x_i\}} d(x_i, x_j) \qquad (2)$$

where $\lambda_S(x_i)$ and $\lambda_S(x_k)$ are the class labels of x_i, x_k respectively, and x_k is the object with minimum Euclidean distance to x_i in the DS. The criterion J is therefore the number of matching labels for the candidates set S in the dissimilarity space.

The GA is an evolutionary method which uses heuristics to converge to better solutions, resembling biological processes such as crossover and mutation. Each potential solution (individual, chromosome) is a set of prototypes of fixed cardinality l codified in a $l - tuple$ of prototypes indexes. Note that, by resorting to this type of solution representation, the parameter that influences the memory requirements is the desired number of prototypes (usually small $l \in [10, 100]$) and not the dimensionality of the EMDS. The GA starts the search in an initial population of individuals randomly generated. In each evolution cycle, it evaluates the population using the fitness function. The population undergoes crossover (with best fitted individuals) and mutation processes until the criteria are met. In our approach, we use uniform crossover since each chromosome reproduces with the best fitted one with a preset probability, usually 0.5, and elitist selection since the best fitted individual is retained for the next generation without undergoing mutation. Besides, the population minus the best fitted individual undergoes mutation with a small preset probability, e.g. 0.05. The probabilities for mutation and crossover are usually set in a way that a good trade-off between "exploitation versus exploration" is obtained. The exploitation means the GA searches in a local region of the space where the last good solution was found (by setting crossover probability), and by the exploration the GA searches in a larger and more global region of the space (by setting mutation probability) to avoid loosing good solutions that may not be in a local neighbourhood of the last good solution.

3 Experimental Analysis

In this section, the multiscale data sets used in our experiments are described. The different approaches for prototype selection are presented. The experimental setup, results and discussion are also provided.

Data. It is worth noting that it is very difficult to collect real-world multiscale data sets where the dissimilarities were proposed by experts as suitable for the problem and where the multiscale information made sense according to them. Three different multiscale data sets were collected for the experiments. They are the Colon, Texture and Chicken Pieces data sets. Their descriptions are as follows. The *Colon* data set represents colon tissue data; it was provided by Dr. Marius Nap from the Atrium Medical Center in Heerlen, The Netherlands. The objects are microscope image patches of size 1024×1024 belonging to four classes: normal, inflamed, adenomatous, and cancer. The Laplacian of different scales was applied to each image patch, and the city-block (L1) distance between the histograms of the response images was used as the dissimilarity measure. The *Texture* data set (Brodatz) was downloaded from [10]. It has 111 images that we consider as classes. The 640×640 images were partitioned into 9 subimages that are used as class objects. The Leung-Malik filter set at different scales was applied to the images, and the Chi square distance between the histograms of the response images was computed. The *Chicken Pieces* data set [11] contains images in binary format representing silhouettes from five different parts of the chicken. From these images the edges are extracted and approximated by segments of different pixel length, and string representations of the angles between the segments are derived. The dissimilarity matrix is composed by edit distances between these strings. A description of the data sets is presented in Table 1.

Table 1. Properties of the multiscale data sets, the last column ($|T|$) refers to the training set cardinality used for the experiments

Data sets	# Classes	# Obj	# scales	$|T|$ in EMDS
Colon	4	375×4	9	100×9
Texture	111	9×111	6	222×6
Chicken pieces	5	446	11	170×11

Experimental Setup. In the experiments our aim is to show that the reduced EMDS obtained by the prototype selection may provide a more compact and discriminative representation of multiscale dissimilarities compared to the space of averaged multiscale dissimilarities, and the DS created by the best individual scale. Note that in the comparison we always use the same dimensionality of the spaces and the same length of vectors codifying the data. For consistency, we compare the same Linear Discriminant classifier (LDC), which is the Bayes classifier assuming normal densities with identical covariance matrices, and the 1-NN in the different spaces and data sets. All the dissimilarity matrices were normalized to avoid scaling problems by setting the mean dissimilarity per scale to 1 using global rescaling.

Since the data sets present a small size, they were 20 times randomly divided into two sets: a training set, that was used as the candidate set of prototypes to optimize the selection criterion and to build the final classifiers in the EMDS;

and a test set, which was only used to compute the final classification error. The prototype selectors executed are: GA in the EMDS (our proposal), random selection in the EMDS (as a baseline for the GA) and recent approaches from the literature on combining non-metric multiscale dissimilarities, also reducing the set of prototypes to ensure comparability of results: GA in the averaged DS, random selection in the averaged DS, random selection in the individual DS for each scale (only for the best performing classifier since for the other one a similar behaviour was found). Different parameters have been proposed for the GA, however, they can be problem-dependent. Thereby we decided to use parameters proposed in previous works on prototype selection [1]: Initial population: 30 individuals or solutions, Probability of reproduction: 0.5, Probability of mutation: 0.05, Stopping condition: 20 generations reached. We analyzed these parameters for our problem and we found that the GA converged to good solutions after 10 generations, but setting 20 generations as stopping condition ensured slightly better results. The results are stable in general for small variations of the parameters, but not for large variations.

Results and Discussion. Figure 1 present the results obtained for the data sets used in our study. Classification errors are presented for increasing numbers of prototypes in multiscale spaces as well as in the individual spaces from the different scales. Standard deviations were not included to maintain the clarity of the plots, but they vary between 0.02 and 0.05 for Chicken Pieces, 0.01 and 0.03 for Colon, and between 0.007 and 0.05 for Texture data set.

For the Colon data set, it can be seen that the EMDS outperforms the averaged DS and the individual scales. Results for the Texture data set show a clear example where the EMDS provides better results than those of the other approaches. In this data set as well as in the Colon data set, the EMDS significantly improves the results of the individual scales. Results for the Chicken Pieces data set show a different behaviour. The averaged DS outperforms the EMDS. We believe that this happens because, in this data set, only four scales present a low classification performance while seven scales perform similarly well. These large number of good performing scales influence the average dissimilarity computation heavier than the four bad ones. However, for the Colon and Texture data sets, the individual scales perform significantly different from each other, and the averaged space suffers from this while the reduced EMDS is able to capture the complementary information for classification. It can also be seen that the proposed selection outperforms the random selection, which usually performs very good for high dimensional DSs.

Note that the proposed approach is less computationally expensive than the combination of the scales by averaging all the DS representations. In the averaging case, the dissimilarities with the prototypes in all the scales must be computed while in our approach only the dissimilarities with the prototypes in the specific scale they were selected are computed. Therefore, for N scales and p prototypes and z number of objects, our approach computes $z \times p$ dissimilarities, while the average approach computes $z \times p \times N$ dissimilarities.

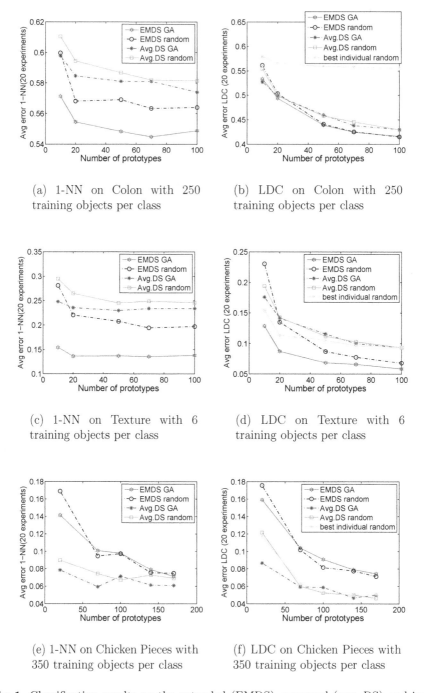

(a) 1-NN on Colon with 250 training objects per class

(b) LDC on Colon with 250 training objects per class

(c) 1-NN on Texture with 6 training objects per class

(d) LDC on Texture with 6 training objects per class

(e) 1-NN on Chicken Pieces with 350 training objects per class

(f) LDC on Chicken Pieces with 350 training objects per class

Fig. 1. Classification results on the extended (EMDS), averaged (avg. DS) and individual dissimilarity spaces for the different data sets

4   Conclusions

In this paper, we proposed a new strategy to represent potentially non-metric multiscale dissimilarity data in a compact and discriminative way. The multiscale representation is achieved by the extended dissimilarity space while the compact representation is achieved by means of a selection criterion optimized by a GA in a way that the most informative examples per scale are selected.

The classification results using the proposed compact multiscale representation outperform results using the representations in individual scales, despite having the same computational cost. In addition, the proposed approach is less computationally expensive than the combination of the scales by averaging all the DS representations, even improving the classification accuracies when the individual scales provide complementary information. Future work will focus on better characterizing the data sets where this approach is useful.

References

1. Plasencia-Calaña, Y.: Prototype selection for classification in standard and generalized dissimilarity spaces. Ph.D. thesis, Delft University of Technology, September 2015
2. Gönen, M., Alpaydin, E.: Multiple kernel learning algorithms. J. Mach. Learn. Res. **12**, 2211–2268 (2011)
3. Liu, Y.M., Ye, L.B., Zheng, P.Y., Shi, X.R., Hu, B., Liang, J.: Multiscale classification and its application to process monitoring. J. Zhejiang Univ. Sci. C **11**, 425–434 (2010)
4. Li, Y., Duin, R.P.W., Loog, M.: Combining multi-scale dissimilarities for image classification. In: Proceedings of the 21th International Conference on Pattern Recognition, ICPR 2012. IEEE Computer Society (2012)
5. Gisbrecht, A., Schleif, F.M.: Metric and non-metric proximity transformations at linear costs. Neurocomputing **167**, 643–657 (2015)
6. Pekalska, E., Duin, R.P.W.: The Dissimilarity Representation for Pattern Recognition: Foundations and Applications (Machine Perception and Artificial Intelligence). World Scientific Publishing Co. Inc., River Edge (2005)
7. Pekalska, E., Duin, R.P.W., Paclík, P.: Prototype selection for dissimilarity-based classifiers. Pattern Recogn. **39**(2), 189–208 (2006)
8. Ibba, A., Duin, R.P.W.: A multiscale approach in combining classifiers in dissimilarity representations. In: Gevers, T., Bos, H., Wolters, L. (eds.) 15th Annual Conference of the Advanced School for Computing and Imaging, ASCI 2009 (2009)
9. Plasencia-Calaña, Y., García-Reyes, E., Orozco-Alzate, M., Duin, R.P.W.: Prototype selection for dissimilarity representation by a genetic algorithm. In: Proceedings of the 2010 20th International Conference on Pattern Recognition, ICPR 2010, pp. 177–180. IEEE Computer Society, Washington (2010)
10. Randen, T.: Brodatz textures. http://www.ux.uis.no/~tranden/brodatz.html
11. Bunke, H., Buhler, U.: Applications of approximate string matching to 2D shape recognition. Pattern Recogn. **26**(12), 1797–1812 (1993)

A Kernel-Based Approach for DBS Parameter Estimation

V. Gómez-Orozco$^{(\boxtimes)}$, J. Cuellar, Hernán F. García, A. Álvarez, M. Álvarez, A. Orozco, and O. Henao

Automatic Research Group, Universidad Tecnológica de Pereira, Pereira, Colombia
{vigomez,jfcuellar,hernan.garcia,andres.alvarez1,
malvarez,aaog,oscarhe}@utp.edu.co

Abstract. The volume of tissue activated (VTA) is commonly used as a tool to explain the effects of deep brain stimulation (DBS). The VTA allows visualizing the anatomically accurate reconstructions of the brain structures surrounding the DBS electrode as a 3D high-dimensional activate/non-activate image, which leads to important clinical applications, e.g., Parkinson's disease treatments. However, fixing the DBS parameters is not a straightforward task as it depends mainly on both the specialist expertise and the tissue properties. Here, we introduce a kernel-based approach to learn the DBS parameters from VTA data. Our methodology employs a kernel-based eigendecomposition from pairwise Hamming distances to extract relevant VTA patterns into a low-dimensional space. Further, DBS parameters estimation is carried out by employing a kernel-based multi-output regression and classification. The presented approach is tested under both isotropic and anisotropic conditions to validate its performance under realistic clinical environments. Obtained results show a significant reduction of the input VTA dimensionality after applying our scheme, which ensures suitable DBS parameters estimation accuracies and avoids over-fitting.

Keywords: Kernels · Volume of tissue activated · Deep brain stimulation

1 Introduction

Deep brain stimulation (DBS) is a surgical therapy used mainly for treatment of a variety of neurological disorders, in patients who do not respond well to medication. In particular, DBS consists of an electrode inserted inside neural tissue of the patient to modulate neural activity with applied electric pulses, which depend on amplitude, pulse-width, frequency, and the electrode characteristics [2]. Although the physiological mechanism of DBS still remains unclear, its clinical effectiveness is evident [5]. A measure of the effects of the DBS resides in the estimation of the volume of tissue activated (VTA), namely, the spatial

© Springer International Publishing AG 2017
C. Beltrán-Castañón et al. (Eds.): CIARP 2016, LNCS 10125, pp. 158–166, 2017.
DOI: 10.1007/978-3-319-52277-7_20

spread of direct neural response to external electrical stimulation as a 3D high-dimensional activate/non-activate image. Nonetheless, fixing the DBS parameters is not a straightforward task since it depends on both the specialist expertise and the brain tissue properties of each patient [6].

The VTA, and its visualization, jointly with reconstructions of the brain structures surrounding the implanted electrode have been proposed as an alternative to accelerate the process of stimulation parameters adjustment, also minimizing the adverse side effects that can occur when they are not carefully adjusted [1]. This system allows the medical specialist to observe the brain structures that are responding directly to the electrical stimulation. Thus, the clinician determines the possible effects of a given stimulation configuration on the patient. Nonetheless, such an approach still involves a heuristic search method (trial and error), requiring high computational load and appropriate expertise by the specialist [6]. So, the problem of computing a set of specific neuromodulation parameters given an objective VTA has been less studied in the DBS literature. In contrast, there is an extensive literature that attempts to estimate the VTA from the stimulation parameters [1,10]. Authors in [7] introduced a machine learning system to predict the VTA from DBS parameter space. It allows the user to select a target to find the correlation between the calculated and the desired VTA. Once the correlation between the VTAs is calculated, the algorithm provides a possible configuration of neurostimulation parameters. However, the approach only operates under isotropic conditions and the system can not represent high stimulation parameter values and/or more than two active contacts. In [4] our group presented an alternative strategy for DBS parameter estimation from a previously specified VTA. We employed a framework based on support vector machines for multi-output regression and classification. However, our strategy is developed only under isotropic conditions.

In this work, we developed a kernel-based approach for DBS parameters estimation from VTA data. In this sense, our data-driven kernel-based scheme comprises mainly two stages *(i)* kernel-based principal components extraction from VTA samples, and *(ii)* DBS parameter estimation using kernel-based multi-output regression and classification. Moreover, our technique is developed under both ideal (isotropic tissue conductivities) and realistic (anisotropic tissue1 conductivities) assumptions. Our aim is to estimate neuromodulation parameters from the planned VTA to support DBS-based treatments. Results obtained show a remarkable reduction of the input VTA dimensionality after applying our feature extraction scheme, ensuring suitable DBS parameter estimation accuracies and avoiding over-fitting. The remainder of this paper is organized as follows: Sect. 2 describes the materials and methods of the approach introduced. Sections 3 and 4 describe the experimental set-up and the results obtained, respectively. Finally, the concluding remarks are outlined in Sect. 5.

2 Materials and Methods

Let $\boldsymbol{X} \in \{0,1\}^{N \times P}$ and $\boldsymbol{Y} \in \mathbb{R}^{N \times Q}$ be a given pair of matrices coding the VTA and the DBS parameter spaces, respectively, holding P axons, Q

stimulation parameters, and N samples. The i-th VTA $\boldsymbol{x}_i \in \mathbb{R}^P$ ($i \in \{1, 2, \ldots, N\}$) is computed from the DBS configuration in row vector $\boldsymbol{y}_i \in \mathbb{R}^Q$. In particular, a six-dimensional DBS parameter vector $\boldsymbol{y} = [D_A \, D_W \, c_0 \, c_1 \, c_2 \, c_3]$ is considered ($Q = 6$), where $D_A \in \mathbb{R}^+$ refers to the amplitude, $D_W \in \mathbb{R}^+$ to the pulse width, and c_r to the r-th contact condition, with $r = \{0, 1, 2, 3\}$. Regarding the VTA simulation, given a DBS parameter vector $\boldsymbol{y}_i$, a finite element method (FEM) is employed to compute the spatial distribution of the extracellular potential [10]. Then, a model of multicompartment myelinated axons is implemented to determine the axonal response to electric stimulation and the VTA is computed as the volume generated by the active axons [1]. The i-th VTA $\boldsymbol{x}_i$ is stored towards axon concatenation, where the element $x_{ip} = 1$ ($p \in \{1, \ldots, P\}$) if the p-th axon is activated by the DBS, otherwise, $x_{ip} = 0$. Figure 1 shows the VTA estimation sketch.

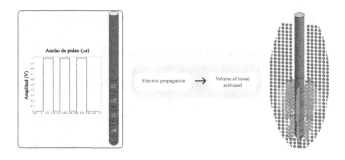

Fig. 1. VTA estimation main sketch. The VTA is composed by elements labeled as active '1' (green dots) and non-active '0' (red dots). (Color figure online)

With the aim to estimate the neuromodulation parameters from a planned VTA in DBS-based treatments, we introduce a data-driven kernel-based scheme to highlight the relevant relations between the VTA and the DBS parameter spaces. Our approach comprises two main stages *(i)* VTA feature extraction through kernel-based eigendecomposition, *(ii)* DBS parameter estimation using kernel-based multi-output regression and classification.

Kernel-Based VTA Feature Extraction. Let $\phi:\{0,1\}^P \to \mathcal{H}$ be a nonlinear mapping function that embeds any $\boldsymbol{x} \in \boldsymbol{X}$ into the element $\phi(\boldsymbol{x}) \in \mathcal{H}$ of the Reproducing Kernel Hilbert Space (RKHS) $\mathcal{H}$. By assuming that the elements in $\mathcal{H}$ are centered ($\sum_{i=1}^{N} \phi(\boldsymbol{x}_i) = 0$), the covariance matrix in the RKHS can be computed as follows: $\boldsymbol{S} = (1/N) \sum_{i=1}^{N} \phi(\boldsymbol{x}_i)\phi(\boldsymbol{x}_i)^\top$, where its eigenvalues $\lambda_m \in \mathbb{R}^+$ and eigenvectors $\boldsymbol{v}_m \in \mathcal{H}$ satisfy $\boldsymbol{S}\boldsymbol{v}_m = \lambda_m \boldsymbol{v}_m$ ($m = \{1, 2, \ldots, M_P\}$, $\boldsymbol{v}_m \neq 0$). Since all solutions $\boldsymbol{v}_m$ lie in the span of $\{\phi(\boldsymbol{x}_i)\}_{i=1}^{N}$ we may consider the equivalent system: $\lambda_m \phi(\boldsymbol{x}_i)^\top \boldsymbol{v}_m = \phi(\boldsymbol{x}_i)^\top \boldsymbol{S}\boldsymbol{v}_m$, and that there exists a coefficient vector set $\{\boldsymbol{\alpha}_m \in \mathbb{R}^N\}_{m=1}^{M_P}$, such that $\boldsymbol{v}_m = \sum_{i=1}^{N} \alpha_i^m \phi(\boldsymbol{x}_i)$ for all $\alpha_i^m \in \boldsymbol{\alpha}_m$. Then, an eigenvalue problem is solved to find $\boldsymbol{\alpha}_m$ as [9]: $N\lambda_m \boldsymbol{\alpha}_m = \boldsymbol{K}\boldsymbol{\alpha}_m$, where

$\boldsymbol{K} \in \mathbb{R}^{N \times N}$ is a kernel matrix holding elements $k_{ij} = \kappa_x(\boldsymbol{x}_i, \boldsymbol{x}_j) = \phi(\boldsymbol{x}_i)^{\top}\phi(\boldsymbol{x}_j)$, being $\kappa_x:\{0,1\}^P \times \{0,1\}^P \rightarrow \mathbb{R}^+$ a positive definite kernel. Due to the binary structure of the VTA, a Gaussian kernel is computed from a Hamming-based distance:

$$\kappa_x(\boldsymbol{x}_i, \boldsymbol{x}_j) = \exp\left(\frac{-d_h^2(\boldsymbol{x}_i, \boldsymbol{x}_j)}{2\sigma_x^2}\right), \tag{1}$$

where $d_h^2(\boldsymbol{x}_i, \boldsymbol{x}_j) = \sum_{p=1}^P (1-\delta(x_{ip}-x_{jp}))$ is the Hamming distance operator, $\delta(\cdot)$ stands for the delta function, and $\sigma_x \in \mathbb{R}^+$ is kernel bandwidth. Afterwards, the kernel principal component extraction (KPCA) $z_i^m \in \mathbb{R}$ of $\phi(\boldsymbol{x}_i)$ onto the m-th basis in $\mathcal{H}$ is computed as:

$$z_i^m = \boldsymbol{v}_m^{\top}\phi(\boldsymbol{x}_i) = \sum_{j=1}^N \alpha_i^m \kappa_x(\boldsymbol{x}_i, \boldsymbol{x}_j) \tag{2}$$

and the feature extraction matrix $\boldsymbol{Z} \in \mathbb{R}^{N \times M_P}$, holding row vectors $\boldsymbol{z}_i \in \mathbb{R}^{M_P}$, is built by concatenating the M_P principal components.

Kernel-Based DBS Parameter Estimation. Given $\boldsymbol{Z}$ we build two kind of vector-valued functions. The former, $f^R:\mathbb{R}^{M_P} \rightarrow \mathbb{R}^2$ estimates the amplitude and pulse-width DBS parameter vector $\boldsymbol{y}_i^R \in \mathbb{R}^2$ in $\boldsymbol{Y}^R \in \mathbb{R}^{N \times 2}$ as: $\hat{\boldsymbol{y}}_i^R = f^R(\boldsymbol{z}_i) = \varphi^R(\boldsymbol{z}_i)\boldsymbol{W}+\boldsymbol{b}$, where $\boldsymbol{W} \in \mathbb{R}^{M_R \times 2}$, $\boldsymbol{b} \in \mathbb{R}^2$, and $\varphi^R:\mathbb{R}^{M_P} \rightarrow \mathbb{R}^{M_R}$. Then, a multi-output support vector regression (MSVR) optimization problem can be defined as follows [8]:

$$\boldsymbol{W}^*, \boldsymbol{b}^* = \arg\min_{\boldsymbol{W},\boldsymbol{b}} \frac{1}{2}\sum_{m=1}^2 \|\boldsymbol{w}_m\|^2 + \gamma_R \sum_{i=1}^N \varsigma(u_i), \tag{3}$$

where $\boldsymbol{w}_m \in \mathbb{R}^{M_R}$ is the m-th column vector of $\boldsymbol{W}$, $\gamma_R \in \mathbb{R}^+$ is a regularization parameter, $u_i = (\boldsymbol{e}_i^{\top}\boldsymbol{e}_i)^{1/2}$, $\boldsymbol{e}_i = \boldsymbol{y}_i^R-\hat{\boldsymbol{y}}_i^R$, and $\varsigma(u_i) = (u-\epsilon)^2$ if $u_i \geq \epsilon$, otherwise, $\varsigma(u_i) = 0$ ($\epsilon \in \mathbb{R}^+$). Writing Eq. (3) in terms of $\boldsymbol{\xi}_m \mathbb{R}^N$, with $\varphi = [\varphi^R(\boldsymbol{z}_1), \ldots, \varphi^R(\boldsymbol{z}_N)]^{\top} \in \mathbb{R}^{N \times M_R}$ and $\boldsymbol{w}_m = \varphi^{\top}\boldsymbol{\xi}_m$, a dual problem can be solved as:

$$\hat{\boldsymbol{y}}_i^R = \boldsymbol{k}_i^R \boldsymbol{\Xi}+\boldsymbol{b} \tag{4}$$

where $\boldsymbol{\Xi} \in \mathbb{R}^{N \times 2}$ is a weighting matrix with column vectors $\boldsymbol{\xi}_m$ and $\boldsymbol{k}_i^R \in \mathbb{R}^N$ is a row vector holding elements: $k_{ij}^R = \kappa_z(\boldsymbol{z}_i, \boldsymbol{z}_j)$, $(i,j \in \{1,2,\ldots,N\})$, being $\kappa_z:\mathbb{R}^{M_P} \times \mathbb{R}^{M_P} \rightarrow \mathbb{R}^+$ a Gaussian kernel function:

$$\kappa_z(\boldsymbol{z}_i, \boldsymbol{z}_j) = \exp\left(\frac{-d_e^2(\boldsymbol{z}_i, \boldsymbol{z}_j)}{2\sigma_z^2}\right); \tag{5}$$

notation $d_e(\cdot, \cdot)$ stands for the Euclidean distance and $\sigma_z \in \mathbb{R}^+$. Then, an iteratively reweighted least squares procedure is used to find $\boldsymbol{\Xi}, \boldsymbol{b}$ [8].

Regarding the latter function, $f^C:\mathbb{R}^{M_P} \rightarrow \{-1,0,1\}$, which allows computing the DBS contact configuration vector $\boldsymbol{y}_i^C \in \{-1,0,1\}^4$ in $\boldsymbol{Y}^C \in \{-1,0,1\}^{N \times 4}$, we built a soft margin support vector classifier (SVC) over $\boldsymbol{Z}$ to compute the r-th contact value as:

$$\hat{y}_{ir}^C = f_r^C(\boldsymbol{z}_i) = \sum_{j=1}^N \varrho_j^r y_{jr}^C \kappa_z(\boldsymbol{z}_i, \boldsymbol{z}_j) + a_r, \tag{6}$$

where $\varrho_j^r \in \mathbb{R}$ is the weight of training sample j for the r-th classifier and $a_r \in \mathbb{R}$ is a bias term. So, each classifier is solved as a quadratic optimization from the well-known SVC dual formulation (for details see [9]).

3   Experimental Set-Up

We built two VTA databases generated by 1000 randomly selected combinations of realistic stimulation parameters, thereby, $c_r\{-1, 0, 1\}$, $D_A \in [0.5, 5.5][V]$, and $D_W \in [60, 450]$ $[\mu s]$. Such parameter value ranges are relevant in the context of VTA estimation for the Medtronic ACTIVA-RC stimulator.[1] The first database is built for both monopolar and bipolar conditions (one or two active contacts), under the assumption of an isotropic tissue medium (ITM), which is the most commonly used in clinical practice settings. The second database comprises isotropic and anisotropic tissue medium conditions (IATM), namely, 500 VTAs are computed for each of them. An extracellular potential model is executed for both databases using the COMSOL Multiphysics 4.2 FEM toolbox. Therefore, a model where the electric conductivity of the brain tissue is assumed to be homogeneous and isotropic is employed for ITM, meanwhile an anisotropic conductivity one is used for IATM. Such anisotropic conductivities are obtained from magnetic resonance imaging by means of diffusion tensors. For concrete testing, a DTI30 dataset is considered[2] with the RESTORE (Robust Estimation of Tensor by Outlier Rejection) algorithm, and then, linearly transformed to conductivity tensors. After that, a model of multicompartment myelinated axons is implemented by using NEURON 7.3 as a Python module, to determine axonal response to the electric stimulation. Nevertheless, solving the gold standard approach for VTA estimation is computationally expensive. So, we use a Gaussian Process classifier (GPC) to emulate the multicompartment myelinated axonal model [3]. In this sense, the multicompartment axon model is executed by a random sample set estimated from the whole axonal population, aiming to simulate the axonal response to the electric stimulation by training the GPC.

For all the experiments, we performed a training-testing validation scheme with 30 repetitions, where 80% of the samples are used as training set and the remaining 20% as testing set. Two kind of systems are tested. The former, high-dimensional kernel learning (HDKL), does not include the KPCA stage for learning relevant components from the VTAs. Instead, the MSVR and the SVC are applied directly from the input set X by applying a Gaussian kernel from a Hamming-based distance. The latter, low-dimensional kernel learning (LDKL), includes the KPCA stage as feature extraction. For both systems the MSVR algorithm is implemented according to an open source code[3]. Furthermore, the kernel bandwidth values are fixed as: $\sigma_x = \mathrm{med}(\mathrm{d}_h(\boldsymbol{x}_i, \boldsymbol{x}_j))$ and $\sigma_z = \mathrm{med}(\mathrm{d}_e(\boldsymbol{z}_i, \boldsymbol{z}_j))$, respectively, where $\mathrm{med}(\cdot)$ stands for the median operator. Moreover, the number of projected features in KPCA is computed from the set $M_R = \{3, 4, \ldots, 30\}$.

---

[1]   Available at www.aann.org/pdf/cpg/aanndeepbrainstimulation.pdf.

[2]   Available at www.cabiatl.com/CABI/resources/dti-analysis/.

[3]   http://isp.uv.es/soft_regression.htm.

Additionally, the MSVR free parameters are fixed within the following ranges based on the system performance: $\sigma_z = [0.5\mathrm{med}(\mathrm{d}_e(\boldsymbol{z}_i, \boldsymbol{z}_j)), 1.5\mathrm{med}(\mathrm{d}_e(\boldsymbol{z}_i, \boldsymbol{z}_j))]$, $\epsilon = [0.5, 2]$, and $\gamma_R = [0.5, 2]$. Alike, the SVC regularization parameter γ_C is tuned from: $\gamma_C = [1, 10, 50, 100, 1000]$. For all the datasets provided, we uniformly subsample each VTA vector to obtain the input dimensionalities $P = \{17.405, 8.703, 3481\}$ and $P = \{17.924, 8.897, 3559\}$ in ITM and IATM, respectively.

4   Results and Discussion

Figure 2(a) and (b) show the kernel-based eigendecomposition projections for both VTA datasets studied. Each dot represents a different neurostimulation configuration, where its size is given by the contact condition, and the color represents the amplitude value. After visual inspection of the ITM results, we note that the kernel-based feature extraction allows differentiating between active and non-active points. Since ITM is computed for isotropic conditions a smooth data structure is revealed by the KPCA projection. In fact, the projected space allows coding the DBS amplitude information, which probes the capability to encode high-dimensional sample relations. Nevertheless, some overlaps are found due to different combinations of the amplitude and pulse width values, which lead to similar VTAs. Regarding the IATM visual inspection results, the achieved projection highlights more overlaps in terms of contact state and amplitude value in comparison to the ITM results. The fact that both isotropic and anisotropic conditions are studied in the same dataset leads to complex data relations between VTAs, however, a bottom to top amplitude increases is presented.

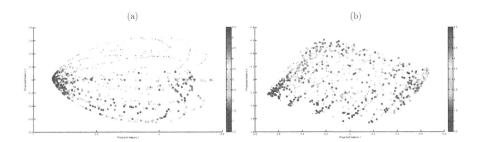

Fig. 2. Kernel-based eigendecomposition results. The dots color represents the amplitude value and the size the contact settings: active (big dot), inactive (small dot). (a) ITM dataset-Contact 0. (b) IATM dataset-Contact 0. (Color figure online)

Now, Tables 1 and 2 summarize the DBS parameter estimation accuracies obtained for both considered kernel-based approaches (HDKL and LDKL). The embedding dimensionality in LDKL is varied in the range $M_P = \{5, 10, 15, 20, 25, 30, 35, 40, 45, 50, 55, 60, 65, 70\}$ and the best result is presented for each provided input dimensionality value. As seen, the HDKL approach

Table 1. DBS parameter estimation results for the ITM dataset. Percentage of training-set selected as support vector is shown in parentheses.

DBS par.	HDKL $P = 17405$	LDKL ($M_P = 40$) $P = 17405$	HDKL $P = 8703$	LDKL ($M_P = 40$) $P = 8703$	HDKL $P = 3481$	LDKL ($M_P = 40$) $P = 3481$
D_A [V]	87.62 ± 1.43 (44.12 ± 9.93)	$\mathbf{91.78 \pm 0.84}$ $\mathbf{(72.50 \pm 4.41)}$	87.49 ± 1.26 39.42 ± 5.08	$\mathbf{91.76 \pm 0.86}$ $\mathbf{(72.44 \pm 3.53)}$	87.28 ± 1.11 44.26 ± 9.96	$\mathbf{91.82 \pm 0.88}$ $\mathbf{(72.20 \pm 4.50)}$
D_W [μs]	$\mathbf{84.65 \pm 1.19}$ $\mathbf{(44.12 \pm 9.93)}$	85.64 ± 0.89 (72.50 ± 4.41)	$\mathbf{84.83 \pm 1.02}$ $\mathbf{(39.42 \pm 5.08)}$	85.55 ± 0.96 (72.44 ± 3.53)	$\mathbf{84.50 \pm 1.10}$ $\mathbf{(44.26 \pm 9.96)}$	85.64 ± 0.90 (72.20 ± 4.50)
c_0	97.47 ± 0.86 (21.07 ± 3.88)	$\mathbf{97.70 \pm 0.96}$ $\mathbf{(13.27 \pm 2.95)}$	$\mathbf{97.25 \pm 1.14}$ $\mathbf{(22.16 \pm 3.91)}$	95.95 ± 3.43 (14.46 ± 2.99)	$\mathbf{97.37 \pm 0.90}$ (20.61 ± 3.70)	97.25 ± 1.11 $\mathbf{(13.91 \pm 3.37)}$
c_1	94.92 ± 1.63 (23.37 ± 4.03)	$\mathbf{97.47 \pm 1.70}$ $\mathbf{(15.28 \pm 3.60)}$	94.84 ± 1.69 (20.87 ± 0.69)	$\mathbf{96.57 \pm 3.47}$ $\mathbf{(15.79 \pm 4.88)}$	94.92 ± 1.59 (23.94 ± 4.35)	$\mathbf{97.55 \pm 0.98}$ $\mathbf{(14.63 \pm 2.68)}$
c_2	95.68 ± 1.44 (23.97 ± 3.67)	$\mathbf{98.12 \pm 0.82}$ $\mathbf{(14.12 \pm 2.41)}$	95.11 ± 1.19 (25.98 ± 5.03)	$\mathbf{96.92 \pm 3.39}$ $\mathbf{(15.42 \pm 2.51)}$	95.71 ± 1.28 (24.06 ± 3.98)	$\mathbf{96.75 \pm 4.23}$ $\mathbf{(16.99 \pm 6.83)}$
c_3	97.65 ± 0.99 (21.10 ± 4.10)	$\mathbf{98.53 \pm 0.76}$ $\mathbf{(13.28 \pm 2.10)}$	96.62 ± 1.10 (20.53 ± 4.03)	$\mathbf{96.38 \pm 2.88}$ $\mathbf{(13.34 \pm 2.70)}$	97.50 ± 1.02 (20.53 ± 4.13)	$\mathbf{97.70 \pm 1.34}$ $\mathbf{(12.37 \pm 2.25)}$
$Average$	93.00 ± 1.26 (26.73 ± 5.12)	94.87 ± 0.99 (25.69 ± 3.09)	92.68 ± 1.24 (25.79 ± 3.75)	93.85 ± 2.49 (26.29 ± 3.32)	92.90 ± 1.17 (26.68 ± 5.22)	94.48 ± 1.54 (26.02 ± 3.93)

Table 2. DBS parameter estimation results for the IATM dataset. Percentage of training-set selected as support vector is shown in parentheses.

DBS par.	HDKL $P = 17793$	LDKL ($M_P = 40$) $P = 17793$	HDKL $P = 8897$	LDKL ($M_P = 40$) $P = 8897$	HDKL $P = 3559$	LDKL ($M_P = 40$) $P = 3559$
D_A [V]	87.74 ± 1.14 (40.08 ± 6.22)	$\mathbf{89.31 \pm 0.91}$ $\mathbf{(73.08 \pm 15.38)}$	87.57 ± 1.14 39.92 ± 5.74	$\mathbf{89.28 \pm 0.91}$ $\mathbf{(74.09 \pm 15.18)}$	87.89 ± 1.17 (43.10 ± 9.57)	$\mathbf{89.45 \pm 0.87}$ $\mathbf{(75.96 \pm 14.93)}$
D_W [μs]	82.91 ± 1.39 (40.08 ± 6.22)	$\mathbf{85.13 \pm 1.45}$ $\mathbf{(73.08 \pm 15.38)}$	82.95 ± 1.50 39.92 ± 5.74	$\mathbf{85.15 \pm 1.45}$ $\mathbf{(74.09 \pm 15.18)}$	82.66 ± 1.48 (43.10 ± 9.57)	$\mathbf{85.15 \pm 1.43}$ $\mathbf{(75.96 \pm 14.93)}$
c_0	$\mathbf{86.78 \pm 2.09}$ $\mathbf{(38.64 \pm 3.57)}$	84.15 ± 2.22 (33.42 ± 5.16)	$\mathbf{86.21 \pm 2.11}$ $\mathbf{(38.17 \pm 3.47)}$	83.53 ± 2.23 (34.34 ± 5.68)	$\mathbf{85.53 \pm 2.34}$ $\mathbf{(38.87 \pm 3.78)}$	83.47 ± 2.13 (33.90 ± 6.43)
c_1	$\mathbf{80.10 \pm 2.80}$ $\mathbf{(51.40 \pm 4.60)}$	77.74 ± 2.38 (43.60 ± 5.35)	$\mathbf{79.51 \pm 2.97}$ $\mathbf{(52.14 \pm 4.78)}$	77.67 ± 2.82 (41.98 ± 4.38)	$\mathbf{79.63 \pm 3.08}$ $\mathbf{(50.07 \pm 3.05)}$	76.60 ± 2.95 (44.65 ± 5.74)
c_2	77.69 ± 3.00 (54.43 ± 4.01)	$\mathbf{77.50 \pm 2.38}$ $\mathbf{(48.40 \pm 8.14)}$	77.72 ± 2.88 (54.35 ± 4.46)	$\mathbf{77.75 \pm 2.25}$ $\mathbf{(46.09 \pm 7.40)}$	$\mathbf{77.10 \pm 2.58}$ $\mathbf{(54.81 \pm 4.85)}$	76.67 ± 2.46 (47.79 ± 7.39)
c_3	$\mathbf{82.45 \pm 2.32}$ $\mathbf{(43.27 \pm 3.42)}$	82.16 ± 2.41 (37.93 ± 5.23)	82.51 ± 2.43 (43.16 ± 2.95)	82.19 ± 2.07 (38.94 ± 5.56)	$82.14 \pm 2.12)$ (44.39 ± 4.18)	$\mathbf{82.11 \pm 2.19}$ $\mathbf{(40.51 \pm 5.88)}$
$Average$	82.94 ± 2.12 (45.57 ± 4.36)	82.67 ± 1.96 (47.29 ± 7.85)	82.75 ± 2.17 (45.55 ± 4.28)	82.60 ± 1.96 (47.09 ± 7.64)	82.49 ± 2.13 (46.25 ± 5.09)	82.24 ± 2.01 (48.56 ± 8.07)

obtains slightly better results in terms of system accuracy in comparison to the LDKL methodology. However, the LDKL extracts a representation space of $M_P = 40$ for both ITM and IATM, without affecting significantly the DBS parameter estimation results. So, our kernel-based eigendecomposition is able to code a high dimensional VTA space in a few number of relevant feature. Hence, LDKL reduces the number of required support vectors by avoiding over-fitting. Overall, LDKL performance is over 94% and 82% in ITM and IATM, respectively, where the highest results are related to the c_0 and c_3 contacts and the lowest ones to c_1 and c_2. The latter can be explained by their central position along the DBS device. So, the activation of the volume around the c_1 and c_2 positions can be affected by the activity of c_0 and c_3. In this sense, similar VTAs can provide different DBS configurations, especially, for axons positions around the center of the stimulation device.

5   Conclusions

In this study, we proposed a novel kernel-based approach to estimate the DBS parameters from VTA data. The data-driven estimation introduced comprises two main stages: *(i)* kernel-based feature extraction from VTA samples, and *(ii)* DBS parameter estimation using kernel-based multi-output regression and classification. In this sense, we carried out a KPCA algorithm from pair-wise Hamming distances between VTAs to extract relevant features from high-dimensional and activated/non-activated-valued data. Then, a MSVR and SVC are trained in the projected space to learn the DBS configuration. The problem we describe in the paper has not been studied in deep in the literature. As we mention in the introduction, there is an extensive literature that attempts to estimate the VTA from stimulation parameters. However, attempting to solve the problem of computing a set of specific neuromodulation parameters given a desired VTA has been much less studied. The proposed approach is tested under both isotropic and anisotropic conditions to validate its performance under realistic clinical environments. According to the results achieved, a significant reduction of the VTA space is obtained based on the kernel-based eigendecomposition analysis, which avoids system over-fitting and ensures stable estimations. As future work, authors plans to develop different kernel-based eigendecomposition, besides the KPCA over the Hamming distances, aiming to enhance the system performance in challenging VTA configurations. Moreover, a pre-image extension of the introduced kernel-based extraction will be carried out for VTA reconstruction tasks.

Acknowledgments. This study was developed under grant provided by the project 111065740687 funded by Colciencias. H.F. García is funded by Colciencias under the program: Formación de alto nivel para la ciencia, la tecnología y la innovación – Convocatoria 617 de 2013.

References

1. Butson, C.R., Cooper, S.E., et al.: Patient-specific analysis of the volume of tissue activated during deep brain stimulation. Neuroimage **34**(2), 661–670 (2007)
2. Da Cunha, C., Boschen, S.L., et al.: Toward sophisticated basal ganglia neuromodulation: review on basal ganglia deep brain stimulation. Neurosci. Biobehav. Rev. **58**, 186–210 (2015)
3. De La Pava, I., Gómez, V., Álvarez, M.A., Henao, Ó.A., Daza-Santacoloma, G., Orozco, Á.A.: A Gaussian process emulator for estimating the volume of tissue activated during deep brain stimulation. In: Paredes, R., Cardoso, J.S., Pardo, X.M. (eds.) IbPRIA 2015. LNCS, vol. 9117, pp. 691–699. Springer, Heidelberg (2015). doi:10.1007/978-3-319-19390-8_77
4. Gómez, V., Alvarez, M.A., et al.: Estimation of the neuromodulation parameters from the planned volume of tissue activated in deep brain stimulation. Revista Facultad de Ingeniería, Universidad de Antioquia **79**, 9–18 (2016)
5. Grimaldi, G., Manto, M.: Mechanisms and Emerging Therapies in Tremor Disorders. Springer Science & Business Media, Heidelberg (2012)

6. Hariz, M.: Deep brain stimulation: new techniques. Parkinsonism Relat. Disord. **20**, S192–S196 (2014)

7. Lujan, J.L., Chaturvedi, A., et al.: System and method to estimate region of tissue activation, 19 November 2013, U.S. Patent No. 8,589,316

8. Sánchez-Fernández, M., de Prado-Cumplido, M., et al.: SVM multiregression for nonlinear channel estimation in multiple-input multiple-output systems. IEEE Trans. Sig. Process. **52**(8), 2298–2307 (2004)

9. Scholkopf, B., Smola, A.J.: Learning with Kernels. MIT Press, Cambridge (2002)

10. Yousif, N., Purswani, N., et al.: Evaluating the impact of the deep brain stimulation induced electric field on subthalamic neurons: a computational modelling study. J. Neurosci. Methods **188**(1), 105–112 (2010)

Consensual Iris Segmentation Fusion

Dailé Osorio Roig[✉] and Eduardo Garea Llano

Advanced Technologies Application Center (CENATAV), 7a ♯ 21406 e/214 y 216,
Rpto. Siboney, Playa, 12200 La Habana, Cuba
{dosorio,egarea}@cenatav.co.cu

Abstract. Recent works have shown that fusion at segmentation level
has contributed to the robustness in iris recognition compared with the
one obtained from a single segmentation, due to different segmentation
algorithms can produce different segmentations of a same eye image. Nevertheless, the fusion process may consider mistakes as important information and include it in the resulting merged image since the analysis
in these methods is made pixel to pixel and the spatial relationships
between pixel structures are not taken into account. In this paper, the
idea of the consensus segmentation as a new method of segmentation
fusion that ensure the spatial relationships between pixel structures is
introduced. This proposal is based in the philosophy of the clustering
ensemble algorithms, treating the iris segmentations as a set of superpixels. The experimental results on a benchmark database (UBIRIS v1)
show promising results.

Keywords: Consensual segmentation · Iris recognition · Clustering
ensemble · Fusion

1 Introduction

The segmentation stage in an iris recognition system has received attention in
many researches. The main aim of these researches is to increase the iris recognition rates. The errors occurred in segmentation stage are propagated on the rest
of stages of the recognition process. Therefore, this process turns out to be one of
the very difficult, especially when the eye images are captured on less controlled
environments, for example: at a distance, on the move, under visible light, among
others. The Noisy Iris Challenge Evaluation (NICE) [19] and Multiple Biometrics Grand Challenge (MBGC)[1] have shown the importance of the robustness
of the iris segmentation in last generation of biometric systems [19]. In recent
years, some authors have focused the fusion at segmentation level (FSL) with
the aim to improve accuracy of an iris recognition system. The segmentation
fusion combines the information of several independent segmentations from a
same image obtaining a single merged segmentation. This merged segmentation
contains more information than each independent segmentation.

---

[1] Multiple Biometric Grand Challenge, http://face.nist.gov/mbgc/.

© Springer International Publishing AG 2017
C. Beltrán-Castañón et al. (Eds.): CIARP 2016, LNCS 10125, pp. 167–174, 2017.
DOI: 10.1007/978-3-319-52277-7_21

Uhl and Wild [16] introduced the concept of multi-segmentation fusion to combine the results of several segmentations. The authors evaluated two fusion algorithms using the CasiaV4-Interval database; the segmentation of the iris region was manually obtained. The authors experimentally demonstrated that the accuracy in the recognition process for different feature extraction algorithms was increased, but they did not show results of the segmentation combination built by automatic segmentation methods. In order to determine the most suitable frame from each video sequence of the MBGC iris video database, Colores-Vargas *et al.* [1] evaluated seven fusion methods to extract the texture information in normalized iris templates. The experimental results showed that the principal component analysis (PCA) method presents the better performance. Sanchez-Gonzalez *et al.* [12] used a *Sum-Rule* fusion algorithm in FSL. The authors used three automatic methods to segment the iris region. The fusion was performed after the step of normalization. The recognition rates were also increased. Garea Llano *et al.* [6] evaluated four fusion methods in FSL using two iris segmentation algorithms as basis. The authors used the same experimental scheme that was proposed by Sanchez-Gonzalez *et al.* [12] modifying the input image. This input image was captured by different sensors. Wild *et al.* [19] revealed auto corrective properties of augmented model fusion on masks before FSL in most of the tested cases. The authors proposed a scanning iris masks before the merging process. Although, the recognition rates were also increased, the authors did not take into account the quality of the mask fusion when boundaries are frequently overestimated. Furthermore, non-convexity of the mask can lead to sample points which are attributed to the wrong boundary. In general the major of these fusion methods are based in the analysis pixel to pixel and the spatial relationships between pixel structures are not taken into account. This implies that the contextual information between the pixel structures is not analyzed.

In the data clustering problems, there are not clustering algorithms that effectively work for every data sets [4]. If some of them are applied to a dataset, different clusters will be obtained. One solution may be to find the best of them, but in the last years, the idea of combining individual segmentations has gained interest showing good results, for example: Vega-Pons *et al.* [18] formalized the segmentation ensemble problem and introduced a new method to solve it, which is based on the *kernel clustering ensemble* philosophy. The proposal was experimented on Berkeley image database and compared to several state-of-the art clustering ensemble algorithms. Franek *et al.* [4] addressed their work to the parameter selection problem by applying general ensemble clustering methods in order to produce consensus segmentation. Other works [5,13] evaluated some clustering ensemble algorithms on image segmentation problems obtaining good results. A similar problem can be seen in the context of the iris segmentation. There is not a segmentation method able to correctly work on different conditions of iris image capturing. Besides several segmentation algorithms could produce very different segmentations of a same iris image. Therefore, we believe it is necessary to introduce the idea of the consensus segmentation based in ensemble clustering algorithms as a new fusion method. It will to allow a better performance compared with the independent segmentations obtained by different segmentation algorithms from a same eye image.

The main contribution of this work is to introduce the proposal of consensus segmentation based in ensemble clustering as a new fusion method for iris recognition at the segmentation level. The aim of this contribution is to replace the analysis pixel to pixel in the fusion process of a iris recognition system at the segmentation level by the analysis of the spatial relationships between pixel structures for a major impact in the merged image. The consensus segmentation is obtained by *weighted median partition* approach [18], modeling in a super-pixel set the segmentations obtained by different algorithms.

The remainder of this paper is organized as follows. In Sect. 2 we present the proposed general scheme and method for iris consensual segmentation fusion. In Sect. 3 we present and discuss the experimental results. Finally, we present the main conclusions and future work of this paper.

2   Iris Consensual Segmentation Fusion by *Segmentation Ensemble via Kernels*

Let $I(x,y)$ be an iris image and $\mathbb{S} = \{S_1, S_2, \ldots, S_n\}$ a set of segmentation masks of I obtained by different automatic iris segmentation algorithms. Then it is possible to model the consensus segmentation S^* as an ensemble clustering problem. For building S^*, two main steps are performed (see Fig. 1). In the first step, the elements of $\mathbb{S}$ are represented as binary matrices with values of $0s$ (black color) and $1s$ (white color) and a matrix G (see Fig. 1) is obtained from $\mathbb{S}$. Each column of G is generated partitioning each segmentation $S_i \in \mathbb{S}$ on clusters of pixels or super-pixels. A super-pixel [18] is a connected component in the image, formed by pixels that were grouped into the same cluster in each segmentation S_i. This representation allows to decrease the count of objects (pixels) of $\mathbb{S}$ and

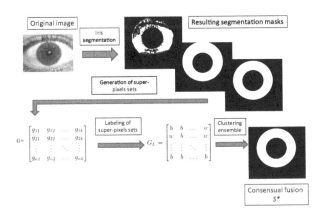

Fig. 1. Scheme for consensual iris segmentation fusion

to maintain the spatial relationship between pixel structures in the image. In the matrix G, gij is the super-pixel j of the initial segmentation S_i. All super-pixels $(g_{ij} : i = 1 \ldots n, j = 1 \ldots k)$ in a column j are compound by pixels located in the same region but in each one of the segmentations S_i; pixels inside a cluster have the same color. If a pixel set forms a super-pixel containing black pixels, then this cluster is labeled with the **b** character, otherwise it will be labeled with the **w** character. The matrix G_L shows a possible labeling of G: The partition set P_i is represented by the concatenation of the labels in the row i, i.e., P_i would be the union of $g_{11} \cup g_{12} \cup, \ldots, \cup g_{1k}$ on G represented on G_L as: $b - b - \ldots - w$. Once obtained P_i for each S_i, in the second step, the consensus segmentation S^* is built. For this process we propose the use of a clustering ensemble algorithm.

Clustering ensemble methods combine partitions of the same dataset into a final consensus clustering. Let $\mathbb{O} = \{o_1, o_2, \ldots, o_m\}$ be a set of objects, $\mathbb{P} = \{P_1, P_2, \ldots, P_n\}$ a clustering ensemble, where each $P_i \in \mathbb{P}$ is a partition of the set $\mathbb{O}$ and $P_{\mathbb{O}}$ the set of all possible partitions of $\mathbb{O}$. The consensus partition P^* is defined through the *median partition* problem [18] and obtained by equation:

$$P^* = arg \max_{P \in P_{\mathbb{O}}} \sum_{i=1}^{n} \Gamma(P, P_i), \tag{1}$$

where Γ is a similarity function defined between partitions, since this problem can be solved by minimizing the dissimilarity between partitions [18].

In the last years, several clustering ensemble methods haven been proposed for *media partition* based problem. With the aim to obtain the consensus segmentation S^*, we propose the use of the WPCK method introduced in [17], because this algorithm uses partitions generated by any kind of clustering algorithm with any initialization parameter in order to build the consensus partition and its low computational cost, lower than the most of the state-of-the-art methods [18]. In WPCK, the theoretical consensus partition is defined[2] as:

$$P^* = arg \max_{P \in P_{\mathbb{O}}} \sum_{i=1}^{n} \tilde{k}(P, P_i), \tag{2}$$

where, $\tilde{k}$ is a positive definite *kernel* function [18]. This similarity function measures the significance of all possible subsets of $\mathbb{O}$ for two partitions (P_i, P_j).

Given that $\tilde{k}$ is a *kernel* function, then problem defined by Eq. 2 can be mapped into *Reproducing Kernel Hilbert Space* associated to $\tilde{k}$, where the exact theoretical solution in this space is defined as:

$$\psi = \frac{\sum_{i=1}^{n} \phi(P_i)}{\| \sum_{i=1}^{n} \phi(P_i) \|}, \tag{3}$$

where ϕ is the function that allows mapping the partitions of $P_{\mathbb{O}}$ into the Hilbert Space [18].

---

[2] The *median partition* problem is defined using weights. In this work, we assume that all weights are equal to 1.

Based on the above, then P^* can be expressed as:

$$P^* = arg \min_{P \in P_0} \|\phi(P) - \psi\|^2, \tag{4}$$

where $\|\phi(P) - \psi\|^2$ can be rewritten in terms of similarity function as:

$$\|\phi(P) - \psi\|^2 = 2 - 2 \sum_{i=1}^{n} \tilde{k}(P, P_i) \tag{5}$$

Taking into account that we use the super-pixel representation, a segmentation $S_i \in \mathbb{S}$ can be defined as the composition of several regions composed by super-pixels. In [18], several region properties of S_i were defined, proving that the set of all segmentations $\mathbb{S}_I$ of the Image I is a subset of the all possible partitions of the super-pixel sets of the image. Then, given a set of initial segmentations $\mathbb{S} = \{S_1, S_2, \ldots, S_n\}$, these are mapped to the Hilbert Space and then, using the Eq. 4, the consensus segmentation S^* can be built of the following way:

$$S^* = arg \min_{S \in \mathbb{S}_I} 2 - 2 \sum_{i=1}^{n} \tilde{k}(S, S_i), \tag{6}$$

The Eq. 6 expresses how close is the consensus segmentation S^* to any segmentation $S \in \mathbb{S}_I$. In order to solve the previous equation, we propose the use of simulated annealing meta-heuristic selecting as first state the closest segmentation $S \in \mathbb{S}_I$ as theoretical consensus segmentation. In this work, we experimented with the simulated annealing parameters and processing defined in [18].

3   Experimental Design

In this section, we show the results oriented to explore the capacity of the proposed fusion method to increase the recognition rates. The experiments were conducted on an international iris dataset and focused on assess how the proposed consensual fusion method improves the recognition rates of the independent segmentations obtained by automatic segmentation algorithms. The iris collection used in our experiments was UBIRIS.v1 [10], a dataset comprised of 1877 images collected from 241 persons in two distinct sessions. This database incorporates images with several noise factors (contrast, reflections, luminosity, focusing, occlusion by eyelids and eyelashes), simulating less constrained image acquisition environments. In the experiments, three initial segmentations were obtained by automatic segmentation algorithms: Contrast-Adjusted Hough Transform (CHT) [8], a traditional sequential (limbic-after pupillary) method based on circular Hough Transform (HT) and contrast-enhancement; Weighted Adaptive Hough and Ellipsopolar Transform (WHT) [15], a two-stage adaptive multiscale HT segmentation technique using elliptical models and Viterbi-based Segmentation Algorithm (VIT) [14], a circular HT-based method with boundary refinement. The motivation for selecting these algorithms was their public availability as open source software for reproducibility and also because, they proved good results in the iris segmentation. From the set of initial segmentations

(CHT-WHT-VIT) the consensual segmentation fusion method S^* was obtained. To obtain S^*, two functions were used to measure the similarity between two iris segmentations: *Rand Index* (RI) [11] and *Kernel* ($\tilde{k}$) [18]. RI is a *positive definite kernel* function proposed in [18] to measure the similarity between two image segmentations. Then using the similarity function defined by Eq. 6, S^* can be computed as:

$$S^* = arg \min_{S \in \mathbb{S}_I} 2 - 2 \sum_{i=1}^{n} RI(S, S_i), \tag{7}$$

$\tilde{k}$ is the similarity function mentioned in the Sect. 2. In general, $\tilde{k}$ and RI take values in the range $[0, 1]$; values close to zero imply the minimum difference between two segmentations. In order to normalize the segmentations, Daugmans rubber sheet normalization [2] was used.

In order to assess the effectiveness of our proposal and possible influence of the type of features used for recognition, four feature extraction methods were experimented. The 2D version of Gabor filters (Daugman) [3]; The wavelet transform (Ma) [7]; The 1D Log-Gabor wavelets (Masek) [8] and the features derived from the zero crossings of the differences between 1D DCT coefficients (Monro) [9]. The results in terms of equal error rate (EER) were obtained by computing and comparing the Hamming distances in the verification task. First, the consensual segmentation fusions were obtained using $\tilde{k}$ and RI functions. Then, we perform the verification process for each obtained consensual fusion and for each initial segmentation (WHT, CHT and VIT) using the four mentioned feature extraction methods and comparing each segmentation algorithm separately and when their results are fused.

3.1   Results and Discussion

Figure 2 reports the results in terms of EER for each one of the automatic segmentation results and their consensual fusion obtained by the two similarity functions RI and $\tilde{k}$ in the UBIRIS.v1 database.

The results in the Fig. 2 show that under less controlled conditions of UBIRIS v1 database, $S^* - RI$ achieved better recognition rate on the results of the

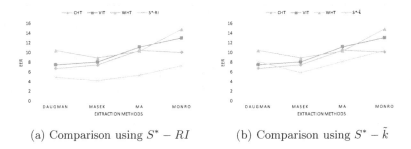

(a) Comparison using $S^* - RI$ (b) Comparison using $S^* - \tilde{k}$

Fig. 2. Recognition accuracy by EER on UBIRIS.v1 database

initial segmentations separately. $S^* - \tilde{k}$ also achieved better recognition rate for Masek and Ma extraction methods. In general $S^* - \tilde{k}$ does not exceed the results obtained by $S^* - RI$. We think that these results are due to the fact that $S^* - RI$ reached a good approximation of the global optimum. However in the case of $S^* - \tilde{k}$ when the Daugman feature extraction algorithm is used, an improvement of the results achieved by the method VIT is not obtained. This may mean that the nature of the feature extraction method combined with the conditions of the database because a negative influence on the similarity measure used. But this should be a matter to investigate in future studies. Besides we can say that despite $S^* - \tilde{k}$ did not reach a good approximation of the global optimum in most of the cases (for example in Daugman for VIT and CHT), but the meta-heuristic could find a close segmentation to theoretical consensus solution in the Hilbert space (ψ), improving at least in one of the initial segmentations (CHT, VIT, WHT). The results show that the idea of fusion by consensual segmentation is promising in the sense of the effect it has on the increased efficacy in recognition. Another advantage is that the fusion process is simplified because the analysis is done at the level of super-pixel unlike traditional fusion methods that perform pixel by pixel analysis. Another advantage is that the spatial relationships between pixel structures are preserved.

4   Conclusions

In this paper we proposed the Consensual Iris Segmentation Fusion. Our proposal allows obtaining a consensus from the initial segmentations produced by automatic segmentation algorithms. Our proposal is based on the *weighted median partition* problem and it uses the super-pixel representation of the iris image in order to overcome some drawbacks as, the volume of calculations to be performed for pixel to pixel image analysis and loss of spatial relation between pixel structures. Experimental results show that the proposed approach is promising. It was demonstrated the robustness of the approach for the degraded conditions of UBIRIS v1 database where the recognition results were improved in the most of the cases. Future work will aim to analyze the influence of similarity measures used on the results of the fusion in the sense to compare their results with the results of the ideal segmentation. Another line of research will be directed to propose new similarity measures closer to the nature of the iris. The combination of the proposed method with other proposed fusion methods can be another future line of research.

References

1. Colores-Vargas, J.M., García-Vázquez, M., Ramírez-Acosta, A., Pérez-Meana, H., Nakano-Miyatake, M.: Video images fusion to improve iris recognition accuracy in unconstrained environments. In: Carrasco-Ochoa, J.A., Martínez-Trinidad, J.F., Rodríguez, J.S., Baja, G.S. (eds.) MCPR 2013. LNCS, vol. 7914, pp. 114–125. Springer, Heidelberg (2013). doi:10.1007/978-3-642-38989-4_12

2. Daugman, J.: How iris recognition works. IEEE Trans. Circuits Syst. Video Technol. **14**(1), 21–30 (2004)
3. Daugman, J.G.: High confidence visual recognition of persons by a test of statistical independence. IEEE Trans. Pattern Anal. Mach. Intell. **15**(11), 1148–1161 (1993)
4. Franek, L., Abdala, D.D., Vega-Pons, S., Jiang, X.: Image segmentation fusion using general ensemble clustering methods. In: Kimmel, R., Klette, R., Sugimoto, A. (eds.) ACCV 2010. LNCS, vol. 6495, pp. 373–384. Springer, Heidelberg (2011). doi:10.1007/978-3-642-19282-1_30
5. Fred, A.L., Jain, A.K.: Combining multiple clusterings using evidence accumulation. IEEE Trans. Pattern Anal. Mach. Intell. **27**(6), 835–850 (2005)
6. Garea Llano, E., Colores-Vargas, J.M., Garcia-Vazquez, M.S., Zamudio Fuentes, L.M., Ramirez-Acosta, A.A.: Cross-sensor iris verification applying robust fused segmentation algorithms. In: 2015 International Conference on Biometrics (ICB), pp. 17–22. IEEE (2015)
7. Ma, L., Tan, T., Wang, Y., Zhang, D.: Personal identification based on iris texture analysis. IEEE Trans. Pattern Anal. Mach. Intell. **25**(12), 1519–1533 (2003)
8. Masek, L., et al.: Recognition of human iris patterns for biometric identification. Ph.D. thesis, Masters thesis, University of Western Australia (2003)
9. Monro, D.M., Rakshit, S., Zhang, D.: DCT-based iris recognition. IEEE Trans. Pattern Anal. Mach. Intell. **29**(4), 586–595 (2007)
10. Proença, H., Alexandre, L.A.: UBIRIS: a noisy iris image database. In: Roli, F., Vitulano, S. (eds.) ICIAP 2005. LNCS, vol. 3617, pp. 970–977. Springer, Heidelberg (2005). doi:10.1007/11553595_119
11. Rand, W.M.: Objective criteria for the evaluation of clustering methods. J. Am. Stat. Assoc. **66**(336), 846–850 (1971)
12. Sanchez-Gonzalez, Y., Chacon-Cabrera, Y., Garea Llano, E.: A comparison of fused segmentation algorithms for iris verification. In: Bayro-Corrochano, E., Hancock, E. (eds.) CIARP 2014. LNCS, vol. 8827, pp. 112–119. Springer, Heidelberg (2014). doi:10.1007/978-3-319-12568-8_14
13. Singh, V., Mukherjee, L., Peng, J., Xu, J.: Ensemble clustering using semidefinite programming with applications. Mach. Learn. **79**(1–2), 177–200 (2010)
14. Sutra, G., Garcia-Salicetti, S., Dorizzi, B.: The viterbi algorithm at different resolutions for enhanced iris segmentation. In: 2012 5th IAPR International Conference on Biometrics (ICB), pp. 310–316. IEEE (2012)
15. Uhl, A., Wild, P.: Weighted adaptive hough and ellipsopolar transforms for real-time iris segmentation. In: 2012 5th IAPR International Conference on Biometrics (ICB), pp. 283–290. IEEE (2012)
16. Uhl, A., Wild, P.: Fusion of iris segmentation results. In: Ruiz-Shulcloper, J., di Baja, G.S. (eds.) CIARP 2013. LNCS, vol. 8259, pp. 310–317. Springer, Heidelberg (2013). doi:10.1007/978-3-642-41827-3_39
17. Vega-Pons, S., Correa-Morris, J., Ruiz-Shulcloper, J.: Weighted partition consensus via kernels. Pattern Recogn. **43**(8), 2712–2724 (2010)
18. Vega-Pons, S., Jiang, X., Ruiz-Shulcloper, J.: Segmentation ensemble via kernels. In: 2011 First Asian Conference on Pattern Recognition (ACPR), pp. 686–690. IEEE (2011)
19. Wild, P., Hofbauer, H., Ferryman, J., Uhl, A.: Segmentation-level fusion for iris recognition. In: 2015 International Conference of the Biometrics Special Interest Group (BIOSIG), pp. 1–6. IEEE (2015)

Depth Estimation with Light Field and Photometric Stereo Data Using Energy Minimization

Doris Antensteiner$^{(\boxtimes)}$, Svorad Štolc, and Reinhold Huber-Mörk

Austrian Institute of Technology, Digital Safety and Security,
2444 Seibersdorf, Austria
{doris.antensteiner,svorad.stolc,reinhold.huber-moerk}@ait.ac.at

Abstract. Through the fusion of light fields and photometric stereo, two state-of-the-art computational imaging approaches, we improve the 3D reconstruction of objects. Light field imaging techniques observe a scene from different angles, which results in a strong absolute depth estimation of the scene. Photometric stereo uses multiple illuminations to reconstruct the surface of objects, which allows estimating fine local details of objects, even when no surface structure is present. We combine both approaches within a minimization algorithm, which exhibits an accurate absolute depth with a high sensitivity to fine surface details.

Keywords: Light fields · Photometric stereo · 3D vision · Energy minimization

1 Introduction

Light field cameras capture a scene from different viewpoints. We solve a correspondence problem to find the correct disparities, and thereby depth values, between those acquisitions. It became popular to analyse the correspondence in light field data using the epipolar image plane (EPI) structure which was originally introduced by Bolles et al. [3], an example is shown in Fig. 1(c). Light fields can be captured by a light field camera, e.g. a plenoptic camera [1] where usually a microlens array is placed in front of the camera sensor to acquire angular dependent reflectance information, or by a multiple camera array [2] where each camera in the array has a different viewing perspective on the scene. Depth reconstructions with light field cameras are globally accurate, but fail in reconstructing fine surface details. Photometric stereo reconstructs surface normals making use of varying illumination directions. It was shown by Woodham [7], that the surface orientation could be recovered from Lambertian surfaces, when at least 3 independent illumination directions are used. From the surface normal vectors a depth map can be reconstructed, which is locally precise, but globally inaccurate.

© Springer International Publishing AG 2017
C. Beltrán-Castañón et al. (Eds.): CIARP 2016, LNCS 10125, pp. 175–183, 2017.
DOI: 10.1007/978-3-319-52277-7_22

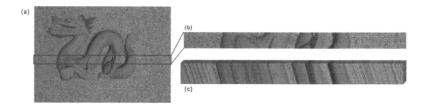

Fig. 1. (a) The Stanford dragon [6] rendered with POV-Ray, (b) Zoomed region around the red dotted line, where (c) EPI stacks with 9 different horizontal viewpoints are formed, each EPI stack is constructed with a certain illumination direction. (Color figure online)

Globally accurate methods were previously combined with approaches that show a high local accuracy, e.g. by combining depth maps with surface normals taking into account the frequency domain [10,11], Kadambi et al. [9] combined polarization normals to enhance depth maps, or by combining the depth from RGB-D cameras with shape-from-shading techniques [8].

The paper is organized as follows. A combined photometric light field setup is introduced in Sect. 2. In Sect. 3 we describe our light field and photometric stereo fusion approach based on energy minimization. Qualitative and quantitative results are provided in Sect. 4, using simulated data. In Sect. 5 we draw conclusions and discuss further work.

2   Acquisition Setup

The light field representation we are considering consists of two directional and two spatial dimensions and thereby provides 4D information about the propagation of light in space. Our light field acquisition setup, as shown in Fig. 2, consists of two components. Namely a light dome which is a half-sphere with 32 light sources and a light field camera. The lights are arranged in circular patterns at three height levels around the object and the light field camera is placed in the domes' top center. Thereby the scene is captured from slightly different viewpoints for each illumination direction. Figure 2(c) shows the top view of the dome in respect to the domes' directions ω_x and ω_y, red dots indicate the viewing perspectives from the light field camera.

The physical acquisition setup is shown in Fig. 2(a). In order to quantify the performance of our proposed algorithm we simulate the physical setup with a virtual one where objects are rendered using POV-Ray [4]. This approach allows rendering objects of arbitrary shapes and surface structures as well as generating corresponding ground truth disparity maps with floating point accuracy.

Captured light fields are analysed in the EPI domain. All light field images which are acquired with a specific illumination direction form a light field stack. A cut through this stack shows linear structures, where the slope angles correspond to a defined distance to the camera of the corresponding object point

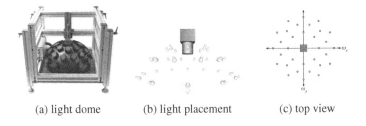

(a) light dome   (b) light placement   (c) top view

Fig. 2. Photometric light field acquisition setup. (Color figure online)

in the scene. Figure 1 shows an image stack from one image line, constructed using 9 camera viewpoints in a row. Photometric information is obtained from different illumination directions in the light dome. We use the central viewpoint of the light field to estimate the surface normals of an object.

3   Combination of Light Fields and Photometric Stereo

In this section we discuss the depth estimation using light field depth and surface normals estimated by photometric stereo. We construct an energy term using both components in order to achieve an improved depth map. Figure 3 shows the acquired light field data where the scene is represented from certain camera viewpoints and the acquired photometric stereo data where the scene is represented by a fixed viewpoint and illuminated from several angles. While the light field depth result shows a strong absolute accuracy but lacks fine details, photometric stereo shows fine surface structures with an absolute depth offset. The fusion of both results in a strong absolute depth accuracy, while we can also reconstruct fine surface details.

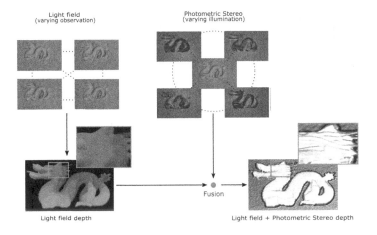

Fig. 3. Fusion of light field and photometric stereo.

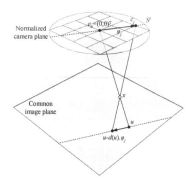

Fig. 4. Light field image parametrisation.

3.1 Light Field Depth Estimation

The light field depth is estimated using EPI stacks, as shown in Fig. 1. We use a parametrisation as illustrated in Fig. 4. The 4D light field is denoted as $L(u, \varphi)$, where $u := (x, y)^T$ denotes the spatial coordinate and φ_j denotes the directional coordinate at the camera position c_j. The maximum camera baseline is normalized to a unit circle S^1.

$$\hat{d}(u) = \underset{d}{\arg\min} \sum_{j=1}^{n} \sum_{u \in \Omega} |L(u - d(u) \cdot \varphi_j, \varphi_j)) - L(u, \varphi_0)|$$

We find the best disparity hypothesis $\hat{d}(u)$, by measuring the difference in irradiance values at a position u from the central reference view $L(u, \varphi_0)$ of the camera c_0 to the camera view $L(\cdot, \varphi_j)$ of a camera c_j at position $(u - d(u) \cdot \varphi_j)$. We want to find the disparity $d(u) \cdot \varphi_j$ for each pixel point $u \in \Omega$, where $\Omega \subset \mathbb{R}$ denotes the image plane. The point x in space maps to an image point u of the camera c_0. In a camera c_j the point u is shifted along the corresponding epipolar line (illustrated as a dotted line) by the disparity $(d(u) \cdot \varphi_j)$. The shift represents the distance of the object point to the common image plane, and thereby the depth value of our point u. Note that the orientation of the epipolar line depends on the camera position vector φ_j.

An object point which lies directly on the common image plane shows no disparities between the views, thereby the depth is equal to the distance of that plane to the camera. Our disparity estimation, and thereby the estimated shift of image points, has a high confidence when those irradiance values are similar. In the EPI stack similar values form slope lines, their angle represents the disparity value of the object point. The results for our disparity hypotheses are represented in a confidence stack, which provides us with an initial depth estimation. We use that cost stack in further processing steps.

3.2   Surface Normals from Photometric Stereo

Photometric stereo methods describe the surface properties of a scene, using several illumination directions and a static camera viewpoint. Our light sources are placed on a sphere around the object. These changes in illumination direction are resulting in pixel intensity changes, depending on the orientation and reflectance characteristics of the object. We describe the irradiance values I using a reflectance map R as in [7]:

$$I(x, y) = R(p(x, y), q(x, y)), \text{where}$$

$$p = \frac{\delta}{\delta x} z(x, y), \text{and} \, q = \frac{\delta}{\delta y} z(x, y)$$

represent two points in the gradient space. We assume a Lambertian reflectance model for our scenes and a constant albedo, therefore we can solve the irradiance as follows:

$$I = \rho \cdot L \cdot N_{PS}.$$

Given the illumination direction $L = [L_1, ..., L_n]$, our observed pixel irradiance vector $I = [I_1, ..., I_n]^T$, and $n = 32$ light sources, we estimate the vector $M = \rho * N_{PS}$, where N_{PS} represents the surface normal unit vector and ρ the albedo. By integrating normal vectors, with e.g. the algorithm of Frankot and Chelappa [5], a depth map can be estimated. It is well known [11,12] that the resulting depth values show a reliable relative depth accuracy, but have an absolute depth offset. The surface normal vectors in our energy term are used to refine the surface structure of our depth result.

3.3   Energy Minimization

The objects' surface (Fig. 5a) is reconstructed using the depth from light field (Fig. 5b) combined with the surface normal vectors N_{PS} from photometric stereo (Fig. 5c). We construct an energy term $E(Z)$ depending on the estimated depth map Z, with a weight λ_{PS}:

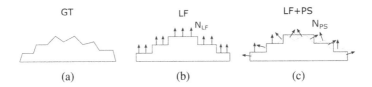

Fig. 5. Energy components: (a) ground truth surface, (b) light field depth with corresponding surface normals, (c) the light field depth map (red) is adapted using photometric stereo surface normals (blue). In 3D estimation, the depth map is close to the light field depth, while the surface orientation is close to the photometric stereo surface normals. (Color figure online)

$$E(Z) = E_{LF}(Z) + \lambda_{PS} \cdot E_{PS}(Z).$$

The solution to our minimization problem is found for both terms analytically by a primal gradient descent algorithm, with a learning rate α:

$$E_{n+1}(Z) = E_n(Z) - \alpha \cdot \left[\frac{\partial E_{LF}}{\partial z} + \lambda_{PS} \cdot \frac{\partial E_{PS}}{\partial z} \right].$$

The light field depth constraint E_{LF} favours depth solutions which have a high confidence from the light field slope analysis (see Sect. 3.1). While the surface orientation constraint E_{PS} favours solutions, where the surface orientation is similar to the surface normal vectors from photometric stereo (see Sect. 3.2).

Light Field Depth Constraint E_{LF}. The light field depth energy depends on the difference of the irradiance values in the center reference view $L(u, \varphi_0)$, compared to the camera view $L(\cdot, \varphi_j)$, as described in Sect. 3.1. The disparity is proportional to the distance of a point to the common image plane.

$$E_{LF}(d) = \sum_{j=1}^{n} \sum_{u \in \Omega} |L(u - d(u) \cdot \varphi_j, \varphi_j)) - L(u, \varphi_0)|$$

Solutions where the distribution of irradiance values along a slope in the EPI stack shows similar values, and thereby solutions with a strong confidence for a defined disparity, will be favoured in the final result. This constraint enforces a strong absolute depth accuracy.

Surface Orientation Constraint E_{PS}. For the surface orientation constraint we minimize the distance between the surface normals N_{PS} and the surface normals $N(Z) = \nabla Z$ from the currently estimated depth map Z.

$$E_{PS}(Z) = \frac{1}{2} \sum_{i=1}^{n} ||N_{PS} \cdot \sqrt{1 + \nabla Z_{ij}} - \begin{pmatrix} \nabla Z_{ij} \\ 1 \end{pmatrix} ||^2$$

Before, we optimized the relative depth of each pixel. With the surface orientation constraint we are now enforcing the reconstruction of fine surface structure details in our scene.

4   Experimental Results

We performed experiments on several rendered objects in our virtual acquisition environment. Images were taken from 9×9 virtual cameras with different viewpoints and with 32 illumination sources. Qualitative examples are shown in Fig. 6. The first column shows a central acquisition of the object. The second column shows the objects' depth map estimated with light field data. The third column shows the depth map estimated with our combined minimization approach. Quantitative results are displayed in Table 1. It is shown that the RMS disparity error improves significantly when photometric stereo normals are combined with the light field depth estimation.

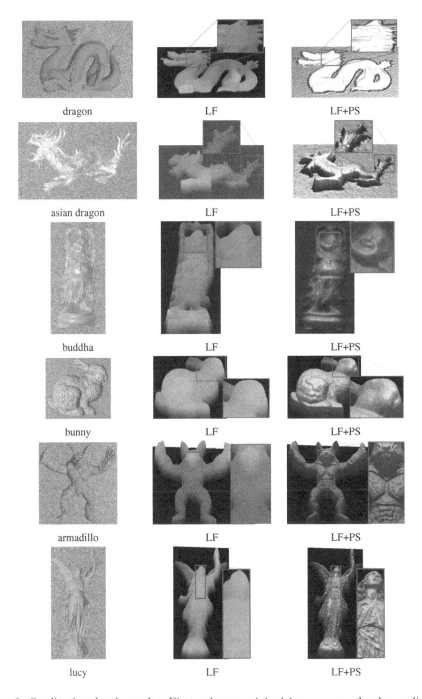

Fig. 6. Qualitative depth results. First column: original image, second column: light field depth estimation, third column: our combined depth result.

Table 1. Quantitative results for the scenes shown in Fig. 6. The table shows the RMS disparity error to the ground truth depth of the objects, both for the initial light field result and for the combination of light field with photometric stereo.

Scene	Dragon	Asian dragon	Buddha	Bunny	Armadillo	Lucy	Average
LF	1.12	1.11	0.56	0.90	0.68	0.40	0.80
LF+PS	0.51	0.50	0.33	0.54	0.59	0.33	0.47

5   Summary and Outlook

We discussed light field and photometric stereo methods for the 3D reconstruction of objects. On the one hand light field depth estimation results in a reliable absolute depth but misses fine details in the surface structure. On the other hand surface normals estimated from photometric stereo allow for the reconstruction of fine surface details but only contain relative depth information. The combination of both methods results in an improved depth reconstruction, which shows a strong absolute depth accuracy as well as fine surface details. To combine these approaches we constructed an energy term, which we minimized using a primal gradient descent approach. For our experimental results we rendered data in a virtual dome setup, which we constructed as shown in Fig. 2. We compared our combined light field and photometric stereo approach with the previously used light field depth estimation in a quantitative and qualitative way. The quantitative results (see Table 1) show a significant improvement of the disparity estimation of the scene from an average RMS error of 0.80 to an average of 0.47. Our qualitative results (see Fig. 6) show an eminent improvement in the reconstruction of fine surface details, which occur e.g. in faces, clothes, armour or skin structures. Further work will cover embedding the algorithm in an industrial setup and evaluating the performance of the physical setup applied to real world objects with different materials and surface structures.

References

1. Ng, R., Levoy, M., Brédif, M., Duval, G., Horowitz, M., Hanrahan, P.: Light field photography with a hand-held plenoptic camera. Technical report CSTR 2005-02, Stanford University (2005)
2. Wilburn, B., Joshi, N., Vaish, V., Talvala, E.-V., Antunez, E., Barth, A., Adams, A., Horowitz, M., Levoy, M.: High performance imaging using large camera arrays. ACM Trans. Graph. **24**(3), 765–776 (2005)
3. Bolles, R.C., Baker, H.H., Marimont, D.H.: Epipolarplane image analysis: an approach to determining structure from motion. Int. J. Comput. Vis. **1**(1), 7–55 (1987)
4. POV-Ray. http://www.povray.org. Accessed 3 June 2016
5. Frankot, R.T., Chellappa, R.: A method for enforcing integrability in shape from shading algorithms. IEEE Trans. Pat. Anal. Mach. Intell. **10**, 439–451 (1988)
6. The Stanford 3D Scanning Repository. http://graphics.stanford.edu/data/3Dscanrep/. Accessed 3 June 2016

7. Woodham, R.J.: Photometric method for determining surface orientation from multiple images. Opt. Eng. **19**(1), 139–144 (1980)
8. Yu, L.F., Yeung, S.K., Tai, Y.W., Lin, S.: Shading-based shape refinement of RGB-D images for consumer depth cameras. In: Proceedings of the Computer Vision and Pattern Recognition (CVPR), pp. 1415–1422 (2013)
9. Kadambi, A., Taamazyan, V., Shi, B., Raskar, R.: Polarized 3D: high-quality depth sensing with polarization cues. In: Proceedings of the International Conference on Computer Vision (ICCV) (2015)
10. Nehab, D., Rusinkiewicz, S., Davis, J., Ramamoorthi, R.: Efficiently combining positions and normals for precise 3D geometry. ACM Trans. Graph. **24**(3), 536–543 (2015)
11. Antensteiner, D., Štolc, S., Huber-Mörk, R.: Depth estimation using light fields and photometric stereo with a multi-line-scan framework. In: Proceedings of Austrian Association for Pattern Recognition (AAPR), vol. 24, no. 3 (2016)
12. Du, H., Goldman, D.B., Seitz, S.M.: Binocular photometric stereo. In: Proceedings of British Machine Vision Conference (BMVC) (2011)

Distributed and Parallel Algorithm for Computing Betweenness Centrality

Mirlayne Campuzano-Alvarez[(⊠)] and Adrian Fonseca-Bruzón

Center for Pattern Recognition and Data Mining, Santiago de Cuba, Cuba
{mirlayne,adrian}@cerpamid.co.cu

Abstract. Today, online social networks have millions of users, and continue growing up. For that reason, the graphs generated from these networks usually do not fit into a single machine's memory and the time required for its processing is very large. In particular, computing a centrality measure, like betweenness, is expensive on these graphs. For addressing this challenge, in this paper we present a parallel and distributed algorithm for computing betweenness. Also, we develop a heuristic for reducing the overall time, obtaining a speedup over 80x in the best cases.

Keywords: Online social network · Betweenness centrality · MPI

1 Introduction

Nowadays, online social networks have gained a huge popularity among people all around the world. An important task in Social Networks Analysis (SNA) is to discover the most prominent users into a social network by using betweenness centrality. This measure determines the frequency that a node or an edge acts as a bridge in the shortest paths in the network.

However, in spite of the massive importance of betweenness, its computation is expensive in large networks. For that reason, several strategies have been proposed in the literature to accelerate this process. Some approaches make a preprocessing of the graph by identifying vertexes with *strategical positions* which can be removed or merged, thus reducing the number of nodes to visit when the shortest paths are calculated [6].

However, due to the fast grow up of online social networks, nowadays previous techniques are not enough, because the time cost can reach several days and the main memory of a single machine is not enough to support the network representation.

To tackle those problems, in this paper we propose a parallel and distributed algorithm for computing betweenness using MPI. Also, we propose a modification that allows our algorithm speed up when it is used in sparse networks. The experimental results show that our proposal is suitable to process large networks and it can be adapted according to the available resources.

© Springer International Publishing AG 2017
C. Beltrán-Castañón et al. (Eds.): CIARP 2016, LNCS 10125, pp. 184–191, 2017.
DOI: 10.1007/978-3-319-52277-7_23

The rest of the paper is organized as follows. First, we expose some definitions related with graph and betweenness concepts. Then, is given a brief description of previous works, including an explanation of the sequential algorithms used. Next, we explain our proposal. Finally, we describe the experimental environment, present the results and make an analysis of them.

2   Background

Commonly, social networks are represented by a graph $G = (V, E)$ where V is a set of nodes, $n = |V|$ and E a set of edges, $m = |E|$. In this paper we use unweighted and undirected graphs. Let d_s be the distance array of the shortest paths from the source vertex s to each node of the graph. We denote $\sigma_{st}(v)$ as the number of the shortest paths from s to t that pass through v and σ_{st} the number of the shortest paths from s to t, $\forall t \in V$.

Betweenness is considered a medial centrality measure, because all walks passing through a node are considered [1]. Formally, it is defined as:

$$Cb(v) = \sum_{\substack{s \neq v, v \neq t \\ s, v, t \in V}} \frac{\sigma_{st}(v)}{\sigma_{st}} \tag{1}$$

The betweenness computation involves two main steps: determining the shortest paths and computing the betweenness values. The first step implies to determine a Directed Acyclic Graph (DAG) for a selected source node s. The second step, consists in the computation of the betweenness value for each node in the DAG previously computed. This process must be repeated taking every node of graph as source node.

According to Eq. 1, betweenness is costly to compute because requires $\mathcal{O}(n^3)$ run-time. In large networks this implies that compute exact betweenness of all nodes can take hours, even days. For that reason, have been proposed algorithms and methods to reduce the time required for its computing.

The best known sequential algorithm was proposed by Brandes in [3]. The author defined the dependency of a node v for a given source s as is shown in Eq. 2, where $P_s(w)$ is the set of predecessors of w in the DAG rooted by s:

$$\delta_{s\bullet}(v) = \sum_{w:v \in P_s(w)} \frac{\sigma_{sv}}{\sigma_{sw}} * (1 + \delta_{s\bullet}(w)) \tag{2}$$

Then, the betweenness value of a node v will be the sum of $\delta(v)$ for each $s \in V, s \neq v$. Using Eq. 2, the running time of the algorithm was reduced to $\mathcal{O}(nm)$ for unweighted graphs. But this reduction in the time cost is not enough for large social networks.

In [2] was proposed a new algorithm which speed up the Brandes's. This method take advantage of the sparsity of social networks identifying the nodes of degree 1. The authors point out that those that had none value of betweenness, however their presence is influential in the betweenness value of their only neighbor. Thus, they compute the betweenness value of the neighbors of the "hanging" nodes once. Then, those nodes can be removed and the process is repeated for residual graph until there are not more nodes with degree 1. The betweenness value of the removed nodes is computed as follows: $Cb(v) = Cb(v) + (n - p(v) - p(u) - 2) * (p(u) + 1)$, where p is an array which store in the position $p(i)$ the number of nodes removed from the sub-tree rooted i.

To keep correctness of Brandes's algorithm they modified the dependency formula as is shown below:

$$\delta_{s\bullet}(v) = \sum_{w:v \in P_s(w)} \frac{\sigma_{sv}}{\sigma_{sw}} * (1 + \delta_{s\bullet}(w) + p(w)) \tag{3}$$

Finally, the betweenness value of the remaining nodes of the graph is computed as follows:

$$Cb(w) = Cb(w) + \delta(w) * (p(s) + 1), w \neq s \tag{4}$$

In [4] was proposed a distributed algorithm to compute betweenness using successor sets instead of predecessor sets. They used the Δ-stepping algorithm during the shortest path traversal [5], allowing use their proposal for weighted and unweighed graphs as well. This is a space efficient approach incorporated at present in the Parallel Boost Graph Library. However, as the authors themselves point out on the experimental results, the algorithm is faster than the sequential version when are available 16 processors or more. Besides, they do not propose any heuristic to distributed nodes, so they need to search across processors where are store vertexes every time they send a message.

3   Our Proposal

In this paper, we propose a solution to reduce the time required for computing the shortest paths using a distributed schema. First, we distribute the nodes of the graph using a round robin strategy across the available processors. Then, we determine the DAGs rooted by the subset of nodes which belong to every processor. Next, we determine the number of the shortest paths that pass through the nodes that lie in the DAGs previously determined. Finally, we compute the dependency values and the betweenness contribution of the nodes reached in the DAGs.

In our case, we opted for distributing the network nodes across the p available processors. Each node is tagged with a unique id ($0 \leq id < n$) and they will be allocated to the processor r if $id \mod p = r$. Using this strategy, each processor stores a subset of nodes of the graph and we are able to know in which processor the nodes are allocated.

When the shortest paths are computed, it is necessary to visit all the reachable nodes. Due to the vertexes are distributed in multiple processors, it is necessary to send messages requesting the needed nodes and receive those nodes and their neighbors. This situation might imply a lot of communications and thus an increase of the time cost compared with the sequential version. To avoid that problem, we propose the use of two threads on each processor. The first one do the real computation, and also it demands and receives the needed nodes. The second thread attends requests, builds the array of neighbors and send the nodes in demand along its neighbors to the processor which need them. Thus, we overlap some computations and communications. In this way the over all time cost is reduced because the overlapping of both the computation and communications process.

As we model the network as an unweighted and undirected graph, we use the BFS algorithm to determine the shortest paths. This algorithm performs a breadth search starting at a root node s_i, by means, it visits the nodes by levels and do not pass to next level until are visited all nodes at the current level. This process must be repeated taking each node of the graph as root. As we have the graph distributed across multiple processors, we are able to discover in parallel the DAG corresponding to a source node s on each processor. However, we notice that various s_i at the same processor might need the same nodes in the same level of BFS algorithm. For that reason, we propose to expand the DAGs corresponding to several s_i at the same time. The number of DAGs simultaneously expanded is controlled by the parameter ch. This strategy reduces communications which leads thrift of time.

Nevertheless, we need to store the DAGs corresponding to each source node and this implies keep loaded in memory a lot of information, because apart from the $S_s(v)$, we need the σ and d_s arrays and a stack S, which stores the nodes in non-increased order from the root, for each source node. Due to the amount of information we need to store, is necessary to minimize the number of structures used in our algorithms. For that reason, we opted for obtaining the number of the shortest paths σ later, because this structure is not needed to discover the shortest paths and it can be deduced from the successors array.

We can deduce the nodes that will be visited at next level from the array d_s of the following way: the first thread only requests those nodes which are not visited yet and that are going to be reachable at next level. This is made by checking the arrays d_s, which are updated before a node is visited. And this process is repeated until there are no more vertexes to process.

Once there have been determined the DAGs, the dependency value of each node is calculated. Then, we compute the dependency value and finally a partial betweenness. That procedure is made in each processor in parallel.

Finally, when the contribution to betweenness of all nodes of the graph has been computed and stored in the δ arrays, we are able to compute the final score. To get the betweenness value of each node, we perform a reduction of the partial values.

3.1   Modification for Sparse Networks

As social networks often are distinguished by their sparsity, this property can be taking in advantage to speed up the betweenness computation. This was made by using the idea of the SPVB algorithm, in [2]. In our approach, we modified our previous distributed algorithm by adding a preprocessing step. First, we look up, in parallel, for all connected components, the number of vertex in each component and which of them belongs to each vertex. Then, we determine, in parallel the nodes of degree 1 for each connected component. For those nodes, we compute the betweenness value that they aport to its unique neighbor and finally those nodes are removed. With the residual graph, we repeat this step, until there are not more nodes of degree 1. Also, we removed the nodes of degree 0. When is finished the pre-processing step, we compute the shortest paths as we explained before and employed Eqs. 3 and 4 to keep correctness of betweenness values.

3.2 Analysis

We consider the worse of the cases to make our analysis, and we focus in the stage of computing the shortest paths, because is there where we propose the major changes. As we process several source nodes at the same time, this process depends of the parameter ch, and the minimum value that it can take is 1, which means that the process must be repeated by taking a single source node s each time, i.e. n/p times for each processor. Then, to select the nodes in the shortest paths, visit the n nodes of the graph at most $diam$ times, where $diam$ is the diameter of the network.

The amount of memory consumed on each processor depends of the parameter ch, because we almost use the same structures than the Brandes's algorithm, but for several source nodes at the same time. Also, as we employ threads, the processes share all the data.

4   Experiments

For validating our proposal, we conducted several experiments with different datasets corresponding to real networks. We analyze the performance of our proposal in comparison with Brandes's algorithm. Also, we make an analysis of the rate between computing time and communications time in our method in relation to the number of processor employed.

Table 1. Datasets from standford large network collection

Dataset	Nodes	Edges	Average	% Hang
soc-Slashdot0922	82168	948464	11.54	2.19
Email-Enron	36692	183831	10.02	31.09
soc-Epinions	75879	508837	6.71	50.84
Email-EuAll	265214	420045	1.58	86.34

To implement our algorithms we use the C++ language programming. To run the experiments, we employed a cluster of 13 machines, each of them with: 4 GB of RAM DDR2, processor DualCore Opteron 2.66 GHz.

We use four well known datasets from Stanford Large Network Dataset Collection[1]. These datasets were selected attending their features, especially the density of the network, as is shown in Table 1. The column Average indicates the average degree of each node. As can be observed, the datasets present different density of edges per nodes. The column % Hang indicates the percent of nodes that can be removed from the graph when the modification for sparse networks is applied.

Table 2. Speedup

| Datas | soc-Slashdot0922 | | Email-Enron | | soc-Epinions | | Email-EuAll | |
Procs	DBB	DSPVB	DBB	DSPVB	DBB	DSPVB	DBB	DSPVB
2	0.75	2.06	0.99	1.86	0.57	2.78	0.38	20.92
3	1.07	2.99	1.36	2.65	0.80	4.00	0.58	26.82
4	1.41	3.98	1.74	3.39	1.00	5.08	0.64	33.60
5	1.68	4.78	2.14	4.10	1.20	6.17	0.70	39.51
6	1.99	5.75	2.49	5.01	1.36	7.58	0.77	45.43
7	2.22	6.50	2.86	5.60	1.55	8.59	0.83	53.45
8	2.59	7.49	3.16	6.25	1.71	9.61	0.92	60.29
9	2.87	8.55	3.56	6.99	1.92	11.01	0.99	61.59
10	3.15	9.28	3.86	7.60	2.10	11.89	1.03	71.94
11	3.41	10.19	4.14	8.36	2.33	12.94	1.16	77.83
12	3.66	10.99	4.48	8.99	2.50	14.11	1.27	80.58
13	3.96	11.72	4.86	9.59	2.68	15.07	1.34	86.29

To evaluate our proposal, we set $ch = 50$ and we use the speedup as measure. Speedup is defined as the ratio between the sequential time and the parallel time. In Table 2 we show the results, we tagged as DBB (Distributed Brandes Betweenness), to our distributed algorithm based in the Brandes one; and DSPVB (Distributed the Shortest Paths Vertex Betweenness) is the modification based in the idea of the SPVB algorithm. As can be observed, the speedup of our DSPVB proposal always gets better behavior than the sequential algorithm and the speedup increases when we augment to number of processors in all cases. In other hands, we noted that when the graph, which represent the social network, is sparse is better to used the DSPVB algorithm, because it can achieve a speed up over 80x, as happened with Email-Euall dataset. It is important to highlight this result, because a special feature of social networks is their sparsity. However, a better performance depends of the network sparsity as well as the number of nodes eliminated on preprocessing step.

---

[1] https://snap.stanford.edu/data/.

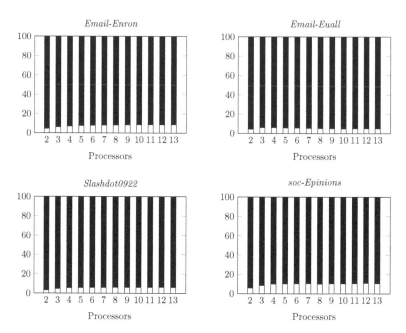

Fig. 1. Time per cent of communications (white bars) and processing (black bars)

We also study the behavior of our proposal with respect to time consuming when it is processing and when it is communicating. In each chunk of 50 sources nodes, from which we expanded the DAGs at the same time in each processor, we measured both times and we computed the average per chunk and the results are shown in the Fig. 1. To establish a comparison, we only show the result of our first proposal, because the other one has a similar behavior.

As we can be observed for all the datasets, the processing time is bigger than the communication time, because in all cases the average time that processors wait for send o receive messages never is greatest than 15 %. That proves that

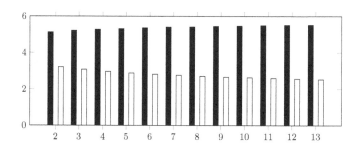

Fig. 2. Time log_{10} of Boost (black bars) and DBB (white bars).

our strategies for reducing the number of message actually work, also proves that we reach to overlap communications and processing by using two threads.

We conducted an experiment to compare our DBB algorithm with the one proposed in [4]. In the Fig. 2, we only show the results of the Email-Enron dataset set, because is the smallest one and there is high difference of time. We can notice that our algorithm is faster than the one implemented in the Boost Library.

5 Conclusions and Future Work

In this paper we presented a distributed and parallel algorithm to compute betweenness centrality. Also, we proposed a version which take advantage of the sparsity of social networks to remove iteratively vertexes of degree 1 and thus reduce the amount of computation. In the experimental results we showed that our proposal is more efficient than the sequential Brandes algorithm. In addition, we overlap communication and processing, thus reducing the delay in the messages interchange due to the use of two threads by processor. Moreover, it necessary to highlight that in our experimental results we observe that our DSPVB algorithm had better performance than the sequential one when we use 2 processors or more for all datasets. Also, in both algorithms, DBB and DSPVB, the speedup increases when we augment the number of processors, which means that our algorithms scale well.

Besides, we proposed to work in a strategy to reduce the memory consumption by not storing the successor arrays for each source node. Moreover, we believe that is possible to reduce the processing time employing a global array for each processor to indicate, during the computing the shortest paths step, the nodes that are going to be visited at next level, instead of traverse the distance arrays.

References

1. Aggarwal, C.C. (ed.): Social Network Data Analytics. Springer, Heidelberg (2011)
2. Baglioni, M., Geraci, F., Pellegrini, M., Lastres, E.: Fast exact computation of betweenness centrality in social networks. In: Proceedings of the 2012 International Conference on Advances in Social Networks Analysis and Mining (ASONAM 2012), pp. 450–456. IEEE Computer Society (2012). http://dx.doi.org/10.1109/ASONAM.2012.79
3. Brandes, U.: A faster algorithm for betweenness centrality. J. Math. Sociol. **25**, 163–177 (2001)
4. Edmonds, N., Hoefler, T., Lumsdaine, A.: A space-efficient parallel algorithm for computing betweenness centrality in distributed memory. In: HiPC, pp. 1–10. IEEE Computer Society (2010)
5. Meyer, U., Sanders, P.: Δ-stepping: a parallel single source shortest path algorithm. In: Bilardi, G., Italiano, G.F., Pietracaprina, A., Pucci, G. (eds.) ESA 1998. LNCS, vol. 1461, pp. 393–404. Springer, Heidelberg (1998). doi:10.1007/3-540-68530-8_33
6. Sariyüce, A.E., Saule, E., Kaya, K., Çatalyürek, Ü.V.: Shattering and compressing networks for betweenness centrality. In: SIAM Data Mining Conference (SDM). SIAM (2013)

A New Parallel Training Algorithm
for Optimum-Path Forest-Based Learning

Aldo Culquicondor[1]([✉]), César Castelo-Fernández[1], and João Paulo Papa[2]

[1] Escuela de Ciencia de la Computacion,
Universidad Catolica San Pablo, Arequipa, Peru
{aldo.culquicondor,ccastelo}@ucsp.edu.pe
[2] Computer Science Department, Sao Paulo State University - UNESP, Bauru, Brazil
papa@fc.unesp.br

Abstract. In this work, we present a new parallel-driven approach to speed up Optimum-Path Forest (OPF) training phase. In addition, we show how to make OPF up to five times faster for training using a simple parallel-friendly data structure, which can achieve the same accuracy results to the ones obtained by traditional OPF. To the best of our knowledge, we have not observed any work that attempted at parallelizing OPF to date, which turns out to be the main contribution of this paper. The experiments are carried out in four public datasets, showing the proposed approach maintains the trade-off between efficiency and effectiveness.

Keywords: Optimum-path forest · Parallel algorithms · Graph algorithms

1 Introduction

Pattern recognition is becoming even more important mainly due to the increasing needs from different applications to extract meaningful information from their data. Additionally, the problem gets worse since data is growing fast in both size and complexity. Humans have an innate ability to recognize patterns, but this is rather difficult to replicate on computers. Several techniques have been developed to address this issue, being the most popular ones Artificial Neural Networks (ANNs) [6] and Support Vector Machines (SVMs) [3]. Recently, a new framework to the design of graph-based classifiers named Optimum Path Forest (OPF) has been introduced in the scientific community. Such framework comprises supervised [10–12], semi-supervised [1,2] and unsupervised learning algorithms [14]. As the main advantages, we shall observe some OPF variants are parameterless, and they do not make assumptions about separability of samples [7].

We refer to OPF as a single classifier in this paper, but it is in fact a framework to the design of graph-based classifiers. This means the user can design his/her own optimum-path forest-driven classifier by configuring three main

C. Beltrán-Castañón et al. (Eds.): CIARP 2016, LNCS 10125, pp. 192–199, 2017.
DOI: 10.1007/978-3-319-52277-7_24

modules: (i) *adjacency relation*, (ii) *methodology* to estimate prototypes, and (iii) *path-cost function*. Since OPF models the problem of pattern recognition as a graph partition task, it requires an adjacency relation to connect nodes (i.e. feature vectors extracted from dataset samples). Further, OPF rules a competition process among prototype samples, which are the most representative samples from each class. Therefore, a careful procedure to estimate them would be wise. Finally, in order to conquer samples, prototypes must offer them rewards, which are encoded by the path-cost function. In this paper, we are considering the OPF classifier proposed by Papa et al. [11,12], which employs a full connectedness graph, and a path-cost function that computes the maximum arc-weight along a path. For the sake of clarity, we shall refer to this version as OPF only.

Although OPF has obtained recognition results comparable or even more accurate than SVMs and ANNs in a number of different applications, as well as it has been usually much faster for training, it can be time-consuming for very large datasets. Although OPF is parameterless, its training phase takes $\theta(n^2)$, where n stands for the number of training samples. Truly speaking, this is not that bad, since SVMs usually require a considerably higher computational load. However, there is still room for improvements, and that is the main contribution of this paper: to introduce a different data structure that allows the OPF parallelization. As a matter of fact, the proposed approach is able to produce equivalent results to the ones obtained by original OPF classifier concerning accuracy, though up to five times faster using a simple personal computer hardware.

The remainder of this paper is organized as follows. Section 2 reviews OPF theoretical background, and Sect. 3 presents the modifications that led to the new parallel training algorithm. Section 4 discusses the experiments, and Sect. 5 states conclusions and future works.

2 Supervised Classification Based on Optimum-Path Forest

Let $\mathcal{Z}$ be a dataset whose correct labels are given by a function $\lambda(x)$, for each sample $x \in \mathcal{Z}$. Thus, $\mathcal{Z}$ can be partitioned into a training ($\mathcal{Z}_1$), validation ($\mathcal{Z}_2$) and testing ($\mathcal{Z}_3$) set. Also, we can derive a graph $\mathcal{G}_1 = (\mathcal{V}_1, \mathcal{A}_1)$ from the training set, where $\mathcal{A}_1$ stands for an adjacency relation known as *complete graph*, i.e. one has a full connectedness graph where each pair of samples in $\mathcal{Z}_1$ is connected by an edge. Additionally, each node $\mathbf{v}_i^1 \in \mathcal{V}_1$ concerns the feature vector extracted from sample $x_i^1 \in \mathcal{Z}_1$. All arcs are weighted by the distance among their corresponding graph nodes. A similar definition can also be applied to the validation and test sets.

The OPF proposed by Papa et al. [12] comprises two distinct phases: (i) *training* and (ii) *testing*. The former step is based upon $\mathcal{Z}_1$, meanwhile the test phase aims at assessing the effectiveness of the classifier learned during the previous phase over the testing set $\mathcal{Z}_3$. Additionally, a *learning* algorithm was proposed to improve the quality of samples in $\mathcal{Z}_1$ by means of an additional set $\mathcal{Z}_2$. Roughly speaking, the idea is to train an OPF classifier over $\mathcal{Z}_1$ and then classify $\mathcal{Z}_2$.

Further, we replace non-prototype samples in $\mathcal{Z}_1$ by misclassified samples in $\mathcal{Z}_2$, and the very same process is executed once again (i.e. training over $\mathcal{Z}_1$ and classification over $\mathcal{Z}_2$). The above procedure is executed until the accuracy between consecutive iteration does not change.

2.1 Training

The training step aims at building the optimum-path forest upon the graph $\mathcal{G}_1$ derived from $\mathcal{Z}_1$. Essentially, the forest is the result of a competition process among prototype samples that ended up partitioning $\mathcal{G}_1$. Let $\mathcal{S} \subseteq \mathcal{Z}_1$ be a set of prototypes, which can be chosen at random or using some other specific heuristic. Papa et al. [12] proposed to find the set of prototypes that minimizes the classification error over $\mathcal{Z}_1$, say that $\mathcal{S}^* \subseteq \mathcal{Z}_1$. Such set can be found by computing a Minimum Spanning Tree $\mathcal{M}$ from $\mathcal{G}_1$, and then marking as prototypes each pair of samples (x_1, x_2), adjacent in $\mathcal{M}$, such that $\lambda(x_1) \neq \lambda(x_2)$.

Further, the competition process takes place in $\mathcal{Z}_1$, where nodes in $\mathcal{S}^*$ try to conquer the remaining samples in $\mathcal{Z}_1 \setminus \mathcal{S}^*$. Basically, such process is based on a reward-compensation procedure, where the prototype that offers the *minimum cost* is the one that will conquer the sample. The reward is computed based on a path-cost function, which should be *smooth* according to Falcão et al. [5]. Therefore, Papa et al. [12] proposed to use f_{max} as the path-cost function, defined as follows:

$$f_{max}(\langle s \rangle) = \begin{cases} 0 & \text{if } \mathbf{s} \in \mathcal{S}^* \\ +\infty & \text{otherwise,} \end{cases}$$
$$f_{max}(\pi_s \cdot (\mathbf{s}, \mathbf{t})) = \max\{f_{max}(\pi_s), d(\mathbf{s}, \mathbf{t})\}, \qquad (1)$$

where $\pi_s \cdot (\mathbf{s}, \mathbf{t})$ stands for the concatenation between path π_s and arc $(\mathbf{s}, \mathbf{t}) \in \mathcal{A}_1$. Also, a path π_s is a sequence of adjacent and distinct nodes in $\mathcal{G}_1$ with terminus at node $\mathbf{s} \in \mathcal{Z}_1$.

In short, by computing Eq. 1 for every sample $\mathbf{s} \in \mathcal{Z}_1$, we obtain a collection of optimum-path trees (OPTs) rooted at $\mathcal{S}^*$, which then originate an optimum-path forest. A sample that belongs to a given OPT means it is more strongly connected to it than to any other in $\mathcal{G}_1$.

2.2 Testing

In the testing step, each sample $\mathbf{t} \in \mathcal{Z}_3$ is classified individually as follows: $\mathbf{t}$ is connected to all training nodes from the optimum-path forest learned in the training phase, and it is evaluated the node $\mathbf{s}^* \in \mathcal{Z}_1$ that conquers $\mathbf{t}$, i.e. the one that satisfies the following equation:

$$C(\mathbf{t}) = \arg\min_{\mathbf{s} \in \mathcal{Z}_1} \max\{C(\mathbf{s}), d(\mathbf{s}, \mathbf{t})\}. \qquad (2)$$

The classification step simply assigns the label of $\mathbf{s}^*$ as the label of $\mathbf{t}$. Notice that a similar procedure to classify $\mathcal{Z}_2$ can be employed, too.

3   Parallel-Driven Optimum-Path Forest Training

In this section, we present the proposed approach based on parallel programming to speed up the naïve OPF training algorithm, hereinafter called POPF. Since standard OPF makes use of a *priority queue* implemented as a binary heap, it does not support multiple access at the same time. Therefore, POPF uses a simpler data structure along with a slightly different (parallel) training process, which is based on three main assumptions, as discussed below.

The first observation concerns the optimum-path computation process for each $\mathbf{t} \in \mathcal{Z}_1$, which is independent to other samples. On the other hand, costs need to be updated on a data structure so that a new sample can be selected for the next iteration, in order to expand the optimum path computed already. For this purpose, LibOPF [13] uses a binary heap as suggested in [5]. However, such data structure is not prepared for concurrent updates, i.e. if one attempts to perform the computation of $f_{max}(\mathbf{t})$ for each $\mathbf{t} \in \mathcal{Z}_1$, a *mutex* would be required for each update process in the heap. However, this approach would not scale well if one increases the number of threads. Furthermore, this data structure introduces a $\mathcal{O}(\log(n))$ overhead in each update, where $n = |\mathcal{Z}_1|$.

The second observation concerns the graph, which is fully connected, implying that, at each iteration, all nodes need to be explored. Therefore, the computation of f_{max} for all $s \in \mathcal{Z}_1$ takes $\theta(n)^2$ operations in total. In order to overcome such quadratic complexity, we can implement the priority queue as a standard array, but exploring the set of nodes in parallel and performing a parallel linear-search. At each iteration, each thread $\delta_i, \forall i = 1, \ldots, m$, explores a subset $\mathcal{W}_{(s,i)}$, such that $\mathcal{W}_s = \mathcal{W}_{(s,1)} \cup \cdots \cup \mathcal{W}_{(s,m)}$ is the set of neighbors of s, thus performing two tasks[1]: (1) to update the costs of each $t \in \mathcal{W}_{(s,i)}$ according to f_{max} using arc $(\mathbf{s}, \mathbf{t})$, and (2) to compute the node $s^{(*,i)} \in \mathcal{W}_{(s,i)}$ with minimum cost for $\mathcal{W}_{(s,i)}$. Afterwards, the main thread finds the node s^* with minimum cost among all $s^{(*,i)}, \forall i = 1, \cdots, m$. Such node s^* will be the first one to come out of the priority queue in the next iteration. Therefore, by using m threads, the overall time complexity of the training algorithm is reduced to $\theta(n^2/m)$.

Finally, the third observation is related to the Prim's algorithm, which is used to calculate the Minimum Spanning Tree over $\mathcal{Z}_1$. As a matter of fact, we can use the very same OPF algorithm with a different path-cost function to compute the MST. Therefore, the aforementioned ideas can be applied to compute the MST too, taking advantage of parallelism in all the steps of the training process.

Algorithm 1 summarizes the ideas presented in this section. Note that even though parallelization takes place during the searching process for the best predecessor only, it is be better to start all threads only once at the beginning of the algorithm. The proposed approach was efficiently implemented using OpenMP [4], a well-known API for shared-memory parallel programming. OpenMP pragmas used in the implementation are included as comments.

---

[1] Notice $\mathcal{W}_{(s,i)}$ stands for the set of neighbours of node s in charge of thread δ_i.

Algorithm 1. POPF Training Algorithm

Input: A λ-labeled training set $\mathcal{Z}_1$, and number m of threads.
Output: Optimum path forest P_1, cost map C_1, label map L_1 and cost-ordered
       set Z_1'

1   $Z_1' \leftarrow \emptyset$
2   $\mathcal{S}^* \rightarrow SelectPrototypes(\mathcal{Z}_1)$
3   **foreach** $\mathbf{s} \in \mathcal{Z}_1 \setminus S$ **do** $C_1(\mathbf{s}) \leftarrow +\infty$
4   **foreach** $\mathbf{s} \in \mathcal{S}^*$ **do** $C_1(\mathbf{s}) \leftarrow 0$, $P_1(\mathbf{s}) \leftarrow nil$, $L_1(\mathbf{s}) \leftarrow \lambda(\mathbf{s})$
5   $\mathbf{s} \leftarrow$ any element from $\mathcal{S}^*$
6   **in parallel for threads** $i = 1, 2, \ldots, m$ # omp pragma: parallel
7   **while** $s \neq nil$ **do**
8 | **in main thread** Insert s in Z_1' # omp pragma: master
9 | $s_i \leftarrow nil$
10 | **synchronization barrier**   # omp pragma: barrier
11 | **split among threads,**   # omp pragma: for
12 | **foreach** $t \in Z_1$ *where* $t \neq s$ **do**
13 | | **if** $C_1(t) > C_1(s)$ **then**
14 | | | $cst \leftarrow \max\{C_1(s), d(s,t)\}$
15 | | | **if** $cst < C_1(t)$ **then** $P_1(t) \leftarrow s$, $L_1(t) \leftarrow L_1(s)$, $C_1(t) \leftarrow cst$
16 | | **if** $s_i = nil$ **or** $C_1(t) < C_1(s_i)$ **then** $s_i \leftarrow t$
17 | **collect** $\arg\min_{s_i} C(s_i)$ in s # omp pragma: critical
18 | **synchronization barrier**
19   **return** classifier $[P_1, C_1, L_1, Z_1']$

4   Experiments and Results

In this section, we present the methodology used to assess the robustness of the proposed approach, as well as the experimental results. Table 1 presents the description of the datasets used in this work, which were taken from UCI Machine Learning Repository [8]. We intentionally chose datasets with numeric features, to avoid extra pre-processing, and with different orders of magnitude, to better describe the scalability of our approach.

Table 1. Description of the datasets and percentages used for $\mathcal{Z}_1$, $\mathcal{Z}_2$ and $\mathcal{Z}_3$.

Dataset	Short description	Instances	Features	Classes	$|\mathcal{Z}_1|$	$|\mathcal{Z}_2|$	$|\mathcal{Z}_3|$
Letter recog.	Character image features	20,000	16	26	20%	40%	40%
Shuttle	Distinguish shuttles	43,500	9	7	20%	40%	40%
MiniBooNE	Distinguish electron neutrinos	130,064	50	2	20%	40%	40%
SUSY	Distinguish supersymmetric particles	5,000,000	18	2	2%	2%	98%

We compared POPF against naïve OPF using a microcomputer equipped with a 3.1 GHz Intel Core i7 processor, 8 GB of RAM, and running Linux 3.16. The programs were compiled using GCC 5.0, which implements OpenMP 4 specification. Also, we varied the number of threads concerning POPF according to the maximum concurrency allowed by the processor. For each experiment, we executed a hold-out-based partition of the dataset over 10 executions for mean accuracy and computational load computation purposes.

Table 2 presents the results regarding execution time, number of learning iterations and classification accuracy for $\mathcal{Z}_3$ – as defined by Papa et al. in [12], where POPF-m stands for POPF executed with m threads. Clearly, POPF maintained OPF accuracy for all number of threads, meaning the classifier obtained through the proposed approach preserves the same properties of the original one. Only a slight variation concerning MiniBooNE dataset can be observed. This is explained by the fact that same-cost samples could be stored in a different order in $\mathcal{Z}_1'$, which will change the evaluation order during the classification process over $\mathcal{Z}_3$ and will assign a different class when ties occur.

Table 2. Comparison against OPF and POPF with different number of threads.

Dataset	Technique	Learning time (sec)	# iter.	Accuracy	S	E
Let. rec.	OPF	15.52 ± 0.14	7	95.64%	1	1
	POPF-2	9.21 ± 0.08	8	95.64	1.69	0.84%
	POPF-4	5.07 ± 0.01	8	95.64	3.06	0.77%
	POPF-8	3.42 ± 0.02	8	95.64	4.54	0.57%
Statlog	OPF	31.45 ± 0.32	4	95.69%	1	1
	POPF-2	16.57 ± 0.15	4	95.69%	1.90	0.95
	POPF-4	9.03 ± 0.09	4	95.69%	3.48	0.87
	POPF-8	5.98 ± 0.08	4	95.69%	5.26	0.66
Miniboone	OPF	$1,442.96 \pm 6.66$	8	79.37%	1	1
	POPF-2	755.20 ± 19.04	8	79.50%	1.91	0.96
	POPF-4	438.81 ± 9.16	8	79.50%	3.29	0.82
	POPF-8	273.95 ± 7.68	8	79.50%	5.27	0.66
SUSY	OPF	$13,149.8 \pm 50.0$	9	66.55%	1	1
	POPF-2	$6,398.9 \pm 32.5$	9	66.55%	2.06	1.02
	POPF-4	$3,784.3 \pm 21.2$	9	66.55%	3.47	0.87
	POPF-8	$2,592.8 \pm 115.7$	9	66.55%	5.07	0.63

In Table 2 we also include parallel performance measures: speedup (S) – measuring gain in running time – and efficiency (E) – measuring thread utilization. They are defined as follows [9]:

$$S = \frac{T_s}{T_p} \quad \text{and} \quad E = \frac{S}{m} = \frac{T_s}{m \cdot T_p}, \tag{3}$$

where T_s and T_p stand for the execution time of traditional and parallel OPF, respectively, and m denotes the number of threads.

We can observe that maximum speedup is obtained using 8 threads, being about five times faster than traditional OPF. Furthermore, speedup improves when the size of the dataset increases. Another worth noticing observation is that for the largest dataset, when using 2 threads, the efficiency obtained was greater than 100%. This confirms that POPF is considerably more efficient than traditional OPF, not only because of the parallel implementation, but also due to its asymptotic improvement. Figure 1 presents charts for S and E.

Regarding the overall parallel efficiency, it is important to stress the efficiency results obtained for both 4 and 8 threads. On one hand, we obtained an efficiency between 77% and 87% considering 4 threads, which is an outstanding result for any parallel implementation. On the other hand, the efficiency considering 8 threads was between 57% and 66%, which is a good thread utilization considering the fact that the processor used has only 4 physical cores (implementing 8 threads with HyperThreading® technology).

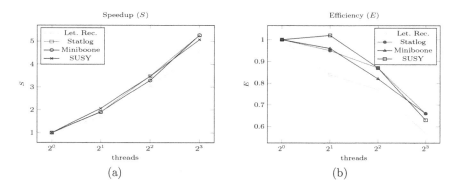

Fig. 1. Parallel performance measures for POPF learning process: (a) speedup (S); and (b) efficiency (E).

5   Conclusions and Future Work

In this work, we were able to parallelize the OPF training algorithm, and we demonstrated its efficiency concerning classification tasks. This new approach is based on three important observations: (i) the optimum-path computation process for each training sample is independent to each other; (ii) the full connectedness training graph allows us to replace the binary heap by a simple list (suitable for parallelization); and (iii) the computation of the MST during the training phase can also be performed in parallel. These changes allow to reduce the asymptotic complexity of the implementation and also turns the parallelization feasible.

We have observed that POPF preserves the accuracy of the original algorithm, but it is able to perform the learning phase at least five times faster using commodity hardware. Thus, an OPF with hundreds of thousands of nodes can be calculated in less than an hour. As such, POPF allows to perform classification of very large datasets when timing restrictions are present, and it brings closer the possibility of performing nearly real-time classification for reasonable sized-datasets even on a single computer or mobile device.

However, such real-time implementation still needs improvements in the classification algorithm. Thus, we are considering the use of spatial data structures to index the optimum path-forest obtained during training, so that a fewer amount of nodes are considered in classification, thus improving its running time.

Acknowledgments. The authors would like to thank Capes PROCAD grant 2966/2014, CNPq grants #306166/2014-3 and #470571/2013-6, FAPESP grant #2014/16250-9 and Universidad Católica San Pablo (UCSP) for their support.

References

1. Amorim, W.P., Falcão, A.X., Carvalho, M.H.: Semi-supervised pattern classification using optimum-path forest. In: 27th SIBGRAPI Conference on Graphics, Patterns and Images, pp. 111–118 (2014)
2. Amorim, W.P., Falcão, A.X., Papa, J.P., Carvalho, M.H.: Improving semi-supervised learning through optimum connectivity. Pattern Recogn. **60**, 72–85 (2016). http://www.sciencedirect.com/science/article/pii/S0031320316300668
3. Cortes, C., Vapnik, V.: Support-vector networks. Mach. Learn. **20**(3), 273–297 (1995)
4. Dagum, L., Enon, R.: OpenMP: an industry standard API for shared-memory programming. IEEE Comput. Sci. Eng. **5**(1), 46–55 (1998)
5. Falcão, A.X., Stolfi, J., de Alencar Lotufo, R.: The image foresting transform: theory, algorithms, and applications. IEEE Trans. Pattern Anal. Mach. Intell. **26**(1), 19–29 (2004)
6. Haykin, S., Network, N.: A comprehensive foundation. Neural Netw. **2**, 2004 (2004)
7. Haynes, S.D., Stone, J., Cheung, P.Y.K., Luk, W.: Video image processing with the sonic architecture. Computer **33**(4), 50–57 (2000)
8. Lichman, M.: UCI machine learning repository (2013). http://archive.ics.uci.edu/ml
9. Pacheco, P.: An Introduction to Parallel Programming. Elsevier, Burlington (2011)
10. Papa, J.P., Falcão, A.X.: A new variant of the optimum-path forest classifier. In: Bebis, G., et al. (eds.) ISVC 2008. LNCS, vol. 5358, pp. 935–944. Springer, Heidelberg (2008). doi:10.1007/978-3-540-89639-5_89
11. Papa, J.P., Falcão, A.X., De Albuquerque, V.H.C., Tavares, J.M.R.: Efficient supervised optimum-path forest classification for large datasets. Pattern Recogn. **45**(1), 512–520 (2012)
12. Papa, J.P., Falcao, A.X., Suzuki, C.T.: Supervised pattern classification based on optimum-path forest. Int. J. Imaging Syst. Technol. **19**(2), 120–131 (2009)
13. Papa, J., Falcão, A., Suzuki, C.: LibOPF: a library for the design of optimum-path forest classifiers (2014). Software version 2.1 http://www.ic.unicamp.br/~afalcao/LibOPF
14. Rocha, L.M., Cappabianco, F.A.M., Falcão, A.X.: Data clustering as an optimum-path forest problem with applications in image analysis. Int. J. Imaging Syst. Technol. **19**(2), 50–68 (2009)

Face Composite Sketch Recognition by BoVW-Based Discriminative Representations

Yenisel Plasencia-Calaña[✉], Heydi Méndez-Vázquez,
and Rainer Larin Fonseca

Advanced Technologies Application Center, 7ma A # 21406, Playa, Havana, Cuba
{yplasencia,hmendez}@cenatav.co.cu, rlarinf@gmail.com

Abstract. Face sketches are one of the main sources used for criminal investigation. In this paper, we propose a new approach for the recognition of facial composite sketches. We propose the use of discriminative representations as a way to bridge the modality gap between sketches and mug-shot photos. The intermediate representation is based on the bag-of-visual-words (BoVW) approach using dense SIFT features on multiple scales. Next, a discriminative representation is computed on top of the intermediate representation. Experimental results show that the discriminative representations outperform state-of-the-art approaches for this task in composite sketch datasets for both a close-set scenario as well as an open-set recognition scenario.

Keywords: Composite sketch recognition · Discriminative representations · Bag-of-visual-words · Metric learning · Modality gap

1 Introduction

Human faces are inherently linked to people identity. Due to this, faces are extensively used to recognize individuals. In forensic scenarios, there are many cases where law enforcement agencies have no photos of a suspect and only a facial sketch made with the help of an eyewitness or victim is available. Recently, automatic facial sketch recognition methods have attracted great attention due to its promising application in subject identification from police mug-shot databases. Facial sketches constitute not a proof but only an approximation to the identity of the subject. However, the use of this approximation in sketch-photo recognition allows reducing the list of potential candidates or suspects.

The manual creation of facial sketches is a challenging task and depends on both, eyewitness and specialists. Independently of this fact, humans can easily recognize facial sketches, even if there are great differences between photos and sketches. Nonetheless, facial sketches are not easily recognized by standard automatic face recognition methods due to significant differences with facial photos. This problem has been defined in the literature as modality gap [15]. The preferred approach to face the modality gap has been to transform facial photos into facial sketches in order to perform the matching operation for the same

© Springer International Publishing AG 2017
C. Beltrán-Castañón et al. (Eds.): CIARP 2016, LNCS 10125, pp. 200–207, 2017.
DOI: 10.1007/978-3-319-52277-7_25

image modality. However, this may loose important discriminative information. We think that this transformation can be avoided if a suitable feature representation that can cope directly with this problem is obtained.

It was shown in [7] that software-generated composite sketches are more effective than hand-drawn sketches for automatic photo-sketch recognition. However there are only few works on this topic. To the best of our knowledge, the first work on using software-generated composites for face recognition was presented by Yuen and Man [16]. It used a combination of local and global features, and it required human intervention for the recognition phase. Other relevant works are the ones by Klare and Jain. Their first proposal [4] was inspired by the component-based manner in which the composite sketches are created. Block-based multi-scale local binary patterns (MLBP) are extracted from facial regions corresponding to 76 facial landmarks, and similarities between the same components for photos and sketches are obtained and combined by score fusion. The second one [6] is a holistic method that extracts SIFT and MLBP features from uniform regions across the face, and learned optimal subspaces for each patch using linear discriminant analysis (LDA). The projected features are concatenated into single vectors, which are compared by using L^2 distance measure. Their last work [8] combines both strategies introducing some modifications and parameters tuning that boosts the performance. More recently, the works by Mittal et al. [11–13] gradually improve the matching accuracy when recognizing composite sketches. Their most recent approach is based on deep learning and transfer learning [13]. First, a deep network is trained using 30 000 face photographs, next the network is updated with information from composite sketch-photo pairs. This is a very interesting approach since it compensates the small amount of sketches available for training.

Despite the promising results obtained by deep networks for this task, they are particularly difficult to train for our problem if there is no outside data available, and only a small amount of photo-sketches pairs is available. However, another promising direction is discriminative or metric learning methods, which have shown a significantly good performance in many problems, and we believe they can be suitable to bridge the modality gap caused by differences between mug-shots and sketches. Therefore, in this paper we focus on learning distances or discriminative representations on top of an intermediate representation based on quantized features, which does not require such a large amount of training data as the deep networks. As intermediate representation we proposed to use densely sampled SIFT features or dense SIFT, quantized by a visual dictionary. This dictionary-based representation compensates geometry differences caused by the component-based manner in which composites are created since it is moderately robust to image distortions. In addition, it was shown in previous works that SIFT features were able to achieve some robustness in front of the modality gap for face-sketch recognition [5].

The remaining part of the paper is organized as follows: Sect. 2 presents the main components of the proposed approach, the BoVW model and the discriminative representations. Experimental results are reported and discussed in Sect. 3. Concluding remarks are presented in Sect. 4.

2   Proposal: BoVW-Based Discriminative Representation

BoVW Representation. We consider the dense SIFT (DSIFT) descriptors as intermediate features for our Bag-of-Visual-Words (BoVW) representation [10]. The image representation by dense SIFT avoids the interest points detection step of the standard SIFT which is computationally expensive, and provides sparse and potentially unreliable points. Instead, DSIFT samples the image using a dense grid with a user-defined size of spatial bins, where the bins are located at a fixed scale and orientation, with a user-defined sampling density step. Note that the representation by DSIFT returns a bag of descriptors. A visual dictionary is created by kmeans clustering and the DSIFT features are vector quantized into K visual words [10]. Next, a histogram of assignment frequencies for each center is computed. This process is repeated for different Gaussian smoothing from coarse to fine (e.g. using 4 different standard deviations) and sizes of spatial bins. Finally, histograms are concatenated in a high dimensional representation. The dense SIFT applied at several resolutions is called Pyramid Histogram of Visual Words (PHOW). This representation is further reduced by principal component analysis (PCA) to a small number of dimensions (e.g. 20) in order to avoid overfitting and to ensure that finding the discriminative projections is practical. After BoVW features are obtained, we create discriminative representations by means of metric learning or other similar methods that learn discriminative projections or similarities. The schematic view of our proposal is shown in Fig. 1.

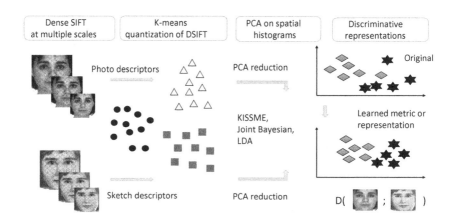

Fig. 1. Schematic view of our proposal

Although the discriminative methods follow the same principle, they rely on different criteria and therefore we briefly describe the foundations for each of them. All the discriminative methods presented here take advantage of the label information either as class labels or as genuine and impostor labels as in a verification setting. We believe that these methods may help to emphasize the discriminative information needed to pull together sketches and mug-shots from the same class while pulling apart those of different classes.

LDA. The Linear Discriminant Analysis (LDA) is one of the oldest methods to find discriminative projections, which still is widely used for its good performance. Here we use the Fisherfaces version of the method [1], which uses PCA before the LDA for regularization. The method learns a projection such that it maximizes the between or inter-class scatter over the within or intra-class scatter. Let the between class scatter be: $S_B = \sum_{i=1}^{C} N_i(\mu_i - \mu)(\mu_i - \mu)^T$ and the within class scatter $S_W = \sum_{i=1}^{C} \sum_{x_j \in X_i}(x_j - \mu_i)(x_j - \mu_i)^T$, where μ is the mean of all objects in the dataset, μ_i is the mean of each class, C the number of classes, and N_i is the total number of objects of class X_i. The optimal projection is:

$$\hat{W} = arg \max_{W} \frac{|W^T S_B W|}{|W^T S_W W|} = [w_1 \ w_2 \ldots w_M], \tag{1}$$

where $w_i, i = 1 : M$ corresponds to the generalized eigenvectors of $S_B w_i = \lambda_i S_W w_i$. There are at most $C - 1$ eigenvectors, due to this, the linear projections generated by these eigenvectors are of dimension $C - 1$ at most.

KISS Metric Learning (KISSME). The idea of metric learning methods in general is to learn a Mahalanobis distance of the form: $(x - y)^t M(x - y)$, where M corresponds to a weight matrix to be learnt by the metric learning methods. This distance in the original space is equivalent to the squared Euclidean distance in the discriminative space where the linear projections of the data are found by $\tilde{x} = Lx$, where L is related to M by $M = L^t L$. Note that the squared Euclidean distance in the original space can be retrieved by using M as the Identity matrix. The KISSME method models the commonalities of genuine and impostor pairs [9]. From a statistical inference sense, the optimal statistical decision about the similarity of a pair (x, y) can be obtained by:

$$r(x, y) = \log \frac{P(x, y|H0)}{P(x, y|H1)}, \tag{2}$$

where we test the hypothesis H0 that a pair is similar versus the alternative H1 that a pair is dissimilar. The method cast the problem in the space of differences with zero mean. Therefore, by using $d_{xy} = x - y$ we have that:

$$\delta(d_{xy}) = \log \frac{P(d_{xy}|H0)}{P(d_{xy}|H1)}. \tag{3}$$

By assuming Gaussianity of the difference space they rewrite Eq. 3 in terms of Gaussian distributions and the parameters θ_0 and θ_1 corresponding to probability density functions for the hypothesis H0 and H1 are estimated from covariance matrices from genuine ($\sum_{y_{ij}=1}$) and impostor ($\sum_{y_{ij}=0}$) pairs. The maximum likelihood estimate of the Gaussian is equivalent to minimize the Mahalanobis distance from the mean by least squares. In this way, relevant directions for each set of genuine and impostor pairs are found, and after reformulation and simplification M is found by clipping the spectrum from the eigendecomposition of $\hat{M} = (\sum^- 1_{y_{ij}=1} - \sum^- 1_{y_{ij}=0})$ to ensure positive-definiteness.

Joint Bayesian. This method was proposed in [2] for face recognition, and it has been used extensively since then especially on top of learned representations by convolutional networks. The Joint Bayesian is based on the idea of the Bayesian face recognition method, but instead of modeling the distances in a 1D space, it models the joint distributions of two samples (x, y) before computing the distance between the samples. The previous method has the problem that by modeling the differences instead of the joint distributions it may reduce class separability since differences lie in a 1D line and therefore may overlap. In the Joint Bayesian, by assuming a face prior, each face is the summation of two independent Gaussian latent variables: the intrinsic variability for identity and intrapersonal variable for intrapersonal variation: $x = I + E$, where the samples have zero mean. The Joint formulation with prior is also Gaussian with zero mean and after algebraic operations the log likelihood ratio is obtained, which can be thought of as a measure of similarity.

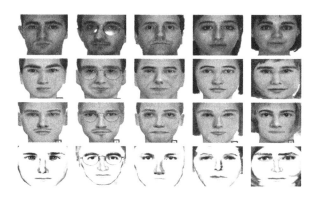

Fig. 2. Images from the PRIP-VSGC database, composites of the subjects with photos in the first row are shown per column. Rows from 2 to 4 correspond to composites by: (2) American user, (3) Asian users, (4) Identi-Kit.

3   Experimental Analysis

We present the results of the proposed discriminative representations for sketch recognition. For the implementation of the BoVW approach we used the VLFeat library [14] while the metric and similarity learning codes are taken from the author's website and for LDA the PRtools library was used [3]. In order to evaluate the proposed discriminative representations for sketch recognition, we use the PRIP Viewed Software-Generated Composite (PRIP-VSGC) [4] dataset (see Fig. 2). It contains photographs from 123 subjects from the AR database and three composites created for each subject using FACES (American and Asian users) and Identi-Kit softwares. Both the mug-shots and sketches are normalized to 150×150. Parameters used for the BoVW representation are: SIFT patch sizes = 4, 6, 8, 10. Standard deviations for Gaussian blur: sigma = sizes/magnification factor, where the magnification factor is 6, and the step size is 3. The number of words of the visual dictionary is 600 and PCA reduces the data to 20 dimensions for the datasets.

Table 1. Rank-10 identification accuracies for state-of-the-art methods and our methods for the different datasets

Method	Faces Am.	Faces As.	Identi-Kit
COTS	11.3 ± 2.1	7.2 ± 2.2	8.1 ± 2.1
Autoencoder + SVM	23.1 ± 1.8	16.0 ± 2.2	18.5 ± 2.0
Autoencoder + NN	46.6 ± 1.7	41.8 ± 1.9	46.5 ± 1.3
DBN + SVM	25.3 ± 2.1	24.0 ± 1.3	21.7 ± 0.9
DBN + NN	49.3 ± 2.6	43.3 ± 2.1	45.3 ± 2.1
Autoencoder + DBN + SVM	38.7 ± 1.6	32.0 ± 2.2	31.6 ± 2.4
Mittal et al. [12]	51.9 ± 1.2	42.6 ± 1.2	45.3 ± 1.5
Mittal et al. [13]	36.0 ± 2.9	26.6 ± 2.5	32.7 ± 2.7
Mittal et al. [13]	56.0 ± 2.1	48.1 ± 1.7	52.0 ± 2.4
Ours (Joint Bayesian)	68.8 ± 2.3	58.1 ± 1.6	**53.0 ± 3.0**
Ours (KISSME)	68.5 ± 2.7	**58.4 ± 1.3**	49.8 ± 2.1
Ours (LDA)	**70.6 ± 2.3**	57.0 ± 1.5	52.8 ± 4.0

We compare our proposal with the most recent approach using deep learning and transfer learning in [13], as well as other sate-of-the-art methods compared in their work, including a commercial-off-the-shelf (COTS) software, FaceVACS. We replicate their experimental protocol, in which 48 photo-sketch pairs for training and 75 for testing are randomly selected five times. The average accuracies and standard deviations are shown in Table 1.

Table 2. Rank-20 and rank-40 identification accuracies on an extended gallery set of 2400 photos

Data	Rank	Mittal et al. [13]	Joint Bayesian	KISSME	LDA
Faces Am.	20	24.6	36.8	30.6	**48.2**
	40	58.8	51.8	40.2	**65.2**
Faces As.	20	15.4	28.5	33.3	**49.8**
	40	52	41.6	43.7	**68.0**
Identi-Kit	20	15.6	20.3	21.0	**32.8**
	40	48.3	28.5	31.5	**53.3**

From the results in Table 1, it can be seen that the proposed approaches outperform the other methods in the closed-set scenario where only 75 images are used for the gallery. We compare our methods in an open-set scenario, where we add up to 2400 mug-shot images for the gallery set, in order to compare our proposal with that in [13], in terms of accuracies at ranks 20 and 40. Results are shown in Table 2. It can be seen that the LDA also outperforms the method in [13] in this setting. Besides, our methods are able to obtain significantly higher accuracies for lower rankings, which may be very useful in real-world applications where it is more convenient for the specialist to check a small list of potential candidates or suspects.

Our intuition behind the better performance of the LDA when more distractors are added to the database is that the LDA uses the most discriminative information of the three compared methods. Besides, the previous PCA projection to a very small dimensional space provides a regularization that is highly beneficial for the generalization (to unseen data) of the method. The KISSME receives the discriminative information in form of genuine and impostor pairs which for small training sets may not generalize as well as the LDA which receives the class memberships. However, for larger training sets there is a possibility that the KISSME may generalize better and avoid overtraining since it does not receive such specific information as the LDA. The Joint Bayesian estimates Gaussian distributions that with small training sets may not be very accurately estimated, which may be problematic in front of distractors.

4   Conclusions

We proposed the use of discriminative representations for the problem of face-sketch recognition. The intermediate representation is achieved by means of multiscale dense SIFT quantized in a visual dictionary and metric learning or other discriminative methods are applied on top of the representation. The results obtained by the proposed approach are very competitive, providing even similar or better results than deep learning-based methods. Besides, the length of the vector representation used for classification of composites after PCA is only 20, which is impressively small and very suitable for real-world applications. In fact, we found that when using more dimensions for the composite sketches the results deteriorate. This shows the intrinsic low dimensionality of composite sketch representation. In addition, it must be taken into account that, for the learning methods, a set of photo-sketch pairs is needed. In our experiments we confirmed that a small-sized set is sufficient for achieving good results, but a larger set must lead to better results.

References

1. Belhumeur, P.N., Hespanha, J.P., Kriegman, D.J.: Eigenfaces vs. fisherfaces: recognition using class specific linear projection. IEEE Trans. Pattern Anal. Mach. Intell. **19**(7), 711–720 (1997)

2. Chen, D., Cao, X., Wang, L., Wen, F., Sun, J.: Bayesian face revisited: a joint formulation. In: Fitzgibbon, A., Lazebnik, S., Perona, P., Sato, Y., Schmid, C. (eds.) ECCV 2012. LNCS, vol. 7574, pp. 566–579. Springer, Heidelberg (2012). doi:10.1007/978-3-642-33712-3_41

3. Duin, R., Juszczak, P., Paclik, P., Pekalska, E., de Ridder, D., Tax, D., Verzakov, S.: PR-Tools4.1, a matlab toolbox for pattern recognition. Technical report, Information and Communication Theory Group: Delft University of Technology, The Netherlands (2007). http://www.prtools.org/

4. Han, H., Klare, B., Bonnen, K., Jain, A.K.: Matching composite sketches to face photos: a component-based approach. IEEE Trans. Inf. Forensics Secur. **8**(1), 191–204 (2013)

5. Klare, B., Jain, A.K.: Sketch-to-photo matching: a feature-based approach. SPIE Defense, Security, and Sensing, vol. 7667 (2010)

6. Klare, B., Jain, A.K.: Heterogeneous face recognition using kernel prototype similarities. IEEE Trans. Pattern Anal. Mach. Intell. **35**(6), 1410–1422 (2013)

7. Klum, S., Han, H., Jain, A.K., Klare, B.: Sketch based face recognition: forensic vs. composite sketches. In: International Conference on Biometrics, ICB 2013, Madrid, Spain, 4–7 June 2013, pp. 1–8 (2013)

8. Klum, S., Han, H., Klare, B., Jain, A.K.: The facesketchid system: matching facial composites to mugshots. IEEE Trans. Inf. Forensics Secur. **9**(12), 2248–2263 (2014)

9. Köstinger, M., Hirzer, M., Wohlhart, P., Roth, P.M., Bischof, H.: Large scale metric learning from equivalence constraints. In: 2012 IEEE Conference on Computer Vision and Pattern Recognition, Providence, RI, USA, 16–21 June 2012, pp. 2288–2295 (2012)

10. Lazebnik, S., Schmid, C., Ponce, J.: Beyond bags of features: spatial pyramid matching for recognizing natural scene categories. In: Proceedings of the 2006 IEEE Computer Society Conference on Computer Vision and Pattern Recognition, CVPR 2006, vol. 2. pp. 2169–2178. IEEE Computer Society, Washington, DC (2006)

11. Mittal, P., Jain, A., Singh, R., Vatsa, M.: Boosting local descriptors for matching composite and digital face images. In: 2013 IEEE International Conference on Image Processing, pp. 2797–2801, September 2013

12. Mittal, P., Jain, A., Goswami, G., Singh, R., Vatsa, M.: Recognizing composite sketches with digital face images via SSD dictionary. In: IEEE International Joint Conference on Biometrics, IJCB 2014, Clearwater, FL, USA, 29 September–2 October 2014, pp. 1–6 (2014)

13. Mittal, P., Vatsa, M., Singh, R.: Composite sketch recognition via deep network - a transfer learning approach. In: International Conference on Biometrics, ICB 2015, Phuket, Thailand, 19–22 May 2015, pp. 251–256 (2015)

14. Vedaldi, A., Fulkerson, B.: VLFeat: an open and portable library of computer vision algorithms. In: Proceedings of the 18th ACM International Conference on Multimedia, MM 2010, pp. 1469–1472. ACM, New York (2010)

15. Wang, X., Tang, X.: Face photo-sketch synthesis and recognition. IEEE Trans. Pattern Anal. Mach. Intell. **31**(11), 1955–1967 (2009)

16. Yuen, P.C., Man, C.H.: Human face image searching system using sketches. IEEE Trans. Syst. Man Cybern. Part A: Syst. Hum. **37**(4), 493–504 (2007)

Efficient Sparse Approximation of Support Vector Machines Solving a Kernel Lasso

Marcelo Aliquintuy[1], Emanuele Frandi[2], Ricardo Ñanculef[1(✉)],
and Johan A.K. Suykens[2]

[1] Department of Informatics, Federico Santa María University, Valparaíso, Chile
{maliq,jnancu}@inf.utfsm.cl
[2] ESAT-STADIUS, KU Leuven, Leuven, Belgium
{efrandi,johan.suykens}@esat.kuleuven.be

Abstract. Performing predictions using a non-linear support vector machine (SVM) can be too expensive in some large-scale scenarios. In the non-linear case, the complexity of storing and using the classifier is determined by the number of support vectors, which is often a significant fraction of the training data. This is a major limitation in applications where the model needs to be evaluated many times to accomplish a task, such as those arising in computer vision and web search ranking.

We propose an efficient algorithm to compute sparse approximations of a non-linear SVM, i.e., to reduce the number of support vectors in the model. The algorithm is based on the solution of a Lasso problem in the feature space induced by the kernel. Importantly, this formulation does not require access to the entire training set, can be solved very efficiently and involves significantly less parameter tuning than alternative approaches. We present experiments on well-known datasets to demonstrate our claims and make our implementation publicly available.

Keywords: SVMs · Kernel methods · Sparse approximation · Lasso

1 Introduction

Non-linear support vector machines (SVMs) are a powerful family of classifiers. However, while in recent years one has seen considerable advancements on scaling kernel SVMs to large-scale problems [1,9], the lack of sparsity in the obtained models, i.e., the often large number of support vectors, remains an issue in contexts where the run time complexity of the classifier is a critical factor [6,13]. This is the case in applications such as object detection in images or web search ranking, which require repeated and fast evaluations of the model. As the sparsity of a non-linear kernel SVM cannot be known *a-priori*, it is crucial to devise efficient methods to impose sparsity in the model or to sparsify an existing classifier while preserving as much of its generalization capability as possible.

Recent attempts to achieve this goal include *post-processing approaches* that reduce the number of support vectors in a given SVM or change the basis used

© Springer International Publishing AG 2017
C. Beltrán-Castañón et al. (Eds.): CIARP 2016, LNCS 10125, pp. 208–216, 2017.
DOI: 10.1007/978-3-319-52277-7_26

to express the classifier [3,12,15], and *direct methods* that modify the SVM objective or introduce heuristics during the optimization to maximize sparsity [2,6,7,11,14]. In a recent breakthrough, [3] proposed a simple technique to reduce the number of support vectors in a given SVM showing that it was asymptotically optimal and outperformed many competing approaches in practice. Unfortunately, most of these techniques either depend on several parameter and heuristic choices to yield a good performance or demand significant computational resources. In this paper, we argue how these problems can be effectively circumvented by sparsifying an SVM solving a simple Lasso problem [4] in the kernel space. Interestingly, this criterion was already mentioned in [12], but was not accompanied by an efficient algorithm neither systematically assessed in practice. By exploiting recent advancements in optimization [4,5,9], we devise an algorithm that is significantly cheaper than [3] in terms of optimization and parameter selection but is competitive in terms of the accuracy/sparsity tradeoff.

2   Problem Statement and Related Work

Given data $\{(\mathbf{x}_i, y_i)\}_{i=1}^m$ with $\mathbf{x}_i \in X$ and $y_i \in \{\pm 1\}$, SVMs learn a predictor of the form $f_{\mathbf{w},b}(\mathbf{x}) = \text{sign}(\mathbf{w}^T \phi(\mathbf{x}) + b)$ where $\phi(\mathbf{x})$ is a feature vector representation of the input pattern $\mathbf{x}$ and $\mathbf{w} \in \mathcal{H}, b \in \mathbb{R}$ are the model parameters. To allow more flexible decision boundaries, $\phi(\mathbf{x})$ often implements a non-linear mapping $\phi : X \to \mathcal{H}$ of the input space into a Hilbert space $\mathcal{H}$, related to X by means of a *kernel function* $k : X \times X \to \mathbb{R}$. The kernel allows to compute dot products in $\mathcal{H}$ directly from X, using the property $\phi(\mathbf{x}_i)^T \phi(\mathbf{x}_j) = k(\mathbf{x}_i, \mathbf{x}_j), \forall \mathbf{x}_i, \mathbf{x}_j \in X$. The values of $\mathbf{w}, b$ are determined by solving a problem of the form

$$\min_{\mathbf{w},b} \tfrac{1}{2}\|\mathbf{w}\|_{\mathcal{H}}^2 + C \sum_{i=1}^m \ell\left(y_i(\mathbf{w}^T \phi(\mathbf{x}_i) + b)\right)^p, \tag{1}$$

where $p \in \{1, 2\}$ and $\ell(z) = (1 - z)_+$ is called *the hinge-loss*. It is well-known that the solution $\mathbf{w}*$ to (1) can be written as a linear combination of the training patterns in the feature space $\mathcal{H}$. This leads to the "kernelized" decision function

$$f_{\mathbf{w},b}(\mathbf{x}) = \text{sign}\left(\mathbf{w}^{*T} \phi(\mathbf{x}) + b^*\right) = \text{sign}\left(\sum_{i=1}^n y_i \beta_i^* k(\mathbf{x}_i, \mathbf{x}) + b^*\right), \tag{2}$$

whose run time complexity is determined by the number n_{sv} of examples such that $\beta_i^* \neq 0$. These examples are called *the support vectors* (SVs) of the model. In contrast to the linear case ($\phi(\mathbf{x}) = \mathbf{x}$), kernel SVMs need to explicitly store and access the SVs to perform predictions. Unfortunately, it is well known that in general, n_{sv} grows as a linear function of the number of training points [6,13] (at least all the misclassified points are SVs) and therefore n_{sv} is often too large in practice, leading to classifiers expensive to store and evaluate. Since n_{sv} is the number of non-zero entries in the coefficient vector β^*, this problem if often referred in the literature as the *lack of sparsity* of non-linear SVMs.

Methods to address this problem can be categorized in two main families: *post-processing* or *reduction methods*, which, starting with a non-sparse classifier, find a more efficient predictor preserving as much of the original predictive accuracy as possible, and *direct methods* that modify the training criterion (1) or introduce heuristics during its optimization to promote sparsity. The first category include methods selecting a subset of the original support vectors to recompute the classifier [12,15], techniques to substitute the original support vectors by arbitrary points of the input space [10] and methods tailored to a specific class of SVM [8]. The second category includes offline [6,7,11], as well as online learning algorithms [2,14]. Unfortunately, most of these techniques either incur in a significant computational cost or depend on several heuristic choices to yield a good performance. Recently, a simple, yet asymptotically optimal reduction method named ISSVM has been presented in [3], comparing favorably with the state of the art in terms of the accuracy/sparsity tradeoff. The method is based on the observation that the hinge loss of a predictor $f_{\mathbf{w},b}$ can be approximately preserved using a number of support vectors proportional to $\|\mathbf{w}\|_{\ell_2}$ by applying sub-gradient descent to the minimization of the following objective function

$$g_{\text{ISSVM}}(\tilde{\mathbf{w}}) = \max_{i:h_i>0} \left(h_i - y_i \left(\tilde{\mathbf{w}}^T \mathbf{x} + b \right) \right) , \tag{3}$$

where $h_i = \max(1, y_i(\mathbf{w}^T\mathbf{x} + b))$. Using this method to sparsify an SVM $f_{\mathbf{w},b}$ guarantees a reduction of n_{sv} to at most $\mathcal{O}(\|\mathbf{w}\|_{\ell_2})$ support vectors. However, since different levels of sparsification may be required in practice, the algorithm is equipped with an additional projection step. In the course of the optimization, the approximation $\tilde{\mathbf{w}}$ is projected into the ℓ_2-ball of radius δ, where δ is a parameter controlling the level of sparsification. Unfortunately, the inclusion of this projection step and the weak convergence properties of sub-gradient descent makes the algorithm quite sensitive to parameter tuning.

3   Sparse SVM Approximations via Kernelized Lasso

Suppose we want to sparsify an SVM with parameters $\mathbf{w}_*, b_*$, kernel $k(\cdot, \cdot)$ and support set $S = \{(\mathbf{x}_{(i)}, y_{(i)})\}_{i=1}^{n_{\text{sv}}}$. Let $\phi : X \to \mathcal{H}$ be the feature map implemented by the kernel and $\phi(\mathbf{S})$ the matrix whose i-th column is given by $\phi(\mathbf{x}_{(i)})$. With this notation, $\mathbf{w}_*$ can be written as $\mathbf{w}_* = \phi(\mathbf{S})\alpha_*$ with $\alpha_* \in \mathbb{R}^{n_{\text{sv}} 1}$. In this paper, we look for approximations of the form $\mathbf{u} = \phi(\mathbf{S})\alpha$ with sparse $\mathbf{u}$. Support vectors such that $u_{(i)} = 0$ are pruned from the approximation.

Our approximation criterion is based on two observations. The first is that the objective function (3) can be bounded by a differentiable function which is more convenient for optimization. Importantly, this function also bounds the expected loss of accuracy incurred by the approximation. Indeed, the following result (whose proof we omit for space constraints) holds:

---

[1] Note that we have just re-indexed the support vectors in (2) to make the model independent of the entire training set and defined $\alpha_j = y_j\beta_j$ for notational convenience.

Proposition 1. *Consider an SVM implementing the decision function* $f_{\mathbf{w},b}(\mathbf{x}) = sign(\mathbf{w}^T \phi(\mathbf{x}) + b)$ *and an alternative decision function* $f_{\mathbf{u},b}(\mathbf{x}) = sign(\mathbf{u}^T \phi(\mathbf{x}) + b)$, *with* $\mathbf{u} \in \mathcal{H}$. *Let* $\ell(z)$ *be the hinge loss. Then,* $\exists M > 0$ *such that*

(i) $g_{ISSVM}(\mathbf{u}) \le M \|\mathbf{u} - \mathbf{w}\|_{\mathcal{H}}$,   (ii) $E\left(\ell\left(yf_{\mathbf{u}}(\mathbf{x})\right) - \ell\left(yf_{\mathbf{w}}(\mathbf{x})\right)\right) \le M \|\mathbf{u} - \mathbf{w}\|_{\mathcal{H}}$.

The result above suggests that we can substitute $\mathbf{w} \in \mathcal{H}$ in the original SVM by some $\mathbf{u} \in \mathcal{H}$ such that $\|\mathbf{u} - \mathbf{w}_*\|^2$ is small. However, the obtained surrogate does need to be sparse. Indeed, minimizing $\|\mathbf{u} - \mathbf{w}_*\|^2$ in $\mathcal{H}$ trivially yields the original predictor $\mathbf{w}_*$ which is generally dense. We thus need to restrict the search to a family of sparser models. Our second observation is that a well-known, computationally attractive and principled way to induce sparsity is ℓ_1-norm regularization, i.e., constraining $\mathbf{u}$ to lie in a ball around 0 with respect to the norm $\|\mathbf{u}\|_{\ell_1} = \sum_i |u_i|$. Thus, we approach the task of sparsifying the SVM by solving a problem of the form

$$\min_{\boldsymbol{\alpha} \in \mathbb{R}^{n_{sv}}} \tfrac{1}{2}\|\phi(\mathbf{S})\boldsymbol{\alpha} - \mathbf{w}_*\|^2 \text{ s.t. } \|\boldsymbol{\alpha}\|_{\ell_1} \le \delta, \tag{4}$$

where δ is a regularization parameter controlling the level of sparsification. The obtained problem can be easily recognized as a kernelized Lasso with response variable $\mathbf{w}_*$ and design matrix $\phi(\mathbf{S})$. By observing that

$$\begin{aligned}
\|\mathbf{w}_* - \phi(\mathbf{S})\boldsymbol{\alpha}\|^2 &= \mathbf{w}_*^T \mathbf{w}_* - 2\boldsymbol{\alpha}_*^T \phi(\mathbf{S})^T \phi(\mathbf{S})\boldsymbol{\alpha} + \boldsymbol{\alpha}^T \phi(\mathbf{S})^T \phi(\mathbf{S})\boldsymbol{\alpha} \\
&= \mathbf{w}_*^T \mathbf{w}_* - 2\boldsymbol{\alpha}_*^T \mathbf{K}\boldsymbol{\alpha} + \boldsymbol{\alpha}^T \mathbf{K}\boldsymbol{\alpha} = \mathbf{w}_*^T \mathbf{w}_* - 2\mathbf{c}^T \boldsymbol{\alpha} + \boldsymbol{\alpha}^T \mathbf{K}\boldsymbol{\alpha},
\end{aligned} \tag{5}$$

where $\mathbf{c} = \mathbf{K}\boldsymbol{\alpha}_*$, it is easy to see that solving (4) only requires access to the kernel matrix (or the kernel function):

$$\min_{\boldsymbol{\alpha} \in \mathbb{R}^{n_{sv}}} g(\boldsymbol{\alpha}) = \tfrac{1}{2}\boldsymbol{\alpha}^T \mathbf{K}\boldsymbol{\alpha} - \mathbf{c}^T \boldsymbol{\alpha} \text{ s.t. } \|\boldsymbol{\alpha}\|_{\ell_1} \le r. \tag{6}$$

This type of approach has been considered, up to some minor differences, by Schölkopf *et al.* in [12]. However, to the best of our knowledge, it has been largely left out of the recent literature on sparse approximation of kernel models. One possible reason for this is the fact that the original proposal had a high computational cost, making it unattractive for large models. We reconsider this technique arguing that recent advancements in Lasso optimization make it possible to solve the problem efficiently using high-performance algorithms with strong theoretical guarantees [4]. Importantly, we show in Sect. 5 that this efficiency is not obtained at the expense of accuracy, and indeed the method can match or even surpass the performance of the current state-of-the-art methods.

Algorithm. To solve the kernelized Lasso problem, we adopt a variant of the Frank-Wolfe (FW) method [5], an iterative greedy algorithm to minimize a convex differentiable function $g(\boldsymbol{\alpha})$ over a closed convex set Σ, specially tailored to handle large-scale instances of (6). This method does not require to compute the matrix $\mathbf{K}$ beforehand, is very efficient in practice and enjoys important convergence guarantees [5,9], some of which are summarized in Theorem 1. Given an

---

Algorithm 1. SASSO: SPARSIFICATION OF SVMS VIA KERNEL LASSO.

---

1 $\boldsymbol{\alpha}^{(0)} \leftarrow \mathbf{0}$, $\mathbf{g}^{(0)} = \mathbf{c}$.
2 **for** $k = 1, 2, \ldots$ **do**
3   Find a descent direction: $j_*^{(k)} \leftarrow \arg\max_{j \in J} |g_j^{(k)}|$, $t_*^{(k)} \leftarrow t \operatorname{sign}\left(g_{j_*}^{(k)}\right)$.
4   Choose a step-size $\lambda^{(k)}$, e.g. by a line-search between $\boldsymbol{\alpha}^{(k)}$ and $\mathbf{u}^{(k)}$.
5   Update the solution: $\boldsymbol{\alpha}^{(k+1)} \leftarrow (1 - \lambda^{(k)})\boldsymbol{\alpha}^{(k)} + \lambda^{(k)} t_*^{(k)} \mathbf{e}_{j_*^{(k)}}$.
6   Update the gradient: $g_j^{(k+1)} \leftarrow (1 - \lambda^{(k)})g_j^{(k+1)} + \lambda^{(k)} t_*^{(k)} K_{j j_*^{(k)}} \ \forall j \in [n_{\mathrm{sv}}]$.
7 **end**

---

iterate $\boldsymbol{\alpha}^{(k)}$, a step of FW consists in finding a descent direction as

$$\mathbf{u}^{(k)} \in \underset{\mathbf{u} \in \Sigma}{\arg\min} \, (\mathbf{u} - \boldsymbol{\alpha}^{(k)})^T \nabla g(\boldsymbol{\alpha}^{(k)}), \tag{7}$$

and updating the current iterate as $\boldsymbol{\alpha}^{(k+1)} = (1 - \lambda^{(k)})\boldsymbol{\alpha}^{(k)} + \lambda^{(k)}\mathbf{u}^{(k)}$. The step-size $\lambda^{(k)}$ can be determined by an exact line-search (which can be done analytically for quadratic objectives) or setting it as $\lambda^{(k)} = 1/(k+2)$ as in [5].

In the case of problem (6), where Σ corresponds to the ℓ_1-ball of radius t in $\mathbb{R}^{n_{\mathrm{sv}}}$ (with vertices $\mathcal{V} = \{\pm t\mathbf{e}_i : i = 1, 2, \ldots, n_{\mathrm{sv}}\}$) and the gradient is $\nabla g(\boldsymbol{\alpha}) = \mathbf{K}\boldsymbol{\alpha} - \mathbf{c}$, it is easy to see that the solution of (7) is equivalent to

$$j* = \arg\max_{j \in [n_{\mathrm{sv}}]} \left| \phi(\mathbf{s}_j)^T \phi(\mathbf{S})\boldsymbol{\alpha} + c_j \right| = \arg\max_{j \in [n_{\mathrm{sv}}]} \left| \sum_{i:\alpha_i \neq 0} \alpha_i K_{ij} + c_j \right|. \tag{8}$$

The adaptation of the FW algorithm to problem (6) is summarized in Algorithm 1 and is referred to as SASSO in the rest of this paper.

Theorem 1. *Consider problem (6) with $r \in (0, \|\boldsymbol{\alpha}^*\|_{\ell_1})$. Algorithm 1 is monotone and globally convergent. In addition, there exists $C > 0$ such that*

$$\|\mathbf{w}_* - \phi(\mathbf{S})\boldsymbol{\alpha}^{(k)}\|^2 - \|\mathbf{w}_* - \phi(\mathbf{S})\boldsymbol{\alpha}^{(k+1)}\|^2 \leq C/(k+2). \tag{9}$$

Tuning of b. We have assumed above that the bias b of the SVM can be preserved in the approximation. A slight boost in accuracy can be obtained by computing a value of b which accounts for the change in the composition of the support set. For the sake of simplicity, we adopt here a method based on a validation set, i.e., we define a range of possible values for b and then choose the value minimizing the misclassification loss on that set. It can be shown that it is safe (in terms of accuracy) to restrict the search to the interval $[b_{\min}, b_{\max}]$ where

$$b_{\min} = \inf_{\mathbf{x} \in S: \mathbf{w}^T \mathbf{x} > 0} - \mathbf{w}^T \mathbf{x}, \qquad b_{\max} = \sup_{\mathbf{x} \in S: \mathbf{w}^T \mathbf{x} < 0} - \mathbf{w}^T \mathbf{x}.$$

4   Experimental Results

We present experiments on four datasets recently used in [2,3] to assess SVM sparsification methods: Adult (a8a), IJCNN, TIMIT and MNIST. Table 1 summarizes the number of training points m and test points t for each dataset. The SVMs to sparsify were trained using SMO with a RBF kernel and parameters set up as in [2,3]. As discussed in Sect. 2, we compare the performance of our algorithm with that of the ISSVM algorithm, which has a publicly available C++ implementation [3]. Our algorithms have been also coded in C++. We executed the experiments on a 2 GHz Intel Xeon E5405 CPU with 20 GB of main memory running CentOS, without exploiting multithreading or parallelism in computations. The code, the data and instructions to reproduce the experiments of this paper are publicly available at https://github.com/maliq/FW-SASSO.

We test two versions of our method, the standard one in Algorithm 1, and an aggressive variant employing a *fully corrective* FW solver (where an internal optimization over the current active set is carried out at each iteration, see

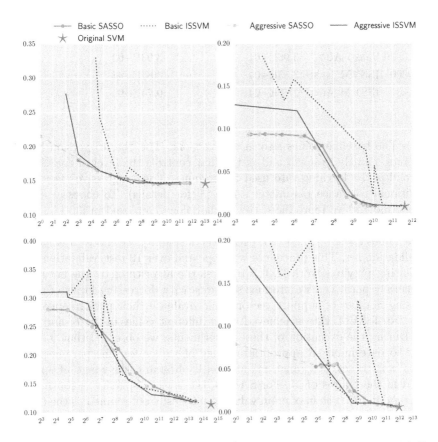

Fig. 1. Test accuracy (y axis) versus sparsity (number of support vectors, x axis). From top-left to bottom-right: Adult, IJCNN, TIMIT and MNIST datasets.

Table 1. Time required to build the sparsity/accuracy path. We report the total time incurred in parameter selection (training with different parameters and evaluation in the validation set) and the average training time to build a path.

Dataset	Method	Average train time (secs)	Total time (secs) train & val.
Adult	SASSO BASIC	2.08E+02	2.66E+02
$m = 22696$	SASSO AGG.	1.99E+02	2.51E+02
$t = 9865$	ISSVM BASIC	2.78E+03	1.97E+04
	ISSVM AGG.	1.33E+03	4.71E+04
IJCNN	SASSO BASIC	8.20E+01	2.74E+02
$m = 35000$	SASSO AGG.	4.34E+01	1.57E+02
$t = 91701$	ISSVM BASIC	4.23E+03	3.47E+04
	ISSVM AGG.	5.35E+03	1.98E+05
TIMIT	SASSO BASIC	1.57E+02	6.22E+02
$m = 66831$	SASSO AGG.	1.46E+02	5.24E+02
$t = 22257$	ISSVM BASIC	1.02E+04	7.68E+04
	ISSVM AGG.	6.99E+03	2.49E+05
MNIST	SASSO BASIC	4.03E+01	4.16E+02
$m = 60000$	SASSO AGG.	3.96E+01	3.93E+02
$t = 10000$	ISSVM BASIC	3.53E+04	3.83E+04
	ISSVM AGG.	2.77E+04	9.74E+05

e.g. [5]). The baseline comes also in two versions. The "basic" version has two parameters, namely ℓ_2-norm and η: the first controls the level of sparsity, and the second is the learning rate used for training. The "aggressive" version has an additional tolerance parameter ϵ (see [3] for details). To choose values for these parameters, we reproduced the methodology employed in [3], i.e., for the learning rate we tried values $\eta = 4^{-4}, \ldots, 4^2$ and for ϵ (in the aggressive variant) we tried values $\epsilon = 2^{-4}, \ldots, 1$. For each level of sparsity, we choose a value based on a validation set. This procedure was repeated over 10 test/validation splits.

Following previous work [2,3], we assess the algorithms on the entire sparsity/accuracy path, i.e., we produce solutions with decreasing levels of sparsity (increasing number of support vectors) and evaluate their performance on the test set. For ISSVM, this is achieved using different values of the ℓ_2-norm parameter. During the execution of these experiments, we observed that it is quite difficult to determine an appropriate range of values for this parameter. Our criterion was to set this range manually till obtaining the range of sparsities reported in the figures of [3]. For SASSO, the level of sparsity is controlled by parameter δ in (4). The maximum value for δ is easily determined as the ℓ_1-norm of the input SVM and the minimum value as 10^{-4} times the former. To make comparison fair, we compute 10 points of the path for all the methods.

Results in Fig. 1 show that the sparsity/accuracy tradeoff path obtained by SASSO matches that of the (theoretically optimal) ISSVM method [3], and often tends to outperform it on the sparsest section of the path. However as it can be seen from Table 1, our method enjoys a considerable computational advantage over ISSVM: on average, it is faster by 1–2 orders of magnitude, and the overhead due to parameter selection is marginal compared to the case of ISSVM, where the total time is one order of magnitude larger than the single model training time. We also note that the aggressive variant of SASSO enjoys a small but consistent advantage on all the considered datasets. Both versions of our method exhibit a very stable and predictable performance, while ISSVM needs the more aggressive variant of the algorithm to produce a regular path. However, this version requires considerable parameter tuning to achieve a behavior similar to that observed for SASSO, which translates into a considerably longer running time.

5 Conclusions

We presented an efficient method to compute sparse approximations of non-linear SVMs, i.e. to reduce the number of support vectors in the model. The algorithm enjoys strong convergence guarantees and it is easy to implement in practice. Further algorithmic improvements could also be obtained by implementing the stochastic acceleration studied in [4]. Our experiments showed that the proposed method is competitive with the state of the art in terms of accuracy, with a small but systematic advantage when sparser models are required. In computational terms, our approach is significantly more efficient due to the properties of the optimization algorithm and the avoidance of cumbersome parameter tuning.

Acknowledgments. E. Frandi and J.A.K. Suykens acknowledge support from ERC AdG A-DATADRIVE-B (290923), CoE PFV/10/002 (OPTEC), FWO G.0377.12, G.088114N and IUAP P7/19 DYSCO. M Aliquintuy and R. Ñanculef acknowledge support from CONICYT Chile through FONDECYT Project 1130122 and DGIP-UTFSM 24.14.84.

References

1. Bottou, L., Lin, C.J.: Support vector machine solvers. In: Bottou, L., Chapelle, O., DeCoste, D., Weston, J. (eds.) Large Scale Kernel Machines. MIT Press (2007)
2. Cotter, A., Shalev-Shwartz, S., Srebro, N.: The kernelized stochastic batch perceptron. In: Proceedings of the 29th ICML, pp. 943–950. ACM (2012)
3. Cotter, A., Shalev-shwartz, S., Srebro, N.: Learning optimally sparse support vector machines. In: Proceedings of the 30th ICML, pp. 266–274. ACM (2013)
4. Frandi, E., Nanculef, R., Lodi, S., Sartori, C., Suykens, J.A.K.: Fast and scalable Lasso via stochastic Frank-Wolfe methods with a convergence guarantee. Mach. Learn. **104**(2), 195–221 (2016)
5. Jaggi, M.: Revisiting Frank-Wolfe: projection-free sparse convex optimization. In: Proceedings of the 30th ICML, pp. 427–435. ACM (2013)

6. Joachims, T., Yu, C.N.J.: Sparse kernel SVMs via cutting-plane training. Mach. Learn. **76**(2–3), 179–193 (2009)
7. Keerthi, S.S., Chapelle, O., DeCoste, D.: Building support vector machines with reduced classifier complexity. J. Mach. Learn. Res. **7**, 1493–1515 (2006)
8. Mall, R., Suykens, J.A.K.: Very sparse LSSVM reductions for large scale data. IEEE Trans. Neural Netw. Learn. Syst. **26**(5), 1086–1097 (2015)
9. Ñanculef, R., Frandi, E., Sartori, C., Allende, H.: A novel Frank-Wolfe algorithm. Analysis and applications to large-scale SVM training. Inf. Sci. **285**, 66–99 (2014)
10. Nguyen, D., Ho, T.: An efficient method for simplifying support vector machines. In: Proceedings of the 22nd ICML, pp. 617–624. ACM (2005)
11. Nguyen, D.D., Matsumoto, K., Takishima, Y., Hashimoto, K.: Condensed vector machines: learning fast machine for large data. IEEE Trans. Neural Netw. **21**(12), 1903–1914 (2010)
12. Schölkopf, B., Mika, S., Burges, C.J., Knirsch, P., Müller, K.R., Rätsch, G., Smola, A.J.: Input space versus feature space in kernel-based methods. IEEE Trans. Neural Netw. **10**(5), 1000–1017 (1999)
13. Steinwart, I.: Sparseness of support vector machines. J. Mach. Learn. Res. **4**, 1071–1105 (2003)
14. Wang, Z., Crammer, K., Vucetic, S.: Breaking the curse of kernelization: budgeted stochastic gradient descent for large-scale SVM training. J. Mach. Learn. Res. **13**(1), 3103–3131 (2012)
15. Zhan, Y., Shen, D.: Design efficient support vector machine for fast classification. Pattern Recogn. **38**(1), 157–161 (2005)

Metric Learning in the Dissimilarity Space to Improve Low-Resolution Face Recognition

Mairelys Hernández-Durán[(✉)], Yenisel Plasencia-Calaña,
and Heydi Méndez-Vázquez

Advanced Technologies Application Center, 7ma A, #21406 Playa, Havana, Cuba
{mhduran,yplasencia,hmendez}@cenatav.co.cu

Abstract. Standard face recognition methods based on a feature representations are not suitable for low-resolution environments. Therefore low-resolution face recognition is still an unsolved problem where the best approaches still obtain very low recognition rates. In this paper, we propose a low-resolution face recognition method using the dissimilarity representation. In addition, we propose the use of metric learning methods to replace the standard Euclidean distance in the dissimilarity space. The effectiveness of our proposal is tested on two different data sets, one of them is the SCface database which is very challenging since the images were collected from surveillance cameras.

Keywords: Low-resolution · Metric learning · Dissimilarity space · Face recognition

1 Introduction

In real world applications such as video-surveillance, captured faces are often of low-resolution (LR). At these environments the obtained LR face image loses important details which are discriminative between persons mainly due to the distance among subjects and camera. These images can also present facial variations such as pose and expression; thus, it represents a challenge for a recognition task. Low-resolution face recognition (LRFR) methods try to cope with a classification problem between LR test images and high-resolution (HR) gallery images, causing the dimensional mismatch problem. Therefore, the dimensional mismatch between gallery/probe pairs and the lack of facial features are some of the main challenges related to LRFR [1]. Some authors have tried to find resolution-robust feature representation [2], but this is a difficult task because most of the effective features used in HR face recognition such as texture and color, may fail with LR images. The performance of traditional methods in the LR case suggests that current feature representation approaches are not suitable to cope with LRFR [3]. To improve the results, it becomes a priority to explore alternatives to the feature-based representation.

A representation based on dissimilarities between objects [4] is advantageous in situations where it is easier to define dissimilarities rather than features. The

© Springer International Publishing AG 2017
C. Beltrán-Castañón et al. (Eds.): CIARP 2016, LNCS 10125, pp. 217–224, 2017.
DOI: 10.1007/978-3-319-52277-7_27

dissimilarity space (DS) representation has been successfully used in many difficult task such as person re-identification [5]. Based on the success of previous works [4], we proposed the use of dissimilarity representations, as an alternative for LRFR. We believe that more discriminative information for classification can be obtained if the LR images are analyzed in the context of dissimilarities with other images. However, previous works assumed standard or designed dissimilarities, and not dissimilarities automatically learned for a given problem. Some researchers have shown that the classification could be greatly improved by learning a suitable distance metric. Then, we consider improving the original DS representation by using metric learning methods on top of it. Metric learning can provide a way to adapt a distance function to the given task.

In this work, the standard Euclidean distance in the DS is replaced by a learned metric, i.e., a Mahalanobis metric. We compared our proposal with some state-of-the-art representative methods based on feature vector representations. To address the dimensional mismatch, we used the best performing strategy proposed in [6], where the HR images are down-scaled and then up-scaled and the LR images are up-scaled to the same resolution. The proposal was evaluated on different face database including the SCFace [7], which is a very difficult database because it emphasizes the challenges of face recognition in surveillance environments.

The paper is organized as follows. Section 2 presents the related work on LRFR, the DS representation and the metric learning approach. Section 3 presents our proposal based on DS with automatically learned metrics to cope with the classification problem of LR facial images. Experiments and discussion are presented in Sect. 4, and concluding remarks are provided in Sect. 5.

2   Related Work

With the growing demands on surveillance applications, extensive efforts have been made on LRFR research. However, it remains an open issue due to the challenges posed by LR. Furthermore, the different resolutions between gallery and probe images lead to the so-called dimensional mismatch. To cope with this problem, different approaches have been used such as unified feature space (CLPM) [8]. This approach is used to project HR and LR images into a common space, which seems feasible to cope with the dimensional mismatch. However, it is not straightforward to find an optimal inter-resolution space and the transformations process may introduce noise. Several methods have used super-resolution (SR) techniques. However, these kind of methods mostly focus on obtaining a good visual reconstruction rather than a higher recognition rate. Current approaches mainly include feature vector representation for addressing LRFR. Resolution-robust feature representation has been considered for the LR case. Multidimensional scaling (MDS) [9] is a representative method, in which the relationships between LR and HR are explored taking into account the dimensional mismatch problem. Many authors have been working on this idea trying to find a common or inter-resolution space to project LR and their corresponding HR images on it [10].

A dissimilarity representation between objects is an alternative solution. Based on the idea proposed in [4], the dissimilarities are considered as the connection between perception and higher-level knowledge, which are key elements in the process of human recognition and categorization. By using the differences with prototypes for creating the representations we may be able to emphasize relevant information for discrimination among the classes, which, otherwise, by only analyzing the image, may be difficult to express in a feature representation. Following up on [6], they proposed the DS for LRFR but they only used the standard Euclidean distance. We believe that the use of a suitable distance metric can improve the classification accuracy. For example, in [11] they showed that it is possible to improve the K-NN classification accuracy using suitable distance metrics. The goal of metric learning algorithms is to take advantage of prior information in form of labels over standard similarity measures.

Compared with previous approaches this work is different in some aspects. We proposed the use of a dissimilarity based representation using learned metrics to achieve more discriminative distances in LRFR. In particular, our proposal is an alternative representation to feature space (FS) based on dissimilarities between objects and also introducing metric learning to replace the standard Euclidean distance in the DS.

3   Proposed Approach: Dissimilarity Space and Metric Learning for LRFR

A general scheme of the proposed strategy can be found in Fig. 1. In the following we will describe in more details the dissimilarity space construction and the metric learning approaches.

Fig. 1. General scheme of our proposal

3.1   Dissimilarity Space

Duin and Pekalska [4] proposed the DS as an Euclidean vector space in which it is possible to use several statistical classifiers. Although it has been used to solve a number of problems [12,13] their advantages to solve the dimensional mismatch in LR case, has not been explored yet. The proximity information is intuitively more discriminative than the features or the composition of each object independently. Based on its advantages, we consider the use of the dissimilarity space to achieve a more discriminative relational representation of the LR images. Let X be the space of objects, let $R = \{r_1, r_2, ..., r_k\}$ be the set of prototypes such that $R \in X$, and let $d : X \times X \to \mathbb{R}^+$ be a suitable dissimilarity measure for the problem. For a training set $T = \{x_1, x_2, ..., x_l\}$ such that $T \in X$, a mapping $\phi_R^d : X \to \mathbb{R}^k$ defines the embedding of training and test objects in the DS by the dissimilarities with the prototypes:

$$\phi_R^d(x_i) = [d(x_i, r_1) \ d(x_i, r_2) \ ... \ d(x_i, r_k)]. \tag{1}$$

3.2 Metric Learning Approach

In this section, we introduce the general idea of metric learning for kNN classification and review some previously studied approaches: LMNN, which directly attempts to optimize k-NN classification error; another method based on the Linear Discriminant Analysis (LDA) [14]; and the KISS metric learning method.

Large Margin Nearest Neighbor (LMNN): A mapping $D : X \times X \to \mathbb{R}_0^+$ over a vector space X is defined as a metric if for all the vectors $\vec{x_i}, \vec{x_j}, \vec{x_k} \in X$ it satisfies some properties such as symmetry and triangular inequality [15]. It is possible to obtain a family of metrics on X by computing Euclidean distances after performing a linear transformation $x = Lx$. These metrics compute quadratic distances that can be expressed in terms of the square matrix $M = L'L$. Thus, any matrix M formed in this way from a real-valued matrix L is guaranteed to be positive semidefinite, refers to the Mahalanobis metric. In LMNN, the distances are viewed as generalizations of Euclidean distances, i.e., Euclidean distances are recovered by setting M to be equal to the identity matrix. The idea is based on the observation that the kNN classification could have a good performance for a sample of the data if its k-nearest neighbors share the same label. By increasing the number of training samples with this property they learned a linear transformation of the input space that precedes kNN classification using Euclidean distances. Their approach has the advantage of improving the original Euclidean distance from a classification perspective and in some cases to provide a lower-dimensional embedding of the data.

Linear Discriminant Analysis: Different ways have been proposed to estimate Mahalanobis distance metrics to compute distances in k-NN classification. One of such methods is Eigen decomposition. This approach has been used to discover informative linear transformations of the input space, which can be seen as inducing a Mahalanobis distance metric in the original space. LDA is a representative Eigenvector method. It operates in a supervised setting and uses the class labels of the inputs to derive informative linear projections. In the context of metric learning, LDA computes a linear projection L that maximizes the amount of between-class variance relative to the amount of within-class variance. The linear transformation L is chosen to maximize the ratio of between-class to within-class variance, subject to the constraint that L defines a projection matrix. The traditional LDA algorithm is still attractive compared to several recently developed metric learning algorithms [16].

Keep It Simple and Straightforward Metric Learning (KISSME):
Another strategy is to learn an optimal distance measure for genuine and impostor pairs. Koestinger et al. [17] proposed an effective method to learn the distance metric based on a likelihood-ratio test. The equivalence constraints are considered natural inputs to metric learning methods because similarity functions mainly establish a relation between pairs of points. KISSME [17] computes the covariance matrix of similar and dissimilar pairs, and uses the difference of the inverse covariance matrix as a projection matrix. It does not rely on complex iterative optimization, which is an advantage for practical applications. It applies the log likelihood ratio test of two Gaussian distributions for metric learning, and so a simplified closed-form solution can be derived.

4 Experimental Evaluation

We present the results of the proposed scheme for low-resolution face recognition. Two different database were considered for the experiments: the SCFace database [7] and the Labeled Faces in the Wild (LFW) [18]. On all of our experiments, the test images were obtained by down-scaling the original images using a bicubic interpolation at different sizes. A bicubic interpolation was also applied in the up-scaling process to obtain high resolution images. The standard Euclidean distance in the DS was replaced by a learned metric, and the linear discriminant classifier (LDC), which assumes equal covariance matrices for the classes, was used. We computed local binary patterns (LBP) on local blocks of the geometrically normalized images. Histograms were computed on each block and concatenated. The dissimilarity measure was computed on top of a feature representation. Particularly, we created the dissimilarity space using chi-square distance between LBP histograms, since it is a more discriminative measure for histograms.

4.1 Experiments and Discussion on SCFace Database

The SCface database [7] was particularly designed for simulating video-surveillance scenarios, thus it is the most suitable to evaluate the low resolution problem. It consists of 4160 images from 130 people taken in uncontrolled environment. Three different distances, namely 4.20 m (distance1), 2.60 m (distance2), and 1.00 m (distance3); each one with five cameras (cam1, cam2, cam3, cam4, cam5) were used to capture the images. The illumination was uncontrolled and the captured images were different in terms of quality, type and resolution. Example images for distances 2 and 3 appear in Fig. 2.

In order to compare our method with existing approaches we follow the protocol in [19], where the images from distance 3 were normalized to 48×48 pixels as HR images, while the corresponding LR images of 16×16 pixels were obtained from distance 2. Besides, 80 subjects were selected for training and the remaining 50 subjects were used for testing. The experiment was repeated 5 times using 200 PCA components, which provided the best results. The results in terms

Fig. 2. Some examples of SCFace database

of Recognition Rates are reported in Table 1. The standard deviation is also presented. As it can be seen in Table 1, in general the proposed scheme achieves relatively high and stable recognition rates when compared with other state-of-the-art algorithms reported in [19]. In particular, the best result is obtained using the LDA metric learning, with a significantly higher recognition rate.

Table 1. Recognition rates in the SCFace database

Method	Recognition Rate(RR)
Proposal in [19]	43.24
CLPM [8]	29.12
SDA [20]	40.08
CMFA [21]	39.56
Our proposal with KISSME (average)	47.20 ± 0.0087
Our proposal with LMNN (average)	50.80 ± 0.0056
Our proposal with LDA (average)	$\mathbf{56.40} \pm 0.0027$

4.2   Experiments and Discussion on LFW Database

In order to corroborate the obtained results on another dataset and to compare the proposed used of metric learning over the dissimilarity space, we conduct experiments on LFW database [18]. It contains 13233 labeled faces from 5749 people. A subset of the database consisting of 3 832 images belonging to 178 subjects was used during the experiments, by selecting the subjects with 8 or more images. The data is challenging, as the faces are detected in the wild, taken from Yahoo! News. The images have different variations such as pose, scale, clothing, expression, focus, resolution and others. Some example images are shown in Fig. 3

All images were geometrically normalized by the center of the eyes to the LR of 16×16 pixels and to the HR of 48×48 pixels. We randomly divided the data set into two sets for training and testing of equal size five times. In this experiment we compare the standard Euclidean distance to the learned metric in the DS. The obtained results in terms of error rates are shown in Table 2. From the results in Table 2 it can be seen that learning a Mahalanobis metric to replace the Euclidean distance improves the classification in a DS by a great margin.

Fig. 3. Some examples of LFW database

Table 2. Error rates in the LFW database

Classifier	DS	LDA
LDC	0.59	**0.45**

5   Conclusions

In this paper we presented the use of metric learning to learn a Mahalanobis distance metric for LRFR in the dissimilarity space. This learned metric enforces objects for the same class to be closer while objects from different classes are pulled apart. Unlike current methods for LR case, which mostly consider the features space, we proposed a new representation space based on dissimilarities between objects and we improved the classification in this space with metric learning. We evaluated our proposal on two challenging datasets. Experiments showed improvements over previously reported methods. Therefore, the improvement of representations based on relational information seems to be a promising research line for future works.

References

1. Wang, Z., Miao, Z., Wu, Q.J., Wan, Y., Tang, Z.: Low-resolution face recognition: a review. Vis. Comput. **30**(4), 359–386 (2014)
2. Ren, C.X., Dai, D.Q., Yan, H.: Coupled kernel embedding for low-resolution face image recognition. IEEE Trans. Image Process **21**(8), 3770–3783 (2012)
3. Hennings-Yeomans, P.H., Baker, S., Kumar, B.V.: Simultaneous super-resolution and feature extraction for recognition of low-resolution faces. In: IEEE Conference on Computer Vision and Pattern Recognition, CVPR 2008, pp. 1–8. IEEE (2008)
4. Duin, R., Pekalska, E.: The Dissimilarity Representations for Pattern Recognition: Foundations and Applications. World Scientific, Singapore (2005)
5. Satta, R., Fumera, G., Roli, F.: Fast person re-identification based on dissimilarity representations. Pattern Recogn. Lett. **33**(14), 1838–1848 (2012)
6. Hernández-Durán, M., Cheplygina, V., Plasencia-Calaña, Y.: Dissimilarity representations for low-resolution face recognition. In: Feragen, A., Pelillo, M., Loog, M. (eds.) SIMBAD 2015. LNCS, vol. 9370, pp. 70–83. Springer, Heidelberg (2015). doi:10.1007/978-3-319-24261-3_6
7. Grgic, M., Delac, K., Grgic, S.: SCface-surveillance cameras face database. Multimed. Tools Appl. **51**(3), 863–879 (2011)
8. Li, B., Chang, H., Shan, S., Chen, X.: Low-resolution face recognition via coupled locality preserving mappings. IEEE Signal Process. Lett. **17**(1), 20–23 (2010)

9. Biswas, S., Aggarwal, G., Flynn, P.J., Bowyer, K.W.: Pose-robust recognition of low-resolution face images. IEEE Trans. Pattern Anal. Mach. Intell. **35**(12), 3037–3049 (2013)

10. Xing, X., Wang, K.: Couple manifold discriminant analysis with bipartite graph embedding for low-resolution face recognition. Signal Process. **125**, 329–335 (2016)

11. Ding, Z., Suh, S., Han, J.J., Choi, C., Fu, Y.: Discriminative low-rank metric learning for face recognition. In: Automatic Face and Gesture Recognition (FG), 11th IEEE International Conference and Workshops on. vol. 1, pp. 1–6. IEEE (2015)

12. Orozco-Alzate, M., Castellanos-Domínguez, C.: Nearest feature rules and dissimilarity representations for face recognition problems. Face Recognition; International Journal of Advanced Robotic Systems, pp. 337–356. Vienna, Austria (2007)

13. Li, Y., Duin, R.P., Loog, M.: Combining multi-scale dissimilarities for image classification. In: Proceedings of the 2012 21th International Conference on Pattern Recognition, ICPR 2012. IEEE Computer Society (2012)

14. Hastie, T., Tibshirani, R.: Discriminant adaptive nearest neighbor classification. IEEE Trans. Pattern Anal. Mach. Intell. **18**(6), 607–616 (1996)

15. Weinberger, K.Q., Saul, L.K.: Distance metric learning for large margin nearest neighbor classification. J. Mach. Learn. Res. **10**, 207–244 (2009)

16. Liao, S., Lei, Z., Yi, D., Li, S.Z.: A benchmark study of large-scale unconstrained face recognition. In: Biometrics (IJCB), 2014 IEEE International Joint Conference on, pp. 1–8. IEEE (2014)

17. Koestinger, M., Hirzer, M., Wohlhart, P., Roth, P.M., Bischof, H.: Large scale metric learning from equivalence constraints. In: 2012 IEEE Conference on Computer Vision and Pattern Recognition (CVPR), pp. 2288–2295. IEEE (2012)

18. Huang, G.B., Ramesh, M., Berg, T., Learned-Miller, E.: Labeled faces in the wild: a database for studying face recognition in unconstrained environments. Technical Report 07–49, University of Massachusetts, Amherst (2007)

19. Shi, J., Qi, C.: From local geometry to global structure: learning latent subspace for low-resolution face image recognition. IEEE Signal Process. Lett. **22**(5), 554–558 (2015)

20. Zhou, C., Zhang, Z., Yi, D., Lei, Z., Li, S.Z.: Low-resolution face recognition via simultaneous discriminant analysis. In: Biometrics (IJCB), 2011 International Joint Conference on, pp. 1–6. IEEE (2011)

21. Siena, S., Boddeti, V.N., Vijaya Kumar, B.V.K.: Coupled marginal fisher analysis for low-resolution face recognition. In: Fusiello, A., Murino, V., Cucchiara, R. (eds.) ECCV 2012. LNCS, vol. 7584, pp. 240–249. Springer, Heidelberg (2012). doi:10.1007/978-3-642-33868-7_24

Video Temporal Segmentation Based on Color Histograms and Cross-Correlation

Anderson Carlos Sousa e Santos and Helio Pedrini[✉]

Institute of Computing, University of Campinas, Campinas, SP 13083-852, Brazil
helio@ic.unicamp.br

Abstract. Several fields of knowledge generate and consume massive volumes of videos, such as entertainment, telemedicine, surveillance and security. The rapid growth in the demand for multimedia content has driven the development of fast and scalable mechanisms for storing, retrieving and transmitting video sequences. The automatic temporal segmentation is a fundamental process in the analysis of video content. This work proposes and evaluates an adaptive video shot detection based on color histograms and normalized cross-correlation. Experiments conducted on several video sequences demonstrate that the combination of these two features achieve high accuracy rates.

Keywords: Multimedia content · Video transition · Shot detection · Temporal segmentation · Frame dissimilarities

1 Introduction

Advances in data acquisition technologies have enabled users to record and share videos through a number of portable devices, such as cell phones, tablets, and digital cameras. Due to this steady increase in multimedia contents, a challenging task is to develop efficient mechanisms for storing, indexing, retrieving and transmitting such large amounts of data.

Video summarization [3] consists in automatically generating a short version of a video sequence, allowing the user to quickly evaluate the relevance of its content by means of only a set of representative frames. As a temporal video segmentation process [4,6], some challenges associated with the video summarization include camera motion, varying lighting conditions, video genres, and subjectivity in the evaluation process.

The main contribution of this work is the proposition and evaluation of a video shot segmentation method based on the combination of inter-frame dissimilarity vectors of color histogram distances and block-based normalized cross-correlation between image pixel intensities. In addition, an adaptive local threshold strategy is defined to automatically detect the boundary frames. Experiments conducted on public video sequences demonstrate that the proposed method achieves high accuracy rates.

© Springer International Publishing AG 2017
C. Beltrán-Castañón et al. (Eds.): CIARP 2016, LNCS 10125, pp. 225–232, 2017.
DOI: 10.1007/978-3-319-52277-7_28

This paper is organized as follows. Section 2 briefly presents some relevant concepts and works related to the topic under investigation. Section 3 describes the proposed shot video detection methodology. Section 4 presents and discusses some of the results obtained with the proposed method. Finally, Sect. 5 concludes our work and includes some future work suggestions for improving the proposed method.

2   Background

Due to the advances in multimedia technology and large availability of digital content, there is an increasing demand for robust mechanisms for storing, indexing, browsing and retrieving video data. An open research problem is the automatic construction of a compact and meaningful representation of massive video sequences to help users understand the most important information of their content [9].

Temporal segmentation of a video into semantic units is a crucial stage in the analysis of video contents, whose process is known as shot boundary detection. A video shot consists of one or more frames generated contiguously to form a continuous action in time and space. A video summary can be constructed from a set of keyframes that represent the shots. In this context, two categories of transitions between shots are commonly defined: abrupt and gradual transitions. An abrupt transition corresponds to a cut between one frame of a shot and its adjacent frame in the next shot, whereas a gradual transition represents a smooth change over several frames.

Several video shot boundary detection approaches have been proposed in the literature [1,2,5,7]. Two main steps are commonly performed in the cut detection methods: (i) a similarity or dissimilarity measure is initially computed for each pair of consecutive frames and (ii) a cut is detected if the measure is higher than a specified threshold.

3   Methodology

The proposed video cut detection method is based on two different dissimilarities between consecutive frames: the Bhattacharyya distance between color histograms and the inverse normalized cross-correlation between the intensity image blocks. The resulting metrics are combined with a simple mean fusion and submitted to an adaptive thresholding technique that detects the relative high disparity and classifies the frames as part of a shot transition or not. These main steps are illustrated in Fig. 1.

3.1   Histogram-Based Dissimilarity

In order to calculate the inter-frame dissimilarity, a quantized color histogram (CH) is extracted from each frame and the distance between two consecutive frames is calculated with the Bhattacharyya distance, as defined in Eq. 1.

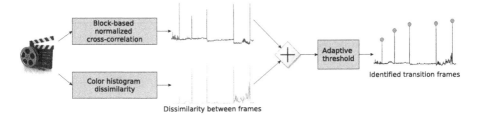

Fig. 1. A flowchart of the proposed video cut detection method.

$$d(H_i, H_{i-1}) = \sqrt{1 - \frac{1}{\sqrt{\overline{H}_i \cdot \overline{H}_{i-1} \cdot N^2}} \sum_b^B \sqrt{H_i(b) \cdot H_{i-1}(b)}} \qquad (1)$$

where $\overline{H}_k = \frac{1}{N} \sum_j H_k(j)$ and $H_i(b)$ is the probability of frame i having a pixel that falls into the color bin b.

3.2 Block-Based Cross-Correlation

The negative normalized cross-correlation (NCC) is a dissimilarity measure over the intensity image, as stated in Eq. 2.

$$d(f_i, f_{i-1}) = -\frac{1}{N} \sum_{x,y} \frac{(f_i(x,y) - \overline{f}_i)(f_{i-1}(x,y) - \overline{f}_{i-1})}{\sigma_{f_i} \sigma_{f_{i-1}}} \qquad (2)$$

where $\overline{f}_i$ is the average of f_i and σ_{f_i} is the standard deviation.

In order to avoid sensitivity to local changes between frames and presence of noise, each video frame is divided into non-overlapping blocks and the negative cross-correlation is calculated for each pair of corresponding blocks. Algorithm 1 summarizes the main steps for the block-based cross-correlation.

The block with the minimum dissimilarity is chosen since a significant change in it implies that all other blocks also changed.

3.3 Fusion

A combination of the dissimilarity vectors of the histogram-based distance and the block cross-correlation is performed to minimize the individual errors and uncertainty. Prior to the fusion process, a z-score normalization followed by a min-max scaling is applied to both vectors. The resulting dissimilarity vector constitutes a weighted mean between each position. Equation 3 summarizes the process.

$$D = \omega \cdot D_{CH} + (1 - \omega) \cdot D_{B\text{-}NCC} \qquad (3)$$

Algorithm 1. Block-based cross-correlation dissimilarity

---

    input : video V, number of blocks K
    output: dissimilarity vector D

1 $D \leftarrow \emptyset$
2 **for** $f_i \in V$ **do**
3 | divide the video frame into K blocks
4 | $NCC \leftarrow \emptyset$
5 | **for** $k \in K$ **do**
6 | | $NCC_k \leftarrow d(f_{k_i}, f_{k_{i-1}})$ // Equation 2
7 | $D \leftarrow \min(NCC)$
8 **return** D

---

where D_{CH} and $D_{B\text{-}NCC}$ are the dissimilarity vectors for the color histogram and the block-based cross-correlation, respectively, D is the final vector of dissimilarities between frames, whereas ω is the weight applied to each dissimilarity measure.

3.4   Adaptive Thresholding

The thresholding over the dissimilarity vector is performed locally through a moving window. Since the goal is to find *peaks* in the frame dissimilarities, this stage is similar to an outlier detection process.

A local median M is calculated for each moving window of size m with center at i. The frames i and $i-1$ are considered as boundary transition frames if their dissimilarity is equal to or greater than the median plus an α value ($d_i \geq M + \alpha$). Furthermore, it needs to be the maximum point within the window to ensure that only the dominant peak is labeled as transition, avoiding redundancy. Figure 2 illustrates the behavior of the proposed thresholding method.

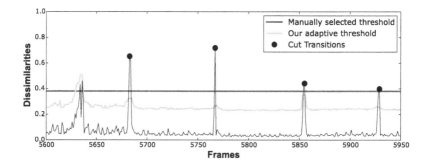

Fig. 2. Adaptive threshold over temporal dissimilarities.

4   Experimental Results

Experiments were conducted on two different annotated data sets. The first one, referred here to as VIDEOSEG'2004 [10], contains 10 video sequences within a diversity of genres, such as news, commercial, movies, cartoons, television shows, as well as other challenging scenarios with low quality digitization, low lighting conditions, fast motions and production effects. The second data set is a shot boundary test collection for the TRECVID'2002 [8]. It consists of 18 videos, where most of them are documentaries and amateur films with low quality, noise and production artifacts, varying in length, date of creation and production style.

The evaluation protocol follows the TRECVID guidelines, such that the results are assessed in terms of precision, recall and their harmonic mean (F_{score}). Equations 4 and 5 express the precision and recall measures, respectively, for a video V with a detection set S.

$$\text{Precision} = \frac{\sum\limits_{f_i \in V} S(i) \in Cut \wedge i \in True\ Cut}{\sum_{f_i \in V} S(i) \in Cut} \tag{4}$$

$$\text{Recall} = \frac{\sum\limits_{f_i \in V} S(i) \in Cut \wedge i \in True\ Cut}{\sum_{f_i \in V} i \in True\ Cut} \tag{5}$$

The F_{score} measure is defined as

$$F_{score} = 2\,\frac{\text{Precision} \times \text{Recall}}{\text{Precision} + \text{Recall}} \tag{6}$$

The adaptive thresholding parameters were empirically determined and applied to all videos in both data sets. In our experiments, values of $\alpha = 0.2$ and window size $m = 7$ achieved the best performance. For the histogram, a quantization with 32 bins for each RGB channel (totalizing 32,768 colors) was defined. The number of blocks K applied to the cross-correlation was set to 16, generating a 4×4 grid on the frames. For the fusion, a constant weight $\omega = 0.5$ demonstrated to be the best overall value in both data sets.

Table 1 shows the results for the VIDEOSEG'2004 data set. The described approaches and a baseline available for the data set [10] based on feature tracking were compared with the proposed fusion method.

It is noticeable that our fusion strategy for color histogram and block-based cross-correlation outperforms the respective individual methods and the provided baseline. The block-based normalized cross-correlation performs poorly in comparison to the global approach, once the videos in the VIDEOSEG'2004 data set have different frame dimensions and the block partitioning can discard important information. Nevertheless, such factors did not affect the performance of our fusion method.

Table 2 shows the results for the TRECVID'2002 data set. It is possible to observe that the proposed fusion outperforms all other approaches. Moreover,

Table 1. Video cut detection results (VIDEOSEG'2004).

Method	Precision (%)	Recall (%)	F_{score} (%)
Color Histogram (CH)	84.66	85.71	84.94
Normalized Cross-Correlation (NCC)	88.94	90.96	88.72
Block-based NCC (B-NCC)	97.44	69.80	75.35
Feature tracking [10]	87.40	96.10	90.80
Proposed fusion (B-NCC + CH)	98.21	91.60	**94.17**

Table 2. Video cut detection results (TRECVID'2002).

Method	Precision (%)	Recall (%)	F_{score} (%)
Color Histogram (CH)	78.34	91.50	83.72
Normalized Cross-Correlation (NCC)	75.05	94.99	80.45
Block-based NCC (B-NCC)	82.46	89.60	85.08
Proposed fusion (B-NCC + CH)	91.60	94.76	**92.93**

the block-based cross-correlation achieves superior performance than the global cross-correlation.

Figure 3 presents the official results, provided by TRECVID'2002, for each participant in the competition. Number in parentheses represent the number of submissions for each team. Our results are presented in a similar manner to allow an adequate comparison.

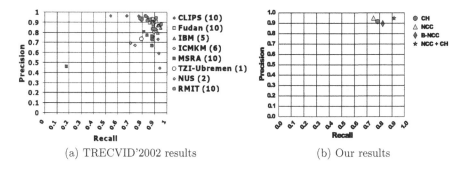

(a) TRECVID'2002 results   (b) Our results

Fig. 3. Precision/Recall performance for (a) participants in the TRECVID'2002 and (b) methods described in this work.

Our proposed method outperforms the majority of the submissions both in precision and recall. Furthermore, even the methods without combination are competitive to those submitted to TRECVID'2002. This corroborates the advantage of our adaptive thresholding method.

Figure 4 illustrates the inter-frame dissimilarities for a video section from TRECVID'2002. It is possible to observe that the number of false positives detected by the color histogram and block-based normalized cross-correlation are reduced through the fusion strategy.

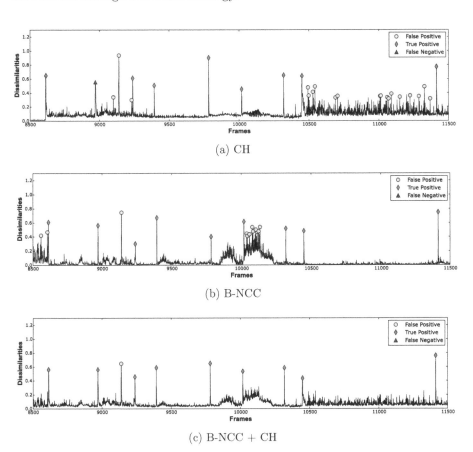

(a) CH

(b) B-NCC

(c) B-NCC + CH

Fig. 4. Frame dissimilarities for a video section from TRECVID'2002.

5   Conclusions and Future Work

This work proposed an adaptive video cut detection method based on the combination of color histograms and cross-correlation. Furthermore, a local thresholding strategy is used to search for relative significant peaks.

Although the inter-frame dissimilarity measures are simple, both the fusion and adaptive thresholding approaches produced significant improvements in the experimental results obtained on two different data sets containing challenging videos. The proposed method also proved to be very competitive compared to the best submissions to TRECVID'2002 shot boundary competition.

As directions for future work, we intend to extend the method to address video gradual transitions and the automatic determination of weights in the fusion process.

Acknowledgments. The authors are thankful to FAPESP (grant #2011/22749-8) and CNPq (grant #307113/2012-4) for their financial support.

References

1. Apostolidis, E., Mezaris, V.: Fast shot segmentation combining global and local visual descriptors. In: IEEE International Conference on Acoustics, Speech and Signal Processing, pp. 6583–6587. IEEE (2014)
2. Birinci, M., Kiranyaz, S.: A perceptual scheme for fully automatic video shot boundary detection. Sig. Process.: Image Commun. **29**(3), 410–423 (2014)
3. Cirne, M.V.M., Pedrini, H.: Summarization of Videos by Image Quality Assessment. In: Shao, L., Shan, C., Luo, J., Etoh, M. (eds.) Progress in Pattern Recognition, Image Analysis, Computer Vision, and Applications. Advances in Pattern Recognition, pp. 901–908. Springer, Heidelberg (2014). doi:10.1007/978-1-84996-507-1_10
4. Jiang, H., Zhang, G., Wang, H., Bao, H.: Spatio-temporal video segmentation of static scenes and its applications. IEEE Trans. Multimedia **17**(1), 3–15 (2015)
5. Jiang, X., Sun, T., Liu, J., Chao, J., Zhang, W.: An adaptive video shot segmentation scheme based on dual-detection model. Neurocomputing **116**, 102–111 (2013)
6. Petersohn, C.: Temporal Video Segmentation. Jörg Vogt Verlag, Niederwaldstr (2010)
7. Tippaya, S., Sitjongsataporn, S., Tan, T., Chamnongthai, K., Khan, M.: Video shot boundary detection based on candidate segment selection and transition pattern analysis. In: IEEE International Conference on Digital Signal Processing, pp. 1025–1029, July 2015
8. TRECVID: TRECVID Data Availability (2016). http://trecvid.nist.gov/trecvid.data.html
9. Veltkamp, R., Burkhardt, H., Kriegel, H.P.: State-of-the-Art in Content-based Image and Video Retrieval, vol. 22. Springer Science & Business Media, Heidelberg (2013)
10. Whitehead, A., Bose, P., Laganiere, R.: Feature based cut detection with automatic threshold selection. In: Enser, P., Kompatsiaris, Y., O'Connor, N.E., Smeaton, A.F., Smeulders, A.W.M. (eds.) CIVR 2004. LNCS, vol. 3115, pp. 410–418. Springer, Heidelberg (2004). doi:10.1007/978-3-540-27814-6_49

Automatic Classification of Herbal Substances Enhanced with an Entropy Criterion

Victor Mendiola-Lau[1]([✉]), Francisco José Silva Mata[1], Yoanna Martínez-Díaz[1], Isneri Talavera Bustamante[1], and Maria de Marsico[2]

[1] Advanced Technologies Application Center, Havana, Cuba
{vmendiola,fjsilva,ymartinez,italavera}@cenatav.co.cu
[2] Universita degli Studi de Roma, Rome, Italy
demarsico@di.uniroma.it

Abstract. This paper presents a novel automatic pattern recognition system for the classification of herbal substances, which comprises the analysis of chemical data obtained from three analytical techniques such as Thin Layer Chromatography (TLC), Gas Chromatography (GC) and Ultraviolet Spectrometry (UV), composed of the following stages. First, a preprocessing stage takes place that ranges from the TLC plate image conversion into a spectrum to the normalization and alignment of spectral data for all techniques. Then, a hierarchical clustering procedure is applied for each technique with the goal of discovering groups or classes that provide evidence concerning the different existing types. Next, an entropy-based template selection step for each group was introduced to exclude the less significant samples, thus allowing to improve the quality of the training set for each technique. In this manner, each class is now described by a set of key prototypes that allows the field expert to have a more accurate characterization and understanding of the phenomenon. Moreover, an improvement of the computational complexity for training and prediction tasks of the Support Vector Machines (SVM) is also achieved. Finally, a SVM classifier is trained for each technique. The experiments conducted show the validity of the proposal, showing an improvement of the classification results on each technique.

Keywords: Herbal substance · Clustering · Entropy · Template selection

1 Introduction

The identification of types of substances is a common task in analytical chemistry. This process entails not only recognizing types of substances or its components, but also discovering features that allow to distinguish among them. These differences arise commonly from types of processing, growing conditions, and geographic location. Recently, in analytical chemistry, the application of computational methods in the analysis of data resulting from analytical techniques, has allowed to increase the speed and accuracy of the system response, thus assisting the field expert in the decision taking process [7]. Analytical techniques such as

© Springer International Publishing AG 2017
C. Beltrán-Castañón et al. (Eds.): CIARP 2016, LNCS 10125, pp. 233–240, 2017.
DOI: 10.1007/978-3-319-52277-7_29

Thin Layer Chromatography (TLC) provide a very economical and straightforward choice to identify chemicals, such as plant extracts or mixtures. However, in order to reach safer conclusions, this technique is usually used in combination with other analytical techniques such as Ultraviolet Spectrometry (UV) and Gas Chromatography (GC) with the purpose of combining information and thus, having a more robust criterion to issue a verdict on the identity of an unknown substance or its type. The identification and classification of herbal substances is a complex problem given the fact that these are composed of a mix of many active compounds, which is where our study is focused. In this paper, we use Cannabis samples as representatives of these types of substances, though our proposal can be applied to any kind of mixed substances.

To our knowledge, no work in the literature proposes an automatic system able to integrate all the process steps: from data acquisition arising from several chemical analytical techniques, to obtaining suitable classification models that allow a proper characterization of the chemical class composition, based on the information provided by each technique. Figure 1 summarizes the main steps of our proposal for a completely automatic identification process, which represents a step forward in the solution of these complex problems regarding herbal substances identification.

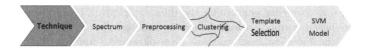

Fig. 1. Proposed methodology for the identification process.

The paper is organized as follows. Section 2 describes the processing applied to spectral data. Next, Sect. 3 discusses the class discovery strategy. Section 4 proposes entropy-based criterion used to select the best templates. In Sect. 5, the classification procedure is presented. Discussion of experimental results is carried at Sect. 6. Conclusions and future work are outlined in Sect. 7.

2   Data Preprocessing

Acquired data for UV and GC analytical techniques have a spectral nature (see Fig. 2 (center) and (right)). However, in the case of the TLC technique, each sample is obtained in an image as spots linearly distributed on a narrow band or lane. Each spot corresponds to the presence of an active compound of the mix. After a photometric normalization of the image lane, a projective integral operation was applied to obtain a spectrum as shown in Fig. 2 (left). This new spectral representation for the TLC technique is more compact and robust than an image and at the same time, the field expert can analyze altogether the three techniques at once.

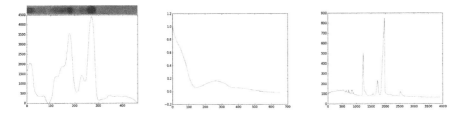

Fig. 2. The corresponding spectra of a substance for different analytical techniques: TLC (left), UV (center) and GC (right).

Next, high frequency and low amplitude noise are removed to prevent them from being confused with real peaks of substances components. Smoothing the spectra through spline interpolation [8] (see Fig. 3) was used. Chromatographic methods (TLC and GC) introduce experimental or instrumental errors. These so-called artifacts are mainly shifted peaks, co-elution of peaks, background off-set, displacement of the baseline and scaling effects [9]. For these reasons, an alignment process is needed, which consists of adjusting the correspondence of concentration peaks of each substance, in accordance with their retention factors, represented as time. This process must be performed with respect to a suitable reference spectrum selected for each technique. The main selection criterion was to use the pattern that best correlates with others patterns, i.e. the one providing the maximum cumulative product of correlation coefficients. As a secondary criterion, the presence of peaks corresponding to basic chemical components of the substances is taken into account [9]. In order to deal with the problems of scaling and baseline shifts, the Multiplicative Scatter Correction (MSC) was employed [4]. Classical alignment algorithms such as the Correlation Optimized Warping (COW) and the Dynamic Time Warping (DTW) [10] were evaluated, where the better results were achieved by COW.

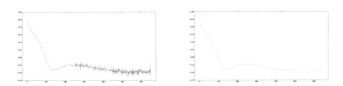

Fig. 3. Smoothing of the UV spectra by spline interpolation.

3   Discovery of Classes

Since the number of existing classes is unknown, an unsupervised clustering technique such as *hierarchical clustering analysis* (HCA) with a complete linkage criterion was applied. A metric to select the optimal partition was not employed, as the final partition was determined by means of an exploratory analysis assessed by a field expert, where the presence and relation among the concentrations of active chemical components were analyzed. Given its precision and resolution capability, GC data was used as *ground truth* for the exploratory analysis performed on the UV and TLC data sets.

Two substances can be considered similar regarding the correlation of their composition, i.e., they have the same compounds present but also with a similar relative concentrations among them. For this reason, two correlation distances [6,11] were selected as measures of similarity aiming to capture this behavior.

Figure 4 shows the clustering results of four of our classes, each class having more or less the same amount of samples for the same technique. The clusters labeled as 1 and 4 agreed in a 100 % of samples for the three techniques. Similarly, clusters labeled as 2 and 3 agreed only in 90 % and 75 % of the samples respectively. Moreover, it can be noticed that each cluster, regardless of the analytical technique it belongs to, contains spectra with similar shape.

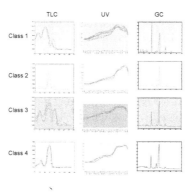

Fig. 4. Visual inspection of our final clustering results.

4   Entropy-Based Template Selection

In many works, models derived and somehow related with Information Theory have been used. Fields range from information communication to biometrics. Particularly, strategies for template selection based on the concept of entropy have been proposed [2], achieving very good results in identifying the most representative templates in a given set. The core concept relies on the computation of the entropy of a gallery of templates, which can be useful for the template selection task. This concept can be extended to our scenario, where each labeled group is considered as class of a given substance.

Given a gallery G of templates, this can be considered as the union of subgalleries G_k where every $g_{i,k} \in G_k$, with $i = 1, 2, \cdots, |G_k|$, belong to the same class k. Therefore, $G = \cup_{k=1}^{K} G_k$ and $G_k \cap G_h = \varnothing$, $\forall k \neq h$. A similarity measure s associates a real scalar value to a pair of templates. The comparison of a probe v with a gallery template $g_{i,k}$ can be denoted by $s_{i,v} = s(v, g_{i,k})$ which can be normalized to be a value in the real interval $[0, 1]$ [2]. If we had a distance measure or a dissimilarity coefficient $d_{i,v} = d(v, g_{ui,k})$, it is possible to consider the similarity value as $s_{i,v} = 1 - d_{i,v}$. Assuming that template v was correctly assigned to the class k, each similarity value $s_{i,v}$ can be interpreted as the probability that template v conforms to $g_{i,k}$, that is $s_{i,v} = p(v \approx g_{i,k})$.

According to this interpretation, a proper normalization of $s_{i,v}$ is needed, to have it ranging in the interval $[0, 1]$. Also, $\sum_{i=1}^{|G_k|} s_{i,v} = 1$ holds given the fact that v was correctly assigned to G_k. It is now possible to compute the (probe-related) entropy of the probability distribution obtained applying (4) to the whole G_k with respect to a probe v as follows:

$$H(G_k, v) = -\frac{1}{\log_2 |G_k|} \sum_{i=1}^{|G_k|} s_{i,v} \log_2 (s_{i,v}), \tag{1}$$

where $\frac{1}{\log_2(|G_k|)}$ is a normalization factor, corresponding to the maximum entropy obtained when (4) has the same value for all the templates in the gallery G_k. In this way, the value obtained for $G(G_k, v)$ is normalized in the range $[0, 1]$. Finally, the entropy for the gallery G_k is computed by considering each template $g_{j,t}$ in turn as a probe v:

$$H(G_k) = -\frac{1}{\log_2 |Q|} \sum_{q_{i,j} \in Q} s_{i,j} \log_2 (s_{i,j}), \tag{2}$$

where Q represents the set of pairs $q_{i,j} = (q_{i,k}, q_{j,k})$ of elements in G_k such that $s_{i,j} > 0$. Once again, values are in the range $[0, 1]$ irrespective of the size of the gallery. $H(G_k)$ represents a measure of heterogeneity for G_k and it is worth noticing that having a common range for $H(G_k)$ allows for comparison of the representativeness of different galleries. Given a gallery G_k, the computation of $H(G_k)$ can be used to select a subset of representative samples out of it. The procedure described in [2] is based on an ordering of the gallery templates according to a representativeness criterion.

After performing sample ordering according to the above procedure, several criteria may be used to select those templates that guarantee a suitable representativeness of the gallery. A possible criterion is the *one-shot* selection proposed in [3]. With this strategy, the entropy difference is computed between each template and the preceding one and all templates are sorted according to this difference value. Finally, the k first templates are chosen as the k best representative templates. Another suitable strategy in this direction could be to select the *top-percentage* of templates according to the entropy difference, which has the additional advantage of allowing to control the resulting data set size.

It is worth underlining that the same technique, depending on the application requirements at hand, can be used for either single sub-galleries G_k, as in the present case, or for an overall gallery G of templates.

5   SVM Classification

Partitions obtained after the selection of the most representative templates are used for training SVM classifiers, one for each technique. Two substances can be considered *similar*, when their components have similar relative concentrations. The standard correlation coefficient and Spearman correlation coefficient are commonly used to account for spectral similarity [1]. Due to the effectiveness of these correlation distances in determining the classes for training and as an effective measure of dissimilarity between the spectra, it was decided to use correlation kernels for the implementation of SVM classifiers [5]. For the correlation kernel based on the standard correlation coefficient, Eq. 3 and Eq. 4 were exploited, where $\bar{X}$ is the sample mean of data matrix $\mathbf{X}$. On the other hand, for the correlation kernel based on the Spearman correlation coefficient, Eq. 3 and Eq. 5 [11] were exploited, where n corresponds to the amount of spectral bands and x_i, y_i correspond to the i-th spectral bands.

$$corr(\mathbf{X}) = \begin{pmatrix} corr(X_1, X_1) & corr(X_1, X_2) & ... & corr(X_1, X_n) \\ corr(X_2, X_1) & corr(X_2, X_2) & ... & corr(X_2, X_n) \\ \vdots & \vdots & \ddots & \vdots \\ corr(X_n, X_1) & corr(X_n, X_2) & ... & corr(X_n, X_n) \end{pmatrix}, \tag{3}$$

$$corr(X, Y) = \frac{(X - \bar{X})^T (Y - \bar{X})}{\sqrt{(X - \bar{X})(X - \bar{X})^T} \sqrt{(Y - \bar{X})^T (Y - \bar{X})}}, \tag{4}$$

$$spea(X, Y) = \frac{n \sum_{i=1}^{n} x_i y_i - (\sum_{i=1}^{n} x_i)(\sum_{i=1}^{n} y_i)}{\sqrt{[n \sum_{i=1}^{n} x_i^2 - (\sum_{i=1}^{n} x_i)^2][n \sum_{i=1}^{n} y_i^2 - (\sum_{i=1}^{n} y_i)^2]}}, \tag{5}$$

The correlation kernels are based on the correlation matrix defined above and the kernel values for any X_i and X_j are defined as:

$$K_{corr}(X_i, X_j) = e^{(1-corr(X_i, X_j))}, \quad K_{spea}(X_i, X_j) = e^{(1-spea(X_i, X_j))}. \tag{6}$$

6   Experimental Results and Discussion

The data sets obtained from the proposed template selection strategies as well as the original ones, were used as training and validation sets in order to assess their descriptive power. Table 1 shows the best template selection strategies compared with the original data sets (used as baseline) for each analytical technique: GC (142 samples), UV (143 samples) and TLC (92 samples). Column 2 shows the template selection strategies analyzed for each technique. In Column 3, the size of optimized data sets yielded by each selection strategy is provided. Finally, Columns 4 and 5 show the classification accuracy (CA) for a 10-fold cross validation procedure obtained for the correlation measures used at the different stages of the process: class discovery, entropy-based template selection and building a precomputed kernel for the SVM classifier.

Our system obtained a very good classification accuracy for the template selection strategies analyzed, almost always improving the classification results

Table 1. Improvement on classification and training set reduction of our strategy over the original data sets.

Technique	Selection strategy	No. samples	CA (Correlation)	CA (Spearman)
GC	baseline	142	95.07 %	94.37 %
	one-shot	**58**	**96.61 %**	**98.28 %**
	top-25 %	38	92.11 %	92.11 %
	top-50 %	**74**	**98.65 %**	93.24 %
	top-75 %	**111**	**97.30 %**	**95.50 %**
UV	baseline	143	96.50 %	95.10 %
	one-shot	**52**	**98.08 %**	**96.23 %**
	top-25 %	38	92.11 %	89.47 %
	top-50 %	**73**	**97.26 %**	**95.89 %**
	top-75 %	**110**	**97.27 %**	**96.36 %**
TLC	baseline	92	96.74 %	89.13 %
	one-shot	**80**	**98.75 %**	**96.77 %**
	top-25 %	22	95.45 %	81.82 %
	top-50 %	**44**	**97.73 %**	**93.18 %**
	top-75 %	**66**	**96.97 %**	**90.91 %**

compared to those obtained for the original data sets, therefore showing the validity of the proposal. The only selection strategy that could not achieve better results was the selection of the 25 % most representative templates of each class (top-25 %), this is due to the fact too few templates were selected, thus excluding others with relevant information for the class description. As can be seen, an additional advantage of the proposed methodology is a significant reduction of the training sets, thus improving the computational efficiency in the training and classification processes. As shown, the *one-shot* variant is usually a very good starting point, but it is also advisable to apply a complementary strategy for template selection. In some cases, the best results were obtained for the *top-percentage* strategy, which not only selected the same templates than the one-shot strategy, but also other templates contribute to improve the classification accuracy. Also, it is important to stress that the measure for spectral comparison based on the standard correlation coefficient showed a higher performance.

Due to differences in cluster size, the *one-shot* criterion was modified to select templates with an associated entropy difference greater than a specific threshold, instead of choosing a fixed k per cluster. Under these conditions, this strategy depends on a threshold, which could result in excluding templates that contribute with valuable information or the opposite, selecting templates that alter the class description or contribute with redundant information. On the other hand, the *top-percentage* criterion does not suffer from this problem with cluster sizes, but it is still necessary to define an optimal percent of representative templates.

7   Conclusions and Future Work

The main contribution of this work is the introduction of a novel pattern recognition system designed to cope with the problem of herbal substances classification and identification by combining different analytical techniques. As previously stated, the entropy-based template selection strategy is beneficial not only from a classification enhancement point of view, but also for the great impact in the process of class description for the field expert. Also, the measure used for spectral comparison based on the standard correlation coefficient proved to be more suitable than the one based on the Spearman correlation coefficient. The results obtained show that the proposal achieved an outstanding classification accuracy, specially considering the complexity of the classification of herbal substances. Moreover, the size of the training sets and the computational effort necessary in the training and classification tasks was reduced significantly. As future work, a thorough analysis should be carried concerning the selection of an optimal threshold for the one-shot selection strategy and the optimal percentage of templates to choose per cluster. Also, additional studies regarding a suitable combination of the independent classifiers would be of great interest.

References

1. Bodis, L.: Quantification of spectral similarity: towards automatic spectra verification. Ph.D. thesis, Babes-Bolyai University (2007)
2. De Marsico, M., Nappi, M., Riccio, D., Tortora, G.: Entropy-based template analysis in face biometric identification systems. Sig. Image Video Process. **7**(3), 493–505 (2013)
3. De Marsico, M., Riccio, D., Vazquez, H.M., Calana, Y.P.: GETSEL: gallery entropy for template selection on large datasets. In: International Joint Conference on Biometrics (IJCB), pp. 1–8 (2014)
4. Helland, I.S., Næs, T., Isaksson, T.: Related versions of the multiplicative scatter correction method for preprocessing spectroscopic data. Chemometr. Intell. Lab. Syst. **29**(2), 233–241 (1995)
5. Jiang, H., Ching, W.K.: Correlation kernels for support vector machines classification with applications in cancer data. Comput. Math. Methods Med. (2012)
6. Kumar, V., Chhabra, J.K., Kumar, D.: Performance evaluation of distance metrics in the clustering algorithms. J. Comput. Sci. **13**(1), 38–52 (2014)
7. Muda, A.K., Choo, Y.H., Abraham, A., Srihari, S.N.: Computational Intelligence in Digital Forensics: Forensic Investigation and Applications. Springer, Heidelberg (2014)
8. Schumaker, L.: Spline Functions: Basic Theory. Cambridge University Press, Cambridge (2007)
9. Skov, T.H.: Mathematical resolution of complex chromatographic measurements. Ph.D. thesis (2008)
10. Tomasi, G., Savorani, F., Engelsen, S.B.: icoshift: an effective tool for the alignment of chromatographic data (2011)
11. Zwillinger, D., Kokoska, S.: CRC Standard Probability and Statistics Tables and Formulae. CRC Press, Boca Raton (1999)

Extended LBP Operator to Characterize Event-Address Representation Connectivity

Pablo Negri[1,2]($^{(\boxtimes)}$)

[1] CONICET, Dorrego 2290, Buenos Aires, Argentine
pnegri@uade.edu.ar
[2] Universidad Argentina de la Empresa (UADE), Lima 717, Buenos Aires, Argentina

Abstract. Address-Event Representation is a flowering technology that can change the visual perception of the computer vision world. This paper proposes a methodology to associate the input data from this kind of sensors. A new descriptor computed using an extended LBP operator seeks to characterize the connectivity of the asynchronous incoming events in a two dimensional space. Those features can be organized on histograms and combined with others descriptors, as histograms of oriented events. They can be the input of traditional classifiers to detect or recognize objects from the scene.

1 Introduction

A new paradigm for visual sensing was introduced in 2006 as the first Event-Driven Dynamic Vision Sensor (DVS) [7]. This sensor is inspired by the asynchronous Address Event Representation (AER), first introduced by Mahowald [9], and by the Kramer's transient detector concept [6]. It consists of a grid of pixels (also called "silicon retinae"), capturing changes of illumination at the focal plane. When a such event occurs, it is transmitted as an information tuple, indicating the pixel position on the grid, the time stamp and the polarity of the event. Thus, this sensor transmits a continuous flow of new events, instead of a 2D frame. This kind of vision sensor is then considered as "frameless".

The DVS visual data representation launch a new branch on the computer vision field. Traditional methodologies have to be modified in order to exploit the new information, and a new theoretical model must be developed to adapt this paradigm. Recent works tackle Visual Flow [1], Corner Detection [3] and Object Recognition [14,15] using the asynchronous temporal flow.

If the DVS camera is fixed, the images have the particularity that only moving objects are captured. In fact, the generated events correspond to the edges of this object. Static objects do not generate events. In some way, the results are similar to the Movement Feature Space [11,12] which constructs a dynamic background model using the boundaries of objects. If the temporal window for learning the background modes is fixed to some milliseconds: a moving object that stops, enters automatically to the background model. This paper seeks to generate a family of features using a histogram representation of the AER data to characterize objects shape.

© Springer International Publishing AG 2017
C. Beltrán-Castañón et al. (Eds.): CIARP 2016, LNCS 10125, pp. 241–248, 2017.
DOI: 10.1007/978-3-319-52277-7_30

The proposed descriptors are inspired on Histograms of Oriented Gradient (HOG) [4] and Local Binary Patterns (LBP) [13]. They are two of the most useful features employed nowadays on Computer Vision. The HOGs organize the gradient information in the image using histograms. The LBP operator performs a simple analysis (binary) about the relationship between the gray scale values of neighbor pixels. In detection problems, these features obtain good performances attributable to their tolerance to monotone illumination changes [10] and their simple computation. Also, they are considered as complementary features [2].

In this article, the HOG features would be arranged in a similar way as the Histogram of Oriented Level Lines (HO2L) proposed in [11]. The descriptors are organized as histograms using the orientation of the events. These orientations would be computed using the plane fitting methodology of [3]. The second contribution consists in an extended LBP operator. It seeks to characterize the connectivity of the AER data generated by the edges of the moving objects.

Next section details the steps to adapt the data provided by the DVS, followed by Sect. 3 where the histogram of oriented events and the extended LBP operator will be presented. Section 4 discusses the results and concludes the paper.

2   Features Generation

2.1   Dynamic Vision Sensor Data

A DVS reproduces the behavior of biological retinas by capturing asynchronous light changes on a 128×128 pixel grid [8]. Each change generates an event $e = (p, t, pol)$, p being the spatial location on the grid, t the event time stamp, and pol defines the polarity. Polarity is a binary ON/OFF output. ON polarity captures an increase on the illumination, and OFF polarity is obtained when illumination decreases. Figure 1(a) shows the operational principle diagram of the DVS Address-Event Representation from [8].

Generally, descriptors transform visual information evaluating relationships between neighbors pixels. These associations are organized on mathematical representations as: filters, histograms, etc. Because the DVS datasets [5] available

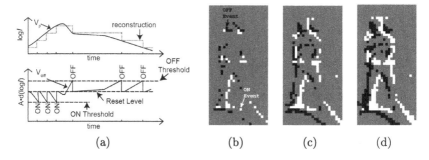

Fig. 1. (a) DVS principle of ON and OFF polarity events generation from [8], (b-d) "Street scene with cars and people walking" sample dataset from [5].

online do not provide pixels intensities or colors (new versions of DVS devices will supply this information), special methodologies should be implemented to study the pixels/event relationships.

Figures 1(b), (c) and (d), show a surveillance sample dataset downloaded from [5]. For visualization purposes the event flow composed of N events is mapped to a 2D matrix E_N which has the same size asf the retina. $E_N(p) = 1$ if the pixel p corresponds to an incoming event with ON polarity (white color). $E_N(p) = -1$ when the polarity of the event at pixel p is OFF (black color). The others pixels of E_N are set to zero (gray color). On Fig. 1(b) is represented a set (window) of $N = 100$ events. The set also corresponds to a temporal delay of 13.24 milliseconds (ms). As can be seen, the available information is not enough to recognize the person. On Fig. 1(c) and (d) $N = 300$ and $N = 500$ events, we respectively have a temporal windows of 40.7 and 67.7 ms. The number of events is now sufficient to identify a person using appropriate features.

Thus, to generate a descriptor using the DVS information, each activated pixel will evaluate its activated neighbors, within the N events on the window, and analyze their orientations and connectivity.

2.2   Events Orientation

Benosman et al. propose in [1] a regularization method to fit a plane, which considers a small spatio-temporal neighborhood $\Omega(\mathbf{e})$ around an incoming event $\mathbf{e}$. The normal component of the resulting plane is considered as a velocity vector. The direction of the normal vector defines the orientation of the event. In [3] event corners consist of those event positions where at least two valid fitting planes intersect. Clady's work uses a maximum number of events within the vicinity of the event $\mathbf{e}$. It is an interesting approach, because opening a temporal window forces to choose a fixed threshold. This threshold would not be appropriate to the dynamic of different objects in the scene. Then, events belonging to different objects/edges could be considered in the vicinity of $\mathbf{e}$ when fitting the plane to obtain the event orientation.

The orientations returned by Benosman's algorithm have values in a range of $[0, 2\pi]$. These values are converted from orientations to directions in a range between $[0, \pi]$, and discretized to integer value between $[1, V]$. This methodology maps the events from E_N to a matrix D_N.

2.3   Events Connectivity

The LBP operator was initially designed for texture recognition [13]. Let I be a gray-scale image, and the pixel $p \in I$. Thus, $LBP(p)$ assigns a label to p analyzing the gray values of their 8-connectivity neighborhood. To do this, and considering the pixels q_i a 8-connectivity neighbors of p, LBP uses an intermediate function $s(p, q_i)$ defined as:

$$s(p, q) = \begin{cases} 1, & \text{if } I(p) - I(q_i) \geq 0 \\ 0, & \text{if } I(p) - I(q_i) < 0 \end{cases} \tag{1}$$

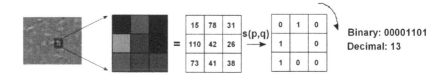

Fig. 2. *LBP* original operator computed on a *p* image pixel.

where $I(x)$ refers to the gray-scale value at position x. The label returned for the operator is obtained as: $LBP(p) = \sum_{i=0}^{8} s(p, q_i) \cdot 2^i$. Figure 2 is an example of the *LBP* computation of the $s(p, q)$ function and the output decimal label: $LBP(p) = 13$. The *LBP* operator is applied to I, giving a label to each pixel.

The patterns '0000000' (0 transition), '00110000' (2 transit.), and '11000111' (2 transit.) are considered as uniform. Others binary labels, '11011001' (4 transit.), and '01010001' (6 transit.) are not uniform. Another particularity of the *LBP* operator, is that the patterns are circular: '00110000' is the same as '11000000'. Thus, in [13] they find 9 uniform unique patterns.

To extend the LBP operator and characterize events connectivity, the Eq. 1 is modified to be adapted to the new data. The neighborhood around the central point p would be evaluated using the equality condition:

$$s(p, q) = \begin{cases} 1, \text{ if } M(p) = M(q_i) \\ 0, \text{ otherwise} \end{cases} \tag{2}$$

The matrix $M(p)$ can be $E_N(p)$ when the polarity of the event is considered to analyze the connectivity, or $D_N(p)$ when the direction of the events is considered.

Figure 3 presents the patterns chosen to identify connectivity on DVS events and defines the operator $eLBP(p)$ around an activated pixel. Two transition patterns capture information of the extrema of a segment (patterns '18' and '21') or possible edges with more than one-pixel width ('1', '2', '3', '5', '8', '13'). Four

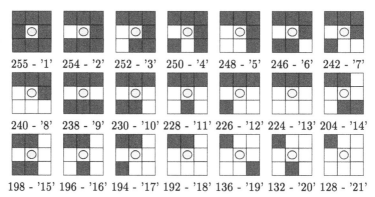

Fig. 3. Extended LBP patterns to characterize events connectivity. It shows the binary code of each pattern and the corresponding label 'x' to identify the pattern.

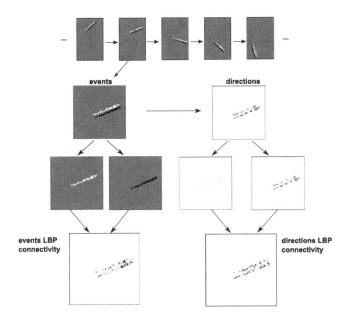

Fig. 4. Connectivity analysis using the extended LBP approach on the events generated by a rotating bar.

transition patterns characterize different configurations of possible connected edges. In total were defined 21 canonical patterns.

There are two possible analysis to study events neighborhood. Figure 4 shows an example of the AER data generated by the rotation of a bar [3]. The 2D representation was obtained histogramming events using a temporal window of $DT = 90\,\text{ms}$. On the left of the figure, the extended LBP operator was applied on the events associated by their polarity. The connectivity is analyzed over the events with ON or OFF polarities. On the right side of the figure, the connectivity analysis characterizes the connectivity of the pixels with the same direction value. Each LBP label gets a specific color on the figure.

Algorithm 1 examples the computation of the extended LBP operator on the directions of the AER data. The inputs of the algorithm are the 2D matrix D

Algorithm 1. Extended LBP Computation on Events

Require: D Events Directions, C extended pattern list, V directions, $L = \oslash$
Ensure: Extended LBP matrix L.
  1: **for** $i = 1$ **to** V **do**
  2:    $\forall\, p\ /\ D(p) = i$
  3:    a) $e = eLBP(p)$
  4:    b) $c \leftarrow C[e]$
  5:    c) $L(p) = c$
  6: **return** L

---

with the direction values at each event. The directions are quantified in V values, and the list C gives the corresponding label to each LBP pattern. The output of the algorithm is the matrix L which gives to each event the eLBP label.

3   Grouping Features as Histograms

To characterize the shape of a moving object in the scene using the two dimensional features (directions and eLBP patterns), the histograms give a measure of the relationship between neighbor pixels.

Using the directions of the events, the construction of the histograms is similar to the HO2L proposed on [11]. Each bin of the histogram corresponds to one direction and accumulates the number of pixels having this value inside a region of the image. The histogram is normalized using the total number of pixels.

The histogram of the eLBP operator is computed in a similar way, and each bin corresponds to one pattern. The position of the pattern on the histogram follows the labels' numbers given on Fig. 3. The other labels which do not belong to one of the 21 canonical patterns get the label 22, and are only considered for histogram normalization by using the total number of pixels of the region.

Figure 5 shows both families of histograms computed on the Poker sequence, which was kindly provided by Dr. Bernabé Linares-Barranco [15]. The events of the sequence were collected inside a window of $N = 150$ events. Using the fitting plane algorithm, a orientation is associated to each event on E_N. Those orientations are then switched to directions $(0 - \pi)$ and quantified in 4 integer values.

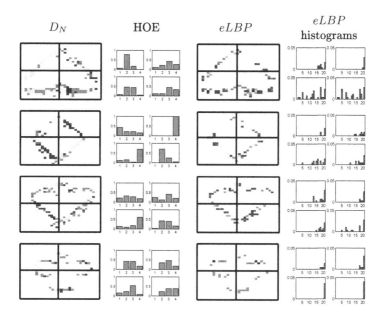

Fig. 5. HOE and histograms of extended LBP patterns obtained on the Poker Dataset.

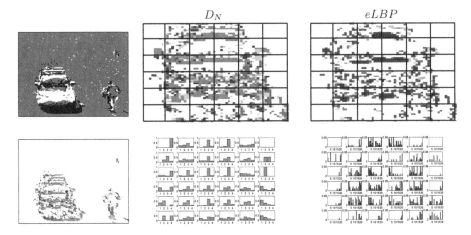

Fig. 6. Histograms analysis of the vehicle on the "Street scene with cars and people walking" dataset.

Figure 5 discriminates events directions with different colors. The image region with the card suits is divided in four patches. Inside each patch, a histogram h_i with V bins, where each bin counts the number of events with the associated direction. The total histogram characterizing the shape is found by concatenating them: $h = \{h_1, h_2, h_3, h_4\}$. This representation is denominated Histograms of Oriented Events (HOE). A similar analysis is performed on the Extended LBP matrix L inside each patch obtaining the histogram $l = \{l_1, l_2, l_3, l_4\}$. Figure 5 presents both features families.

It is also shown on Fig. 6 the same analysis inside an event window on the "Street scene with cars and people walking" dataset. Here, the analysis was performed on the vehicle dividing its region of interest on a grid of 6×6 patches.

4 Discussions and Conclusions

This paper proposes a histogram based feature family that shows promising discriminant properties, and can be employed later on shapes recognition.

As can be seen on Fig. 5, each HOE corresponding to a different poker sign, h^{heart}, h^{spade}, $h^{diamant}$ and h^{clug}, defines the shape in a discriminant way. An analysis of the eLBP histograms, will give a measure of how well the shape is defined within the events window. Uniform eLBP histograms will describe shapes with connected events. On the other hand, when the histogram has highest values at bin 21, as the *club* sign shows, the shape is not completely defined and the event window E_N has numerous isolated events.

Once the histograms h and l are obtained on a event window, they can be the input of classifiers (Boost, SVM, MLP, etc.) in order to perform the detection or recognition of a given class of object. Future research will be conducted on the implementation of shape recognition systems using both Histograms of Oriented Events (HOE) and the Extended LBP operator.

Acknowledgements. This work was funded by PID Nro. P16T01 (UADE, Argentine).

References

1. Benosman, R., Clercq, C., Lagorce, X., Ieng, S.H., Bartolozzi, C.: Event-based visual flow. IEEE Trans. Neural Netw. Learn. Syst. **25**, 407–417 (2013)
2. Cevikalp, H., Triggs, B.: Efficient object detection using cascades of nearest convex model classifiers. In: CVPR, pp. 3138–3145 (2012)
3. Clady, X., Ieng, S., Benosman, R.: Asynchronous event-based corner detection and matching. Neural Netw. **66**, 91–106 (2015)
4. Dalal, N., Triggs, B.: Histograms of oriented gradients for human detection. In: CVPR, vol. 1, pp. 886–893 (2005)
5. JAER: Aer dataset. http://sourceforge.net/p/jaer/wiki/AERdata/. Accessed 30 May 2015
6. Kramer, J.: An integrated optical transient sensor. CS **49**(9), 612–628 (2002)
7. Lichtsteiner, P., Posch, C., Delbrück, T.: A 128 x 128 120 db 30 mW asynchronous vision sensor that responds to relative intensity change. In: ISSCC (2006)
8. Lichtsteiner, P., Posch, C., Delbruck, T.: A 128 * 128 120 db 15μs latency asynchronous temporal contrast vision sensor. JSSC **43**(2), 566–576 (2008)
9. Mahowald, M.: An Analog VLSI System for Stereoscopic Vision. Springer Science & Business Media, Heidelberg (1994)
10. Martínez-Díaz, Y., Méndez-Vázquez, H., Plasencia-Calaña, Y., García-Reyes, E.B.: Dissimilarity representations based on Multi-Block LBP for face detection. In: Alvarez, L., Mejail, M., Gomez, L., Jacobo, J. (eds.) CIARP 2012. LNCS, vol. 7441, pp. 106–113. Springer, Heidelberg (2012). doi:10.1007/978-3-642-33275-3_13
11. Negri, P., Goussies, N., Lotito, P.: Detecting pedestrians on a movement feature space. PR **47**(1), 56–71 (2014)
12. Negri, P., Lotito, P.: Pedestrian detection using a feature space based on colored level lines. In: Alvarez, L., Mejail, M., Gomez, L., Jacobo, J. (eds.) CIARP 2012. LNCS, vol. 7441, pp. 885–892. Springer, Heidelberg (2012). doi:10.1007/978-3-642-33275-3_109
13. Ojala, T., Pietikainen, M., Maenpaa, T.: Multiresolution gray-scale and rotation invariant texture classification with local binary patterns. PAMI **24**(7), 971–987 (2002)
14. Orchard, G., Meyer, C., Etienne-Cummings, R., Posch, C., Thakor, N., Benosman, R.: Hfirst: a temporal approach to object recognition. To Appear in PAMI (2015)
15. Pérez-Carrasco, J., Zhao, B., Serrano, C., Acha, B., Serrano-Gotarredona, T., Chen, S., Linares-Barranco, B.: Mapping from frame-driven to frame-free event-driven vision systems by low-rate rate coding and coincidence processing-application to feedforward convnets. PAMI **35**(11), 2706–2719 (2013)

Definition and Composition of Motor Primitives Using Latent Force Models and Hidden Markov Models

Diego Agudelo-España$^{(\boxtimes)}$, Mauricio A. Álvarez, and Álvaro A. Orozco

Faculty of Engineering, Universidad Tecnológica de Pereira,
La Julita, Pereira, Colombia
{dialagudelo,malvarez,aaog}@utp.edu.co

Abstract. In this work a different probabilistic motor primitive parameterization is proposed using latent force models (LFMs). The sequential composition of different motor primitives is also addressed using hidden Markov models (HMMs) which allows to capture the redundancy over dynamics by using a limited set of hidden primitives. The capability of the proposed model to learn and identify motor primitive occurrences over unseen movement realizations is validated using synthetic and motion capture data.

Keywords: Movement representation · Motor primitives · Latent force models · Hidden Markov models · Switched models · Multi-output GPs

1 Introduction

Motor primitives have been traditionally defined based on the theory of dynamical systems (DMP) and they have been extensively used to learn motor behavior [5]. In this work, a novel parameterization of motor primitives is proposed relying on the latent force model (LFM) framework [2]. The proposed solution represents a contribution along the same line of the switched dynamical latent force model (SDLFM) introduced in Álvarez et al. [1]. However, the main difference lies in the composition mechanism for different LFMs because in the SDLFM these primitives are articulated via switching points which become hyper-parameters of a Gaussian process covariance matrix. This covariance matrix grows quadratically on the length of the movement time series and as a consequence, the method is unscalable for long realizations of movement. HMM are used for composing different LFMs which allows to have a fixed-length covariance matrix for each primitive, and a simpler representation of the sequential dynamics of motor primitives. HMMs have been used before in the context of motor primitives [4,9] to combine them either sequentially or simultaneously.

This paper is organized as follows. In Sect. 2 the probabilistic formulation of the motor primitive representation and the composition mechanism is introduced along with the inference algorithm. In Sect. 3 we show experimental results over synthetic and real data, with the corresponding discussion. Finally, some conclusions are presented in Sect. 4.

© Springer International Publishing AG 2017
C. Beltrán-Castañón et al. (Eds.): CIARP 2016, LNCS 10125, pp. 249–256, 2017.
DOI: 10.1007/978-3-319-52277-7_31

2   Materials and Methods

2.1   Latent Force Models

The definition of motor primitives is done using the LFM framework which was introduced in Álvarez et al. [2] motivated by the idea that for some phenomena a weak mechanistic assumption underlies a data-driven model. The mechanistic assumptions are incorporated using differential equations. Similarly to the DMP approach, a primitive is defined by a second order dynamical system described by

$$\frac{d^2 y_d(t)}{dt} + C_d \frac{dy_d(t)}{dt} + B_d y_d(t) = \sum_{q=1}^{Q} S_{d,q} u_q(t),\tag{1}$$

where C_d and B_d are known as the damper and spring coefficients respectively, and $\{y_d(t)\}_{d=1}^{D}$ is the set of D outputs of interest. In order to keep the model flexible enough to fit arbitrary trajectories even under circumstances where the mechanistic assumptions are not rigorous fulfilled [3], a forcing term is added and it is shown in the right side of Eq. (1). This forcing term is governed by a set of Q latent functions $\{u_q(t)\}_{q=1}^{Q}$ whose contribution to the outputs dynamics is regulated by a set of constants $\{S_{d,q}\}$ which are known as the sensitivities.

The main difference of the LFM approach in contrast with the classical DMP is that it assumes Gaussian process priors with RBF covariance over the latent forcing functions. As a consequence of this assumption, it turns out that outputs are jointly governed by a Gaussian process as well and the corresponding covariance function can be derived analytically. The cross-covariance between outputs $y_p(t)$ and $y_r(t)$ under the dynamical model of Eq. (1) can be computed with

$$k_{y_p,y_r}(t,t') = \sum_{q=1}^{Q} \frac{S_{p,q} S_{r,q} l_q \sqrt{\pi}}{8 \omega_p \omega_r} k_{y_p,y_r}^{(q)}(t,t'),\tag{2}$$

where $k_{y_p,y_r}^{(q)}(t,t')$ represents the covariance between outputs $y_p(t)$ and $y_r(t)$ associated to the effect of the $u_q(t)$ latent force and its exact form, along with definitions for constants ω_p, ω_r and l_q can be found in [2]. Remarkably, many of the desired properties of a motor primitive representation such as co-activation, modulation, coupling and learnability naturally arise as a consequence of the probabilistic formulation [6] which also allows to quantify the uncertainty over the learned movements.

2.2   Hidden Markov Models

Formally, an HMM models a sequence of observations $\mathbf{Y} = \{\mathbf{y}_1, \mathbf{y}_2, \dots, \mathbf{y}_n\}$ by assuming that the observation at index i (i.e. $\mathbf{y}_i$) was produced by an emission process associated to the k-valued discrete hidden state z_i and that the sequence of hidden states $\mathbf{Z} = \{z_1, z_2, \dots, z_n\}$ was produced by a first-order Markov process. Therefore, the complete-data likelihood for a sequence of length n can be written as

$$p(\mathbf{Y}, \mathbf{Z}|\mathbf{A}, \boldsymbol{\pi}, \boldsymbol{\theta}) = p(z_1|\boldsymbol{\pi})p(\mathbf{y}_1|z_1, \boldsymbol{\theta}) \prod_{i=2}^{n} p(z_i|z_{i-1}, \mathbf{A})p(\mathbf{y}_i|z_i, \boldsymbol{\theta}), \qquad (3)$$

where $\mathbf{A} = \{a_{j,j'}\}$ denotes the hidden state transition matrix, $\boldsymbol{\pi} = \{\pi_j\}$ is the initial hidden state probability mass function and $\boldsymbol{\theta}$ represents the set of emission parameters for each hidden state. The problem of how to estimate the HMM parameters $\zeta = \{\mathbf{A}, \boldsymbol{\pi}, \boldsymbol{\theta}\}$ is well-known and solutions for particular choices of emission processes have been proposed [8]. However, the use of novel probabilistic models as emission processes bring new challenges from the perspective of probabilistic inference and it also broadens the horizon of potential applications.

2.3   HMM and LFM

In this work HMMs are used differently by introducing a hybrid probabilistic model as emission process to represent motor primitives, namely the latent force models. The proposed model is based on the idea that movement time-series can be represented by a sequence of non-overlapping latent force models. This is motivated by the fact that movement realizations have some discrete and non-smooth changes on the forces which govern the movements. These changes can not be modeled by a single LFM because it generates smooth trajectories. Moreover, the use of HMM and LFM emissions enables us to capture the existing redundancy in dynamics over a movement trajectory since the whole trajectory is explained by a limited set of hidden primitives.

Formally, the overall system is modeled as an HMM where the emission distribution for each hidden state is represented by a LFM. Therefore the complete-data likelihood still fulfills the equation in (3) but the emission process is performed as follows

$$p(\mathbf{y}_i|z_i, \boldsymbol{\theta}, \boldsymbol{\chi}) = \mathcal{N}(\mathbf{y}_i|f(\boldsymbol{\chi}), I\sigma^2), \qquad f(t) \sim \mathcal{GP}(0, k_{y_p, y_r}(t, t'; \boldsymbol{\theta}_{z_i})), \quad (4)$$

where $k_{y_p, y_r}(.,.)$ represents the second order LFM kernel already defined in Eq. (2) with hyper-parameters given by $\boldsymbol{\theta}_{z_i}$ and σ^2 denotes the noise variance. Notice that the HMM framework allows the observable variables to have a different dimensionality with respect to the latent variables, in this case $\mathbf{y}_i$ is a continuous multivariate vector whereas z_i is univariate and discrete. It should also be noticed that there is an additional variable χ conditioning the emission process, this variable denotes the set of sample locations where the LFMs are evaluated and this set is assumed to be independent of the hidden variable values.

2.4   Learning the Model

A common approach for estimating HMMs parameters is maximum likelihood via the expectation-maximization (EM) algorithm which is also known as the Baum-Welch algorithm. It can be shown that the E-step and the update equations for the parameters associated to the hidden dynamics $\{\mathbf{A}, \boldsymbol{\pi}\}$ are unchanged by the use of LFMs as emission processes and their exact form can be found in [8]. In order to update the emission process parameters $\boldsymbol{\theta}$, only one term of the $Q(\zeta, \zeta^{old})$ equation of the EM algorithm must be taken into account. This term is basically a weighted sum of Gaussian log-likelihood functions and gradient ascent methods can be used for optimizing and updating the emission parameters in the M-step.

3   Experimental Results

3.1   Synthetic Data

A particular instance of the model was used for sampling 20 trajectories with 20 segments each. This data-set was divided in two parts. One half was used for learning the motor primitives and the rest was used for validation (i.e. motor primitive identification). Notice that, as the model was formulated in Sect. 2.3, it does not necessarily generate continuous trajectories, thus the synthetic trajectories were generated in such a way that the constant mean of each segment is equal to the last value of the last segment which produces real-looking realizations. To generate the synthetic trajectories, we consider an HMM with three hidden states with transition matrix $\mathbf{A}$ and initial state probability mass function $\boldsymbol{\pi}$. The same parameters inferred from training data are shown alongside (i.e. $\mathbf{A}^*$ and $\boldsymbol{\pi}^*$).

$$\mathbf{A} = \begin{bmatrix} 0.8 & 0.1 & 0.1 \\ 0.6 & 0.3 & 0.1 \\ 0.3 & 0.2 & 0.5 \end{bmatrix}, \boldsymbol{\pi} = \begin{bmatrix} 0.1 \\ 0.3 \\ 0.6 \end{bmatrix}, \quad \mathbf{A}^* = \begin{bmatrix} 0.83 & 0.08 & 0.09 \\ 0.63 & 0.27 & 0.1 \\ 0.27 & 0.25 & 0.48 \end{bmatrix}, \boldsymbol{\pi}^* = \begin{bmatrix} 0.0 \\ 0.4 \\ 0.6 \end{bmatrix}.$$

Regarding the emission process, a LFM with 4 outputs ($D = 4$), a single latent force ($Q = 1$) and sample locations set $\chi = \{0.1, \ldots, 5.1\}$ with $|\chi| = 20$ was chosen. The sensitivities were fixed to one and they were not estimated. The actual emission parameters and the corresponding inferred values are depicted in Table 1, where the spring and damper constants are indexed by the output index.

Table 1. LFM emission parameters

	Hidden state 1	Hidden state 2	Hidden state 3
Spring const.	{3., 1, 2.5, 10.}	{1., 3.5, 9.0, 5.0}	{5., 8., 4.5, 1.}
Damper const.	{1., 3., 7.5, 10.}	{3., 10., 0.5, 0.1}	{6., 5., 4., 9.}
Lengthscale	10	2	5
Spring const.*	{3.21, 1.05, 2.67, 11.09}	{0.67, 1.72, 3.39, 2.26}	{6.92, 10.66, 6.32, 2.3}
Damper const.*	{1.2, 3.36, 8.39, 9.61}	{0.5, 2.62, 1.07, 0.27}	{6.09, 5.54, 4.47, 9.76}
Lengthscale*	85.77	180.48	159.96

It can be argued that there is a similar pattern with respect to the original emission values, particularly for hidden states 1 and 3. The hidden state 2 exhibits a more diverse pattern in comparison to its original parameters. Nevertheless, when the inferred LFM covariances are plotted (see Fig. 2) for each hidden state, it is easy to see that the underlying correlations between outputs were successfully captured by the inferred emission processes. The poor length-scale estimation can be explained by the fact that the range covered by the sample locations set χ (i.e. from 0.1 to 5.1) is reduced for this property to be noticeable and inferable.

A sample synthetic trajectory can be seen in Fig. 1. This is a trajectory used to validate the model's capability to detect motor primitives (i.e. hidden states) given the inferred model parameters, which is achieved using the Viterbi algorithm [8]. The resulting hidden state sequence is shown on the top of Fig. 1, and it turned out to be exactly the same sequence used for generation, which was assumed to be unknown.

The Viterbi algorithm was executed for each validation trajectory and the resulting hidden state sequences were compared against the actual values used

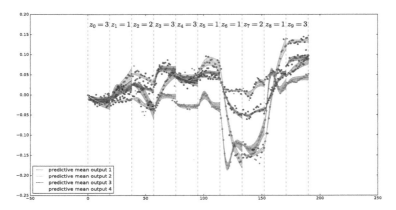

Fig. 1. Primitives identification over a synthetic trajectory. In the top, the most probable hidden state sequence $\{z_0, z_1, \ldots, z_9\}$ given by the inferred model is shown. The predictive mean after conditioning over the hidden state sequence and observations is also depicted with error bars accounting for two standard deviations.

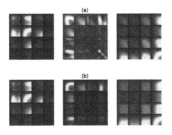

Fig. 2. Actual and inferred toy experiment covariance matrices for the 3 hidden states. Top row **(a)**: LFM covariance matrices used for generating the synthetic trajectories. Bottom row **(b)**: estimated LFM covariance matrices.

during the data-set generation. The correct hidden state was recovered with a success rate of 95% failing only in 10 out of 200 validation segments.

3.2   Real Data

The real-data experiment consists in inferring a set of motor primitives from a group of realizations of the same particular behavior and assessing whether or not, a similar pattern is recovered over the inferred activated motor primitives for unseen realizations of the same behavior. To achieve this, the CMU motion capture database (CMU-MOCAP) is used. The chosen behavior is the walking action because it exhibits a rich redundancy over dynamics as a consequence of the cyclic nature of gait. Specifically, the subject No. 7 was used with trials $\{01, 02, 03, 06, 07, 08, 09\}$ for training and trials $\{10, 11\}$ for validation. In order to take advantage of the multiple-output nature of LFMs a subset of four joints was selected for the validation. The chosen joints are both elbows and both knees since they are relevant for the walking behavior and their potential correlations might be exploited by the model.

The model formulation given in Sect. 2 implies fixing some model parameters a priori such as the number of hidden primitives, the number of latent forces and the sample locations set χ used in the emission process in Eq. 4. Using a number of hidden states equals to three was enough for capturing the dynamics of the chosen behavior. Experiments with a higher number of hidden states were carried out with equally good results but the local variations associated to each observation started to be captured by the model thanks to the higher number of available hidden states. Similarly, the use of a high number of latent forces makes the model more flexible from the point of view of regression at the expense of favoring overfitting. By using three latent forces a good trade-off is obtained between the regression capability and the intended behavior generalization at the motor primitive level. Finally, the sample locations set χ was defined to cover the interval $[0.1, 5.1]$ with 20 sample locations equally spaced (i.e. $|\chi| = 20$). This choice was motivated by the functional division of a gait cycle into eight phases [7]: initial contact, loading response, mid stance, terminal stance, pre-swing, initial swing, mid swing, and terminal swing. Having $|\chi| = 20$ a complete gait cycle over the selected observations is made up of roughly seven segments which is in accordance with the functional subdivision given that the initial contact can be considered an instantaneous event.

In Fig. 3 the identified motor primitives are shown along the resulting fit. The resulting Viterbi sequences over training and validation observations are shown in Table 2. From this table a cyclic sequential pattern over the identified motor primitives can be observed with a period of seven segments as expected. The first three segments of a gait cycle correspond to activations of the hidden state one, and the remaining gait cycle segments are generally explained by the hidden state two, although the last cycle segment exhibits high variability. Remarkably, the discussed pattern was also obtained over the unseen trajectories (i.e. No. 10, 11) which suggests that the sequential dynamics of the motor primitives associated to the walking behavior of subject seven were learned by the proposed model.

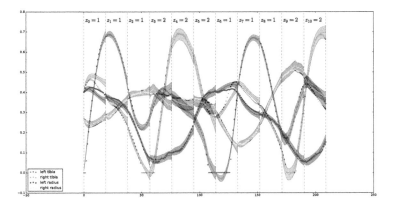

Fig. 3. Primitives identification over a real walking trajectory. In the top the most probable hidden state sequence $\{z_0, z_1, \ldots, z_{10}\}$ given by the inferred model is shown. The predictive mean with error bars is depicted for the four joints.

Table 2. Motor primitives identified over walking realizations.

Trial number	Motor primitives identified
01	1, 1, 1, 2, 2, 2, 1, 1, 1, 1, 2, 2, 2, 1, 1, 1
02	1, 1, 1, 2, 2, 2, 1, 1, 1, 1, 2, 2, 2, 1, 1, 1
03	1, 1, 1, 2, 2, 2, 2, 1, 1, 1, 2, 2, 2, 2, 2, 1, 1, 1, 2, 2, 2
06	1, 1, 1, 2, 2, 2, 2, 1, 1, 1, 2, 2, 2, 2, 1, 1
07	1, 1, 1, 2, 2, 2, 2, 1, 1, 1, 2, 2, 2, 2, 2, 1, 1, 1, 2
08	1, 1, 1, 2, 2, 2, 2, 2, 1, 1, 1, 1, 2
09	1, 1, 1, 2, 1, 2, 1, 1, 1, 2, 2
10*	1, 1, 1, 2, 2, 2, 1, 1, 1, 2, 2
11*	1, 1, 1, 2, 2, 2, 2, 1, 1, 1, 2, 2, 2, 2, 2, 1, 1, 1, 2

4   Conclusions

In this work, a novel probabilistic parameterization of motor primitives and their sequential composition is proposed relying on LFMs and HMMs. We showed how to estimate the model's parameters using the EM algorithm and, through synthetic and real data experiments, the model's capability to identify the occurrence of motor primitives after a training stage was successfully validated.

For future work, alternative formulations which explicitly include the switching points as parameters are suggested to increase the model's flexibility. A further step in the validation might involve using it as a probabilistic generative model in a classification task as in the case of identifying pathological walking.

Acknowledgments. This work has been supported by the project *Human-motion Synthesis through Physically-inspired Machine Learning Models* funded by Universidad

Tecnológica de Pereira, with code 6-15-3. The authors would also like to thank Maestría en Ingeniería Eléctrica, from Universidad Tecnológica de Pereira, for additional funding.

References

1. Álvarez, M.A., Peters, J.R., Lawrence, N.D., Schölkopf, B.: Switched latent force models for movement segmentation. In: Advances in Neural Information Processing Systems, pp. 55–63 (2010)
2. Álvarez, M.A., Luengo, D., Lawrence, N.D.: Latent force models. In: International Conference on Artificial Intelligence and Statistics, pp. 9–16 (2009)
3. Álvarez, M.A., Luengo, D., Lawrence, N.D.: Linear latent force models using Gaussian processes. IEEE Trans. Pattern Anal. Mach. Intell. **35**(11), 2693–2705 (2013)
4. Chiappa, S., Peters, J.R.: Movement extraction by detecting dynamics switches and repetitions. In: Advances in Neural Information Processing Systems, pp. 388–396 (2010)
5. Ijspeert, A.J., Nakanishi, J., Hoffmann, H., Pastor, P., Schaal, S.: Dynamical movement primitives: learning attractor models for motor behaviors. Neural Comput. **25**(2), 328–373 (2013)
6. Paraschos, A., Daniel, C., Peters, J.R., Neumann, G.: Probabilistic movement primitives. In: Advances in Neural Information Processing Systems, pp. 2616–2624 (2013)
7. Perry, J., Davids, J.R., et al.: Gait analysis: normal and pathological function. J. Pediatr. Orthop. **12**(6), 815 (1992)
8. Rabiner, L.R.: A tutorial on hidden Markov models and selected applications in speech recognition. Proc. IEEE **77**(2), 257–286 (1989)
9. Williams, B., Toussaint, M., Storkey, A.J.: Modelling motion primitives and their timing in biologically executed movements. In: Advances in Neural Information Processing Systems, pp. 1609–1616 (2008)

Similarity Measure for Cell Membrane Fusion Proteins Identification

Daniela Megrian[1], Pablo S. Aguilar[2], and Federico Lecumberry[3($\boxtimes$)]

[1] Unidad de Bioinformática, Institut Pasteur de Montevideo, Montevideo, Uruguay
dmegrian@pasteur.edu.uy
[2] Laboratorio de Biología Celular de Membranas, IIBINTECH, CONICET,
Universidad Nacional de San Martín, San Martín, Argentina
[3] Departamento de Procesamiento de Señales, Instituto de Ingeniería Eléctrica,
Facultad de Ingeniería, Universidad de la República, Montevideo, Uruguay
fefo@fing.edu.uy

Abstract. This work proposes a similarity measure between secondary structures of proteins capable of fusing cell membranes and its implementation in a classification system. For the evaluation of the metric we used secondary structures estimated from amino acid sequences of Class I and Class II viral fusogens (VFs), as well as VFs precursor proteins. We evaluated three different classifiers based on k-Nearest Neighbors, Support Vector Machines and One-Class Support Vector Machines in different configurations. This is a first approach to the similarity measure with satisfactory results. It is possible that this method could allow the identification of unknown membrane fusion proteins in other biological models than the proposed in this work.

Keywords: Cell membrane fusion · Viral fusogen · Similarity measure · Support Vector Machines · One-Class Support Vector Machines · k-Nearest Neighbors

1 Introduction

Fusion between cells is needed in many cellular events. Some of the most studied events are myoblasts fusion during muscle formation [1], fusion of gametes during fertilization [2] and fusion between extracellular vesicles (EVs) and target cells [3]. Cellular membranes cannot fuse spontaneously, this process is catalyzed by proteins named fusogens [4]. However, it is still unknown which proteins carry out the fusion mechanism during these events.

One of the best understood fusion mechanisms is the fusion between the membrane of an enveloped virus and the membrane of the target cell. The viral fusion proteins, or viral fusogens, can be grouped in at least three classes according to their structure and mechanism of action. Most of the known viral fusogens belong to Class I and II, that is why these classes are the better characterized. At secondary structure level, Class I viral fusogens present mostly α-helix structure, while Class II are organized mainly in β-sheet [5]. One of the few known

© Springer International Publishing AG 2017
C. Beltrán-Castañón et al. (Eds.): CIARP 2016, LNCS 10125, pp. 257–265, 2017.
DOI: 10.1007/978-3-319-52277-7_32

cellular membrane fusion proteins is EFF protein from C. elegans. This protein is structurally homologous to Class II viral fusogens, also preserving the β-sheet secondary structure organization. In spite of this homology, the amino acid sequence highly differs from Class II viral fusogens sequences [6].

In this work, we intend to develop a similarity measure able to discriminate proteins with fusion capacity in different biological models, based on the proteins secondary structure.

In Sect. 2 we describe the previous attempts to find similarity between secondary structure sequences and its applications, including pattern recognition methods. In Sect. 3 the secondary structure alignment algorithm is explained, as well as the advantages of implementing it to our problem. The description of the data available and the experimental results are in Sect. 4. Section 5 concludes and presents some possible directions of work.

2   Background

The search for protein secondary structures alignment gathered strength with the appearance of reliable tools for secondary structure prediction from amino acid sequences such as described by Cuff et al. [7] and Mc Guffin et al. [8]. These tools return, for each amino acid position, an H (α-helix), E (β-sheet), or C (random coil) character corresponding to the most probable structure in that position, considering the propensities of individual amino acids (Fig. 1).

Most of the literature relative to secondary structure alignment is based on the method proposed by Przytycka et al. [9] called SSEA (Secondary Structure Element Alignment). In this method, the secondary structure of each protein is represented as a summarized and ordered sequence of characters H, E and C

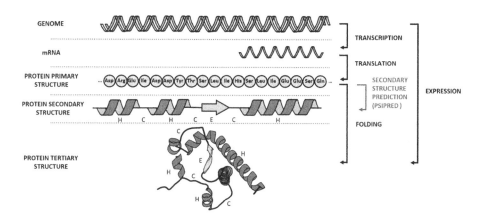

Fig. 1. Context of secondary structure prediction. Gene expression is the process by which information contained in a genome is used to direct protein synthesis. The protein folds into a functional tridimensional molecule at two levels: secondary and tertiary structure.

(Fig. 2). The consecutive repeated characters are collapsed in an element, and the length of the element is stored. SSEA algorithm is analogous to the global alignment algorithm based on dynamic programming proposed by Needleman and Wunsch [10], but using a different score assignment system. When aligning two secondary structures X and Y, a score is calculated for each pair of elements x and y. Each score $S(x, y)$ is defined in Eq. 1, where $L(x)$ and $L(y)$ are the length of the element x and y, respectively. The score is used to fill an alignment matrix as described by Needleman-Wunsch. Besides the score system, the other parameter in an alignment is the gap penalty. A gap comprises an insertion or deletion in a sequence, usually occurring from a single mutational event. SSEA method do not analyze explicitly the role of gap penalty in the alignment.

$$S(x, y) = \begin{cases} \min(L(x), L(y)) & \text{if } x = y \\ \frac{1}{2}\min(L(x), L(y)) & \text{if } (x = \{H, E\} \text{ and } y = C) \text{ or} \\ & \quad (x = C \text{ and } y = \{H, E\}) \\ 0 & \text{if } (x = H \text{ and } y = E) \text{ or} \\ & \quad (x = E \text{ and } y = H) \end{cases} \quad (1)$$

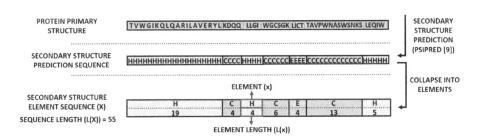

Fig. 2. Secondary Structure Element Alignment (SSEA) concepts. A secondary structure sequence is obtained from an amino acid sequence using Psipred [8]. The consecutive repeated characters are collapsed into elements. Each element is associated with the length of the collapsed characters. The group of ordered elements is a secondary structure element sequence. The sum of lengths of the elements equals the length of the sequence.

The final similarity score in a Needleman-Wunsch alignment corresponds to the last cell score in an alignment matrix and is normalized by the mean of the length of the two sequences. This final score ($d(X, Y)$) is between 0 and 1, the higher the score, the higher the similarity between those two proteins according to the secondary structure. Przytycka et al. proposed and applied this metric to generate a taxonomic tree through a clustering algorithm. The generated tree was compared with trees generated with methods that involve more information, and the taxonomic organization was in agreement. Almost at the same time, Xu et al. [11] used a similar measure to identify two enzymes in Archaea. They also do alignments using a dynamic programming algorithm, but do not collapse consecutive characters.

McGuffin et al. [12] proposed that the prediction of proteins secondary structure and the alignment of its elements allows to detect distant homologs in a better way than methods based on amino acid sequence. Different amino acid sequences may adopt similar tridimensional structures. The capacity to identify distant homologs from the alignment of secondary structure elements was also evaluated by Zhang et al. [13]. The identification of distant homology was accomplished through a method based on Support Vector Machines (SVM), and different metrics were compared. The classification from secondary structure alignments obtained one of the highest accuracy values. Si et al. [14] applied the method to identify proteins with a highly conserved tridimensional domain, called TIM-barrel (triose-phosphate isomerase) allowing to identify this domain in Bacillus subtilis proteome with 99% of accuracy using SVM. SSEA method was also applied successfully by Ni and Zou [15] to the prediction of outer membrane proteins from bacteria. They developed a kernel function based on the metric proposed in SSEA, capable of classifying outer membrane proteins using a method based on SVM with a 97.7% accuracy.

3   Proposed System

3.1   Development of a Similarity Measure

Our bioinformatics search is based on viral fusogens, since these are the only known fusion machineries capable of catalyzing the fusion of membranes outside cells. Because of the amount of information available, we focus on Classes I and II. Owing to the high divergence at sequence level in viral fusogens, algorithms that find similarity between amino acid sequences are frequently not enough to identify similar proteins. To solve this, we evaluate a metric capable of discriminating secondary structure signals between viral fusogens. Our task is to tune up this technique so we can evaluate it later with proteins from other biological models, as the ones described previously.

Viral fusogens are synthesized as inactive precursors (VFPs), that under certain conditions are cleaved, releasing a transmembrane protein with fusion capacity. We refer to the ectodomain as the fusogen (VF) (Fig. 3). Considering the necessity to search fusogens in proteins synthesized as precursors, our algorithm is intended to correctly align VFs with other VFs, but also with VFPs.

Our protein similarity measure is developed based on SSEA but modifies the alignment algorithm and score normalization, and explores the gap penalty incidence. When aligning the secondary structure of a VF and a VFP the algorithm will not consider the local alignment between the VF and the VFP fusogenic region, since Needleman-Wunsch algorithm computes a global alignment. For this reason, we propose to apply a local alignment algorithm, analogous to Smith-Waterman algorithm [16], which allows the correct alignment between fusogens and proteins that contain a fusogen. Thus, we perform secondary structure representations alignments (VFs and VFPs) in pairs, applying SSEA method, substituting the alignment algorithm with Smith-Waterman algorithm. Although the local alignment approach was described by Fontana et al. [17] it was not applied to a specific problem, the tool is no longer publicly available and we did not find any articles that apply this modification.

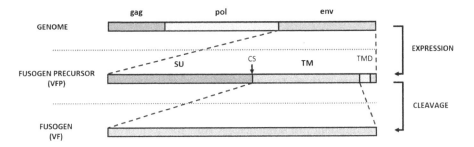

Fig. 3. Viral fusion proteins processing. In this example HIV is presented for its simplicity. Env gene is synthesized as a non-functional precursor protein (VFP) containing a surface protein (SU) and a transmembrane protein (TM). Both proteins are cleaved at the cleavage site (CS) to form an active fusion protein. The released TM is a Class I viral fusion protein. The transmembrane domain separates the protein into an intraviral domain and an ectodomain (VF). The latter carries out the fusogenic activity of the virus.

In SSEA method the final score corresponds to the last cell score in the matrix of a Needleman-Wunsch global alignment. Since we work with a local alignment algorithm, our final score is the maximum score obtained in the dynamic programming matrix as described by Smith-Waterman.

The VFs sequences length is variable between the two classes, and also inside each class. The same happens with the VFPs sequences length, so we would not expect to find a relation between the length of VFs and the respective VFPs. For this reason we could expect that the normalization method applied in SSEA would fail, since the alignment final score is divided between the mean of the pair of proteins sequences length. Thus, we propose another modification to the metric, where the final score is normalized by the mean of the aligned regions length for each pair of proteins.

3.2 Classification

Similarly to viral fusogens, we consider that protein candidates for fusion capacity in other biological models may exist as a part of a precursor. For this reason, besides evaluating the metric when classifying a group of Class I and Class II VFs, we will also evaluate the metric when classifying a group of Class I and Class II VFPs.

Another approach consists of training a One-class SVM (OC-SVM) classifier with Class I VFs and classifying a group of VFs or VFPs as Class I (positive class) or Class II (negative class). This is a first approximation to evaluate the method in order to consider its application for classifying proteins from other biological models as proteins with fusogenic capacity (positive class) and no fusogenic capacity (negative class).

4   Data Description and Experimental Results

4.1   Data Pre-processing

We obtained the amino acid sequences for the VFPs available in the public database UniProt [18]. We selected those proteins labeled as Class I viral fusion protein or Class II viral fusion protein. We obtained 27846 Class I and 1800 Class II sequences, with variable lengths from 446 to 1376 amino acids. We extracted the VF from each protein using the annotations available in UniProt. The lengths varied from 136 to 584 amino acids. From here on, we worked with the VFP and the VF in parallel.

Knowing the redundancy of sequences in UniProt, as a previous step for secondary structure prediction, the sequences were clustered with 99% identity with CD-HIT tool [19]. Thus, we obtained 1769 representative sequences of Class I viral fusogens, and 1103 Class II fusogens. We selected randomly 100 Class I VFs and 100 Class II VFs. We also selected randomly another 100 Class I VFs, 100 Class II VFs, and their corresponding 100 Class I VFPs and 100 Class II VFPs.

For these 600 sequences the secondary structure predictions were calculated with Psipred, using the HHSuite package [20]. This method considers a multiple sequence alignment for each amino acid sequence to improve the accuracy, as evolution provides a closer description of structural tendencies. Finally, we computed similarity matrices for the proposed metric. We analyzed the similarity matrices obtained for a constant gap penalty with values between 0 and -5, and chose to work with a penalty value of -1. This value maximizes the score when comparing sequences of the same class, and minimizes it when comparing different classes.

Training and Classification. SVM method has been widely used in biological sequences analysis. This method uses kernel functions, mapping the problem into a high-dimensional space. This feature allows the construction of a hyperplane that has the highest separation between two classes in the transformed space.

A distance between protein sequences was obtained from the computed similarity with the kernel [15]:

$$k(x, y) = \exp(\gamma \, d(X, Y)). \tag{2}$$

We worked with LIBSVM package [21] for Python. LIBSVM can generate a classifier from the precalculated kernel and estimate the performance.

VFs Classification Training with Two Classes. We trained a SVM classifier with a set of Class I and Class II VFs. The classification was performed with another set of Class I and Class II VFs. The performance of the classifier depends on parameters C and γ. The parameter C affects the flexibility of the classification, allowing some errors, but also penalizing. The parameter γ establishes how far the influence of a sample can reach. The best combination of C and γ was selected using a grid search with 10-fold cross-validation, with C values between 2^{-15} and 2^{15}, and γ values between 5×10^{-4} and 5×10^2 with uniform intervals.

We selected as optimal parameters $C = 2^{-15}$ and $\gamma = 5.4 \times 10^{-1}$. For these parameters, the classification accuracy for Class I and Class II fusogens was 99.0%.

To obtain a second evaluation of the proposed metric, we classified the set of VFs using k-NN as the classification method, for 1, 3 and 7 NNs. The classification accuracy was 98.5%, 98.5% and 97.5% respectively (Table 1).

Table 1. Accuracies obtained for VFs and VFPs classification.

Classification	1-NN	3-NN	7-NN	SVM	OC-SVM
VFs	98.5	98.5	97.5	99.0	92.0
VFPs	98.5	97.5	95.5	90.5	69.5

VFs Classification Training with a Positive Class. SVM classifiers are based in training with samples belonging to two classes (e.g. positive and negative). However, in some situations there are only positive samples for training. This is the case of the problem suggested in this work, as we have a set of training proteins known to be fusogenic, and we intend to select fusogen candidates from a diverse set of proteins. Given the characteristics of the candidate proteins set and the virtually infinite variability a protein can present, it is not possible to create a representative negative samples. Schölkopf et al. [22] described the one-class classification method that allows to train a model with just positive samples.

For this reason, we trained an OC-SVM classifier with Class I VFs as positive samples. We evaluated the classification of a Class I VFs set (distinct from the training set) and a Class II VFs set. For this part, we also used LIBSVM package. The same kernel was applied to the data, and the best combination of parameters was chosen. In this case, parameter ν substitutes parameter C. The meaning of parameter ν is analogous to C meaning, but the values should be between 0 and 1. Similarly to previous part, the best combination of parameters ν and γ were selected using a grid search with 10-fold cross-validation, with ν values between 0.05 and 1, and γ values between 5×10^{-4} and 5×10^2 with uniform intervals.

We selected as optimal parameters $\nu = 0.05$ and $\gamma = 5.4 \times 10^{-1}$. For these parameters, the classification accuracy for Class I and Class II VFs was 92.0% (Table 1).

VFPs Classification Training with Two Classes. In order to evaluate the modified algorithm performance for local alignments, the SVM classifier was evaluated classifying a group of VFPs. It was trained with the same two classes used previously. The classification accuracy for Class I and Class II VFPs was 90.5%. The k-NN classification accuracy for 1, 3 and 7 NNs was 98.5%, 97.5% and 95.5% respectively (Table 1).

VFPs Classification Training with a Positive Class. The classification of Class I and Class II VFPs when training only with Class I VFs resulted in an accuracy of 69.5% (Table 1).

5   Conclusions and Future Work

The developed metric, based on SSEA allowed the satisfactory classification of VFs using the three proposed methods (k-NN, SVM y OC-SVM). The classification of VFPs using k-NN gave a similar accuracy as the obtained for VFs classification. We also obtained an acceptable accuracy when classifying VFPs using a SVM model. However, the performance is reduced considerably when classifying VFPs using an OC-SVM model. It is clear that a reduction of the accuracy is expected since this is the most challenging case where the classifier is trained with VFs and tested with VFPs, and the reduction for the two classes SVM is already greatly reduced. However, the reduction is too abrupt, and further analysis of the dissimilarities between classes would help to understand the reasons of this reduction. In spite of the SVM and OC-SVM classification results, the metric by itself appears to accomplish the objective according to the results obtained for 1-NN classification.

This work was performed on a reduced set, selected randomly from an original set of VFs and VFPs obtained from UniProt. We discarded a significant subset of sequences from the original set as those sequences did not have annotations for the cleavage of VFPs. The first step when reviewing the metric should be to expand the set.

On the other hand, we propose to work in detail on the influence of gap penalties in the metric. Evaluations not presented in this work showed that gap penalty value does not have influence on the performance of k-NN classifiers, but does have influence on SVM and OC-SVM classifiers. In this work, we used a constant gap system, so the gap penalty value is always the same. It would be interesting to evaluate the performance of the metric when working with a linear gap system (dependent on the length of the gap) or with affine gap penalty, in which the gap opening is penalized differently than gap extension.

The set-up of this method could make possible the identification of unknown viral fusogens from the genome or proteome of enveloped viruses. It could also be used to identify proteins with fusogenic capacity in different biological models.

Acknowledgment. This work was partially supported by Agencia Nacional de Investigación e Innovación (DCI-ALA/2011/023-502 "Contrato de apoyo a las políticas de innovación y cohesión territorial"). We would also like to include the double affiliation of Federico Lecumberry (Laboratorio de Procesamiento de Señales, Institut Pasteur de Montevideo, Uruguay).

References

1. Rochlin, K., Yu, S., Roy, S., Baylies, M.K.: Myoblast fusion: when it takes more to make one. Dev. Biol. **341**, 66–83 (2010)

2. Primakoff, P., Myles, D.G.: Penetration, adhesion, and fusion in mammalian sperm-egg interaction. Science **296**, 2183–2185 (2002)
3. van der Pol, E., Bing, A.N., Harrison, P., Sturk, A., Nieuwland, R.: Classification, functions, and clinical relevance of extracellular vesicles. Pharmacol. Rev. **64**, 676–705 (2012)
4. Harrison, S.C.: Viral membrane fusion. Nat. Struct. Mol. Biol. **15**, 690–698 (2009)
5. Kielian, M., Rey, F.A.: Virus membrane-fusion proteins: more than one way to make a hairpin. Nat. Rev. Microbiol. **4**, 67–76 (2006)
6. Perez-Vargas, J., Krey, T., Valansi, C., Avinoam, O., Haouz, A., Jamin, M., Raveh-Barak, H., Podbilewicz, B., Rey, F.A.: Structural basis of eukaryotic cell-cell fusion. Cell **157**, 407–419 (2014)
7. Cuff, J.A., Clamp, M.E., Siddiqui, A.S., Finlay, M., Barton, G.J.: JPred: a consensus secondary structure prediction server. Bioinformatics **14**, 892–893 (1998)
8. McGuffin, L.J., Bryson, K., Jones, D.T.: The PSIPRED protein structure prediction server. Bioinformatics **16**, 404–405 (2000)
9. Przytycka, T., Aurora, R., Rose, G.D.: A protein taxonomy based on secondary structure. Nat. Struct. Biol. **6**, 672–682 (1999)
10. Needleman, S.B., Wunsch, C.D.: A general method applicable to the search for similarities in the amino acid sequence of two proteins. J. Mol. Biol. **48**, 443–453 (1970)
11. Xu, H., Aurora, R., Rose, G.D., White, R.H.: Identifying two ancient enzymes in Archaea using predicted secondary structure alignment. Nat. Struct. Biol. **6**, 750–754 (1999)
12. McGuffin, L.J., Jones, D.T.: Targeting novel folds for structural genomics. Proteins **48**, 44–52 (2002)
13. Zhang, Z., Kochhar, S., Grigorov, M.G.: Descriptor-based protein remote homology identification. Protein Sci. **14**, 431–444 (2005)
14. Si, J.N., Yan, R.X., Wang, C., Zhang, Z., Su, X.D.: TIM-finder: a new method for identifying TIM-barrel proteins. BMC Struct. Biol. **9**, 73 (2009)
15. Ni, Q., Zou, L.: Accurate discrimination of outer membrane proteins using secondary structure element alignment and support vector machine. J. Bioinform. Comput. Biol. **12**, 1450003-1–1450003-12 (2014)
16. Smith, T.F., Waterman, M.S.: Identification of common molecular subsequences. Mol. Biol. **147**, 195–197 (1981)
17. Fontana, P., Bindewald, E., Toppo, S., Velasco, R., Valle, G., Tosatto, S.C.: The SSEA server for protein secondary structure alignment. Bioinformatics **21**, 393–395 (2005)
18. UniProt Consortium: UniProt: a hub for protein information. Nucleic Acids Res. **43**, D204–D212 (2015)
19. Li, W., Godzik, A.: Cd-Hit: a fast program for clustering and comparing large sets of protein or nucleotide sequences. Bioinformatics **22**, 1658–1659 (2006)
20. Soding, J., Biegert, A., Lupas, A.N.: The HHpred interactive server for protein homology detection and structure prediction. Nucleic Acids Res. **33**, W244–W248 (2005)
21. Chang, C.C., Lin, C.J.: LIBSVM: a library for support vector machines. ACM Trans. Intell. Syst. Technol. **2**, 27:1–27:27 (2011)
22. Schölkopf, B., Williamson, R.C., Smola, A.J., Shawe-Taylor, J., Platt, J.C.: Support vector method for novelty detection. In: NIPS, vol. 12, pp. 582–588 (1999)

Abnormal Behavior Detection in Crowded Scenes Based on Optical Flow Connected Components

Oscar E. Rojas$^{(\boxtimes)}$ and Clesio Luis Tozzi

School of Electrical and Computer Engineering, UNICAMP, Av. Albert Einstein,
400, Campinas, SP, Brazil
oscar.rojas87@gmail.com, clesio@dca.fee.unicamp.br

Abstract. This paper presents a new approach for automatic abnormal behavior detection in crowded scenes. Background subtraction algorithm, optical flow and connected component analysis are used to define the optical flow connected components (OFCC). An unsupervised normal behavior model is computed using the main magnitude and direction of each OFCC. Experimental results on the standards UCSD and UMN anomaly detection and localization benchmarks demonstrate the method performance compared to other approaches considering detection rate and processing time.

Keywords: Anomaly behavior detection · Optical flow · Crowded scenes analysis · Anomaly localization

1 Introduction

Abnormal behavior analysis on crowded scenes is an important and growing research field. Video cameras, given their ease installation and low cost, have been widely used for monitoring internal and external areas such as buildings, parks, stadiums etc. With the world's population increasing, the presence of people in common areas has been increasing too. Algorithms for pose detection and action recognition for single or, in some cases, very low density groups of people are extensively treated in the pattern recognition community. Nevertheless, abnormal behavior detection and localization in crowded scenes remains an open problem due to high levels of occlusion and the impractical approach of individual segmentation.

The concept of abnormal behavior is always associated with the scene context, a behavior considered as normal in a scene may be considered abnormal in other. These specific conditions increase the difficulties for automatic analysis and require specific modeling of the abnormal behavior for each particular scene.

In order to build such models many algorithms have been proposed. In [6] optical flow is used to compute interaction forces between adjacent pixels and a model, known as Social Force Model, is created based in a bag of words approach

© Springer International Publishing AG 2017
C. Beltrán-Castañón et al. (Eds.): CIARP 2016, LNCS 10125, pp. 266–273, 2017.
DOI: 10.1007/978-3-319-52277-7_33

for classify frames either normal or abnormal. In [5] dynamic textures (DT) are used to model the appearance and dynamics of normal behavior, samples with a low probabilistic values in the model are labeled as abnormal. In [7] entropy and energetic concepts are used as features to model the probability of finding abnormal behavior in the scene. Natural language processing is used in [10] as classification algorithm for features based on viscous fluid field concepts.

Many algorithms employs machine learning techniques as classification tool. Support Vector Machine (SVM) are used in [8,11] for classify histograms of the orientation of optical flow. Multilayer Perceptron Neural Networks is used in [13]. k-Nearest Neighbors is used in [1] for classify outlier observed trajectories as abnormal behavior. Finally, Fuzzy C-Means are used in [2,3] to derive an unsupervised model for the crowd's trajectory patters.

In general, to construct the feature vector used in many of the algorithms described above, a several set of parameters must be correctly set in order to achieve the performance reported by the authors. Some of the state-of-the-art methods are based in complex probabilistic models which leads high processing time. Despite the processing time per frame is reported only for a very few papers, it is in general high. For example, in [5] the authors reported a test time of 25 s per frame for 160×240 pixel images and in [12] the reported test time per frame is 5 s in videos with 320×240 pixel resolution.

The main contribution for this paper is present a simple but efficient method that reduce the processing time per frames in near real time allowing practical use.

The rest of this paper is organized as follows. Section 2 describes the proposed approach. Section 3 presents the experimental results. Section 4 presents the conclusions.

2   Proposed Method

The general pipeline for the proposed approach is shown in Fig. 1. The five initial modules (1 to 5 in Fig. 1) aims to compute the model features and are the same for the training and test phases. These initial modules are described in Sect. 2.1. In the training phase, represented by module 6, frames with normal

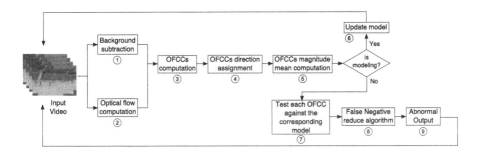

Fig. 1. General pipeline for the proposed method.

behavior are used to update the model as described in Sect. 2.2. In the test phase, represented by module 7, each new sample is compared with the model and classified as normal or abnormal as described in Sect. 2.3. A false positive reduction methodology, represented by module 8, is also described in Sect. 2.3.

2.1   OFCCs Computation

In the training phase a sequences of N frames are used to build a normal behavior model. The algorithm presented in [9] is used to compute the background model and a foreground mask I_{fm} of each frame is obtained.

In order to reduce the noise and de computational load a connected components labeling algorithm is used to obtain the blobs $(b_1, b_2, \ldots, b_n)$ where n is the total number of blobs in the foreground mask I_{fm}. In parallel to foreground extraction the dense optical flow of each frame is computed using [14]. The optical flow vectors are used to obtain the magnitude $m(x, y)$ and direction $\theta(x, y)$ values of each (x, y) point in the input image.

An Optical Flow Connected Component $OFCC_i$ can be defined as the set of values $[m(x, y), \theta(x, y)]$ for all (x, y) points belonging to the i-th blob, as expressed in Eq. 1.

$$OFCC_i = [m(x, y), \theta(x, y)] \; \forall \; (x, y) \in b_i. \tag{1}$$

The main direction $\bar{\theta}_i$ of the i-th OFCC is computed as follows. A histogram of the direction values of $OFCC_i$ is obtained with a fixed bin width of $\Delta\theta = 45°$. The angle associated with the highest bin is used as the main direction $\bar{\theta}_i$ of $OFCC_i$.

The main magnitude $\bar{m}_i$ of $OFCC_i$ is obtained as the statistical mean of the magnitudes values in $OFCC_i$ as shown in Eq. 2.

$$\bar{m}_i = \frac{1}{S} \sum_{k=1}^{S} m(x, y) \mid m(x, y) \in OFCC_i \tag{2}$$

where S is the total number of magnitude values in $OFCC_i$. Finally, the main direction $\bar{\theta}_i$ and the main magnitude $\bar{m}_i$ values of each $OFCC_i$ are used to construct the normal behavior model.

2.2   Normal Behavior Model

In this algorithm the behavioral model is composed of m matrices $(A_1, A_2, \ldots, A_m)$ where m is computed as

$$m = \frac{360}{\Delta\theta} \tag{3}$$

and represents the number of possible values that $\bar{\theta}_i$ can adopt. For instance, if $\Delta\theta = 45°$ then $m = 8$ matrices will be defined.

For each frame in the training video a set of n OFCCs are obtained as described in the previous section. After compute the $\overline{\theta}_i$ and $\overline{m}_i$ values of each $OFCC_i$ the corresponding A matrix number η is obtained as,

$$\eta = \frac{\overline{\theta}_i}{\Delta\theta} \tag{4}$$

Then, the values of the A_η matrix are updated using the next condition

$$A_\eta(x,y) = \begin{cases} \overline{m}_i, & \text{if } \overline{m}_i > A_\eta(x,y) \\ A_\eta(x,y), & \text{otherwise} \end{cases}, \forall\, (x,y) \in b_i. \tag{5}$$

At the end of the training phase each matrix A_η will store the maximum principal magnitude $\overline{m}_i$ in the full training video at each point (x,y) at the direction $\eta * \Delta\theta$.

Figure 2 shows an example of a normal behavior model with $\eta = (1, 2, \ldots, 8)$ matrices for $\Delta\theta = 45°$. A color map was applied to each matrix A_η for better visualization.

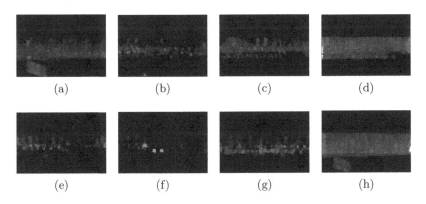

Fig. 2. The magnitude model for eight directions ($\eta = 8$). Each image shows the pixel highest magnitude for direction angles between (a) $(0°–45°]$, (b) $(45°–90°]$, (c) $(90°–135°]$, (d) $(135°–180°]$, (e) $(180°–225°]$, (f) $(225°–270°]$, (g) $(270°–315°]$ and (h) $(315°–360°]$. (Color figure online)

2.3 Abnormality Detection

After all the training frames have been processed and the model is completed, test videos with both normal and abnormal behaviors can be analyzed.

The set of OFCCs and their main directions $\overline{\theta}_i$s are obtained as described in Sect. 2.1 for each video frame. To determine if the $OFCC_i$ is abnormal or not its main direction $\overline{\theta}_i$ is used to find the corresponding A_η matrix with η computed using Eq. 4. Next the maximum value $\hat{a}_\eta$ in A_η within the same region defined by the blob b_i is founded according to

$$\hat{a}_\eta = max(A_\eta(x,y)) \mid (x,y) \in b_i. \tag{6}$$

Then, the comparison between each $m(x,y)$ value in $OFCC_i$ and $\hat{a}_\eta$ is done as follows. If $m(x,y)$ is grater than $\hat{a}_\eta$ then the pixel (x,y) is marked as abnormal, otherwise its marked as normal.

After compare all the magnitude values in $OFCC_i$ an abnormal binary mask image $I_{ab}(x,y)$ with the same size as the input frames, can be use to store the abnormal marked pixels as $I_{ab}(x,y) = 1$ and the normal ones as $I_{ab}(x,y) = 0$.

In order to improve the algorithm performance a FIFO type list with fixed size M is defined and filled up with the latest M binary images $I_{ab}(x,y)$. To consider an OFCC as abnormal it must appear at least a number W of times in the list. The list size M and the number W are user controlled parameters and can be used for sensitivity adjustment, since a higher value of W means a higher alarm delay time.

3   Results and Comparisons

The proposed algorithm was implemented in Qt/C++ using OpenCV on a 2.7 GHz Intel Core i7 PC with 16 GB of RAM. The method was tested in two popular datasets: UMN[1] and UCSD[2]. Figure 3 shows a frame for each of the scenarios in the UMN dataset and the abnormality detected by the proposed approach. The frame size in all UMN videos is 320×240 pixels. The frame size in the UCSDPed1 videos is 238×158 and for UCSDPed2 is 360×240 pixels.

Fig. 3. Examples of normal (*top*) and abnormal (*bottom*) situations in UMN dataset.

Figure 4 shows three examples frames with abnormal behavior for each of the two scenarios in the UCSD dataset.

The proposed method was compared with similar state-of-the-art algorithms including Mixture Dynamic Texture (MDT) [5], Mixture of Optical Flow

---

[1] http://mha.cs.umn.edu/proj_events.shtml.
[2] http://www.svcl.ucsd.edu/projects/anomaly/dataset.htm.

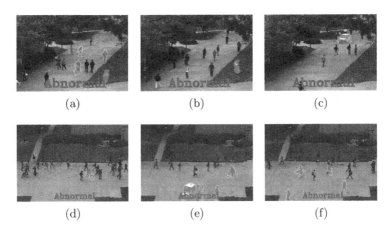

Fig. 4. Example of abnormal behavior detected in the UCSD dataset. UCSDped1 (*top*) and UCSDped2 (*below*).

(MPPCA) [4], Social Force [6], Social Force with MPPCA [4] and the Hierarchical Activity Approach [12]. Figure 5 shows the Receiver Operation Characteristic (ROC) curves for the proposed method and the comparative algorithms, taken from [12]. Table 1 shows the Area Under the ROC curve (AUC) for the five comparative methods and the proposed one. Finally, Fig. 6 shows the processing time per frame for some state-of-the-art algorithms and the proposed in this paper.

The ground truth provided by the UCSD dataset, and used for performance evaluation in all the comparison methods, labels people in wheelchair as abnormal behavior, even when their speed is lower than the speed of walking people. This leads to additional False Negative detected frames because, in the presented algorithm, this situation is not considered as abnormal. A second situation when the output of the presented algorithm differs from the ground truth is when

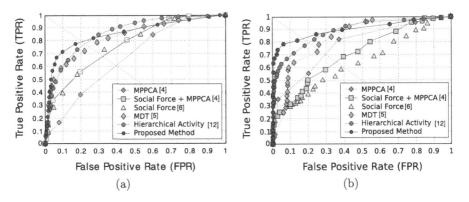

Fig. 5. Quantitative comparison of abnormal behavior detection in (a) UCSDped1 and (b) UCSDped2 against state-of-the-art algorithms.

somebody, in the test phase, walks in a region where no people walked in the training phase. Examples of this type of abnormal detection are shown in Fig. 4(a), (b), (e) and (f). Frames that present only this kind of abnormality are ignored in the comparison results.

Table 1. Comparison of the Area Under Curve of the proposed method compared with the others algorithms.

Algorithm	Area Under Curve (AUC)	
	UCSDped1	UCSDped2
MPPCA [4]	0.59	0.693
Social force [6]	0.675	0.556
Social force + MPPCA [4]	0.668	0.613
MDT [5]	0.818	0.829
Hierarchical activity [12]	0.854	0.882
Proposed method	**0.852**	**0.958**

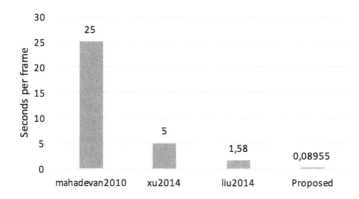

Fig. 6. Comparison of consumed time per frame with others state-of-the-art algorithms. The showed time is for the test phase in UCSDped1.

4   Conclusions

This paper presents a new method for abnormal behavior detection. It's based on optical flow and connected component analysis. From the experimental results it can be concluded that, when compared to other state of the art methods, the proposed method presents better performance in abnormal detection in the UCSDped2 dataset and is very close to the best one in the UCSDped1 but, as shown in Fig. 6 it presents the lowest processing time, near to real-time processing which allows practical use in modern computers.

Acknowledgments. The authors wish to thank Conselho Nacional de Desenvolvimento Científico (CNPq), Brazilian Research Support Foundations, for sponsoring this work.

References

1. Alvar, M., Torsello, A., Sanchez-Miralles, A., Armingol, J.M.: Abnormal behavior detection using dominant sets. Mach. Vis. Appl. **25**(5), 1351–1368 (2014)
2. Chen, Z., Tian, Y., Zeng, W., Huang, T.: Detecting abnormal behaviors in surveillance videos based on fuzzy clustering and multiple Auto-Encoders. In: IEEE International Conference on Multimedia and Expo, pp. 1–6 (2015)
3. Cui, J., Liu, W., Xing, W.: Crowd behaviors analysis and abnormal detection based on surveillance data. J. Vis. Lang. Comput. **25**, 628–636 (2014)
4. Kim, J., Grauman, K.: Observe locally, infer globally: a space-time MRF for detecting abnormal activities with incremental updates. In: Proceedings of International Conference on Computer Vision and Pattern Recognition (CVPR) (2009)
5. Mahadevan, V., Li, W., Bhalodia, V., Vasconcelos, N.: Anomaly detection in crowded scenes. In: IEEE Conference on Computer Vision and Pattern Recognition, pp. 1975–1981 (2010)
6. Mehran, R., Oyama, A., Shah, M.: Abnormal crowd behavior detection using social force model. In: IEEE Conference on Computer Vision and Pattern Recognition, vol. 2, pp. 935–942 (2009)
7. Ren, W.Y., Li, G.H., Chen, J., Liang, H.Z.: Abnormal crowd behavior detection using behavior entropy model. In: International Conference on Wavelet Analysis and Pattern Recognition, pp. 212–221 (2012)
8. Snoussi, H., Wang, T.: Detection of abnormal visual events via global optical flow orientation histogram. IEEE Trans. Inf. Forensics Secur. **9**(6), 988–998 (2014)
9. Stauffer, C., Grimson, W.: Adaptive background mixture models for real-time tracking. In: Proceedings in IEEE Computer Society Conference on Computer Vision and Pattern Recognition, vol. 2 (1999)
10. Su, H., Yang, H., Zheng, S., Fan, Y., Wei, S.: Crowd event perception based on spatio-temporal viscous fluid field. In: IEEE Ninth International Conference on Advanced Video and Signal-Based Surveillance, pp. 458–463 (2012)
11. Wang, T., Snoussi, H.: Histograms of optical flow orientation for visual abnormal events detection. In: IEEE Ninth International Conference on Advanced Video and Signal-Based Surveillance (AVSS), pp. 13–18 (2012)
12. Xu, D., Song, R., Wu, X., Li, N., Feng, W., Qian, H.: Video anomaly detection based on a hierarchical activity discovery within spatio-temporal contexts. Neurocomputing **143**, 144–152 (2014)
13. Zhang, D., Peng, H., Haibin, Y., Lu, Y.: Crowd abnormal behavior detection based on machine learning. Inf. Technol. J. **12**, 1199–1205 (2013)
14. Zivkovic, Z., van der Heijden, F.: Efficient adaptive density estimation per image pixel for the task of background subtraction. Pattern Recogn. Lett. **27**(7), 773–780 (2006)

Identifying Colombian Bird Species from Audio Recordings

Angie K. Reyes[1], Juan C. Caicedo[2], and Jorge E. Camargo[1(✉)]

[1] Laboratory for Advanced Computational Science and Engineering Research,
Universidad Antonio Nariño, Bogotá, Colombia
{angreyes,jorgecamargo}@uan.edu.co
[2] Fundación Universitaria Konrad Lorenz, Bogotá, Colombia
juanc.caicedor@konradlorenz.edu.co

Abstract. This paper presents a methodology to represent, process and visualize large bird audio recordings. The proposed methodology is based on audio signal processing, image processing, and machine learning techniques. We evaluated the methodology in a collection of 6,856 audio recordings belonging to 560 Colombian bird species. Results show that it is possible to build a tool that help ornithologists to process and analyze large collections of bird species in and effective and efficient way.

Keywords: Biosignal processing · Species recognition · Bag of audio features · Visualization · Deep Neural Networks

1 Introduction

Currently, there is a large number of vulnerable species of animals and plants, due to climate change and other human interventions in the natural world. In particular, a total of 3,401 species of birds are endangered worldwide; 68 of them are native from Colombia [4]. Birds are of great importance to keep a healthy balance in most ecosystems, and many biologists make great efforts to track and identify them in specific places. The ability to collect information about the location and presence of birds allows us to understand trends in population changes in the short and long terms, and therefore, make informed decisions to protect them [4].

Audio processing technologies and pattern recognition methods are a promising tool to automate the identification of bird species from sound recordings. Currently, such recordings are analyzed by expert ornithologists who spend a lot of time classifying and organizing audible evidence that some species are still present in their habitats. The automated identification of birds from audio recordings would help biologists and ornithologists to study larger regions of land more frequently. However, this is a challenging task because bird songs are very variable, and are rarely heard without background noise, such as the sound of rivers, wind and crickets.

© Springer International Publishing AG 2017
C. Beltrán-Castañón et al. (Eds.): CIARP 2016, LNCS 10125, pp. 274–281, 2017.
DOI: 10.1007/978-3-319-52277-7_34

There has been long term interest in identifying animal species from audio recordings. One of the first works for automatic bird identification was conducted by Anderson et al. [1] who used spectrograms to compare audio signals. Kogan and Margoliash [6] used Hidden Markov Models and Dynamic Time Warping for recognizing bird songs. Some of these studies have been limited by the small number of involved audios and species, which are usually in the order of tens [10]. In our work, we make an attempt to scale bird species identification to hundreds of species. Feature extraction is a key component of any audio processing system, and MFCC features are widely used for bird identification [2]. These features are usually computed in frames and then some quantization mechanism is used to facilitate matching. Classifiers are also important for bird species identification, and early works attempted the use of Neural Networks (NN) [8]. More recently, Deep Neural Networks (DNN) have been explored on this same task [7]. We also use DNNs in this paper, and compare against the performance of SVMs.

In this paper, we present and evaluate a methodology to process audio recordings of birds in the wild. We use a bag of features approach to build a representation of recordings with variable length using MFCC[1] features. A collection of more than 6,000 recordings of 560 Colombian bird species were processed. We trained two classification models, one based on SVM classifiers, and another based on Deep Neural Networks to compare classification accuracy. Also, the representation of these recordings was used to analyze similarities among bird species using several visualization techniques.

2   Materials and Methods

The goal of our work is to investigate computational tools for processing and analyzing audio recordings of Colombian bird species. We divide these computational tools in three types: (1) audio representation, (2) classification methods, and (3) visualization techniques.

2.1   Dataset

The collection of audio recordings used in this work is part of the *Xeno-canto* project[2], which has more than 192,000 recordings of 9,120 bird species from around the world. We used a subset of the BirdCLEF collection [13], which contains 24,607 recordings mainly from Latin American countries. The BirdCLEF collection has metadata associated to each audio file, including the species and GPS coordinates of the recording. We selected all audios that correspond to Colombian species filtering by geographic location. We used the Google Maps Geo-coding service to check which audio files were recorded in Colombian territory. We found a total of 6,856 recordings with samples of 560 Colombian bird species using this approach.

---

[1] Mel Frequency Cepstral Coefficient.

[2] http://www.xeno-canto.org.

2.2   Feature Extraction and Audio Representation

We construct a bag of features representation for audios of variable length. Starting from a raw input signal, we construct a histogram with the most representative features of the recording after applying noise removal, feature computation, and vector quantization (Fig. 1).

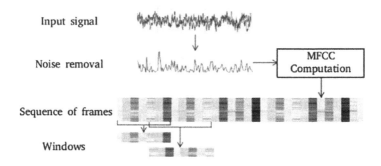

Fig. 1. Pre-processing steps for the construction of the bag of features for audio recordings.

Noise Reduction. We leveraged the SoX software to automatically remove noise from the recordings collection, using the *noisered* filter. This filter removes constant background noise after analyzing the signal and profiling the amount of noise. We set the parameter of amount of noise to 0.5 in our experiments. When the tool is run, background noise is removed, leaving parts of the recording silent. We also applied the same tool to cut those blank spaces of the recording left after noise removal. In particular, any silent interval longer than 0.1 s is removed from the recording. This results in a continuous audio sequence without the usual pauses at the beginning and the end of the recording, and even without intermediate spaces while the bird sings.

MFCC Computation and Windowing. We compute MFCC features [3] using the *Python Speech Features Library*, setting frames of 25 ms with strides of 10 ms, obtaining a 13-dimensional feature vector per frame. Frames are organized in windows, concatenating 10 consecutive feature vectors for a total of 130 features per window. This new feature vector represents a segment of audio of 0.25 s. For a particular audio recording, we collect windows like this each 5 frames, resulting in a dense sample of features.

Bag of Features. We proceed with the construction of a dictionary of audio features, which is used to compute statistics of the distribution of features in a recording. First, we collect a large sample of windows from the training set of audio recordings. Then, these windows are quantized in K groups using the

K-means algorithm. The resulting means are used as the reference dictionary to build histograms for individual recordings.

The bag of features histogram is built for a new recording by first computing MFCC features and windows. Each of the windows is compared against the dictionary using the Euclidean distance to identify the closest element. For each dictionary element, we keep the count of windows in the recording that match that particular element. All these are the frequencies of features, and are useful to encode the contents of an audio with a fixed length histogram representation (K features).

2.3   Automatic Identification of Bird Species

We use a classification approach to identifying bird species from audio recordings. In this context, a classifier is a function that transforms the input signal in a decision of which bird species is more likely to be associated with the audio contents. We evaluated two supervised classification functions in this work: Support Vector Machines (SVMs) and Deep Neural Networks (DNNs).

SVM classifiers can be used to model a multi-class problem following a one-versus-all strategy. We use this strategy, which results in a total of $n(n-1)/2$ individual classifiers that need to be trained. We also use SVMs with linear kernel functions, which can be trained efficiently using Stochastic Gradient Descent, and are also very fast to evaluate during the testing phase.

DNNs are non-linear classifiers that can also be trained and evaluated efficiently. The non-linearity is usually introduced between layers, after the input has been processed with a linear function. Stacking multiple transformations of this type usually results in better performance since intermediate (or hidden) layers learn hierarchically organized patterns that disentangle the complexity of the signal. We evaluated different DNN architectures with up to 5 hidden layers plus a classification layer trained with the softmax function. This architecture is illustrated in Fig. 2.

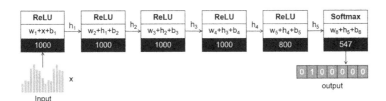

Fig. 2. Architecture of the DNN used in this work: the input is the bag of words histogram representation with dimensionality K. This input is processed by up to 5 hidden layers, each consisting of a linear transformation followed by a ReLU non-linearity. The final output is produced by a softmax layer that computes classification scores. Numbers in the black box of each layer indicate the number of outputs of that layer.

2.4 Visualization of Bird Species

We are interested in visually analyzing the audio bird collection by the use of information visualization techniques. The most frequently used approach is based on dimensionality reduction methods. In our case, we represent a bird audio record in a features vector (histogram), which has D dimensions as it was described in Sect. 2.2. The dimensionality of this vector can be reduced to obtain $d \ll D$ dimensions, which can be visualized in a 2D coordinates system when the 2 most important dimensions are selected.

In this work, we have evaluated PCA [5], Isomap [11], t-SNE [12], and LLE [9] as mechanisms for generating visualizations of the audio collection.

3 Experiments and Results

We present an experimental evaluation on a collection of 6,856 audio recordings of Colombian birds. We start with the pre-processing phase applying noise removal to all audio files, and also computing MFCC features on frames and extracting windows. We removed audio files from the collection if they have less than 8 windows in total after the pre-processing steps. This leaves us with a total of 5,242 audio recordings with examples of 547 bird species. This collection was split in three sets for training, validation and test using 50%, 25% and 25% of the samples respectively, based on stratified sampling.

3.1 Classification Experiments

We train SVM and DNN classifiers to compare performance and evaluate the difficulty of classifying a large number of bird species. We adopt two performance measures to assess the quality of the results: Accuracy and Mean Average Precision. Intuitively, accuracy measures the number of times that a classifier correctly produced the expected output for a set of test examples. This measurement is also known as Top-1 classification precision, since classifiers can usually predict the probability of different outputs, and only the most confident one is used in this performance metric.

In contrast, Mean Average Precision (MAP) evaluates the ranking of predictions made by the classifier and computes the average precision for each prediction. Precision can be thought of as the percent of correct outputs in a partial list of n ranked results. MAP averages this over all examples in the test set.

We approach the question of how many features should be used for this task using the collection of available audio recordings. We construct the bag of features representation varying the number K of clusters produced by the k-means algorithm. A bag of features representation can be adapted to extract representative features from a collection of items, but the exact number of features depends on the data modality, and the amount of examples in the training set.

We evaluated dictionaries of features with 512, 1024, 2048 and 4096 elements, and the results are reported in Fig. 3. In these experiments, we evaluated

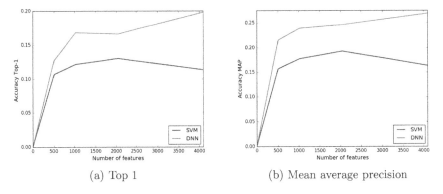

(a) Top 1   (b) Mean average precision

Fig. 3. Performance evaluation of classification methods. The number of features is represented in the x-axis and the performance measure in the y-axis.

both classifiers, SVMs and DNNs, with the best parameters found with cross-validation. The results show that DNNs with 5 hidden layers can produce a higher performance compared to shallow SVMs. This result is consistent regardless of the performance measure: Top-1 and MAP show a similar tendency.

Notice that performance improves as more features are added to the dictionary. SVMs start to decline with too many features, likely because linear combinations of these features cannot inform further the classification decision. However, DNNs robustly improve with more features because multiple layers can make non-linear combinations of the input features, and therefore find more complex patterns. Test results are reported in Table 1.

Table 1. Classification results on the test set. Features are the number of elements in the dictionary, Top-1 and MAP are performance measures.

Classifier	Features	Top-1	MAP
Support Vector Machine	2,048	0.12	0.18
Deep Neural Network	4,096	0.19	0.26

3.2 Visualization Experiments

In Fig. 4 are showed the visualization results for PCA (Fig. 4a), t-SNE (Fig. 4b), Isomap (Fig. 4c) and LLE (Fig. 4d). In this visualization all the bird audio records are visualized, and three selected species are highlighted in color for illustration purpose: green (*axillaris*), blue (*bicolor*) and magenta (*latrans*). Note that each method projects in a different way the audio records. In particular, Isomap is the method that best preserves the similarity of the bird species. LLE projects the audio records in 5 axis but without preserving species similarity. t-SNE uses the visualization space in an optimal way, avoiding the occlusion of data points.

This result is very interesting since we can use different visualization methods to visually analyze the audio bird collection.

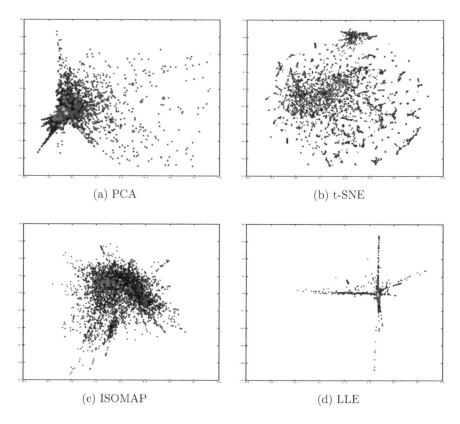

(a) PCA   (b) t-SNE

(c) ISOMAP   (d) LLE

Fig. 4. 2D visualization of the bird audio collection (Color figure online)

4   Conclusion and Future Work

This paper presented a methodology to process and analyze bird audio recordings. We followed a bag of audio features approach to represent audio content. Results showed the potential of the proposed methodology to offer ornithologist new mechanisms that help to process and analyze large collection of bird species using audio content. As future work we want to evaluate the proposed methodology with ornithologists to measure its effectiveness and efficiency.

References

1. Anderson, S.E., Dave, A.S., Margoliash, D.: Template-based automatic recognition of birdsong syllables from continuous recordings. J. Acoust. Soc. Am. **100**(2), 1209–1219 (1996)

2. Chou, C.-H., Liu, P.-H., Cai, B.: On the studies of syllable segmentation and improving MFCCs for automatic birdsong recognition. In: IEEE Asia-Pacific Services Computing Conference, APSCC 2008, pp. 745–750. IEEE (2008)
3. Davis, S.B., Mermelstein, P.: Comparison of parametric representations for monosyllabic word recognition in continuously spoken sentences. IEEE Trans. Acoust. Speech Signal Process. **28**(4), 357–366 (1980)
4. World Bank Group: World Development Indicators 2012. World Bank Publications, Washington (2012)
5. Jolliffe, I.T.: Principal Component Analysis. Springer, New York (1989)
6. Kogan, J.A., Margoliash, D.: Automated recognition of bird song elements from continuous recordings using dynamic time warping and hidden Markov models: a comparative study. J. Acoust. Soc. Am. **103**(4), 2185–2196 (1998)
7. Koops, H.V., Balen, J., Wiering, F.: Automatic segmentation and deep learning of bird sounds. In: Mothe, J., Savoy, J., Kamps, J., Pinel-Sauvagnat, K., Jones, G.J.F., SanJuan, E., Cappellato, L., Ferro, N. (eds.) CLEF 2015. LNCS, vol. 9283, pp. 261–267. Springer, Heidelberg (2015). doi:10.1007/978-3-319-24027-5_26
8. Ranjard, L., Ross, H.A.: Unsupervised bird song syllable classification using evolving neural networks. J. Acoust. Soc. Am. **123**(6), 4358–4368 (2008)
9. Saul, L.K., Roweis, S.T.: Think globally, fit locally: unsupervised learning of low dimensional manifolds. J. Mach. Learn. Res. **4**, 119–155 (2003)
10. Somervuo, P., Härmä, A.: Bird song recognition based on syllable pair histograms. In: Proceedings of IEEE International Conference on Acoustics, Speech, and Signal Processing (ICASSP 2004), vol. 5, p. V-825. IEEE (2004)
11. Tenenbaum, J.B., de Silva, V., Langford, J.C.: A global geometric framework for nonlinear dimensionality reduction. Science **260**, 2319–2323 (2000)
12. Van der Maaten, L., Hinton, G.: Visualizing data using t-SNE. J. Mach. Learn. Res. **9**(2579–2605), 85 (2008)
13. Vellinga, W., Planqu, R., Joly, A., Rauber, A.: LifeCLEF Bird Identification Task 2014 (2014)

GMM Background Modeling Using Divergence-Based Weight Updating

Juan D. Pulgarin-Giraldo[1(✉)], Andres Alvarez-Meza[1],
David Insuasti-Ceballos[1], Thierry Bouwmans[2],
and German Castellanos-Dominguez[1]

[1] Signal Processing and Recognition Group,
Universidad Nacional de Colombia, Manizales, Colombia
jdpulgaring@unal.edu.co
[2] Laboratoire MIA - Université de La Rochelle, La Rochelle, France

Abstract. Background modeling is a core task of video-based surveillance systems used to facilitate the online analysis of real-world scenes. Nowadays, GMM-based background modeling approaches are widely used, and several versions have been proposed to improve the original one proposed by Stauffer and Grimson. Nonetheless, the cost function employed to update the GMM weight parameters has not received major changes and is still set by means of a single binary reference, which mostly leads to noisy foreground masks when the ownership of a pixel to the background model is uncertain. To cope with this issue, we propose a cost function based on Euclidean divergence, providing nonlinear smoothness to the background modeling process. Achieved results over well-known datasets show that the proposed cost function supports the foreground/background discrimination, reducing the number of false positives, especially, in highly dynamical scenarios.

Keywords: Background modeling · GMM · Euclidean divergence

1 Introduction

Intelligent video surveillance systems, devoted to detect and track moving objects, can accomplish unsupervised results using background modeling methodologies, where a representation of the background is estimated and the regions that diverge from it are subtracted and labeled as moving objects named foreground [1]. Afterwards, surveillance systems interpret the activities and behaviors from the foreground objects to support computer vision analysis (e.g. object classification, tracking, activity understanding, among others) [2]. To achieve proper results, background modeling approaches focus on the elaboration of a background model, which suitably represents the pixel dynamics from real-world scenarios [3]. Among developed background modeling approaches, the most used are the ones derived from the conventional pixel-wise Gaussian Mixture Models (GMM) since they provide a trade-off between robustness to real-world video conditions and computational burden [4]. To date, several adapted

© Springer International Publishing AG 2017
C. Beltrán-Castañón et al. (Eds.): CIARP 2016, LNCS 10125, pp. 282–290, 2017.
DOI: 10.1007/978-3-319-52277-7_35

versions have been proposed. In fact, Authors in [5] provide a survey and propose the classification of the most salient GMM-based background modeling approaches.

Regarding the updating rules of the GMM parameters, improvements have been mainly reported related to the learning rate parameter, which aims to adapt the background model by observing the pixel dynamics through time. Originally, from the derivation of the GMM parameter updating rules, a Gaussian kernel term is attained, providing smoothness to the updating rules of the mean and variance parameters, nonetheless, the cost function of the weights is set using a binary ownership value. This updating rule may lead to noisy foreground masks, especially, when the pixel labels are uncertain like in dynamical scenarios. Zivkovic et al. proposed an improvement for the original GMM, which uses Dirchilet priors to update the weight parameters. Nonetheless, this improvement was mainly made to decrease the computational cost and the foreground/background discrimination performance remains similar to the original GMM.

Here, we propose a new cost function for the GMM weights updating. Using Euclidean divergence (ED), we compare the instant and cumulative probabilities of each Gaussian GMM model fitting the pixel input samples. Then, employing Least Mean Squares (LMS), we minimize the ED of obtained probabilities to adjust the weights values through time. By doing so, we provide non-linear smoothness to the whole GMM parameter updating rules, reducing the number of false positives in the obtained foreground masks. The proposed cost function is coupled into the traditional GMM approach, producing a new background modeling approach named ED-GMM, which improves the foreground/background discrimination in the case of real-world scenarios, especially in dynamical environments.

2   Methods

Background Modeling Based on GMM: The probability that a query input pixel $x_t \in \mathbb{R}^C$ ($C \in \mathbb{N}$ is the color space dimension), at time $t \in T$, belongs to a given GMM-based background model is as:

$$p\left(x_t | \mu_t, \Sigma_t\right) = \sum_{m=1}^{M} w_{m,t}\, \mathcal{N}\left(x_t | \mu_{m,t}, \Sigma_{m,t}\right), \tag{1}$$

where $M \in \mathbb{N}$ is the number of Gaussian models of the GMM, $w_{m,t} \in \mathbb{R}$ the weight related to the m-th Gaussian model, $\mathcal{N}\{\cdot, \cdot\}$ with mean value, $\mu_{m,t} \in \mathbb{R}^{1 \times C}$, and covariance matrix $\Sigma_{m,t} \in \mathbb{R}^{C \times C}$. For computational burden alleviation, all elements of the color representation set are assumed as independent and having the same variance value [4]: $\sigma_{m,t}^2 \in \mathbb{R}^+$, $\forall m$, that is, $\Sigma_{m,t} = \sigma_{m,t}^2 I$, being $I \in \mathbb{R}^{C \times C}$ the identity matrix. Afterwards, each query pixel, x_t, is evaluated until it matches a Gaussian model of the GMM. Here, the match occurs whenever a pixel value ranges within 2.5 standard-deviation interval of the Gaussian model. However, if x_t does not match any Gaussian model, the least probable model is replaced by a new one having low initial weight, large initial variance, and mean $\mu_{m,t} = x_t$ [4]. In the positive case that the m-th model matches a new input pixel, its parameters are updated as follows:

$$w_{m,t} = w_{m,t-1} + \alpha(o_{m,t} - w_{m,t-1}) \tag{2a}$$

$$\mu_{m,t} = \mu_{m,t-1} + \rho_{m,t} o_{m,t}(x_t - \mu_{m,t}) \tag{2b}$$

$$\sigma_{m,t}^2 = \sigma_{m,t-1}^2 + \rho_{m,t} o_{m,t}\left((x_t - \mu_{m,t})^\top (x_t - \mu_{m,t}) - \sigma_{m,t-1}^2\right) \tag{2c}$$

where $\alpha \in \mathbb{R}^+$ is the weight learning rate, $o_t \in \{0,1\}$ is a binary number indicating the membership of a sample to a model, and $\rho_{m,t} \in \mathbb{R}^+$ is the mean and variance learning rate set as a version of the α parameter smoothed by the Gaussian kernel $g(x_t; \cdot, \cdot)$, i.e.: $\rho_{m,t} = \alpha g(x_t; \mu_{m,t}, \sigma_{m,t})$. Lastly, the derived models are ranked according to the ratio w/σ to determine the most likely produced by the background, making suitable the further foreground/background discrimination [6].

Enhanced GMM-Based Background Using Euclidean Divergence (ED-GMM): The updating rules of the GMM parameters, used in Eq. (2), can be derived within the conventional Least Mean Square (LMS) formulation framework as follows:

$$\theta_t = \theta_{t-1} - \eta_{\theta_{t-1}} \partial_{\theta_{t-1}} \{\varepsilon_{\theta_{t-1}}^2\}, \tag{3}$$

where $\theta_t \in \{w_t, \mu_t, \sigma_t\}$ is each one of the estimated parameters by the corresponding learning rate:

$$\eta_{\theta_{t-1}} \in \begin{cases} \eta_{w_{t-1}} = \alpha/2 \\ \eta_{\mu_{t-1}} = \alpha \sigma_{t-1}^2/2 \\ \eta_{\sigma_{t-1}} = \alpha g(x_t; \mu_t, \sigma_{m,t})/2 \end{cases} \tag{4}$$

and the following cost functions, respectively:

$$\varepsilon_{\theta_{t-1}} \in \begin{cases} \varepsilon_{w_{t-1}} = o_t - w_{t-1} \\ \varepsilon_{\mu_{t-1}} = g(x_t; \mu_t, \sigma_{m,t}) \\ \varepsilon_{\sigma_{t-1}} = |x_t - \mu_t|^2 - \sigma_{t-1}^2 \end{cases} \tag{5}$$

It is worth noting that the μ_t and σ_t updating rules, grounded on kernel similarities $g(x_t; \cdot, \cdot)$ (Eqs. (4) and (5)), provide smoothness to encode the uncertainty of a pixel belonging whether to the background or foreground. In contrast, the cost function of the weights is set using a binary reference (i.e., membership o_t). This updating rule may lead to noisy foreground masks when the ownership value is uncertain, especially, in environments holding dynamical sources like trees waving, water flowing, snow falling, etc. To cope with this, we propose to set the cost function of w_t using the ED as follows:

$$\varepsilon_{w_{t-1}} = \bar{g}(x_t; \mu_{m,t}, \sigma_{m,t}) - w_{t-1}. \tag{6}$$

The ED allows measuring the difference between two probabilities [7]. So, back into the LMS scheme we aim to minimize the ED between an instant probability determined by the Gaussian kernel $\bar{g}\left(\boldsymbol{x}_t; \boldsymbol{\mu}_{m,t}, \sigma_{m,t}\right)$ and a cumulative probability encoded by w_{t-1}. This is grounded by the fact that, if a model has a high cumulative probability w_{t-1}, means that such model has suitably adapted to the pixel dynamics through time, then, $\bar{g}\left(\boldsymbol{x}_t; \boldsymbol{\mu}_{m,t}, \sigma_{m,t}\right)$ should have a high value too. Since the difference between both is expected to be low, the following updating rule is introduced:

$$w_{m,t} = w_{m,t-1} + \alpha(\bar{g}\left(\boldsymbol{x}_t; \boldsymbol{\mu}_{m,t}, \sigma_{m,t}\right) - w_{m,t-1}) \tag{7}$$

where the kernel term, $\bar{g}\left(\boldsymbol{x}_t; \boldsymbol{\mu}_{m,t}, \sigma_{m,t}\right) = \mathbb{E}\left\{g\left(\boldsymbol{x}_t^c; \boldsymbol{\mu}_{m,t}^c, \sigma_{m,t}\right) : \forall c \in C\right\}$, measures the average similarity along color channels. Also, since we aim to incorporate the information about each new input sample into all the GMM Gaussian models, we exclude the ownership o_t from the weight updating.

3 Experimental Set-Up

Aiming to validate the proposed cost function, the ED-GMM approach is compared against the traditional GMM (GMM1) and the Zivkovic GMM proposed (ZGMM) in [8], which uses Dirichlet priors into the weight updating rules to automatically set the number of Gaussians M. The following three experiments are performed: (i) Visual inspection of the temporal weight evolution to make clear performance of background model and the foreground/background discrimination through time. (ii) Foreground/background discrimination over a wide variety of real-world videos that hold ground-truth sets. (iii) Robustness against variations of the learning rate parameter in foreground/background discrimination tasks. The following datasets are employed for the experiments:

DBa- *Change Detection:* (at http://www.changedetection.net/) Holds 31 video sequences of indoor and outdoor environments, where spatial and temporal regions of interest are provided. Ground-truth labels are background, hard shadow, outside region of interest, unknown motion, and foreground.

DBb- *A-Star-Perception:* (at http://perception.i2r.a-star.edu.sg) Recorded in both indoor and outdoor scenarios, contains nine image sequences with different resolution. The ground-truths are available for random frames in the sequence and hold two labels: background and foreground.

Measures: The foreground/background discrimination is assessed only for two ground-truth labels (*foreground* and *background*) by supervised pixel-based measures: Recall, $r = t_p/(t_p + f_n)$, Precision, $p = t_p/(t_p + f_p)$, and $F_1 = 2pr/(p + r)$. Here, t_p is the number of true positives, f_p is the false positives, and f_n is the false negatives. These values are obtained comparing against the ground-truth. Measures range within [0,1], where the higher the attained measure – the better the achieved segmentation.

Implementation and Parameter Tuning: The ED-GMM algorithm is developed using as basis the C++ BGS library [9]. Parameters are left as default for all the experiments except for task three requiring to vary the learning rate α. We set three mixing models $M = 3$ (noted as *Model1*, *Model2*, and *Model3*). The GMM1 and ZGMM algorithms are also taken from the BGS library.

4   Results and Discussion

Temporal Analysis: We conduct a visual inspection of the temporal evolution of estimated parameters to make clear the contribution of the proposed weight cost function. Testing is carried out on the video *DBa-snowFall* for which a single pixel in the red color channel is tracked as seen in Fig. 1, showing temporal evolution of μ_t (top row) and w_t (bottom row). Also, the inferred foreground/background labels and the ground-truth are shown in subplots Fig. 1(c) and (d) ('1': foreground and '0': background). It can be observed that the estimated μ_t parameter by either GMM1 (see subplot 1(a)) or ED-GMM (subplot 1(b)) is similar for the three considered mixing models. However, the weights estimated by ED-GMM are updated better along time. Particularly, the ED-GMM weight increases around the $500th$ frame, where the *Model2* (in green) is generated (see subplot 1(d)). Then, the model properly reacts to the pixel change occurring close to the $800th$

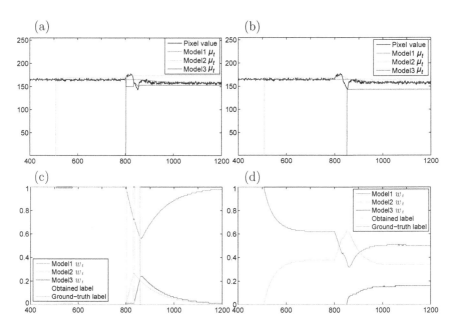

Fig. 1. Temporal evolution of w_t and μ_t for pixel $(150, 250, 1)$ from DBa-snowFall video. (a) GMM1 μ_t, (b) ED-GMM μ_t, (c) GMM1 w_t and label, (d) ED-GMM w_t and label (Color figure online)

frame, obtaining labels corresponding to the ground-truth (background). In contrast, the GMM1 updating rule makes the w_t weight remain almost zero even if the *Model2* gets very close to the pixel value (see subplot 1(c)). As a consequence, this strategy infers wrongly foreground labels.

Performance of the Foreground/Background Discrimination Task: Aiming to check for the generalizing ability of the ED-GMM method, we test 25 videos embracing a wide variety of dynamics. The videos are grouped into two categories *a* and *b*. The former holds videos where the background is mostly static and the latter videos where the background exhibit highly dynamical variations. The total average seen in Table 1 shows that the ED-GMM reaches higher precision

Table 1. Foreground discrimination performance

	Video	GMM1			ZGMM			ED-GMM		
		r	p	F_1	r	p	F_1	r	p	F_1
Category *a*	DBa-abandonedBox	0.34	0.43	0.38	0.32	0.46	0.37	0.35	**0.60**	**0.44**
	DBb-bootstrap	**0.59**	0.56	0.57	0.55	0.60	0.58	0.55	0.61	0.58
	DBa-corridor	0.60	0.74	0.66	0.58	0.79	0.67	0.61	**0.88**	**0.72**
	DBa-diningRoom	0.45	0.88	0.60	**0.57**	0.94	**0.71**	0.45	0.93	0.61
	DBb-hall	0.51	0.65	0.57	**0.54**	**0.73**	**0.62**	0.51	0.69	0.59
	DBa-highway	0.90	0.90	0.90	0.82	0.91	0.86	0.89	0.93	0.91
	DBa-library	0.17	0.91	0.29	0.19	0.94	0.32	0.19	0.93	0.32
	DBa-office	0.29	0.63	0.40	0.28	0.63	0.38	**0.31**	**0.71**	**0.43**
	DBa-park	0.65	0.70	0.67	0.64	0.73	0.68	0.53	**0.80**	0.64
	DBa-parking	0.27	0.78	0.40	0.17	0.69	0.28	0.27	**0.84**	0.41
	DBa-pedestrians	0.94	0.51	0.66	0.93	0.46	0.61	0.90	**0.71**	**0.79**
	DBa-pets2006	**0.69**	0.54	0.61	0.67	0.57	0.62	0.63	**0.74**	**0.68**
	DBb-shoppingMall	0.64	0.53	**0.58**	0.63	0.48	0.55	0.54	**0.58**	0.56
	DBa-sofa	0.42	0.65	0.51	**0.46**	0.64	**0.54**	0.35	**0.84**	0.49
	DBa-streetLight	0.17	0.48	0.25	**0.23**	0.52	**0.32**	0.18	**0.72**	0.29
	DBa-tramStop	0.37	0.87	0.52	0.38	0.87	0.53	0.33	0.88	0.48
	Average	0.50	0.67	0.53	0.50	0.69	0.54	0.47	**0.77**	**0.56**
Category *b*	DBa-boats	0.43	0.24	0.31	**0.46**	0.18	0.26	0.43	**0.27**	**0.33**
	DBa-canoe	0.49	0.31	0.38	0.53	0.32	0.40	**0.63**	**0.35**	**0.45**
	DBa-fountain02	0.86	0.33	0.48	0.85	0.28	0.42	0.82	**0.52**	**0.64**
	DBa-overpass	0.73	0.76	0.74	**0.80**	0.77	0.78	0.78	**0.83**	0.80
	DBa-skating	0.76	0.80	0.78	0.77	0.77	0.77	0.78	**0.89**	**0.83**
	DBa-snowFall	0.73	0.35	0.47	0.71	0.48	0.58	0.66	**0.65**	**0.65**
	DBb-waterSurface	0.55	0.74	0.63	0.52	0.71	0.60	**0.61**	**0.79**	**0.69**
	DBa-wetSnow	**0.64**	0.76	0.69	0.58	0.80	0.67	0.59	0.81	0.68
	DBa-winterDriveway	0.59	0.40	0.48	0.58	0.43	0.49	0.60	**0.52**	**0.55**
	Average	0.64	0.52	0.55	0.64	0.53	0.56	0.66	**0.63**	**0.62**
	Total average	0.55	0.62	0.58	0.55	0.63	0.59	0.54	**0.72**	**0.62**

during the discrimination of the foreground/background labels, decreasing the amount of false positives. This fact is explicable since the proposed weight updating rule (see Eq. (7)) allows the ED-GMM models to adapt faster to changes of the pixel dynamics. The above can be even more remarked for videos with dynamical background sources as seen in the *Category b* in which the precision is improved by 10% comparing against the other two methods. By the other hand, GMM1 and ZGMM attain very similar results, since the main proposal of Zivkovic was focused to reduce computational cost. As a result, the foreground masks attained by ED-GMM have less false positives and are more similar to the ground truth masks as seen in Fig. 2 showing concrete scenarios with highly dynamical background sources relating to snow falling (DBa-snowFall, DBa-winterDriveway) and water flowing (DBa-fountain02, DBb-waterSurface).

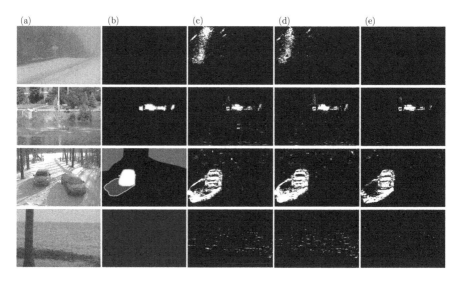

Fig. 2. Foreground masks of interest. (a) Original frame, (b) Ground-truth, (c) GMM1, (d) ZGMM, (e) ED-GMM.

Robustness Against Variation of the Learning Rate Parameter: The influence of the learning rate variation on the foreground/background discrimination is assessed through supervised measures, which are estimated from the videos of Category a: DBa-highway, DBa-office, DBa-pedestrians and DBa-pets2006 and Category b: DBa-boats, DBa-canoe, DBa-fountain02 and DBa-overpass.

Figure 3 shows the obtained supervised measures, averaged over the 10 videos, where the x axis is the logarithm of the employed α rate ranging within: $\{0.0005, 0.001, 0.005, 0.01, 0.03, 0.05, 0.07, 0.1, 0.15, 0.2\}$. It can be seen that the proposed ED-GMM (continuous lines) behaves similar as the traditional GMM1 method (dashed lines) and ZGMM (pointed lines). However, the obtained Precision and $F1$ measures are everywhere higher than the ones reached by GMM1 and

ZGMM. An interval of confidence is found within the interval $\alpha \in 0.005 - 0.01$, where the highest $F1$ measure is reached.

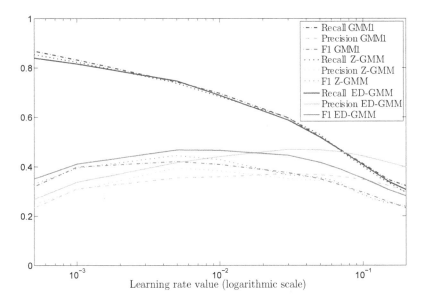

Fig. 3. Supervised measures changing the learning rate value for GMM1, Z-GMM and ED-GMM.

5 Conclusions

We propose a cost function for the GMM weights updating using Euclidean divergence. The proposed cost function is coupled into the traditional GMM, producing a new approach named ED-GMM used to support the background modeling task for videos recorded in highly dynamical scenarios. The Euclidean divergence allows comparing the instant and cumulative probabilities of a GMM model fitting the pixel input samples. Then, employing LMS, we minimize the Euclidean divergence of such probabilities to adjust the weights values through time. Carried out experiments show that ED-GMM reduces the amount of false positives in the obtained foreground masks comparing against traditional GMM and Zivkovic GMM, especially, for videos holding dynamical background sources: water flowing, snow falling and trees waving. Additionally, the proposed cost function demonstrated to be robust when varying the learning rate parameter value, always achieving better results than traditional GMM. Consequently, the proposed cost function can be coupled into more complex GMM based background modeling approaches to improve the foreground/background discrimination. As future work, authors plan to test the proposed cost function using selective updating strategies to improve the discrimination in scenarios holding motionless foreground objects.

Acknowledgment. This work was developed in the framework of the research project entitled *"Caracterización de cultivos agrícolas mediante estrategias de teledetección y técnicas de procesamiento de imágenes"* (36719) under the grants of *"Convocatoria conjunta para el fomento de la investigación aplicada y desarrollo tecnológico"* 2016, as well as by program *"Doctorados Nacionales convocatoria 647 de 2014"* funded by COLCIENCIAS and partial Ph.D. financial support from Universidad Autonoma de Occidente.

References

1. Molina-Giraldo, S., Álvarez-Meza, A.M., García-Álvarez, J.C., Castellanos-Domínguez, C.G.: Video segmentation framework by dynamic background modelling. In: Petrosino, A. (ed.) ICIAP 2013. LNCS, vol. 8156, pp. 843–852. Springer, Heidelberg (2013). doi:10.1007/978-3-642-41181-6_85
2. Maddalena, L., Petrosino, A.: A self-organizing approach to background subtraction for visual surveillance applications. IEEE Trans. Image Process. **17**(7), 1168–1177 (2008)
3. Alvarez-Meza, A.M., Molina-Giraldo, S., Castellanos-Dominguez, G.: Background modeling using object-based selective updating and correntropy adaptation. Image Vis. Comput. **45**, 22–36 (2016)
4. Stauffer, C., Grimson, W.E.L.: Adaptive background mixture models for real-time tracking. In: IEEE Computer Society Conference on Computer Vision and Pattern Recognition, vol. 2. IEEE (1999)
5. Bouwmans, T., El Baf, F., Vachon, B., et al.: Background modeling using mixture of Gaussians for foreground detection - a survey. Recent Pat. Comput. Sci. **1**(3), 219–237 (2008)
6. Hayman, E., Eklundh, J.-O.: Statistical background subtraction for a mobile observer. In: Proceedings of Ninth IEEE International Conference on Computer Vision, vol. 1, pp. 67–74, October 2003
7. Principe, J.: Information Theoretic Learning. Renyi's Entropy and Kernel Perspectives. Springer, New York (2010)
8. Zivkovic, Z.: Improved adaptive Gaussian mixture model for background subtraction. In: Proceedings of the 17th International Conference on Pattern Recognition, ICPR 2004, vol. 2, pp. 28–31, August 2004
9. Sobral, A.: BGSLibrary: an OpenCV C++ background subtraction library. In: IX Workshop de Visão Computacional (WVC 2013), Rio de Janeiro, Brazil, pp. 38–43, June 2013

Bayesian Optimization for Fitting 3D Morphable Models of Brain Structures

Hernán F. García[✉], Mauricio A. Álvarez, and Álvaro A. Orozco

Grupo de Investigación en Automática,
Universidad Tecnológica de Pereira, Pereira, Colombia
{hernan.garcia,malvarez,aaog}@utp.edu.co

Abstract. Localize target areas in deep brain stimulation is a difficult task, due to the shape variability that brain structures exhibit between patients. The main problem in this process is that the fitting procedure is carried out by a registration method that lacks of accuracy. In this paper we proposed a novel method for 3D brain structure fitting based on Bayesian optimization. We use a morphable model in order to capture the shape variability in a given set of brain structures. Then from the trained model, we perform a Bayesian optimization task with the aim to find the best shape parameters that deform the trained model, and fits accurately to a given brain structure. The experimental results show that by using an optimization framework based on Bayesian optimization, the model performs an accurate fitting over cortical brain structures (thalamus, amygdala and ventricle) in comparison with common fitting methods, such as iterative closest point.

Keywords: Bayesian optimization · 3D brain structures · Shape fitting · Morphable model

1 Introduction

Deep brain stimulation surgery (DBS) in Parkinson's disease, is a surgical procedure used to treat the most common neurological symptoms such as stiffness, slowed movement, tremor and walking problems. In this procedure, the neurosurgeons perform an electrical stimulation of the basal ganglia area (thalamus, ventricles and sub-thalamus) by placing a micro-electrode device over this region [1]. However, a misplacement of the final position of this micro-electrode can induce negative effects in the post-surgery, like abnormal postures, loss of speech and even loss of mobility, among others symptoms [2]. That is why the neurosurgeons need to locating with high accuracy, the cortical brain structures related with the basal ganglia area [3]. The most common planning DBS surgery methods include a registration process in which a given brain atlas (volume of labeled brain structures with respect to a healthy patient), is fitted into the magnetic resonance images (MRI) of the patient to be treated. This step allows to the neurosurgeons localize the target area in the MRI volume [4]. The main problem of these approaches is that the brain atlases used to estimate the target brain structures lack of generalizability (since we need to modeling the brain

© Springer International Publishing AG 2017
C. Beltrán-Castañón et al. (Eds.): CIARP 2016, LNCS 10125, pp. 291–299, 2017.
DOI: 10.1007/978-3-319-52277-7_36

volumetry in a given population), and the fitting accuracy of the atlases in the brain volumes is still an open research topic [5]. Therefore, methods that can be able to capture the shape variability over a set of MRI volumes could improve the robustness (generalizability and accuracy) of the target area localization [3].

The model-based approaches such as morphable models (MM), make use of the prior knowledge of the shape variability in a set of images (MRI volumes) and typically finds the best match between the model and the new image (brain structure) [6]. In medical image analysis, the use of prior information (shape contour of the brain structures labeled by a medical specialist), combined with some model that can be able to represent a shape contour, leads to an accurate fitting of a given brain structure [3]. A commonly used method for fitting morphable models, is the Iterative Closest Point (ICP) [6]. This method establishes closest point associations from one shape to another, and finds a rigid transformation that minimizes an Euclidean error between shapes. However, due to the large deformations that the brain volumes present over different patients (i.e. different thickness for ventral brain structures), the performance of the fitting process becomes inaccurate [7]. The reason for this performance, is that the global minimum of the cost function that measures the fitting process (mean square error between the deformed and the target brain structure), often does not corresponds to the optimal fitting (That is, a rigid transform only estimates the scale, rotation and translation parameters, discarding the shape parameters that control the deformations) [8].

Bayesian optimization (BO) provides an elegant approach for the global optimization problem, in which a given cost function is minimized in a probabilistic way [9]. For continuous functions, Bayesian optimization assumes that the unknown function is sampled from a Gaussian process, and maintains a posterior distribution for this function as observations [10]. In the fitting process of the morphable model, these observations are the measure of generalization performance (matching accuracy) under different settings of the hyperparameters that we wish to optimize (shape parameters that control the deformations) [11]. In this paper we propose a Bayesian optimization framework for fitting 3D morphable models of brain structures. We use as morphable model, a point distribution model (PDM) in order to capture the shape variability in a set of brain volumes (brain structures related with Parkinson's disease). From the trained model, we perform a fitting process based on Bayesian optimization in order to find the optimal match between the morphable model and a given brain structure. The main contribution of our work, is based on the Bayesian optimization process that finds the shape parameters that control the deformations of a morphable model in a probabilistic way.

2   Materials and Methods

2.1   Database

In this work we use a MRI database from the Universidad Tecnológica de Pereira (DB-UTP). This database contains recordings of MRI volumes from ten patients with Parkinson's disease. The database was labeled by neurosurgeons from the Institute of Parkinson and Epilepsy of the Eje Cafetero, located in Pereira-Colombia. The database contains $T1$, $T2$ and CT sequences with $1\,\text{mm} \times 1\,\text{mm} \times 1\,\text{mm}$ voxel size and slices of 512×512 pixels. Also a set of detailed label maps related to the brain structures (i.e. basal ganglia area) are provided. Moreover, three-dimensional models of the labeled anatomical structures for each patient are available in the dataset.

2.2   3D Brain Model

We use a 3D brain model based on a PDM to represent the shape of the brain structures. The shape information is captured by the vertexes information (point-cloud data in the $\mathbb{R}^3$) from the mesh data that represents each brain structure. Besides, the model uses statistical information of the shape variation across the training set in order to model the brain volumetry of a given brain structure [6]. The PDM is a parametrized model, $\mathbf{S} = \gamma(\mathbf{b})$, where $\mathbf{S} = [\mathbf{w}_1, \mathbf{w}_2, \ldots, \mathbf{w}_N]$, with $\mathbf{w}_i = [x_i, y_i, z_i]^\top$, $\mathbf{w}_i \in \mathbb{R}^{3\times1}$ representing each landmark (point data in $3D$ space). The vector $\mathbf{b}$ holds the parameters which can be used to vary the shape across the surface and γ defines the function over the parameters. We use N landmarks representing the points of the surface related to a given brain structure, from a training set of L brain meshes, where each shape is $\mathbf{S}^k = [\mathbf{w}_1^k, \mathbf{w}_2^k, \ldots, \mathbf{w}_N^k]$, $\mathbf{S}^k \in \mathbb{R}^{3\times N}$.

In order to eliminate the global transformations for the training shapes, we use Procrustes analysis [6]. The alignment process is carried out by minimizing the square-distance of each shape $\mathbf{S}_k$ with respect to their mean $\bar{\mathbf{S}} = \frac{1}{N}\sum_{k=1}^{L}\mathbf{S}_k$ ($\bar{\mathbf{S}}$ is scaled such that $|\mathbf{S}| = 1$). We use Principal Component Analysis (PCA) to model the shape variations of the brain structures in the training set. The model estimate these variations by computing the eigenvalues and eigenvectors of the covariance matrix of the training set defined by, $\mathbf{C} = \frac{1}{L-1}\sum_{k=1}^{L}(\mathbf{s}_k - \bar{\mathbf{s}})(\mathbf{s}_k - \bar{\mathbf{s}})^\top$, where $\mathbf{s}_k$ and $\bar{\mathbf{s}}$ are the vectorized forms of the shape $\mathbf{S}_k$ and the mean shape $\bar{\mathbf{S}}$ respectively. Les us define ϕ_i and λ_i as the ith eigenvector and eigenvalue of the covariance matrix $\mathbf{C}$. If $\boldsymbol{\Phi}$ holds the t eigenvectors corresponding to the largest eigenvalues, a given plausible shape (similar to the brain structures in the training set) can be estimated by

$$\hat{\mathbf{s}} \approx \bar{\mathbf{s}} + \boldsymbol{\Phi}\mathbf{b}, \tag{1}$$

where $\boldsymbol{\Phi} = (\phi_1|\phi_2|\ldots|\phi_t)$ and $\mathbf{b}$ is a t dimensional vector representing the shape parameters (those who controls the shape variability). The value of t is chosen such that the model represents the 98% of the shape variance [6].[1]

2.3   Bayesian Optimization with Gaussian Process Priors

Since we want to compute the model parameters in a probabilistic way, the goal is to find the minimum of a cost function $f(\mathbf{x})$ (i.e. Euclidean distance

---

[1] The variance of the ith parameter, b_i, across the training set is given by λ_i.

between the ground truth landmarks and the landmarks deformed by the 3D-MM model) on some bounded set $\mathcal{X}$ that controls the shape variations. To this end, Bayesian optimization constructs a probabilistic framework for $f(\mathbf{x})$ with the aim to exploit this model to make predictions of the shape parameters $\mathcal{X}$ evaluated in the cost function [9].[2]

The main components of the Bayesian optimization framework are the prior over the function being optimized and the acquisition function that will allow us to determine the next point to evaluate the cost function [10]. In this work we use Gaussian process prior, due to its flexibility and tractability. A Gaussian Process (GP) is an infinite collection of scalar random variables indexed by an input space such that for any finite set of inputs $\mathbf{X} = \{\mathbf{x}_1, \mathbf{x}_2, \ldots, \mathbf{x}_n\}$, the random variables $\mathbf{f} \triangleq [f(\mathbf{x}_1), f(\mathbf{x}_2), \cdots, f(\mathbf{x}_n)]$ are distributed according to a multivariate Gaussian distribution $\mathbf{f}(\mathbf{X}) \sim \mathcal{GP}(m(\mathbf{x}), k(\mathbf{x}, \mathbf{x}'))$. A GP is completely specified by a mean function $m(\mathbf{x}) = \mathbb{E}[f(\mathbf{X})]$ (usually defined as the zero function) and a positive definite covariance function given by $k(\mathbf{x}, \mathbf{x}') = \mathbb{E}\left[(f(\mathbf{x}) - m(\mathbf{x}))(f(\mathbf{x}') - m(\mathbf{x}'))^{\top}\right]$ (see [12] for further details).

Let us assume that $f(\mathbf{x})$ is drawn from a Gaussian process prior and that our observations are of the form $\{\mathbf{x}_n, y_n\}_{n=1}^{N}$, where $y_n = \mathcal{N}(f(\mathbf{x}_n), v))$ and v is the variance of noise introduced into the function observations. The acquisition function is denoted by $u : \mathcal{X} \in \mathbb{R}^{+}$ and establish the point in $\mathcal{X}$ that is evaluated in the optimization process as $x_{\text{next}} = \arg\max_{\mathbf{x}} u(\mathbf{x})$. Since the acquisition function depends on the GP hyperparameters, $\boldsymbol{\theta}$, and the predictive mean function $\mu(\mathbf{x}, \{\mathbf{x_n}, \mathbf{y_n}\}, \boldsymbol{\theta})$ (as well as the predictive variance function), the best current value is then $x_{\text{best}} = \arg\min_{\mathbf{x}_n} f(\mathbf{x}_n)$.

2.4   Morphable Model Fitting Using Bayesian Optimization

Our approach is based on the Bayesian optimization process for estimating the shape parameters that fit accurately a given brain structure in a probabilistic way. In this work, we used the 3D models of the labeled anatomical structures for the ten patients of the dataset. We trained three PDM models by capturing the shape variability for the thalamus, amygdala and ventricles (due to their importance in the Parkinson's disease). Since the brain structures are non-rigid shapes and their size vary along the training set, we need to decimate all the 3D shapes (with respect to the smallest brain structure) to perform the eigenvalue decomposition. We use *leave-one-out* validation (we train the 3D-MM with $L-1$ brain structures) to measure the fitting accuracy. Besides, we initialize the fitting process by performing rigid iterative closest point (ICP) in order to find the scale, rotation and translation parameters between the model and the given brain structure. The main reason for this prior initialization is to let the BO process explore the shape parameters that deform the PDM model. For the BO process we use as cost function, the Euclidean distance between the landmarks

---

[2] The main reason of the Bayesian optimization framework is to use all of the available information from previous evaluations of $f(\mathbf{x})$.

of the model and the ground truth landmarks of the brain structure. We use the $GPyOpt^3$ toolbox for python. In this work, we report results for the expected improvement (EI), and the probability of improvement (PI) acquisition functions [9]. Figure 1 shows the block diagram of the proposed model used in this work.

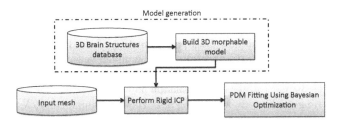

Fig. 1. Block diagram of the proposed model fitting approach based on BO.

3 Results and Discussions

In the following sections we show the results for our BO framework for fitting 3D morphable models of brain Structures. On Sect. 3.1, we show the results of the 3D shape modeling for the brain structures using a PDM as morphable model. Section 3.2 shows the results for the fitting process of the morphable model using BO.

3.1 Training Results for 3D-MM

Figure 2 shows the training process results for the morphable models. From the figure, it can be noticed that for the three brain structures (thalamus, amygdala and ventricle), the morphable models capture the shape variability along the training set. Moreover, the results also show that by modeling the covariance matrix of the training dataset using PCA, the morphable model capture the relevant information in the latent space by analyzing the shape parameters that controls the deformation of a given brain structure. Besides, Fig. 2 shows that by changing the shape parameters (eigenvalues of the PDM) the models deform a given brain structure from thin shapes (upper left corner of the Fig. 2(c)) to curvy shapes (lower right corner of the Fig. 2(c)). This shape variability can be related with the range of ages of the subjects in the database (between 35–65 years), due to the fact that the brain volume decreases their mass over time in patients with Parkinson's disease (thin shapes modeled by the first eigenvalues of the morphable models) [2].

---

[3] *Gpyopt* is a BO framework in python available at http://github.com/SheffieldML/GPyOpt.

(a) PDM for the thalamus (b) PDM for the amygdala (c) PDM for the ventricle

Fig. 2. Effects of varying the first 3 shape parameters of the PDM for the analyzed brain structures. Figures (a), (b) and (c) show the model variation for the most relevant eigenvalues: (top) $b_1 = \{-3\sqrt{\lambda_1}, \ldots, 3\sqrt{\lambda_1}\}$; (middle) $b_2 = \{-3\sqrt{\lambda_2}, \ldots, 3\sqrt{\lambda_2}\}$; (bottom) $b_3 = \{-3\sqrt{\lambda_3}, \ldots, 3\sqrt{\lambda_3}\}$. Each shape parameter (structures for each row in the subfigures) ranges from $-3\sqrt{\lambda_i}$ starting with the left column till $3\sqrt{\lambda_i}$ for the right column.

3.2   Fitting Results Using Bayesian Optimization

From the trained morphable models, we deform each PDM in order to match a target brain structure. The BO process estimate the best shape parameters that fit each of the brain structures (amygdala, thalamus and ventricle). Figure 3

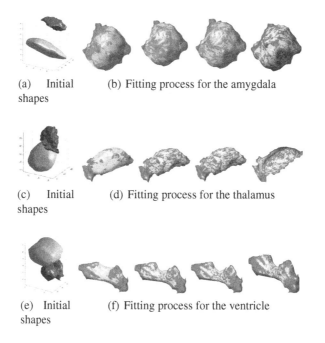

(a)   Initial shapes   (b) Fitting process for the amygdala

(c)   Initial shapes   (d) Fitting process for the thalamus

(e)   Initial shapes   (f) Fitting process for the ventricle

Fig. 3. Fitting process for the PDM models of the amygdala, thalamus and ventricle brain structures (shapes with yellow color are related to the model and those with color magenta depict the target shapes). The figures show the model to be deformed and the target brain structure. Also the figure shows the deformed models at 10, 50, 100 and 200 iterations. (Color figure online)

shows the results for the fitting process using BO. The results show that by using a fully Bayesian treatment of the optimization process, the model can estimate the shape parameters that performs accurately the fitting process between shapes. Figure 3 shows that by adding a prior registration process through rigid ICP, the BO method can explore an exploit the probabilistic search of the shape parameters that minimizes the cost function (Euclidean distance between the deformed model landmarks and the target shape). Besides, the results in Fig. 3(a) prove that the optimization step estimates the plausible shape parameters that deforms a given brain structure, even if the target shape has large curvatures (significant changes over the shape surface, see Figs. 3(c) and (e)). Beside, the Fig. 4 shows the convergence of the BO process for both the EI and the PI acquisition functions. The results show that EI has better global convergence, due to the fact that by using this acquisition function the model fits more accurately the target shape and take less iteration to converge (200 iterations for the EI and more than 300 iterations for the PI). The main reason is that the EI estimates best shape parameters in the exploration step than the PI acquisition function. However, both acquisition functions exploit the probabilistic search of the shape parameters with lower values in the evaluated cost function (mean square error of the Euclidean distance, 52 for EI and 60 for the PI). Finally, Table 1 shows the accuracy of the BO process compared against the common fitting method such as Rigid ICP. The results show that by using a given acquisition function the optimization process improves the selection of those values that control the deformation with high variance (shape parameters in regions not well explored) and values with high mean value (shape parameters worth exploiting that increase the fitting accuracy, MSE error of 10.295 for the amygdala) in comparison with the Rigid ICP which only removes the translation, rotation and scale between two shapes (MSE error of 23.971 for the amygdala).

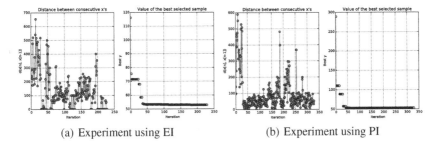

(a) Experiment using EI   (b) Experiment using PI

Fig. 4. Convergence of the BO process for the EI (up) and PI (down) acquisition functions. The figure shows the distance between consecutive selected **x** values (left column), the mean of the current model in the selected sample (middle) and the variance of the model in the selected sample (right column).

Table 1. Accuracy of the BO process for fitting the three morphable models using EI and PI acquisition functions. The table shows the mean square error of the Euclidean distance between the deformed model and the target brain structure.

Brain model	Fitting method		
	Rigid ICP	BO with EI	BO with PI
Amygdala	23.971 ± 4.163	10.295 ± 2.575	13.927 ± 4.665
Thalamus	32.486 ± 4.169	19.907 ± 1.340	22.351 ± 4.170
Ventricle	26.623 ± 4.287	15.924 ± 2.079	16.686 ± 4.289

4   Conclusions and Future Works

In this paper we propose a Bayesian optimization framework for fitting 3D morphable models of brain structures. Our method deforms a trained point distribution model in order to match a given brain structure. Besides, the shape parameters that control the model variations (relevant eigenvalues derived from the training set), are optimized in a probabilistic way. The results show that placing a Gaussian process prior over the function being optimized, the proposed model can estimate the best parameters that perform the fitting process. Moreover, by using the *Expected improvement* as acquisition function, the optimization process improves the matching accuracy by exploiting the model parameters bounded by the eigenvalues of the training set.

As future works, we want to analyze the Bayesian optimization framework in high dimensional problems that includes the whole optimization process of a raw point-cloud data. Finally, we want to analyze a morphable model that adds both shape an appearance information in order to model tissue properties related to a given brain structure.

Acknowledgments. This research is developed under the projects with code $FP44842$-584-2015 and 111065740687, financed by Colciencias and SINCLECLICK SOLUTIONS S.A.S. H.F. García is funded by Colciencias under the program: *Formación de alto nivel para la ciencia, la tecnología y la innovación - Convocatoria 617 de 2013.*

References

1. Hill, D.: Neuroimaging to assess safety and efficacy of AD therapies. Expert Opin. Investig. Drugs **19**(1), 23–26 (2010). PMID: 19947893
2. Ibarretxe-Bilbao, N., Tolosa, E., Junque, C., Marti, M.J.: MRI and cognitive impairment in Parkinson's disease. Mov. Disord. **24**(S2), S748–S753 (2009)
3. Cabezas, M., Oliver, A., Lladó, X., Freixenet, J., Cuadra, M.B.: A review of atlas-based segmentation for magnetic resonance brain images. Comput. Methods Programs Biomed. **104**(3), e158–e177 (2011)

4. Cosa, A., Canals, S., Valles-Lluch, A., Moratal, D.: Unsupervised segmentation of brain regions with similar microstructural properties: application to alcoholism. In: 2013 35th Annual International Conference of the IEEE on Engineering in Medicine and Biology Society (EMBC), pp. 1053–1056, July 2013

5. Sotiras, A., Davatzikos, C., Paragios, N.: Deformable medical image registration: a survey. IEEE Trans. Med. Imaging **32**(7), 1153–1190 (2013)

6. Nair, P., Cavallaro, A.: 3-D face detection, landmark localization, and registration using a point distribution model. IEEE Trans. Multimed. **11**(4), 611–623 (2009)

7. Sjöberg, C., Ahnesjö, A.: Multi-atlas based segmentation using probabilistic label fusion with adaptive weighting of image similarity measures. Comput. Methods Programs Biomed. **110**(3), 308–319 (2013)

8. Sjöberg, C., Johansson, S., Ahnesjö, A.: How much will linked deformable registrations decrease the quality of multi-atlas segmentation fusions? Radiat. Oncol. **9**(1), 1–8 (2014)

9. Snoek, J., Larochelle, H., Adams, R.P.: Practical Bayesian optimization of machine learning algorithms. In: Advances in Neural Information Processing Systems 25: 26th Annual Conference on Neural Information Processing Systems 2012. Proceedings of a Meeting Held 3–6 December 2012, Lake Tahoe, Nevada, United States, pp. 2960–2968 (2012)

10. Srinivas, N., Krause, A., Seeger, M., Kakade, S.M.: Gaussian process optimization in the bandit setting: no regret and experimental design. In: Fürnkranz, J., Joachims, T. (eds.) Proceedings of the 27th International Conference on Machine Learning (ICML 2010), pp. 1015–1022. Omnipress (2010)

11. Wang, Z., Zoghi, M., Hutter, F., Matheson, D., de Freitas, N.: Bayesian optimization in high dimensions via random embeddings. In: International Joint Conferences on Artificial Intelligence (IJCAI) - Distinguished Paper Award (2013)

12. Rasmussen, C.E., Williams, C.: Gaussian Processes for Machine Learning. The MIT Press, Cambridge (2006)

Star: A Contextual Description of Superpixels for Remote Sensing Image Classification

Tiago M.H.C. Santana[1(✉)], Alexei M.C. Machado[2], Arnaldo de A. Araújo[1], and Jefersson A. dos Santos[1]

[1] Department of Computer Science, Universidade Federal de Minas Gerais (UFMG), Belo Horizonte, Brazil
{tiagohubner,arnaldo,jefersson}@dcc.ufmg.br
[2] Department of Computer Science, Pontifícia Universidade Católica de Minas Gerais, Belo Horizonte, Brazil
alexei@pucminas.br

Abstract. Remote Sensing Images are one of the main sources of information about the earth surface. They are widely used to automatically generate thematic maps that show the land cover of an area. This process is traditionally done by using supervised classifiers which learn patterns extracted from the image pixels annotated by the user and then assign a label to the remaining pixels. However, due to the increasing spatial resolution of the images resulting from advances in the acquisition technology, pixelwise classification is not suitable anymore, even when combined with context. Therefore, we propose a new descriptor for superpixels called Star descriptor that creates a representation based on both its own visual cues and context. Unlike the most methods in the literature, the new approach does not require any prior classification to aggregate context. Experiments carried out on urban images showed the effectiveness of the Star descriptor to generate land cover thematic maps.

Keywords: Remote sensing · Thematic maps · Land cover · Contextual descriptor

1 Introduction

Since the Remote Sensing Images (RSIs) became available to the non-academic community, classification has played an essential role to generate new geographic products like thematic maps [1], which in turn, are fundamental for the decision-making process in several areas such as urban planning, environmental monitoring and economic activities. In this process, low-level descriptors are extracted from few image samples, such as pixels, regions, superpixels (a superpixel can be considered as a perceptually meaningful atomic region [2]), etc., which are annotated by the user, and used to train a classifier. Thereafter the generated classifier should be able to annotate the remaining samples in the image. The precision of the resultant thematic map depends on the quality of the descriptors and the training samples selected [3].

© Springer International Publishing AG 2017
C. Beltrán-Castañón et al. (Eds.): CIARP 2016, LNCS 10125, pp. 300–308, 2017.
DOI: 10.1007/978-3-319-52277-7_37

From the very beginning, RSI classification was based on pixel statistics analysis. With the increasing in the spatial resolution of the images, the information from neighboring pixels (either texture or context) was used to improve results. Although this approach has been a dominant paradigm in remote sensing for many years, pixelwise classification does not meet the current increasing demand for faster and more accurate classification anymore [4]. Region-based classification, which aims at capturing information from pixel patterns inside each segmented region of the image, has become more suitable for nowadays' scenario. Nevertheless, the use of the contextual information among regions began to be considered only very recently in RSI processing [3].

The main motivation behind using contextual information is that traditional low-level appearance features, such as color, texture or shape, are limited while capturing the appearance variability of real world objects represented in images. In the presence of factors that modify the acquired image of a scene, such as noise and changes in lighting conditions, the intra-class variance is increased, leading the classifier to many errors. In these scenarios, the coherent arrangement of the elements expected to be found in real world scenes can be used to help describing objects that share similar appearance features, adjusting the confidence of the classifier predictions or correct the results [5].

Existing approaches for contextual description can be divided into three categories [5]: semantic, that is regarded as the occurrence and co-occurrence of objects in scenes; scale, related to the dimension of an object with respect to the others; and spatial, that refers to the relative localization and position of objects in a scene. In addition, context can be regarded as being either global or local. The first available methods were based on fixed and predefined rules [6–8]. More effective approaches used machine learning techniques to encompass contextual relationships [9–11]. A recent trend consists on combining different kinds of context to improve the classification [12], which is nevertheless computationally inefficient and, therefore, little used so far. The main drawback of these methods is the requirement of previous identification of other elements in the image. A way to overcome this deficiency is through feature engineering, which consists in building a representation for image objects/regions that implicitly encodes their context. This approach must somehow include co-occurrences, scale or spatial relationships between descriptors of image elements without labeling them. An example can be found in the work of Lim *et al.* [13] that represents the scene as a tree of regions where the leaves are described by a combination of features from their ancestors. This resulting descriptor encodes context in a top-down fashion. To the best of our knowledge, the only approach of this type in remote sensing was proposed by Vargas *et al.* [3] to create thematic maps. In that work, each superpixel of the image is described through a histogram of visual elements, using the method of Bag of Visual Words (BoVW). Then, contextual information is encoded by concatenating the superpixel description with a combination of the histograms of its neighbors to generate a new contextual descriptor. One of the main drawbacks of this method is the lack of explicit encoding of the relational aspects among the features extracted from adjacent superpixels.

Thereby, this paper presents a novel contextual descriptor for superpixels of RSIs, which builds a representation for the target superpixel in terms of its own visual cues, visual features extracted from its neighbors and pairwise interactions between them. The resulting representation implicitly encodes spatial relationships and co-occurrences of patterns extracted within a neighborhood and, thus, does not require any prior classification to aggregate context.

2   Star Descriptor

Unlike the most methods found in the literature, our approach builds a representation for image segments that implicitly encodes co-occurrences (semantic context) and spatial relations (spatial context) without the need of labeling them. The pipeline to generate the Star descriptor is summarized in Fig. 1. Each step of the proposed approach is further explained in the following.

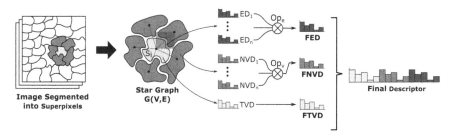

Fig. 1. Process to generate the proposed contextual descriptor for a superpixel s_i. Given a segmented image, the local neighborhood of s_i is modeled as a star graph $G(V,E)$ where s_i is the central vertex (or the root), the superpixels adjacent to it are the leaves and edges link the mass centers of them. A feature descriptor is extracted from s_i and from each of its n neighbors. Every edge is then taken as the diagonal of a rectangle (reddish region) from which a texture descriptor is computed. The n resultant edge descriptors are combined into one of the same dimensionality through some operation Op_e. Likewise, the n neighbor vertex descriptors are used to build only one through Op_v. Lastly, the final contextual descriptor for s_i is composed by concatenating its own vertex descriptor, the final neighborhood vertex descriptor and the final edge descriptor, in this order and after individually normalizing each of them. (Color figure online)

2.1   Segmentation into Superpixels

Firstly, segmentation is applied to delineate objects or object parts in the image from which visual feature descriptors will be extracted. Superpixels are used instead of the traditional regions because some low level descriptors are more discriminative when extracted from regular regions such that provided by superpixel generation methods [2].

Among several methods, Simple Linear Iterative Clustering (SLIC) was chosen for this work because of it found to be more effective according to boundary recall [14]. Since the edge descriptors capture borders between adjacent super-pixels, the ability of SLIC to adhere to the borders of objects in the image can leverage edge descriptors computed by our descriptor.

2.2 Graph Modeling

Given an image segmented into N superpixels, the local neighborhood of each superpixel s_i, $i = 1, \ldots, N$, is regarded as a graph $G(V, E)$ in star topology (see Fig. 1) where V are the superpixels and the edges in E represent adjacency relation between s_i and the other superpixels. Formally, two superpixels s_x and s_y are adjacent if and only if at least one pixel of s_x is 4-connected to a pixel of s_y. In addition, the target superpixel s_i is the central vertex (or root), each of its n neighbor superpixels ns_j, $j = 1, \ldots, n$, are the leaves and there is an edge e_k, $k = 1, \ldots, n$ linking the mass centers of s_i and every ns_j. Such a graph modeling provides a clear understanding of the proposed descriptor in terms of the level of context taken into account and the types of context exploited (spatial relations and co-occurrences between the s_i and a pattern of neighborhood).

2.3 Vertex Descriptors

A visual feature descriptor is computed within every superpixel in a given local neighborhood modeled as a star graph $G(V, E)$. More formally, a feature vector - referred to as target vertex descriptor (TVD) - is extracted from the target superpixel s_i. Likewise, a neighbor vertex descriptor (NVD_j) is built for ns_j, $j = 1, \ldots, n$, as can be seen in Fig. 1. Notice that the same algorithm is used for both TVD and every NVD_j.

Although the only restriction for the vertex descriptor chosen is that it must represent every superpixel by a fixed-size numerical vector, we propose to use two types: low level global color/texture descriptors and BoVW for mid level representation. In the former approach, a global descriptor is extracted from each superpixel taking it as it were a whole image. To account for size differences among them, the resultant feature vector is normalized. The second way was proposed by Vargas *et al.* [3]: dense grid sampling is applied and low level color/texture descriptors are computed from each 5×5 local patch around the selected pixels; the extracted feature vectors are used to conform the codebook using the k-means clustering algorithm; hard assignment is used to assign the closest visual word to each pixel of the grid; a histogram is then computed for every superpixel by taking into account the grid pixels within it; finally, a normalization is applied to each histogram, which is divided by the number of grid pixels inside its respective superpixel.

2.4 Edge Descriptors

The edge descriptor proposed by Silva *et al.* [15] was used to better capture the patterns found in the borders of neighbors, since it directly represents the transition across the frontiers of two adjacent superpixels by extracting texture descriptors around the edge. More precisely, given a local neighborhood represented as a star graph $G(V, E)$, the k-th edge descriptor (ED_k) is computed by extracting a low level texture descriptor within the rectangle formed by taking e_k as its diagonal (as exemplified by the reddish area nearby the edge in Fig. 1). This process is repeated for each of the n edges in E.

2.5   Final Descriptor Composition

Since the vertex and edge descriptors were extracted, they are combined into only one vertex descriptor and one edge descriptor through some operation. This step is applied to tackle with two issues: due to the large number of feature vectors extracted from each star graph, the computational cost to train a classifier with them would be prohibitively high and the variability in the number of leaves of the graphs would result in a feature vector of non-fixed size if a simple concatenation would be done.

More specifically, an operation Op_v is applied to summarize the n NVDs, resulting in one final neighbor vertex descriptor ($FNVD$). Similarly, the n EDs are combined into just one final edge descriptor (FED) through an operation Op_e. The final target vertex descriptor ($FTVD$) is the TVD itself. Because vertex and edge descriptors lie in different feature spaces, $FTVD$, $FNVD$ and FED are individually normalized using L_2 norm and then concatenated to compose the final descriptor which has $2 * |vertexdescriptor| + |edgedescriptor|$ dimensions.

The only constraint imposed to Op_v and Op_e is that they must summarize p m-dimensional vectors into one of same dimensionality. Concretely, we propose to use three operations commonly found in BoVW pooling step: sum pooling, average pooling and max pooling. These operations are formally defined as follows: let D_j be the j-th m-dimensional feature vector in a sequence $\langle D_1, \ldots, D_p \rangle$, whose components are d_i, $i \in \{1, \ldots, m\}$ as stated in Eq. 1; the i-th component of D_j can be summarized through either sum, average or max pooling, which are respectively showed in Eq. 2.

$$D_j = \{d_i\}_{i \in \{1,\ldots,m\}} \tag{1}$$

$$d_i = \sum_{j=1}^{p} d_{i,j} \qquad d_i = \frac{1}{p}\sum_{j=1}^{p} d_{i,j} \qquad d_i = \max_{j \in \{1,\ldots,p\}} d_{i,j} \tag{2}$$

3   Experimental Protocol

Datasets. The experiments were carried out on an imbalanced multi-class dataset: the grss_dfc_2014 [16]. The dataset consists of a Very High Resolution (VHR) image, spatial resolution of 20 cm, taken in 2013 over an urban area near Thetford Mines in Québec, Canada. This dataset was annotated into seven classes: road, trees, red roof, grey roof, concrete roof, vegetation and bare soil. The grss_dfc_2014 dataset provides a specific subset of the entire image for training a classifier which should be used to generate a thematic map for the whole image. **Setup.** The superpixel segmentation was performed using SLIC with 25,000 regions and 25 of compactness for the training image and 37,000 regions and 25 of compactness for the whole image. The number of regions for the whole image was chosen to be 37,000 because the image is about 50% bigger than the training image. Since color descriptors usually achieve better results in RSI classification [17], the vertices were described by using just one texture - Unser (USR) [17]- and three color descriptors - Border/Interior pixel Classification (BIC), Color Coherence Vector (CCV) and Global Color Histogram (GCH) [17] - as either global descriptor or BoVW with 256 words in the codebook. Histograms of Local Binary Patterns (LBP) were used initially for the edges because it is the original proposal of Silva *et al.* [15]. However, LBP is not one of the best options for RSI classification [17]. For this reason, USR (which is a good trade-off between accuracy and number of dimensions) was also tested. All three operations - sum, average and max pooling - were used to summarize the final vertices and edge descriptor. The extracted contextual descriptors were used to train a Support Vector Machine (SVM) classifier with Radial Basis Function (RBF) kernel and the parameters were determined through grid searching 5-fold Cross-validation in the training set. In order to assess the robustness of Star descriptor to changes in the segmentation scale, a second experiment was carried out varying the number of regions of SLIC for the best configurations of the first experiment. **Baselines.** The first baseline used for comparison is the low/mid level representation for the superpixels without any context, which is referred to as NO-CTXT in Sect. 4. The second baseline is the contextual descriptor proposed by Vargas *et al.* [3] which is the only approach that implicitly encodes context with the purpose of generating thematic maps. Its results are shown under the name VARGAS in the next section. It is worth to mention that the results reported for NO-CTXT used BIC with BoVW and the configuration for VARGAS consists of BIC as global descriptor, which achieved the best accuracy for each of them. **Evaluation metrics.** All results are reported in Sect. 4 in terms of overall accuracy (Ovr.), average accuracy (Avg.) and Kappa index (κ). It is worth to mention that although a single label is assigned to each superpixel, the metrics are calculated in terms of pixels. This is done by assigning the label of the superpixel to every pixel within it.

4   Results

The best results achieved by Star descriptor for each vertex descriptor are reported in Table 1 for LBP and USR as edge descriptors. As can be seen from the table, BIC was the most prominent vertex descriptor when combined with max pooling. In general, the average pooling operation was better to summarize the edge descriptor and the USR descriptor obtained the best results for the proposed descriptor, shown in boldface in Table 1. Notice that although using BoVW usually produces slightly more accurate maps, the highest kappa index was achieved using BIC as a global descriptor.

A comparison between the best configuration of Star found in Table 1 and the baselines is presented in Table 2. The proposed contextual descriptor achieved

Table 1. Best results of Star descriptor on grss_dfc_2014

Edge descriptor:		LBP					USR				
Global/BoVW	Descriptors	Op_v	Op_e	κ	Ovr.	Avg.	Op_v	Op_e	κ	Ovr.	Avg.
BoVW	STAR-BIC	Max	Avg	0.681	0.772	0.812	Max	Avg	0.707	0.795	0.805
	STAR-CCV	Max	Avg	0.638	0.740	0.769	Max	Max	0.679	0.779	0.756
	STAR-GCH	Max	Sum	0.642	0.746	0.757	Max	Avg	0.680	0.780	0.753
	STAR-USR	Max	Avg	0.559	0.684	0.648	Max	Avg	0.522	0.655	0.623
Global	**STAR-BIC**	Max	Avg	0.522	0.664	0.592	**Max**	**Avg**	**0.735**	**0.822**	**0.775**
	STAR-CCV	Max	Avg	0.636	0.742	0.753	Avg	Sum	0.676	0.782	0.737
	STAR-GCH	Max	Avg	0.630	0.738	0.743	Max	Avg	0.677	0.779	0.745
	STAR-USR	Sum	Max	0.555	0.684	0.629	Sum	Sum	0.510	0.648	0.595

better results than all baselines for all metrics used to assess them. Another key observation is that encoding context to describe superpixels improved the accuracy of the automatic generated thematic maps. It is worthwhile to mention that in the experiments carried out, Star descriptor usually performed better than baselines when combined with the most low/mid level descriptors and operations that summarize them.

Results of the second experiment are presented in Fig. 2. As can be seen from the graphic, Star descriptor is more robust to changes in segmentation scale than Vargas' descriptor, whose Kappa index drastically drops for more than 36,997 regions, becoming worse than the baseline without context.

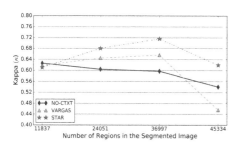

Fig. 2. Comparison of robustness of descriptors to changes in the segmentation scale

Table 2. Comparison between STAR descriptor and baselines

Descriptors	κ	Ovr.	Avg.
NO-CTXT	0.619	0.724	0.767
VARGAS	0.651	0.751	0.766
STAR	**0.735**	**0.822**	**0.775**

5   Conclusion

A new approach for superpixel description which encodes context was proposed in this paper. The Star descriptor builds a representation for each superpixel in terms of its own visual cues and in terms of its context, by taking into account the spatial relations and co-occurrences of visual patterns within the local neighborhood, modeled as a star graph. A classifier is then trained with the resultant feature vectors of the training set and used to generate a thematic map by painting the remaining superpixels according to the labels assigned to them. From the experiments carried out on the grss_dfc_2014 dataset, we found that the use of context improved the quality of the resultant maps and the proposed descriptor achieved better results over the baselines. We believe that it is possible to generate even more representative contextual descriptors by using combinations of low-level descriptors or other operations which preserves the information about the relative positions of the superpixels. In future, we plan to investigate these improvements and assess the Star descriptor using learned features as vertex descriptors.

Acknowledgements. This work was partially financed by CNPq, CAPES, and Fapemig. The authors would like to thank Telops Inc. (Québec, Canada) for acquiring and providing the data used in this study, the IEEE GRSS Image Analysis and Data Fusion Technical Committee and Dr. Michal Shimoni (Signal and Image Centre, Royal Military Academy, Belgium) for organizing the 2014 Data Fusion Contest, the Centre de Recherche Public Gabriel Lippmann (CRPGL, Luxembourg) and Dr. Martin Schlerf (CRPGL) for their contribution of the Hyper-Cam LWIR sensor, and Dr. Michaela De Martino (University of Genoa, Italy) for her contribution to data preparation.

References

1. Wilkinson, G.G.: Results and implications of a study of fifteen years of satellite image classification experiments. IEEE Trans. Geosci. Remote **43**(3), 433–440 (2005)
2. Achanta, R., Smith, K., Lucchi, A., Fua, P., Süsstrunk, S.: SLIC superpixels. Technical report 149300, EPFL, June 2010
3. Vargas, J.E., Falcão, A.X., dos Santos, J.A., Esquerdo, J.C.D.M., Coutinho, A.C., Antunes, J.F.G.: Contextual superpixel description for remote sensing image classification. In: International Geoscience and Remote Sensing Symposium. IEEE (2015)
4. Blaschke, T., Hay, G.J., Kelly, M., Lang, S., Hofmann, P., Addink, E., Feitosa, R.Q., van der Meer, F., van der Werff, H., van Coillie, F., Tiede, D.: Geographic object-based image analysis towards a new paradigm. ISPRS J. Photogramm. **87**, 180–191 (2014)
5. Galleguillos, C., Belongie, S.: Context based object categorization: a critical survey. Comput. Vis. Image Underst. **114**(6), 712–722 (2010)
6. Hanson, A.R., Riseman, E.M.: VISIONS: a computer system for interpreting scenes. In: Hanson, A.R., Riseman, E.M. (eds.) Computer Vision Systems. Academic Press, New York (1978)
7. Strat, T.M., Fischler, M.A.: Context-based vision: recognizing objects using information from both 2D and 3D imagery. IEEE Trans. Pattern Anal. Mach. Intell. **13**(10), 1050–1065 (1991)
8. Fischler, M.A., Elschlager, R.: The representation and matching of pictorial structures. IEEE Trans. Comput. **C−22**(1), 67–92 (1973)

9. Rabinovich, A., Vedaldi, A., Galleguillos, C., Wiewiora, E., Belongie, S.: Objects in context. In: IEEE 11th International Conference on Computer Vision, pp. 1–8, October 2007

10. Torralba, A.: Contextual priming for object detection. Int. J. Comput. Vis. **53**(2), 169–191 (2003)

11. Shotton, J., Winn, J., Rother, C., Criminisi, A.: Textonboost for image understanding: multi-class object recognition and segmentation by jointly modeling texture, layout, and context. Int. J. Comput. Vis. **81**(1), 2–23 (2009)

12. Mottaghi, R., Chen, X., Liu, X., Cho, N.G., Lee, S.W., Fidler, S., Urtasun, R., Yuille, A.: The role of context for object detection and semantic segmentation in the wild. In: Proceedings of the CVPR IEEE, pp. 891–898, June 2014

13. Lim, J.J., Arbelaez, P., Arbelez, P., Gu, C., Malik, J.: Context by region ancestry. In: IEEE International Conference on Computer Vision, pp. 1978–1985, September 2009

14. Achanta, R., Shaji, A., Smith, K., Lucchi, A., Fua, P., Susstrunk, S.: SLIC superpixels compared to state-of-the-art superpixel methods. IEEE Trans. Pattern Anal. Mach. Intell. **34**(11), 2274–2282 (2012)

15. Silva, F.B., Goldenstein, S., Tabbone, S., Torres, R. da S.: Image classification based on bag of visual graphs. In: IEEE International Conference on Image Processing, pp. 4312–4316, September 2013

16. IEEE: GRSS data fusion contest (2014). http://www.grssieee.org/community/technical-committees/data-fusion/

17. dos Santos, J.A., Penatti, O.A.B., Torres, R. da S.: Evaluating the potential of texture and color descriptors for remote sensing image retrieval and classification. In: VISAPP, Angers, France, May 2010

A Similarity Indicator for Differentiating Kinematic Performance Between Qualified Tennis Players

J.D. Pulgarin-Giraldo[1]([⊠]), A.M. Alvarez-Meza[2], L.G. Melo-Betancourt[3],
S. Ramos-Bermudez[3], and G. Castellanos-Dominguez[1]

[1] Signal Processing and Recognition Group, Universidad Nacional de Colombia,
Manizales, Colombia
jdpulgaring@unal.edu.co
[2] Faculty of Engineering, Universidad Tecnológica de Pereira, Pereira, Colombia
[3] Physical Activity and Sport Group, Universidad de Caldas, Manizales, Colombia

Abstract. This paper presents a data-driven approach to estimate the kinematic performance of tennis players, using kernels to extract a dynamic model of each player from motion capture (MoCap) data. Thus, a metric is introduced in the Reproducing Kernel Hilbert Space in order to compare the similarity between models so that the built kernel enhances groups separability: the baseline reference group and the group including players developing their skills. Validation is carried out on a specially constructed database that contains two main testing actions: serve and forehand strokes (carried out on a tennis court). Besides, the classical kinematic analysis is used to compare our kernel-based approach. Results show that our approach allows better representing the performance for each player regarding the ideal group.

Keywords: Multi-channel data · Kernel methods · Kinematics · QKLMS · Similarity indicator

1 Introduction

Human action recognition from video data is a well-established area in computer vision [1]. To date, the main efforts are directed at creating a sufficiently robust dynamic model of human movement accomplished under a priori given actions. All of these models have been validated in predicting the movement and/or action recognition through a certain indicator of performance [2]. However, the models are mostly oriented to detect discrepancies in the performance of players rather than to observe their progress or development of actions.

Aiming at modeling, playing dynamics are usually constructed based on Hidden Markov Models or Gaussian Process from high-dimensional MoCap markers [3,4]. Although these models can provide adequate performance, the involved actions are assumed to be produced in the form of smooth, slow, and cyclical

© Springer International Publishing AG 2017
C. Beltrán-Castañón et al. (Eds.): CIARP 2016, LNCS 10125, pp. 309–317, 2017.
DOI: 10.1007/978-3-319-52277-7_38

movements. In the case of movements having quick actions or responses, development and tuning of either approach may face difficulties, leading to low rate classification. Furthermore, these methods assign high weights to the body segments that involve more information about the posture and gait as the lower limbs, trunk, and pelvis. However, the actions that involve faster movements and smaller body segments are dismissed, such as the upper limbs. To overcome this issue, the kernel methods appear as an option that enables nonlinear dynamics modeling in Hilbert space, and hence, that can enhance the classification and prediction performance.

On the other hand, the efficiency rate may be diminished if the movement analysis is carried out just on a movement that involves more speed or much higher dynamic response and upper limbs [2]. Therefore, the performance evaluation of tennis players should also incorporate the active participation of upper limbs and high-speed response, during the service and forehand strokes. Another aspect to consider is the technique mastered by each player so that its biomechanics plays a key role in stroke production [5]. Each stroke movement has its fundamental mechanical structure, making necessary to analyze the practicing and training player procedures individually. However, the biomechanical analysis is quite extensive and strongly depends on two types of variables: kinetics and kinematics [6]. In this regard, the use of motion capture (MoCap) makes the kinematic analysis lower in cost than kinetic analysis [1].

In practice, the kinematic analysis is of interest to the early analysis of player skills in sports training. Although the kinetic study looks like a more suitable approach due to the calculation of force and torque in specific biomechanical gestures, its involves mass calculation of every single body segment, resulting in a very time consuming and expensive procedure when only MoCap data are employed. As a result, just the professional athletes have access to these tools because of their price. On the other hand, the kinematic analysis is mainly performed using information of linear and angular velocity, as well as the joints and angles of alignment as the initial input [7]. Yet, the majority of approaches do not take into consideration the parameter drift that can describe the progressive development of player performance due to its training improvement. Therefore, there is growing demand for a kinematic analysis at affordable implementation cost but with more flexibility to capture the continuous progress of players.

In this paper, we introduce a data-driven approach to estimate the kinematic performance devoted to training of tennis players that is based on kernels to extract a dynamic model of each player from motion capture (MoCap) data. This methodology allows to see each player performance against other players, his progress, and as a group their similarity in motions executed. This level of interpretation is not achieved with classical kinematics variables studied on tennis training.

2   Kernel Based Multi-channel Data Representation

Let $\boldsymbol{X} \in \mathbb{R}^{N \times M}$ be a multi-channel input matrix that holds M channels and N samples, where $\boldsymbol{x}_i \in \mathbb{R}^{1 \times M}$ is a row vector containing the information of all the provided channels at different time instants, with $i \in \{1, \ldots, N\}$. With the aim at modeling properly each player from MoCap data, we represent the dynamic behavior of its movements through the kernel methods using generalized similarities. To this end, a kernel function is employed as a nonlinear mapping $\varphi : \mathbb{R}^{N \times M} \mapsto \mathcal{H}$, where $\mathcal{H}$ is a Reproducing Kernel Hilbert Space - RKHS. Thus, the kernel based representation allows dealing with nonlinear structures that can not be directly estimated by traditional operators, such as, the linear correlation function. Regarding this, the inner product between two samples $(\boldsymbol{x}_i, \boldsymbol{x}_j)$ is computed in RKHS as $\kappa(\boldsymbol{x}_i, \boldsymbol{x}_j) = \langle \varphi(\boldsymbol{x}_i), \varphi(\boldsymbol{x}_j) \rangle_2$, being $\kappa(\cdot, \cdot)$ a Mercer's kernel. Taking advantage of the so-called kernel *trick*, the kernel function can be computed directly from $\boldsymbol{X}$. Here, we rely on the well-known Gaussian kernel defined as follows:

$$\kappa(\boldsymbol{x}_i, \boldsymbol{x}_j) = \exp\left(-\|\boldsymbol{x}_i - \boldsymbol{x}_j\|_2^2 / 2\sigma^2\right), \tag{1}$$

where $\sigma \in \mathbb{R}^+$ is the kernel band-width. Notation $\|\cdot\|_2$ is the L_2 Euclidean norm.

Therefore, we obtain the similarity matrix $\boldsymbol{S} \in \mathbb{R}^{N \times N}$ that holds elements $s_{i,j} = \kappa(\boldsymbol{x}_i, \boldsymbol{x}_j)$ with $s_{i,j} \in \mathbb{R}^+$. In this sense, matrix $\boldsymbol{S}$ encodes the temporal dynamics of the multi-channel input data.

Due to the Representer Theorem, the nonlinearities of wide range problems can be described as a kernel expansion in terms of the data [8]. That is,

$$f(\boldsymbol{\xi}) = \sum_{i \in N} \alpha_i \kappa(\boldsymbol{x}_i, \boldsymbol{\xi}), \, \boldsymbol{\xi} \in \boldsymbol{X} \tag{2}$$

where N is the number of samples, $\alpha_i \in \mathbb{R}^+$ is built on Kernel least-mean-squares algorithms (kernel least mean square – KLMS) like the one based on solving a least-square problem (KRLS) or the ones using stochastic gradient descent methods [9]. In particular, we employ the Quantized KLMS method (QKLMS) that constructs a dictionary set or *codebook*, noted as C, according to a quantization process in which data points are mapped onto the closest dictionary point. By using QKLMS, the learned mapping before iteration k is as follows:

$$f_{k-1}(\boldsymbol{\xi}) = \sum_{i=1}^{size(C_{k-1})} \alpha_{i,k-1} \kappa(C_{i,k-1}, \boldsymbol{\xi}) \tag{3}$$

where $C_{i,k-1}$ denotes the i-th element (code-vector) that belongs to codebook C_{k-1} and $\alpha_{i,k-1}$ is the coefficient of i-th center.

3   Performance Comparison Expanded to RKHS Representation

The main rationale behind validation is to define a real-valued distance $\mathrm{d}() \in \mathbb{R}^+$ that accomplishes the pairwise comparison of the models estimated for the tested players. Taking into account Eq. 2, we compute the distance between a couple of models, f_n and f_m, as below:

$$\mathrm{d}_f(f_n, f_m) = \varphi(f_n(\boldsymbol{X}^n), f_m(\boldsymbol{X}^m)). \tag{4}$$

where $\boldsymbol{X}^n$ and $\boldsymbol{X}^m$ are the multi-channel input matrices for subjects n and m, respectively. Here, we use the following RKHS-related norm:

$$d_f(f_n, f_m) = \|f_n - f_m\|_2, \tag{5}$$

where $f_n \in \mathcal{H}\{\varphi_n(\cdot); \{\alpha_i^n : \forall i \in size(C^n)\}\}$. Consequently, the evaluation of the euclidean distance proposed in Eq. 5 is as below:

$$d_f(f_n, f_m) = \langle f_n, f_n \rangle_2 - 2\langle f_n, f_m \rangle_2 + \langle f_m, f_m \rangle_2. \tag{6}$$

Because of the codebook in Eq. 3, we have $i \in size(C^n)$ and $j \in size(C^m)$ so that we can rewrite the distance (5) as follows:

$$d_f(f_n, f_m) = \left\langle \sum_i \alpha_i^n \kappa(\boldsymbol{x}_i^n, \cdot), \sum_i \alpha_i^n \kappa(\boldsymbol{x}_i^n, \cdot) \right\rangle_2 - 2\left\langle \sum_i \alpha_i^n \kappa(\boldsymbol{x}_i^n, \cdot), \sum_j \alpha_j^m \kappa(\boldsymbol{x}_j^m, \cdot) \right\rangle_2$$
$$+ \left\langle \sum_j \alpha_j^m \kappa(\boldsymbol{x}_j^m, \cdot), \sum_j \alpha_j^m \kappa(\boldsymbol{x}_j^m, \cdot) \right\rangle_2, \text{ with } \boldsymbol{x}_i^n \in \boldsymbol{X}^n, \boldsymbol{x}_j^m \in \boldsymbol{X}^m.$$

Lastly, we obtain the following closed expression for the Euclidean-distance based indicator:

$$d_f(f_n, f_m) = \sum_{i,i'} \alpha_i^n \alpha_{i'}^n \kappa(\boldsymbol{x}_i^n, \boldsymbol{x}_{i'}^m) - 2\sum_{i,j} \alpha_i^n \alpha_j^m \kappa(\boldsymbol{x}_i^n, \boldsymbol{x}_j^m) + \sum_{j,j'} \alpha_j^m \alpha_{j'}^m \kappa(\boldsymbol{x}_j^m, \boldsymbol{x}_{j'}^m). \tag{7}$$

Thus, the obtained indicator allows encoding the pairwise relationships between all considered models as a functional measure in a RKHS representation.

4   Experimental Setup and Results

4.1   Database

The data were collected from 16 players: four labeled as high performance (HP) and 12 more as regular (RP). All participants belonged to the Caldas-Colombia tennis league and explicitly volunteered to participate in the study, approved by the Ethics Committee of the *Universidad de Caldas*. Each group had the following anthropometric parameters (HP vs RP): age 21 ± 2.7 vs. 16.7 ± 3.9 years, mass 64 ± 14.9 vs. 61.6 ± 9 kg, and height 168.8 ± 8.4 vs. 170.8 ± 7.7 cm. The employed motion capture protocol was Biovision Hierarchy (BVH), stipulating the placement of 34 markers for collecting information of body joints. Optitrack Flex V100 (100 Hz) infrared videography was collected from six cameras to acquire sagittal, frontal, and lateral planes. All subjects were encouraged to hit the ball with the same velocity and action just as they would in a match. They were instructed to hit one series of five serve strokes followed by 12 forehand strokes.

4.2   Baseline Kinematic Analysis

For the purpose of deriving representative and accurate kinematics of the video recorded hits, a total of four serves and six forehand strokes per subject were manually selected and considered for kinematic analysis. The kinematics variables of interest for the serve and forehand were the following: (a) maximum angular displacement of shoulder alignment, hip alignment and wrist extension, (b) maximum velocity of hip, shoulder, elbow, and wrist, lastly (c) maximum angular velocity of elbow extension, trunk rotation, and pelvis rotation [7].

As seen in Table 1, analysis of variance (high performance, regular) estimated for each swing allows detecting some statistical differences and effects in the selected kinematic variables. Thus, the used one-way ANOVA shows that no group differences are achieved by the kinematics variables for the serve hit, assuming a significance level $p < 0.05$. Instead, the hip alignment ($p < 0.01$) and elbow velocity ($p < 0.03$) are significantly different between high performance and regular players in forehand stroke; all these findings are consistent with some related works on this area [5]. However, these selected variables do not clearly distinguish between groups. Indeed, one can not identify any cluster relating

Table 1. Tennis serve and forehand stroke kinematics (mean $\pm$ SD) of high performance and regular players. [†] High performance group tends to be different from regular: hip alignment ($p < 0.01$) and elbow velocity ($p < 0.03$)

Variable	Serve stroke		Forehand stroke	
	High performance	Regular	High performance	Regular
Max. shoulder alignment angle (°)	29.57 (04.01)	25.36 (9.12)	27.43 (8.36)	21.78 (5.55)
Max. hip alignment angle (°)	7.09 (3.96)	4.87 (6.60)	9.97 (2.82)[†]	4.73 (3.69)
Max. wrist extension angle (°)	78.10 (11.03)	85.12 (5.68)	73.48 (10.68)	68.56 (13.76)
Max. hip linear velocity (m/seg)	0.95 (0.71)	0.63 (0.28)	0.46 (0.13)	0.52 (0.15)
Max. shoulder linear velocity (m/seg)	1.89 (0.64)	2.16 (1.68)	1.16 (0.22)	0.93 (0.34)
Max. elbow linear velocity (m/seg)	3.08 (0.66)	2.89 (1.06)	2.07 (0.50)[†]	1.56 (0.37)
Max. wrist linear velocity (m/seg)	4.13 (1.43)	3.37 (1.04)	3.66 (0.96)	2.94 (0.89)
Max. elbow extension velocity (°/seg)	89.67 (0.29)	89.69 (0.33)	88.76 (1.17)	89.61 (0.78)
Max. trunk angular velocity (°/seg)	392.29 (87.58)	425.77 (260.83)	335.98 (67.78)	256.07 (112.05)
Max. pelvis angular velocity (°/seg)	107.49 (26.24)	98.32 (38.36)	90.08 (22.91)	116.67 (49.90)

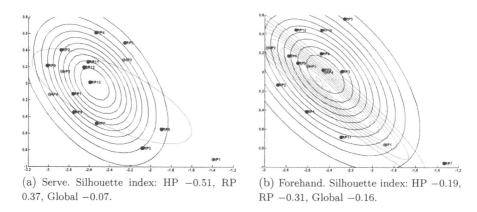

(a) Serve. Silhouette index: HP −0.51, RP 0.37, Global −0.07.

(b) Forehand. Silhouette index: HP −0.19, RP −0.31, Global −0.16.

Fig. 1. 2-D embedding of the 16 players using kinematics characteristics. HP means high performance (blue color) and RP means regular performance (red color). Gaussian densities of each group are observed. (Color figure online)

to the considered groups of players when embedding these ten variables into a 2-D representation using Kernel PCA as shown in Fig. 1(a) and (b). In fact, the comparison between players using this characteristic does not allow to see a compact group of high-performance players which regular players can come closer in performance. Silhouette index is used to measure the cohesion and separation between groups [10]. A silhouette group index near to 1 indicates that the group is well clustered. On the other hand, group index near to −1 is not well clustered. Index close to 0 indicates that a point is on the border of two natural clusters.

4.3 Kernel-Based Performance Analysis

We compute each model $f_n(\cdot)$ as a pairwise kernel-based similarity measure between two samples x_i and x_j. The Gaussian kernel defined in Eq. 1 estimates each pairwise sample relationship, mapping from the original feature space to an RKHS representation. The kernel bandwidth σ is calculated by maximization of information potential variability, which provides better results in this kind of data than the Silverman's rule. QKLMS parameters are set as follows: quantization size $\epsilon_U = 0.95$, step size $\eta = 0.81$, window size $w_s = 30$ and the initial codebooks are built directly from the input time series $X \in \mathbb{R}^{N \times M}$, with $M = 57$ corresponding to 19 joints estimated from .bhv file in x, y and z positions. Each model is validated doing a simple task: predict $z(t + 1)$ coordinate from $x(t)$ and $y(t)$. For the serve models, the mean prediction error is 3.29 ± 4.84, and this amount is 2.56 ± 3.90 for the forehand models. Although the variability is high, it shows a low and regular mean error, which is significant due to the high variability of both: MoCap data and kinematics variables. Besides, our approach works with the full-length videos of two series: five serve strokes and 12 forehand strokes per each player; segmentation and selection of actions are not required in our approach.

Our proposed functional distance d() allows having models that evaluate the performance each other and between groups, where the two groups are well established, and the performance between players is highlighted thanks to the metric proposed. Embedding the coefficients obtained from the QKLMS model of f_n in 2-D with KPCA, a better clustering is achieved for high-performance players. In fact, a clear cluster is described as the high-performance players in the serve action according to its silhouette index (0.92), and the other clusters also exhibit some improvement.

5   Discussion and Concluding Remarks

In the kinematic performance analysis of tennis players, mostly, the following steps are to be accomplished: estimation of the contact with the time ball position, calculation of the segments ranging from starting position till the recovery position, and then, extraction of kinematics parameters for each segment. It is worth noting that segmentation highly influences the performed classifier accuracy, where manual segmentation is often carried out even that MoCap data are widely used. Therefore, efforts to avoid the segmentation stage will promote the automation of supporting tools for improving player skills.

Since there is a strong intersubject variability that should be taken into consideration, we focus on developing an accurate dynamic model individually. This allows to discriminate each player among others much better. Thus, Fig. 2(a) shows an improvement of the performed service action for each one of the players that belong to the high-performance group. Likewise, Fig. 2(b) shows a better performance of forehand action. Also, we consider the progressive performance index to assess the performance evolution along the time. As a result, we find

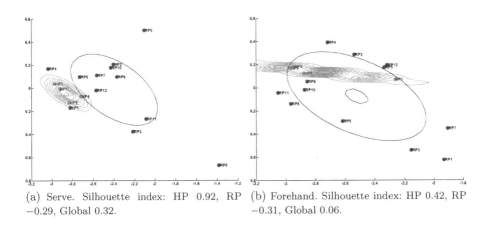

(a) Serve. Silhouette index: HP 0.92, RP −0.29, Global 0.32.

(b) Forehand. Silhouette index: HP 0.42, RP −0.31, Global 0.06.

Fig. 2. 2-D embedding of the 16 players using QKLMS f_n coefficients model for each player. HP means high performance (blue color) and RP means regular performance (red color). Gaussian densities of each group are observed. (Color figure online)

that the performance of fast actions increases, when involving dynamics information of the upper limbs. Though the actions involving upper limbs can be more precisely measured by employing capture techniques like inertial wireless wearable sensors, the proposed methodology can extract from MoCap data a suitable training model that provides good performance of actions involving the movement of upper limbs. Particularly, the developed dynamic model takes advantage of the trajectories derived from the upper limbs, learning from the extracted nonlinear dynamical time series from MoCap data with enough accuracy so that segmentation is not required any more. Namely, no segmentation is required for each action or marker of joint body selection.

To this end, the kernel based training is a data-driven approach that effectively encodes the stochastic behavior of the MoCap trajectories. Here, we use the QKLMS model that seeks for the time relationships among channels, even for actions with big changes between adjacent frames. Consequently, the dynamics model obtained by the kernel-based methodology along with the incorporated similarity indicator allows evaluating more accurately the kinematic performance reached by a subject in the tennis court, having less complexity of implementation. Moreover, the Kernel-based analysis allows to represent the performance reached for each player; this aspect is crucial to individualize the improving process of training skills.

In the future, the authors plan to incorporate more complex dynamics, meaning the introduction of more elaborate measures of similarity. Also, the inclusion of trainers information (like labeled segments) is to be considered.

Acknowledgments. This work is supported by the project 16882 funded by Universidad Nacional de Colombia sede Manizales and Universidad de Caldas, by program "Doctorados Nacionales 2014" funded by COLCIENCIAS, as well as partial Ph.D. financial support from Universidad Autonoma de Occidente.

References

1. Wang, L., Zhao, G., Cheng, L., Pietikainen, M.: Machine Learning for Vision-Based Motion Analysis, 1st edn. Springer, London (2011)
2. García-Vega, S., Álvarez-Meza, A.M., Castellanos-Domínguez, C.G.: MoCap data segmentation and classification using kernel based multi-channel analysis. In: Ruiz-Shulcloper, J., Sanniti di Baja, G. (eds.) CIARP 2013. LNCS, vol. 8259, pp. 495–502. Springer, Heidelberg (2013). doi:10.1007/978-3-642-41827-3_62
3. Ravet, T., Tilmanne, J., d'Alessandro, N.: Hidden Markov model based real-time motion recognition and following. In: Proceedings of 2014 International Workshop on Movement and Computing, MOCO 2014, pp. 82–87. ACM, New York (2014)
4. Wang, J.M., Fleet, D.J., Hertzmann, A.: Gaussian process dynamical models for human motion. IEEE Trans. Pattern Anal. Mach. Intell. **30**(2), 283–298 (2008)
5. Whiteside, D., Elliott, B., Lay, B., Reid, M.: A kinematic comparison of successful and unsuccessful tennis serves across the elite development pathway. Hum. Mov. Sci. **32**(4), 822–835 (2013)
6. Zatsiorsky, V.M.: Kinetics of Human Motion, 1st edn. Human Kinetics, Champaign (2002)

7. Landlinger, J., Lindinger, S., Stoggl, T., Wagner, H., Muller, E.: Key factors and timing patterns in the tennis forehand of different skill levels. J. Sports Sci. Med. **9**, 643–651 (2010)
8. Liu, W., Príncipe, J.C., Haykin, S.: Kernel Adaptive Filtering: A Comprehensive Introduction. Wiley, Hoboken (2010)
9. Van Vaerenbergh, S., Santamaria, I.: A comparative study of kernel adaptive filtering algorithms. In: 2013 IEEE Digital Signal Processing and Signal Processing Education Meeting (DSP/SPE), pp. 181–186, August 2013
10. Theodoridis, S., Koutroumbas, K.: Chapter 16 - Cluster validity. In: Theodoridis, S., Koutroumbas, K. (eds.) Pattern Recognition, 4th edn, pp. 863–913. Academic Press, Boston (2009)

Subsampling the Concurrent AdaBoost Algorithm: An Efficient Approach for Large Datasets

Héctor Allende-Cid[1($\boxtimes$)], Diego Acuña[2], and Héctor Allende[2,3]

[1] Pontificia Universidad Católica de Valparaíso, Avda. Brasil 2241, Valparaíso, Chile
hector.allende@pucv.cl
[2] Universidad Técnica Federico Santa María, Avda. España 1680, Valparaíso, Chile
diego.acuna@usm.cl, hallende@inf.utfsm.cl, hallende@uai.cl
[3] Universidad Adolfo Ibáñez, Padre Hurtado 750, Viña del Mar, Chile

Abstract. In this work we propose a subsampled version of the Concurrent AdaBoost algorithm in order to deal with large datasets in an efficient way. The proposal is based on a concurrent computing approach focused on improving the distribution weight estimation in the algorithm, hence obtaining better capacity of generalization. On each round, we train in parallel several weak hypotheses, and using a weighted ensemble we update the distribution weights of the following boosting rounds. Instead of creating resamples of size equal to the original dataset, we subsample the datasets in order to obtain a speed-up in the training phase. We validate our proposal with different resampling sizes using 3 datasets, obtaining promising results and showing that the size of the resamples does not affect considerably the performance of the algorithm, but the execution time improves greatly.

Keywords: Concurrent AdaBoost · Subsampling · Classification · Machine Learning · Large data sets classification

1 Introduction

In the last decades, ensemble algorithms have been widely used in the Machine Learning and Data Science community, due to its remarkable results. There are both practical and theoretical reasons of why ensembles are preferred over single learners. For example, it is known that a group of learners with similar training performance may have different generalization performance when exposed to sparse data, large volume of data or data fusion. So, the basic idea of building an ensemble is to construct an inference model by combining a set of learning hypotheses, instead of designing the complete map between inputs and responses in a single step. Ensemble based systems have shown to produce favorable results compared to those of single-expert system for a broad range of applications such as financial, medical and social models, network security, web mining or bioinformatics, to name a few [3,8].

© Springer International Publishing AG 2017
C. Beltrán-Castañón et al. (Eds.): CIARP 2016, LNCS 10125, pp. 318–325, 2017.
DOI: 10.1007/978-3-319-52277-7_39

Boosting algorithms, since the mid nineties, have been a very popular technique for constructing ensembles in the areas of Pattern Recognition and Machine Learning (see [2,5–7]). Boosting is a learning algorithm designed to construct a predictor by combining, what are called, weak hypotheses. The AdaBoost algorithm, introduced by Freund and Schapire [5], builds an ensemble incrementally, placing increasing weights on the examples in the data set which appear to be "difficult". Ensemble learning is the discipline which studies the use of a committee of models to construct a joint predictor which improves the performance over a single more complex model.

Currently, most of the modern computers have processors with multiple cores. AdaBoost was proposed in a time were the number of cores per machine was more limited. So, it seems natural to use all the resources available to improve the quality of the inference made by this model. In [1] a concurrent ensemble approach is presented that improves the weight estimation phase, which is one of the most important stages in the AdaBoost algorithm (Concurrent AdaBoost). By using concurrent computation, authors showed that one can effectively improve the generalization ability of the algorithm. In this work, we use this approach not only to improve the estimation of the weights for the resampling stage, but also to improve the time efficiency, without sacrificing generalization performance.

This paper is organized as follows. In Sect. 2, we briefly introduce AdaBoost and Parallel Adaboost approaches. In Sect. 3, we present our proposed model Concurrent AdaBoost with Subsampling. In Sect. 4 we compare the performance of our proposal using different percentages of subsampled data. The last section is devoted to concluding remarks and to delineate future work.

2   Adaptive Boosting and Some Parallel Variants

The AdaBoost Algorithm [5], introduced in 1997 by Freund and Schapire, has its theoretical background based on the "PAC" learning model [13]. The authors of this model were the first to pose the question of whether a weak learner that is slightly better than random guessing can be "boosted" to a strong learning algorithm. The classic AdaBoost takes as an input a training set $\mathcal{Z} = \{(x_1, y_1) \ldots (x_n, y_n)\}$ where each x_i belongs to $\mathcal{X} \subset \mathbb{R}^d$ and each label y_i is in some label set $\mathcal{Y}$ such as $\{-1, 1\}$. Using a set of weak learners, AdaBoosts main idea is to maintain a sampling distribution D_t over the training set where in a sequence of $t = 1 \ldots T$ rounds, D_t is used to train each weak learner. More formally, let $D_t(i)$ be the sampling weight assigned to the example i on round t. In the initial round, $D_t(i) = \frac{1}{n}$ for all i. Then, at each round, the weights of the incorrectly classified examples are increased, so that the following weak learner is forced to focus on the "hard" examples of the training set. The job of each weak learner is to find a hypothesis $h_t : \mathcal{X} \to \{-1, 1\}$ appropriate for the distribution D_t. The goodness of the obtained hypothesis can be quantified as the weighted error:

$$\epsilon_t = Pr_{i \sim D_t}[h_t(x_i) \neq y_i] = \sum_{i:h_t(x_i) \neq y_i} D_t(i). \tag{1}$$

AdaBoost loss function is $\ell(y, f(x)) = \exp(-yf(x))$ where y is the target and $f(x)$ is the approximation made by the model. To obtain the exponential loss function, using the weak hypothesis $h_T \in \{-1, 1\}$, one must solve:

$$(\alpha_t, h_t) = \underset{\alpha, h}{\operatorname{argmin}} \sum_{i=1}^{n} D_i^{(t)} exp(-\alpha y_i h(x_i)), \tag{2}$$

for the weak hypothesis h_t and corresponding coefficient α_t to be added at each step and with $D_i^{(t)} = \exp(-y_i H_{t-1}(x_i))$, where H_{t-1} is the strong Hypothesis without the learner h_t. The solution to Eq. 2 for h_t for any value of $\alpha > 0$ is

$$h_t = \underset{h}{\operatorname{argmin}} \sum_{i=1}^{n} D_i^{(t)} \operatorname{sign}(y_i h(x_i)), \tag{3}$$

In general, weight changes of the boosting procedure can occur in two ways: reweighting (the numerical weights for each example are passed directly to the learner) and resampling (the training set is resampled following the weight distribution creating a new training set). In [12] the authors state that the latter approach gives better results. Nevertheless, the weak learner h_t that is selected with this approach may be sensitive to the resampling technique.

There have been several approaches to use Parallel Computing together with AdaBoost. In [10,11] the authors propose two parallel boosting algorithms, ADABOOST.PL and LOGITBOOST.PL, which facilitate simultaneous participation of multiple computing nodes to construct a boosted ensemble classifier. The authors claim that these proposed algorithms are competitive to the corresponding serial versions in terms of the generalization performance while achieving a significant speedup. In [4] a randomised parallel version of Adaboost is proposed. The algorithm uses the fact that the logarithm of the exponential loss is a function with coordinate-wise Lipschitz continuous gradient, in order to define the step lengths. They provide the proof of convergence for this randomised Adaboost algorithm and a theoretical parallelization speedup factor. The authors in [9] propose an algorithm called BOOM, for boosting with momentum. Namely, BOOM retains the momentum and convergence properties of the accelerated gradient method while taking into account the curvature of the objective function. They describe a distributed implementation of BOOM which is suitable for massive high dimensional datasets. To the best of our knowledge, all proposed algorithms that use parallel computing, try to improve the computation time, instead of improving the generalization performance. The results obtained by these approaches are at most similar to the ones obtained with classic AdaBoost, because they try to approximate the behaviour of the classic algorithm.

3   Subsampling the Concurrent AdaBoost Algorithm

Concurrent computing is a form of computing in which several computations are executed during overlapping time periods concurrently instead of sequentially (one completing before the next starts). In this research the main idea is to work with all the processors of the machine in order to use, in other case, idle processors, to improve the generalization ability and the execution time of the AdaBoost algorithm. In most parallel Adaboost approaches, the multiple cores are used to decrease the computation times, by partitioning the dataset in smaller fractions. These approaches try to get an approximation to the model that would have been obtained it had been trained with classic approaches.

In this proposal, instead of using a single weak learner in each AdaBoost round, the idea is to use all p processors available to subsample, in a parallel fashion, the original data, and also in parallel train several weak learners. With all p weak learners we build an ensemble, that is weighted with its training accuracy, and then with the output of the ensemble update the weights of the examples. This proposal is based in a previous work, where the parallel resampling was performed using the original size of the dataset [1]. In this work, we choose each h^j according to Eq. 3 where $j = 1, \ldots, p$, and then obtain the ensemble output $E_t(x_i)$ for the example x_i at the t-th round as:

$$E_t(x_i) = \text{sign} \left(y_i \sum_{j=1}^{p} \phi^j h_t^j(x_i) \right), \tag{4}$$

where ϕ^j is the training accuracy of weak hypothesis h_t^j and $\sum_{j=1}^{p} \phi^j h_t^j(x_i)$ is the decision of the p weak learners that were trained in parallel. Then, the weighted error ϵ_t is computed considering the output of the ensemble E_t using $\epsilon_t = Pr_{i \sim D_t}[E_t(x_i) \neq y_i]$. With this, we avoid the necessity of having to select a learner explicitly and use an ensemble of learners instead. Algorithm 1 shows our proposal.

Note that in the classic AdaBoost approach, a single weak learner is trained with a resample of the original data, and the updates of the weights of the distributions are changed using the outputs of the single weak hipothesis. In the case of the concurrent approach (see Fig. 1), on each boosting round, p resamples

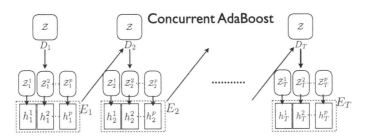

Fig. 1. Concurrent approach. p weak learner h^i per boosting round, where $i = 1, \ldots, p$

obtained in a concurrent fashion are used to train p weak learner to build an ensemble weighted by the training accuracy. This ensemble is then used to update the weights of the distribution.

Algorithm 1. Concurrent AdaBoost

1: **input**: the training data set $\mathcal{Z} = \{(x_1, y_1), \ldots, (x_n, y_n)\}$, where $x_i \in \mathcal{X}$ and $y_i \in \{-1, 1\}$, p the number of parallel processes and T the number of rounds.

2: Set $t = 0$ and initialize the empirical distribution $D_1(i) = \frac{1}{n}$, $\forall (x_i, y_i) \in \mathcal{Z}$.

3: **repeat**

4:     Increment t by one.

5:     Take p bootstrap subsamples $\mathcal{Z}_t^j$ of size s from $\mathcal{Z}$ with distribution D_t, where j is the number of the parallel process, and $j = 1, \ldots, p$. s is a smaller fraction of the original dataset size and it is performed in parallel.

6:     Train p weak learners $h_t^j : \mathcal{X} \to \{-1, 1\}$ with the bootstraped subsamples $\mathcal{Z}_t^j$ as the training sets (in parallel).

7:     Generate an ensemble E_t with all weak learners h_t^j and compute its weighted error ϵ_t as

$$\epsilon_t = Pr_{i \sim D_t}[E_t(x_i) \neq y_i]$$

8:     Compute empirical importance for the t-th ensemble α_t with equation

$$\alpha_t = \frac{1}{2} \ln \left(\frac{1 - \epsilon_t}{\epsilon_t} \right)$$

9:     Update the empirical distribution as

$$D_{t+1}(i) = \frac{D_t(i)}{Z_t} \times e^{(-\alpha_t \beta(i) y_i E_t(x_i))}$$

    where Z_t is the normalization factor.

10:    Classify the training data set $\mathcal{Z} = \{(x_1, y_1), \ldots, (x_n, y_n)\}$ with the strong hypothesis $H_t(x)$ given by:

$$H_t(x) = sign \left(\sum_{t=1}^{t} \alpha_t E_t(x) \right)$$

11: **until** The stopping criterion is met

12: **output**: The strong hypothesis $H_T(x)$

---

4 Experimentation

In this section we validate our proposal with 3 real datasets commonly used for binary classification tasks. The datasets are: SDSS DR7 Quasar Catalog/SDSS SEGUE Stellar Parameter Pipeline data (Quasar/Star)[1], the well-known Breast Cancer dataset and a set of Twitter messages used for Sentiment Analysis tasks.

---

[1] http://classic.sdss.org/dr7/products/spectra.

The Quasar/Star dataset consists of 433043 examples and 10 attributes and the Twitter dataset contains 400 examples and has 59794 attributes (this is only a subset of the entire dataset which has 4000 examples). The latter was obtained in 2011 and is related with the chilean presidential election. The sentiment (positive or negative) from the twitter messages was labeled by 3 journalism students.

The weak learners used were Decision Stumps, and the number of experiments made was 20. The data was split in two groups: 80% training and 20% test. We implemented our proposal in the Python Language (v2.7.6) and used the latest version of the library for parallel processing, Parallel Python[2]. The experiments were run in an Intel i7 2.6 GHz (8 threads) with 16 GB RAM running with Ubuntu 14.04 x64. The number of parallel processes p in each proposal experiment was 7 (odd number of decisions). The performance measures used to report the results are: Accuracy, Area under the ROC curve, F1-score, Precision and Recall.

Table 1. Test results of Quasar, Breast Cancer and Twitter in terms of ROC-AUC, F1-score, Precision and Recall performance measures with different percentages of concurrent subsampling s, using $p = 7$ parallel processes.

Subsample size (percentage)	Training				Test			
	ROC-AUC	F1	Precision	Recall	ROC-AUC	F1	Precision	Recall
Quasar/Star								
1%	**0.967**	**0.959**	**0.979**	**0.940**	**0.965**	**0.957**	**0.979**	0.936
5%	0.965	0.957	0.977	0.938	0.961	0.952	0.976	0.923
10%	0.963	0.954	0.978	0.932	0.963	0.955	0.976	0.934
50%	0.963	0.955	0.978	0.933	0.965	**0.957**	0.978	**0.937**
100%	0.955	0.948	0.984	0.915	0.955	0.948	0.982	0.917
Twitter								
1%	0.523	0.091	**0.917**	0.048	0.499	0.216	0.24	**0.262**
5%	0.572	0.310	0.770	0.203	0.533	0.197	**0.808**	0.125
10%	0.642	**0.520**	0.794	**0.399**	0.534	**0.275**	0.614	0.176
50%	**0.645**	0.487	0.881	0.338	0.522	0.216	0.556	0.134
100%	0.995	0.995	0.992	0.988	0.605	0.659	0.585	0.767
Breast cancer								
1%	-	-	-	-	-	-	-	-
5%	1	1	1	1	**0.971**	**0.966**	**0.977**	**0.955**
10%	1	1	1	1	0.968	0.962	0.971	0.953
50%	1	1	1	1	0.954	0.952	0.943	0.944
100%	1	1	1	1	0.942	0.930	0.945	0.920

---

[2] http://www.parallelpython.com/.

In Table 1 we observe the results of the experimentation over the 3 datasets in terms of ROC-AUC, F1-score, Precision and Recall. We show the results with s equal to 1%, 5%, 10%, 50% and 100% (for subsampling).

Analyzing the results of our experimentation, we observe that the proposed approach has a similar performance on the classification task using smaller resamples compared to using resamples with more data. Event in some cases (Quasar/Star dataset) with the smallest subsample we reach the best performance. This shows that our proposed algorithm is not affected considerably by reducing the size of the resamples. In Table 2 we show the execution time and the test accuracy of the experiments made over all three datasets, using different subsample size percentages. Although there is a slight improvement when the size of the resamples increases, the execution time increases greatly. The best exemplification of this, are the results obtained in the Quasar/Star dataset. The results for 1% and 50% are the same in 20 experiments but the execution time is considerably less.

Table 2. Execution time and test accuracy with $p = 7$ parallel learners in each round.

Dataset	Percentage s	Time [sec]	Test accuracy
Quasar/Star	1%	257	0.980
	5%	606	0.982
	10%	923	0.978
	50%	3645	0.980
	100%	2912	0.975
Twitter	1%	1398	0.512
	5%	7046	0.524
	10%	16685	0.537
	50%	83215	0.555
	100%	113652	0.606
Breast cancer	1%	-	-
	5%	7.54	0.974
	10%	7.56	0.972
	50%	7.64	0.957
	100%	7.67	0.950

5   Conclusions

In this work we introduce a subsampled concurrent approach of the classic AdaBoost algorithm. By using subsamples of the training data, our proposal is able to improve the efficiency in terms of execution speed, with minimum to none accuracy loss, specially in large datasets. From a previous work it is known that by training more than one weak learner per round (using concurrent computation), a significant improvement in the generalization ability of the algorithm can be obtained. By using this result, we showed that via resampling methods we can speed up in a remarkable fashion the execution time of the process, making this algorithm suitable and efficient for problems involving large datasets.

In future work we will try this approach in other boosting algorithms and using new datasets. Also we are planning to formalize the approach in terms of the convergence of the AdaBoost algorithm.

Acknowledgments. This work was supported by the following research grants: Fondecyt Initiation into Research 11150248 and DGIP-UTFSM.

References

1. Allende-Cid, H., Valle, C., Moraga, C., Allende, H., Salas, R.: Improving the weighted distribution estimation for AdaBoost using a novel concurrent approach. Intell. Distrib. Comput. IX **616**(1), 223–232 (2015)
2. Bauer, E., Kohavi, R.: An empirical comparison of voting classification algorithms: bagging, boosting, and variants. Mach. Learn. **36**(1-2), 105–139 (1999)
3. Bergstra, J., Casagrande, N., Erhan, D., Eck, D., Kégl, B.: Aggregate features and AdaBoost for music classification. Mach. Learn. **65**(2), 473–484 (2006)
4. Fercoq, O.: Parallel coordinate descent for the AdaBoost problem. In: 12th International Conference on Machine Learning and Applications (ICMLA), vol. 1, no. 1, pp. 354–358, 4–7 December 2013
5. Freund, Y., Schapire, R.: A decision-theoretic generalization of on-line learning and an application to boosting. J. Comput. Syst. Sci. **55**(1), 119–139 (1997)
6. Kuncheva, L.I., Whitaker, C.J.: Using diversity with three variants of boosting: aggressive, conservative, and inverse. In: Roli, F., Kittler, J. (eds.) MCS 2002. LNCS, vol. 2364, pp. 81–90. Springer, Heidelberg (2002). doi:10.1007/3-540-45428-4_8
7. Liu, H., Tian, H.Q., Li, Y.F., Zhang, L.: Comparison of four AdaBoost algorithm based artificial neural networks in wind speed predictions. Energy Convers. Manag. **92**(1), 67–81 (2015)
8. Markoski, B., Ivanković, Z., Ratgeber, L., Pecev, P., Glusac, D.: Application of AdaBoost algorithm in basketball player detection. Acta Polytechnica Hungarica **12**(1), 189–207 (2015)
9. Mukherjee, I., Canini, K., Frongillo, R., Singer, Y.: Parallel boosting with momentum. Mach. Learn. Knowl. Discov. Databases **8190**(1), 17–32 (2013)
10. Palit, I., Reddy, C.K.: Scalable and parallel boosting with MapReduce. IEEE Trans. Knowl. Data Eng. **24**(10), 1904–1916 (2012)
11. Palit, I., Reddy, C.K.: Parallelized boosting with MapReduce. In: 2010 IEEE International Conference on Data Mining Workshops (ICDMW), vol. 1, no. 1, pp. 1346–1353, 13 December 2010
12. Seiffert, C., Khoshgoftaar, T.M., Van Hulse, J., Napolitano, A.: Resampling or reweighting: a comparison of boosting implementations. In: Proceedings of 20th IEEE International Conference on Tools with Artificial Intelligence (ICTAI 2008), vol. 1, no. 1, pp. 445–451, November 2008
13. Valiant, L.G.: A theory of the learnable. Commun. ACM **27**(11), 1134–1142 (1984)

Deep Learning Features for Wireless Capsule Endoscopy Analysis

Santi Seguí[1,2]([✉]), Michal Drozdzal[3], Guillem Pascual[1], Petia Radeva[1,2],
Carolina Malagelada[4], Fernando Azpiroz[4], and Jordi Vitrià[1,2]

[1] Dept. Matemàtica Aplicada i Anàlisi,
Universitat de Barcelona, Barcelona, Spain
santi.segui@ub.edu
[2] Computer Vision Center (CVC), Barcelona, Spain
[3] Medtronic GI, Yoqneam, Israel
[4] Digestive System Research Unit, Hospital Vall d'Hebron, Barcelona, Spain

Abstract. The interpretation and analysis of wireless capsule endoscopy images is a complex task which requires sophisticated computer aided decision (CAD) systems in order to help physicians with the video screening and, finally, with the diagnosis. Most of the CAD systems for capsule endoscopy share a common system design, but use very different image and video representations. As a result, each time a new clinical application of WCE appears, new CAD system has to be designed from scratch. Therefore, in this paper we introduce a system for small intestine motility characterization, based on Deep Convolutional Neural Networks, which avoids the laborious step of designing specific features for individual motility events. Experimental results show the superiority of the learned features over alternative classifiers constructed by using state of the art hand-crafted features.

1 Introduction

Wireless Capsule Endoscopy (WCE) is a tool designed to allow for inner visualization of entire gastrointestinal tract. It was developed in the mid-1990s and published in 2000 [1]. The invention is based on a swallowable capsule, equipped with a light source, camera, lens, radio transmitter and battery, that is propelled by the peristalsis along all GastroIntestinal (GI) tract, allowing the full visualization of it from inside without pain and sedation.

In order to help physicians with the diagnosis, several WCE-based CAD systems have been presented in the last years. Extensive reviews of these CAD systems can be found in [2,3]. Generally, these systems are designed either for efficient video visualization [4,5] or to automatically detect different intestinal abnormalities such as bleeding [6], polyp [7], tumor [8], ulcer detection [9], motility disorders [10] and other general pathologies [11]. Reviewing these publications one observation can be made: each method uses a different, specially hand-crafted image representation (e.g. intestinal content or bleeding classifiers are based on color analysis, wrinkle-like frames detectors use structure analysis and

C. Beltrán-Castañón et al. (Eds.): CIARP 2016, LNCS 10125, pp. 326–333, 2017.
DOI: 10.1007/978-3-319-52277-7_40

polyps characterization is based on structure and texture analysis). This hand-crafting process implies that each time a new event of physiological interest appears, the method for quantification of these phenomena has to be designed from scratch by an expert and therefore, the process is very time consuming.

Lately, a technology for generic feature learning in computer vision tasks, called Convolutional Neural Network (CNN), appeared and has been shown to obtain very good results in image classification tasks [12]. CNN use a hierarchy of computational layers in order to learn a mapping from an input (image) to an output (class). The main advantage of CNN is the fact that they allow to classify a variety of classes using a single model without the need of feature hand-crafting. However, in order to learn a good model, a relatively large number of labeled images is needed.

In this paper, we present a CAD system with a generic feature-learning app-roach for WCE endoscopy that merges CNN technology with WCE data. In order to be able to train the network, we build a large data set of 120K labeled WCE images. Our hypothesis is that a single, well trained deep network would be generic enough to classify a variety of events present in WCE recording. The proposed method uses as an input not only color information (RGB channels), but also priors in a form of higher-order image information. We show that this system outperforms other well-known computer vision descriptors (e.g. SIFT, GIST, color histogram).

2   Frame-Based Motility Events

We have defined a set of motility interesting events that can be observed in a single WCE frame (see Fig. 1):

- *Turbid:* Turbid frames represent food in digestion. These frames usually present a wide range of green-like colors with a homogeneous texture.
- *Bubbles:* The presence of bubbles is related to agents that reduce surface tension, analogous to a detergent. These frames are usually characterized by the presence of many circular blobs that can be white, yellow or green.
- *Clear blob:* These frames display an open intestinal lumen that is usually visible as a variable-size dark blob surrounded by orange-like intestinal wall.
- *Wrinkles:* These frames are characterized by a star-shape pattern produced by the pressure exert by the nerve system and are often present in the central frames of intestinal contractions.
- *Wall:* These images display the orange-like intestinal wall (frames without the presence of the lumen).
- *Undefined:* These frames correspond to visually undefined clinical events.

As a result, we have built a data set of WCE images consists of frames from 50 WCE recordings of small intestine in healthy subjects obtained using the PillCam SB2 (Given Imaging, Ltd., Yoqneam, Israel). During the labeling process, the frames from all 50 videos were randomized and showed to an expert who labeled all of them (until reaching a minimal number of labeled images, namely 20,000, per each class).

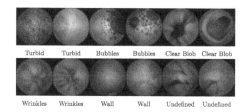

Fig. 1. Exemplary images for each of the category in the database. (Color figure online)

3   Feature Learning Using Convolution Neural Networks

Modern Neural Network architectures date from the beginning of 80s [13]. However, it is not until the very recent years when their use in the computer vision field has fully emerged. Their success has been possible, because of some recent advances related to the training methods of large Convolutional Neural Networks (CNNs) [12] as well as to the availability of very large training datasets. CNNs are a variation of the classical Multilayer Perceptron (MLP), which are inspired by biological models. CNN architecture, compared to the MLP, has a much fewer number of connections and parameters by using prior knowledge, such as using weight sharing among local regions of the image.

3.1   Our Network Architecture

We have considered one basic architecture of them 5-layer deep with one variation. The basic networks called CNN_{RGB} uses as a input the 3 channels representing the RGB image, and the variation called CNN_{RGBHL} uses as a input a 5 channels image composed by concatenating the RGB image, the Hessian (H) and the Laplacian (L) bands. The consideration of priors in a form of L and H is based on the observation that these features showed very good results, when detecting several kinds of WCE frame events [10,14].

More formally, L can be easily computed from image derivatives:

$$L(x,y) = \frac{\partial^2 I}{\partial x^2} + \frac{\partial^2 I}{\partial y^2}$$

H can be derived from the Hessian matrix of every pixel of the image. The Hessian Matrix (HM) is a matrix derived from the second derivatives of the image that summarizes the predominant directions of the local curvatures and their magnitudes in a neighborhood of a pixel:

$$HM(x,y) = \begin{pmatrix} \frac{\partial^2 I}{\partial x^2} & \frac{\partial^2 I}{\partial x \partial y} \\ \frac{\partial^2 I}{\partial x \partial y} & \frac{\partial^2 I}{\partial y^2} \end{pmatrix}$$

Let λ_1 be the largest eigenvalue by absolute value, $|\lambda_1| > |\lambda_2|$. $|\lambda_1|$ shows the strength of the local image curvature, a concept that can be used to represent foldings of the intestinal wall. To build H, we consider for every pixel the map represented by $max(0, \lambda_1)$.

In both cases the lower part of the network is composed of three convolutional layers, along with their corresponding normalization and pooling layers, with 25×25, 5×5 and 3×3 convolution filters respectively. The dimension of their feature map output is 64 in all cases. The higher part of the network is composed of two fully connected layer of 512 neurons. The output layer is a 6 neuron fully connected layer where each neuron represents a different class.

3.2   CNN Training

Original PillCam SB2 images have the resolution of 256×256 pixels. First, we resize all images to 128×128 pixels then we crop a central 100×100 part of the image and, finally, we pre-calculate Laplacians and Hessians. The CNN is trained with the open source Convolutional Architecture for Fast Feature Embedding (CAFFE) environment, presented in [15]. The parameters are initialized with Gaussians (std $= 1$). In order to optimize the network, we use Stochastic Gradient Descent policy with batch size of 128. We start with the learning rate of 0.1 and decrease it every 100k iterations by a factor of 10. The algorithm is stopped after 400k iterations. The system is installed on an Intel Xeon Processor E5–2603 with 64 GB RAM and a Nvidia Tesla K40 GPU. The network training, for every of the proposed network variation, takes approximately 1 day.

4   Experimental Results

In this section, we present the experimental results of the proposed system. First, the database is split into two different sets: training and test set with 100K and 20K of the samples (randomly sampled in a stratified way) from the full database, correspondingly. Second, we evaluate the system quantitatively comparing its performance to the state of the art image descriptors. Finally, in order to provide additional understanding of the proposed system, a qualitative analysis is performed.

4.1   Quantitative Results

Classical image representations (such as GIST, SIFT or COLOR) followed by a Linear Support Vector Machine classifier (available in the sklearn toolbox [16]) are used to establish a baseline in our problem. In particular, we use the following image representations:

– GIST [17]: This is a low dimensional representation of the image based on a set of perceptual dimensions (naturalness, openness, roughness, expansion, ruggedness) that represent the dominant spatial structure of a scene. These dimensions may be reliably estimated using spectral and coarsely localized information.

- COLOR: Each image is described with 128 color words that are learnt with k-means clustering. The training set is a large random sample of (r, g, b) values from WCE videos.
- SIFT [18]: Each image is described with 128 SIFT words that are learnt with k-means clustering. The training set is a large random sample of image frames from WCE videos.
- SIFT + COLOR: A concatenation of SIFT and COLOR descriptor is used.
- GIST + SIFT + COLOR: A concatenation of all above features.

In the experiments, we use the following notations for different CNN modalities: CNN_{RGB} refers to our network trained with RGB images as input and CNN_{RGBHL} refers to the network trained with RGB images concatenated with Hessian and Laplacian.

The following experiment is designed to compare the classification results we can obtain with different image features: GIST, SIFT, COLOR and CNN. As it can be seen in Table 1, the best results are obtained with the CNN_{RGBHL} system with a mean accuracy of 96.22%. It is worth noticing that CNN_{RGBHL} outperforms CNN_{RGB} as well as the classical image representations in all categories, with the best performance for *wall* class (99.1%) and the worst performance for *turbid* class (92.6%). Then, when comparing SIFT, GIST and COLOR, we can observe that GIST descriptor achieves the best results with a mean accuracy of 78.0%. Not surprisingly, GIST descriptor performs well for *wall, wrinkles* and *bubbles*, while COLOR descriptor achieves good results only for the *wall* class. The second best result for our database is obtained by concatenating all classical image representations (GIST with SIFT and COLOR).

Table 1. Comparison of accuracy obtained for frame classification in WCE data. The numbers represent percentages.

	GIST	SIFT	COLOR	SIFT + COLOR	GIST + SIFT + COLOR	CNN_{RGB}	CNN_{RGBHL}
Wall	89.9	49.15	92.2	95.2	90.7	99.0	99.1
Wrinkles	88.2	67.0	36.9	75.7	82.2	95.9	96.1
Bubbles	95.9	39.75	65.4	82.5	91.2	97.1	97.2
Turbid	65.1	41.1	50.2	58.6	80.2	92.2	92.6
Clear blob	70.1	77.3	80.9	77.7	77.4	97.7	97.2
Undefined	58.4	21.25	61.0	39.1	74.7	92.9	92.9
Mean	**78.0**	**50.3**	**64.4**	**71.5**	**82.8**	**96.01**	**96.22**

In order to understand better the results for the most difficult class, the *turbid* class, the confusion matrix for CNN_{RGB} is presented in Table 2. Not surprisingly, the turbid class is miss-classified from time to time as *bubble* class (5.9%), but even an expert can have difficulty in distinguishing turbid from bubbles in some images. Another interesting case is the *undefined* class, which is miss-classified for *clear blob* 3% of the time and for *wall* 1% of the time.

Table 2. Confusion matrix for our method CNN_{RGB}. The numbers represent percentages.

	Wall	Wrinkles	Bubbles	Turbid	Clear blob	Undefined
Wall	99.0	0.0	0.0	0.0	0.0	0.8
Wrinkles	0.7	95.9	0.2	0.5	1.2	1.6
Bubbles	0.1	0.0	97.1	2.6	0.0	0.1
Turbid	0.3	0.1	5.9	92.2	1.0	0.3
Clear blob	0.0	0.3	0.0	0.5	97.7	1.2
Undefined	2.1	1.3	0.1	0.4	3.0	92.9

4.2 Qualitative Results

In this section, a qualitative analysis of the results is performed. First, the filters learned by CNN_{RGB} and CNN_{RGBHL} networks are displayed. Second, a visual evaluation of the learned representation is done. Finally, we show where our system fails.

In order to understand better the differences between CNN_{RGB} and CNN_{RGBHL} architectures, we analyse the filters form the network's first layer. The learnt filters are displayed in Fig. 2. Figure 2(a) shows filters form CNN_{RGB}, while Figs. 2(b)–(d) show the resulting kernels from CNN_{RGBHL}. Note that in both cases, the number of parameters distributed along the filters is constant. As it can be seen, filters from Fig. 2(a) combine both color information and texture. However, if we add additional streams to the network, the system uses RGB channels to learn color information, and Laplacian and Hessian streams to learn the structure and the texture present in the WCE images.

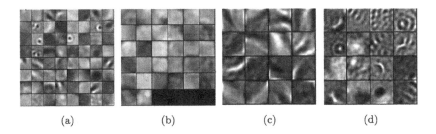

    (a)   (b)   (c)   (d)

Fig. 2. Visualization of the learnt filters: (a) 64 filters from CNN_{RGB}, (b) 32 color filters from CNN_{RGBHL}, (c) 16 Laplacian filters from CNN_{RGBHL}, (d) 16 Hessian filters from CNN_{RGBHL}. All filters are 25×25 pixels. The figure is best seen in color. (Color figure online)

Finally, some failures of the system (CNN_{RGB}) are presented in Fig. 3. Each row shows 10 images from one class that are miss-classified by the system providing additional information about the system's errors. For example, in many cases the frontier between intestinal content and bubbles is not clear (see third and fourth row of Fig. 3), or, while looking at fifth row, it can be seen that some clear blobs have wrinkle-like structure.

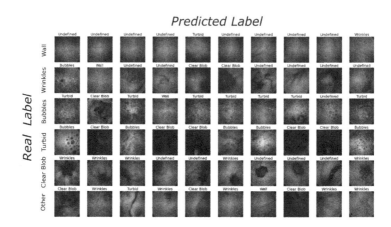

Fig. 3. Images, where the proposed system fails. Each row contains sample images from one event and at the top of the images, the predicted label is printed. (Color figure online)

5   Conclusions

In this paper, we have presented a generic feature descriptor for the classification of WCE images that achieves very good results 96% of accuracy beating the second best method by almost 14%. In order to build our system, several steps needed to be done. First, a large data set of labeled images was built; second, we designed a CNN-like architecture using as an input only color images or color images with some additional information; finally, we performed an exhaustive validation of the proposed method.

Acknowledgements. This work was supported in part by a research grant from Given Imaging Ltd., Yoqneam Israel, as well as by Spanish MINECO/EU Grant TIN2012-38187-C03 and SGR 1219. MD has received funding from the People Programme of the EUs 7th Framework Programme under REA grant agreement no. 607652 (ITN NeuroGut). We gratefully acknowledge the support of NVIDIA Corporation with the donation of a Tesla K40 GPU used for this research.

References

1. Iddan, G., Meron, G., Glukhovsky, A., Swain, P.: Wireless capsule endoscopy. Nature **405**(6785), 417 (2000)
2. Liedlgruber, M., Uhl, A.: Computer-aided decision support systems for endoscopy in the gastrointestinal tract: a review. IEEE Rev. Biomed. Eng. **4**, 73 (2011)

3. Belle, A., Kon, M.A., Najarian, K.: Biomedical informatics for computer-aided decision support systems: a survey. Sci. World J. **2013**, 8 (2013)
4. Mackiewicz, M., Berens, J., Fisher, M.: Wireless capsule endoscopy color video segmentation. IEEE Trans. Med. Imaging **27**(12), 1769–1781 (2008)
5. Drozdzal, M., Seguí, S., Vitrià, J., Malagelada, C., Azpiroz, F., Radeva, P.: Adaptable image cuts for motility inspection using WCE. Comput. Med. Imaging Graph. **37**(1), 72–80 (2013)
6. Fu, Y., Zhang, W., Mandal, M., Meng, M.-H.: Computer-aided bleeding detection in WCE video. IEEE J. Biomed. Health Inf. **18**(2), 636–642 (2014)
7. Mamonov, A.V., Figueiredo, I.N., Figueiredo, P.N., Tsai, Y.H.R.: Automated polyp detection in colon capsule endoscopy. IEEE Trans. Med. Imaging **33**(7), 1488–1502 (2014)
8. Cobrin, G.M., Pittman, R.H., Lewis, B.S.: Increased diagnostic yield of small bowel tumors with capsule endoscopy. Cancer **107**(1), 22–27 (2006)
9. Ciaccio, E.J., Tennyson, C.A., Bhagat, G., Lewis, S.K., Green, P.H.: Implementation of a polling protocol for predicting celiac disease in videocapsule analysis. World J. Gastrointest. Endosc. **5**(7), 313 (2013)
10. Seguí, S., Drozdzal, M., Zaytseva, E., Malagelada, C., Azpiroz, F., Radeva, P., Vitrià, J.: Detection of wrinkle frames in endoluminal videos using betweenness centrality measures for images. IEEE J. Biomed. Health Inf. **18**(6), 1831–1838 (2014)
11. Malagelada, C., Seguí, S., Mendez, S., Drozdzal, M., Vitria, J., Radeva, P., Santos, J., Accarino, A., Malagelada, J., Azpiroz, F., et al.: Functional gut disorders or disordered gut function? Small bowel dysmotility evidenced by an original technique. Neurogastroenterol. Motil. **24**(3), 223–e105 (2012)
12. Krizhevsky, A., Sutskever, I., Hinton, G.E.: Imagenet classification with deep convolutional neural networks. In: Advances in Neural Information Processing Systems, pp. 1097–1105 (2012)
13. Fukushima, K.: Neocognitron: a self-organizing neural network model for a mechanism of pattern recognition unaffected by shift in position. Biol. Cybern. **36**(4), 193–202 (1980)
14. Seguí, S., Drozdzal, M., Vilarino, F., Malagelada, C., Azpiroz, F., Radeva, P., Vitrià, J.: Categorization and segmentation of intestinal content frames for wireless capsule endoscopy. IEEE Trans. Inf. Technol. Biomed. **16**(6), 1341–1352 (2012)
15. Jia, Y.: Caffe: an open source convolutional architecture for fast feature embedding (2013). http://caffe.berkeleyvision.org/
16. Pedregosa, F., Varoquaux, G., Gramfort, A., Michel, V., Thirion, B., Grisel, O., Blondel, M., Prettenhofer, P., Weiss, R., Dubourg, V., Vanderplas, J., Passos, A., Cournapeau, D., Brucher, M., Perrot, M., Duchesnay, E.: Scikit-learn: machine learning in Python. J. Mach. Learn. Res. **12**, 2825–2830 (2011)
17. Oliva, A., Torralba, A.: Modeling the shape of the scene: a holistic representation of the spatial envelope. Int. J. Comput. Vis. **42**(3), 145–175 (2001)
18. Lowe, D.: Object recognition from local scale-invariant features. In: 1999 Proceedings of 7th IEEE International Conference on Computer Vision, vol. 2, pp. 1150–1157 (1999)

Interactive Data Visualization Using Dimensionality Reduction and Similarity-Based Representations

P. Rosero-Montalvo[1,2], P. Diaz[2,3], J.A. Salazar-Castro[4,5], D.F. Peña-Unigarro[5], A.J. Anaya-Isaza[6,7], J.C. Alvarado-Pérez[8,9], R. Therón[8], and D.H. Peluffo-Ordóñez[1(✉)]

[1] Universidad Técnica del Norte, Ibarra, Ecuador
dhpeluffo@utn.edu.ec
[2] Universidad de las Fuerzas Armadas ESPE, Sangolquí, Ecuador
[3] Universidad Nacional de la Plata, Ensenada, Argentina
[4] Universidad Nacional Sede Manizales, Manizales, Colombia
[5] Universidad de Nariño, Pasto, Colombia
[6] Universidad Surcolombiana, Neiva, Colombia
[7] Universidad Tecnológica de Pereira, Pereira, Colombia
[8] Universidad de Salamanca, Salamanca, Spain
[9] Coorporación Universitaria Autónoma de Nariño, Pasto, Colombia

Abstract. This work presents a new interactive data visualization approach based on mixture of the outcomes of dimensionality reduction (DR) methods. Such a mixture is a weighted sum, whose weighting factors are defined by the user through a visual and intuitive interface. Additionally, the low-dimensional representation space produced by DR methods are graphically depicted using scatter plots powered via an interactive data-driven visualization. To do so, pairwise similarities are calculated and employed to define the graph to be drawn on the scatter plot. Our visualization approach enables the user to interactively combine DR methods while provided information about the structure of original data, making then the selection of a DR scheme more intuitive.

Keywords: Data visualization · Dimensionality reduction · Pairwise similarity

1 Introduction

The aim of dimensionality reduction (DR) is to obtain lower dimensional representations of high-dimensional input data keeping -under a pre-established criterion- the structure of data as well as possible. Reaching this aim, entails both the performance of a pattern recognition system and intelligible data representation can be improved [1]. Traditionally, DR methods are designed by following pre-established optimization criteria and design parameters. But they mostly lack of properties like interactivity and controllability, being important characteristics of the field of Information Visualization (InfoVis) [2]. InfoVis

© Springer International Publishing AG 2017
C. Beltrán-Castañón et al. (Eds.): CIARP 2016, LNCS 10125, pp. 334–342, 2017.
DOI: 10.1007/978-3-319-52277-7_41

provides interfaces and graphical ways of representing data making the available information more usable and intelligible for the user. However, it turns out that DR outcomes can be enhanced by taking advantages of some properties of InfoVis methods [3,4]. Following this premise, some approaches have proposed [5,6], making use of interactivity with equalizer-bar like interfaces or geometric interaction models. In general, such approaches implement interesting interactive models but their final visualization lacks the information about structure of the data from the original input space -at least in an easy to understand and/or visual way-.

In this work, we introduce a new visualization approach using an interactive mixture of data representations resultant from DR methods. After performing the DR methods on the input data, a set of lower-dimensional representation spaces are obtained. Particularly, the mixture is done via a weighted sum. In order to give users a sense of the structure of data, we implement a data-driven visualization in addition to the conventional scatter plot. Such a visualization captures the structure of the input data by using a similarity matrix (as well, affinity matrix from graph theory), which captures the degree of similarity or affinity between every pair of data points. The visualization consists of plotting lines (edges) between data points exhibiting the highest value of similarity. Additionally, to provide more sense of interactivity, user can control the number of edges by a varying parameter -working as a slider bar within an interface-. By design, affinity is selected as a Gaussian one so that the structure of local neighbor points can be taken into account. Particularly, low-dimensional spaces are obtained by the state of the art of methods such as: Classical Multidimensional Scaling (CMDS) [2], Laplacian Eigenmaps (LE) [7], Locally Linear Embedding (LLE) [8], Stochastic Neighbor Embedding (SNE), and t-Student-distributed-SNE (t-SNE) [1,7]. To perform the mixture, user can set the weighting factors by picking up values from a equalizer-bar-like interface. To test our visualization approach, we use a 3D artificial spherical shell data set. The quality of resultant representation spaces is quantified by a scaled version of the average agreement rate between K-ary neighborhoods [9]. The proposed mixture may represent every single dimensionality reduction approach as well as it helps users to find a suitable representation of input data within a visual and friendly user interface.

The remaining of the paper is organized as follows: In Sect. 2, Data visualization via dimensionality reduction is outlined. Section 3 introduces the proposed interactive data visualization scheme. Experimental setup and results are presented in Sects. 4 and 5, respectively. Finally, Sect. 6 gathers some final remarks as conclusions and future work.

2   Data Visualization via Dimensionality Reduction

Perhaps, one of the most intuitive ways of visualizing numerical data is through a 2- or 3-dimensional representation of original data, which can be readily represented using a scatter plot. In consequence, dimensionality reduction arises as an Correspondingly, DR is aiming at reaching a low-dimensional data representation, upon which both the classification task performance is improved in terms of accuracy, as well as the intrinsic nature of data is properly represented [10]. So, when performing a DR method, a more realistic and intelligible visualization for the user is expected [1]. More technically, the goal of dimensionality reduction is to embed a high dimensional data matrix $\mathbf{Y} = [\mathbf{y}_i]_{1 \leq 1 \leq N}$ such that $\mathbf{y}_i \in \mathbb{R}^D$ into a low-dimensional, latent data matrix $\mathbf{X} = [\mathbf{x}_i]_{1 \leq 1 \leq N}$ being $\mathbf{y}_i \in \mathbb{R}^d$, where $d < D$ [1,11]. Figure 1 depicts an instance where a manifold, so-called 3D spherical shield, is embedded into a 2D representation, which resembles to an unfolded version of the original manifold.

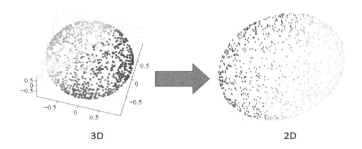

Fig. 1. Dimensionality reduction effect over an artificial (3-dimensional) spherical shell manifold. Resultant embedded (2-dimensional) data is an attempt to unfolding the original data.

3   Interactive Data Visualization Scheme

The proposed visualization approach, here called DataVisSim, involves three main stages: mixture of DR outcomes, interaction, and visualization, as depicted in the block diagram of Fig. 2. One of the most important contributions of this work is that information on the structure of the input high-dimensional space is added to the visual final representation, by using a pairwise-similarity-based scheme.

3.1   Mixture

Let us suppose that the input matrix $\mathbf{Y}$ is reduced by using M different DR methods, yielding then a set of lower-dimensional representations: $\{\mathbf{X}^{(1)}, \cdots, \mathbf{X}^{(M)}\}$. Herein, we propose to perform a weighted sum in the form:

$$\bar{\mathbf{X}} = \sum_{m=1}^{M} \alpha_m \mathbf{X}^m,$$ (1)

where $\{\alpha_1, \cdots, \alpha_M\}$ are the weighting factors. To make the selection of weighting factors intuitive, we use probability values so that $0 \leq \alpha_m \leq 1$ and $\sum_{m1=1}^{M} \alpha_m = 1$, and therefore all matrices $\mathbf{X}^{(m)}$ should be normalized to rely within a hypersphere of ratios.

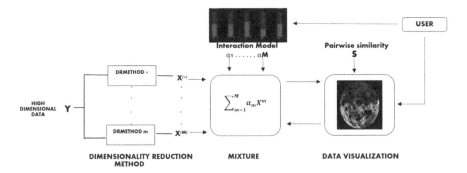

Fig. 2. Block diagram of proposed interactive data visualization using dimensionality reduction and similarity-based representations (DataVisSim). Roughly speaking, it works as follows: first performs a mixture of resultant lower-dimensional representation spaces by taking advantage of conventional implementations of traditional DR methods. The interaction is provided through a interface that enables user to dynamically input the weighting factors for the aforementioned mixture. For visualization, a novel similarity-based approach is used.

3.2   Interaction Model

For the sake of interactivity, the values of every α_m, required to calculate $\bar{X}$ according to Eq. (1), are to be defined by the users using an equalizer-bar available in the interface. Within a friendly-user and intuitive environment, weighting factors can be readily inputted by just picking up values from bars. In order to provide quick views of resultant representation space, as soon as a point is picked up the remaining ones are automatically completed following a uniform density probability function. The same is done in case than more than one value is selected.

3.3   Similarity-Based Visualization

The most used method to visualize 2- or 3-dimensional data is the scatter plot. In this work, we introduce a similarity-based visualization approach with the aim to provide a visual hint about the structure of the high-dimensional input data matrix Y into the scatter plot of its representation in a lower-dimensional space To do so, we use a pairwise similarity matrix $S \in \mathbb{R}^{N \times N}$, such that $S = [s_{ij}]$. In terms of graph theory, entries s_{ij} defines the similarity or affinity between the i-th and j-th data point from Y. Doing so, we can hold the structure of original input space in a topological fashion, specifically in terms of pairwise relationships. For visualization purposes, such a similarity is used to define graphically the relationship between data points by plotting edges. In order to control the amount of edges and make an appealing visual representations, the value of s_{ij} is constrained as $s_{ij} > s_{max}$, being s_{max} a maximum admissible similarity value to be given by the users as well. In other words, our visualization approach consists of building a graph with constrained affinity values.

4   Experimental Setup

Database: In order to visually evaluate the performance of the DataVisSim approach, we use an artificial spherical shell (N = 1500 data points and D = 3), as depicted in Fig. 1.

Parameter Settings and Methods: In order to capture the local structure for visualization, i.e. data points being neighbors, we utilize the Gaussian similarity given by: $s_{ij} = -exp(-0.5\|\boldsymbol{y}_{(i)} - \boldsymbol{y}_{(j)}\|^2/\sigma^2)$. The parameter is a bandwidth value set as 0.1, being the 10% of the hypersphere ratio (applicable once matrices are normalized as discussed in Sect. 3.1. To perform the dimensionality reduction we consider $M = 5$ DR methods, namely: CMDS, LE, LLE, SNE, and t-SNE. All of them are intended to obtain spaces in dimension $d = 2$.

Performance Measure: To quantify the performance of studied methods, the scaled version of the average agreement rate $R_{NX}(K)$ introduced in [9] is used, which is ranged within the interval [0, 1]. Since $R_{NX}(K)$ is calculated at each perplexity value from 2 to $N-1$, a numerical indicator of the overall performance can be obtained by calculating its area under the curve (AUC). The AUC assesses the dimension reduction quality at all scales, with the most appropriate weights.

5   Results and Discussion

Figure 3 shows the scatter plots for the resultant low-dimensional spaces obtained by the considered dimensionality reduction methods, as well as the performed mixture. Quality curves and corresponding scatter of each mixture are shown

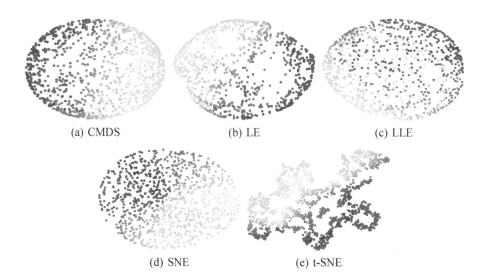

(a) CMDS   (b) LE   (c) LLE

(d) SNE   (e) t-SNE

Fig. 3. The effects of dimensionality reduction of RD methods considered on the 3D sphere. The results are embedded data represented in a bidimensional space.

in Figs. 4, 5, 6, 7 and 8. As seen, $R_{NX}(K)$ measure allows for assessing both the different mixtures and the RD methods independently. Since the area under its curve represents a quality measure of the low-dimensional space, is in turn a visual and intuitive indicator that helps the user to find the best either a single DR method or the proper mixture.

To test the DataVis approach, we implement an interface on Processing software, which allows to easily code visual arts. Then, it results appealing for creating visual analytics interfaces. Figure 9 shows a view of the implemented interface. For the sake of easily handling so that (even non-expert) users may interact with DR methods and their feasible combinations in an intuitive manner

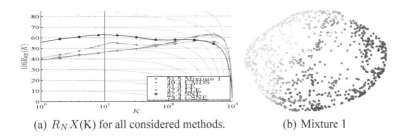

(a) $R_N X(K)$ for all considered methods. (b) Mixture 1

Fig. 4. (a) Performance of the mixture 1 and all methods deemed RD. In (b) the embedded data resulting from mixture 1 are indicated.

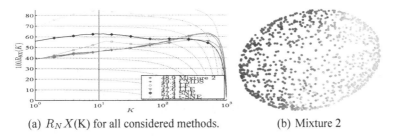

(a) $R_N X(K)$ for all considered methods. (b) Mixture 2

Fig. 5. (a) Performance of the mixture 2 and all methods deemed RD. In (b) the embedded data resulting from mixture 2 are indicated.

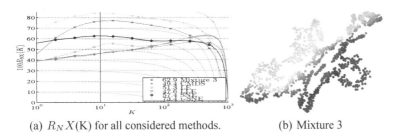

(a) $R_N X(K)$ for all considered methods. (b) Mixture 3

Fig. 6. (a) Performance of the mixture 3 and all methods deemed RD. In (b) the embedded data resulting from mixture 3 are indicated.

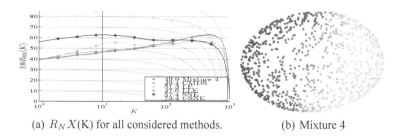

(a) $R_N X$(K) for all considered methods.   (b) Mixture 4

Fig. 7. (a) Performance of the mixture 4 and all methods deemed RD. In (b) the embedded data resulting from mixture 4 are indicated.

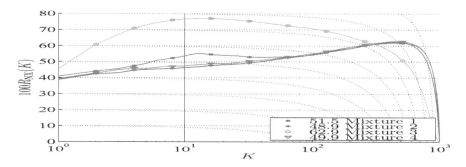

Fig. 8. Performance of all selected mixtures.

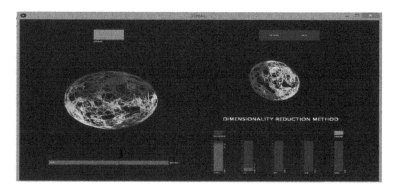

Fig. 9. View of the DataVisSim interface implemented on processing software (https:// sites.google.com/site/intelligentsystemsrg/home/gallery).

using equalizer-like bars. This is possible because of resultant data representations are properly set according to the human perception. As well, the interface incorporates a slider bar to dynamically draw the edges between nodes. This is useful for visual analysis given that it allows to relate the structure of high-dimensional data (original data) within the visualization of the low-dimensional representation space. Therefore, it is provided a powerful tool for making decisions of the most suitable representation of the original data, in other words, the most proper DR methods.

6 Conclusions and Future Work

This work presents a new interactive data visualization approach based on mixture of the outcomes of dimensionality reduction (DR) methods. The core of this approach consists of plotting lines (edges) between data points exhibiting the highest value using a similarity matrix which measure the degree of similarity or affinity between every pair of data points capturing the structure of the input data. Such visualization of a topology can be represented by a data-driven graph in addition to the conventional scatter plot, to provide more sense of interactivity to the user for selecting and/or combining DR methods while providing information about the structure of original data. Correspondingly, data points represent the nodes and an affinity matrix holds the pairwise edge weights. As a future work, other dimensionality reduction methods are to be integrated into data-driven graph, so that a good trade between preservation of data structure and intelligible data visualization can be reached. More mathematical properties will be explored to design data-driven schemes that best approximate the topology data.

References

1. Peluffo-Ordóñez, D.H., Lee, J.A., Verleysen, M.: Short review of dimensionality reduction methods based on stochastic neighbour embedding. In: Villmann, T., Schleif, F.-M., Kaden, M., Lange, M. (eds.) Advances in Self-Organizing Maps and Learning Vector Quantization. AISC, vol. 295, pp. 65–74. Springer, Heidelberg (2014). doi:10.1007/978-3-319-07695-9_6
2. Borg, I., Groenen, P.J.: Modern Multidimensional Scaling: Theory and Applications. Springer Science & Business Media, New York (2005)
3. Dai, W., Hu, P.: Research on personalized behaviors recommendation system based on cloud computing. Indones. J. Electr. Eng. Comput. Sci. **12**, 1480–1486 (2013)
4. Ward, M.O., Grinstein, G., Keim, D.: Interactive Data Visualization: Foundations, Techniques, and Applications. CRC Press, Boca Raton (2010)
5. Peluffo-Ordónez, D.H., Alvarado-Pérez, J.C., Lee, J.A., Verleysen, M., et al.: Geometrical homotopy for data visualization. In: European Symposium on Artificial Neural Networks, Computational Intelligence and Machine Learning (ESANN 2015) (2015)
6. Díaz, I., Cuadrado, A.A., Pérez, D., García, F.J., Verleysen, M.: Interactive dimensionality reduction for visual analytics. In: Proceedings of 22th European Symposium on Artificial Neural Networks, Computational Intelligence and Machine Learning (ESANN 2014), pp. 183–188. Citeseer (2014)
7. Belkin, M., Niyogi, P.: Laplacian eigenmaps for dimensionality reduction and data representation. Neural Comput. **15**, 1373–1396 (2003)
8. Roweis, S.T., Saul, L.K.: Nonlinear dimensionality reduction by locally linear embedding. Science **290**, 2323–2326 (2000)

9. Lee, J.A., Renard, E., Bernard, G., Dupont, P., Verleysen, M.: Type 1 and 2 mixtures of Kullback-Leibler divergences as cost functions in dimensionality reduction based on similarity preservation. Neurocomputing **112**, 92–108 (2013)

10. Bertini, E., Lalanne, D.: Surveying the complementary role of automatic data analysis and visualization in knowledge discovery. In: Proceedings of ACM SIGKDD Workshop on Visual Analytics and Knowledge Discovery: Integrating Automated Analysis with Interactive Exploration, pp. 12–20. ACM (2009)

11. Peluffo-Ordóñez, D.H., Lee, J.A., Verleysen, M.: Generalized kernel framework for unsupervised spectral methods of dimensionality reduction. In: 2014 IEEE Symposium on Computational Intelligence and Data Mining (CIDM), pp. 171–177. IEEE (2014)

Multi-labeler Classification Using Kernel Representations and Mixture of Classifiers

D.E. Imbajoa-Ruiz[1](✉), I.D. Gustin[1], M. Bolaños-Ledezma[1],
A.F. Arciniegas-Mejía[1], F.A. Guasmayan-Guasmayan[1,2],
M.J. Bravo-Montenegro[2], A.E. Castro-Ospina[3], and D.H. Peluffo-Ordóñez[1,4]

[1] Universidad de Nariño, Pasto, Colombia
deivy311@hotmail.com
[2] Universidad Mariana, Pasto, Colombia
[3] Research Center of the Instituto Tecnológico Metropolitano, Medellín, Colombia
[4] Universidad Técnica del Norte, Ibarra, Ecuador

Abstract. This work introduces a multi-labeler kernel novel approach for data classification learning from multiple labelers. The learning process is done by training support-vector machine classifiers using the set of labelers (one labeler per classifier). The objective functions representing the boundary decision of each classifier are mixed by means of a linear combination. Followed from a variable relevance, the weighting factors are calculated regarding kernel matrices representing each labeler. To do so, a so-called supervised kernel function is also introduced, which is used to construct kernel matrices. Our multi-labeler method reaches very good results being a suitable alternative to conventional approaches.

Keywords: Multi-labeler classification · Supervised kernel · Support vector machines

1 Introduction

Typically, supervised pattern recognition systems are trained by using prior knowledge - expressed by labels - given by a single expert labeler or annotator. Nonetheless, for some applications is suitable to consider a set of labelers rather than only one. For instance, a set of specialists diagnosing a patient's pathology [1] or a teacher team assessing the academic performance of a student [2]. Despite it is necessary, considering multiple labelers makes the classification problem difficult since there is no a clearly, identified ground truth. To classify data within this scenarios, multi-labeler strategies have been proposed, which should be able to both compensating the influence of wrong labels regarding the assumed ground truth [3], as well as identifying the good and bad labelers [4,5]. In this connection, support-vector-machines (SVM) based approaches have

---

D.H. Peluffo-Ordóñez—This work is supported by Faculty of Engineering from Universidad Técnica del Norte.

C. Beltrán-Castañón et al. (Eds.): CIARP 2016, LNCS 10125, pp. 343–351, 2017.
DOI: 10.1007/978-3-319-52277-7_42

shown to be a suitable alternative [3,6–8]. Most of the currently available methods make strong assumptions on the resultant labeling vector, introducing then naturally noise over the classification task [6].

In this work, we present a novel approach for data classification within a multi-labeler approach. The classification is done by making a mixture of classifiers trained by using each labeler. Therefore, there is no assumptions made on the estimation of ground truth. Our method just naturally obtains an adjusted objective function defining an improved boundary decision. The proposed approach start by introducing a so-called supervised kernel, which is used to construct kernel matrices - one per labeler. Obtained matrices are incorporated within a variable relevance procedure to estimate their corresponding weighting factors. Finally, a mixture of classifiers trained by the labelers using the cost functions and the aforementioned weighting factors. Particularly, the used mixture is a linear combination. Experiments are carried out using the well-known Iris databases. Labeling vectors are created introducing different percentage of noise into the beforehand ground truth. The simulated labellers are penalized. Five different experiments are performed with several iterations each, to prove stability of our approach. Our multi-labeler method reaches very good results being an alternative to conventional approaches. The great advantage of our method is that classification is performed with no assumptions on the mixture of labeling vectors. Instead, we perform a mixture of classifiers.

This paper is organized as follows: Sect. 2 reviews some related works and outlines basics on support vector machines and their extensions to multi-labeler scenarios. Section 3 describes the operation of the proposed multi-labeler approach in three subsections. Section 4 gathers some results and discussion. Finally, conclusions and final remarks are drawn in Sect. 5.

2   Related Works and Background

Given its versatility and outstanding performance in several applications, many approaches to deal with multi-labeler problems are formulated within support-vector machines (SVM) frameworks. In a previous work [9], a bi-class multi-labeler classifier (BMLC) is introduced. It starts from the simplest formulation for a bi-class or binary SVM-based classifier. For further statements, let us define the ordered pair $\{\boldsymbol{x}_i, \bar{y}_i\}$ to denote the i-th sample or data point, where $\boldsymbol{x}_i$ is its d-dimensional feature vector and $\bar{y}_i \in \{1, -1\}$ is its binary class label. If considering a data set of N samples, all feature vectors can be gathered into a $N \times d$ data matrix $\boldsymbol{X}$ such that $\boldsymbol{X} = [\boldsymbol{x}_1, \ldots, \boldsymbol{x}_N]^\top$, whereas labels into a labeling vector $\bar{\boldsymbol{y}} \in \mathbb{R}^m$. As well, consider k labelers or labeling vectors $\{\boldsymbol{y}^{(1)}, \ldots, \boldsymbol{y}^{(k)}\}$. That said, the labeling vector $\bar{\boldsymbol{y}}$ is a reference vector to be determined. There are some approaches to estimate it. For instance, by calculating the simple average as done in [3]. To pose the classifier's objective function, we assume a latent variable model in the form: $e_i = \boldsymbol{w}^\top \boldsymbol{x}_i + b = \langle \boldsymbol{x}_i, \boldsymbol{w} \rangle + b$, where $\boldsymbol{w}$ is a d-dimensional vector, b is a bias term and notation $\langle \cdot, \cdot \rangle$ stands for Euclidean inner product. As can be readily noted, vector $\boldsymbol{e} = [e_1, \ldots, e_m]$ results from a linear mapping of elements of $\boldsymbol{X}$, which, from a geometrical point of view, is an hyperplane and can thus be seen as a projection vector. By design, if assuming $\boldsymbol{w} \in \mathbb{R}^d$ as an orthogonal vector to the hyperplane, projection vector can be used to encode the class assignment by a decision function in the form $\text{sign}(e_i)$. Alternatively, projection vector can be expressed in matrix terms as $\boldsymbol{e} = \boldsymbol{X}\boldsymbol{w} + b\boldsymbol{1}_m$, being $\boldsymbol{1}_m$ an m-dimensional all ones vector.

Moreover, in order to avoid that data points lie in an ambiguity region for the decision making, the distance between the hyperplane and any data point can be constrained to be at least 1 by fulfilling the condition: $\bar{y}_i e_i \geq 1$, $\forall i$. The distance between data point $\boldsymbol{x}_i$ and hyperplane $\boldsymbol{e}$ can be calculated as: $\text{d}(\boldsymbol{e}, \boldsymbol{x}_i) = \bar{y}_i e_i / ||\boldsymbol{w}||^2$, where $|| \cdot ||$ denotes Euclidean norm. Therefore, since the upper boundary of $\text{d}(\boldsymbol{e}, \boldsymbol{x}_i)$ is $1/||\boldsymbol{w}||^2$, one expect that $\bar{y}_i \simeq e_i$. Then, the classifier objective function to be maximized can be written as: $\max_{\boldsymbol{w}} \bar{y}_i e_i / ||\boldsymbol{w}||^2$; $\forall i$. Accordingly, for accounts of minimization, we can write the problem so: $\min_{\boldsymbol{w}} \frac{1}{2} ||\boldsymbol{w}||^2$, s. t. $\bar{y}_i e_i = 1$, $\forall i$. Notice that previous formulation is attained under the *hard* assumption that $\bar{y}_i = e_i$, and can then be named as hard-margin SVM. By relaxing it, and by adding slack terms, a soft-margin SVM (SM-SVM) can be written as:

$$\min_{\boldsymbol{w},\boldsymbol{\xi}} f(\boldsymbol{w}, \boldsymbol{\xi}|\lambda) = \min_{\boldsymbol{w},\boldsymbol{\xi}} \frac{\lambda}{2} ||\boldsymbol{w}||^2 + \frac{1}{m} \sum_{i=1}^{m} \xi_i^2; \quad \text{s. t.} \quad \xi_i \geq 1 - \bar{y}_i e_i, \quad (1)$$

where λ is a regularization parameter and ξ_i is a slack term associated to data point i.

Binary Approach: Aimed at designing a multi-label classifier, in [6,9], the SM-SVM given in Eq. 1 is modified by adding penalty factors $\theta_t{}_{t=1}^{k}$ and computing $\bar{\boldsymbol{y}}$ as the average of the set of the labeling vectors. This factor is intended to make f increases when adding wrong labels otherwise f should not or insignificantly decrease. In other words, consider a set of k labelers or panelists who singly provide their corresponding labeling vectors. Then, the t-th panelist's quality is quantified by the penalty factor θ_t. Accordingly, incorporating the penalty factors $\boldsymbol{\theta}$, a new binary classification problem is introduced by modifying problem stated in 1, as detailed in [9]. The solution of this problem is conducted by a primal-dual formulation as explained.

Multi-class Approach: Another work [10] naturally extends previous approach to multi-class scenarios by using a one-against-all strategy. Basically, this approach consists of building a number of SVM models - one per class. Applying c times the BMLC approach, a multi-class approach is accomplished. In general, in case of using SVM-approaches, class c is compared with the remaining ones in such a way that it is matched with a positive label, meanwhile the others with a negative label [11]; so that a binary labeling vector per each single class is formed. Concretely, the labeling reference vector $\bar{\boldsymbol{y}}^{(\ell)}$ associated to class ℓ is assumed as a binarized version of labeling vector, as explained in [10]. In this sense, the BMCL is generalized to deal with more that two classes. Consequently, the decision hyperplanes are given by $\{e_i^{(\ell)}, \ldots, e_i^{(\ell)}\}$, where $\boldsymbol{e}^{(\ell)} = \boldsymbol{X}\boldsymbol{w}^{(\ell)} + b^{(\ell)}\boldsymbol{1}_m$.

3   Proposed Multi-labeler Classification Approach

Briefly put, the proposed model works as follows: It involves mainly three stages. We first introduce a so-called supervised kernel, which is used to construct kernel matrices - one per labeler. Such matrices are incorporated within a variable relevance procedure to estimate their corresponding weighting factors. Finally, such weighting factors are used to performing the mixture of classifiers. Following are explained our approach's stages.

3.1   Modified Supervised Kernel

Both the binary and multi-class problems explained in Sect. 2 can be solved by a means of a primal-dual formulation as explained in [10]. Typically, the dual problem is in the form: $\alpha - (1/2)\alpha^\top G\alpha$ where α is the Lagrangian multiplier vector and G is a symmetric and positive semi-definite matrix. As demonstrated in [12], vectors $\alpha^{(\ell)}$ (one per class) pointed out a feasible direction where classes are readily distinguished as the kernel captures the nature of data. Indeed, kernel matrices represent data by means of pairwise similarities between data points. Here, we propose to incorporate the supervised information within the design of a kernel similarly as done in kernelized SVM classifiers [13]. Specifically, we introduce the modified kernel as

$$G_{ij}^{(t)} = \sum_{\ell=1}^{c} y_{\ell i}^{(t)} y_{\ell j}^{(t)} \mathcal{K}(\boldsymbol{x}_i, \boldsymbol{x}_j), \qquad (2)$$

where $\mathcal{K}(\cdot,\cdot)$ is a kernel function and $y_{\ell i}^{(t)}$ stands for the binary label assignment to sample i regarding class ℓ given by labeler t given by: $y_{\ell i}^{(t)} = 1$ if $\boldsymbol{x}_i$ belongs to class ℓ and otherwise -1. The kernel matrix $G^{(t)}$ must be calculated for each labeler to account for multi-labeler settings, i.e. $t \in \{1, \ldots, k\}$.

3.2   Multi-labeler Approach

Our approach may result appealing since it is easy to solve by means of a quadratic programming search, given the form of the dual formulation. However, as BMLC, solution is highly dependent on the chosen reference vector $\bar{\boldsymbol{y}}$ as well as a no new coordinate axis is provided since only one vector α is yielded. Furthermore, to design a multi-labeler approach from this formulation, the quadratic problem should be solved k times (one per labeler). Instead, we propose to perform a mixture of classifiers. Let us define $f^t(\boldsymbol{X})$ the trained cost function by using the labels given by the labeler t. Then, in order to take advantage of the information of the whole set of labelers, we propose a classifier whose cost function is the following mixture:

$$\bar{f}(X) = \sum_{t=1}^{k} \eta_t f^{(t)}(\boldsymbol{X}), \qquad (3)$$

where η_t are the weighting factors to be defined.

3.3   Estimation of Weighting Factors

Now, we are intended to estimate η_t. Here, we propose to estimate the coefficients by using an adapted version of the variable ranking approach proposed in [14]. Similarly as in [15], we start by define a matrix $\boldsymbol{\mathcal{G}} \in \mathbb{R}^{N^2 \times k}$ holding the vectorization of the kernel matrices $\boldsymbol{G}^{(t)}$, as well as a lower-rank representation $\widehat{\boldsymbol{\mathcal{G}}} \in \mathbb{R}^{N^2 \times k}$. Regarding any orthonormal matrix $\boldsymbol{U} = [\boldsymbol{u}^{(1)} \cdots \boldsymbol{u}^{(m)}] \in \mathbb{R}^{k \times m}$, with $m < k$, we can write the lower-rank matrix as $\widehat{\boldsymbol{\mathcal{G}}} = \boldsymbol{\mathcal{G}} \boldsymbol{U}$. Following the variable relevance scheme proposed in [14], we can pose the following optimization problem:

$$\min_{\boldsymbol{U}} \|\boldsymbol{\mathcal{G}} - \widehat{\boldsymbol{\mathcal{G}}}\|_F^2 \quad \text{s. t.} \quad \boldsymbol{U}^\top \boldsymbol{U} = \boldsymbol{I}_m, \tag{4}$$

where $\| \cdot \|_F$ stands for Frobenius norm. Previous problem has a dual version being a maximization problem regarding the variance of $\widehat{\boldsymbol{\mathcal{G}}}$ as: $\max_{\boldsymbol{U}} \text{tr}(\boldsymbol{U}^\top \boldsymbol{\mathcal{G}} \boldsymbol{\mathcal{G}} \boldsymbol{U})$, s.t. $\boldsymbol{U}^\top \boldsymbol{U} = \boldsymbol{I}_m$. Finally, the coefficients η_t for mixture are the ranking values quantifying how much each column of matrix $\boldsymbol{\mathcal{G}}$ (each kernel) contributes to minimizing the cost function given in (4). Again, applying the variable relevance approach presented in [14], we can calculate the ranking vector $\boldsymbol{\eta} = [\eta_1, \ldots, \eta_k]$ using:

$$\boldsymbol{\eta} = \sum_{t=1}^{k} \lambda_t \boldsymbol{u}^{(t)} \circ \boldsymbol{u}^{(t)}, \tag{5}$$

being λ_t and $\boldsymbol{u}^{(t)}$ the t-th eigenvalue and eigenvector of $\boldsymbol{\mathcal{G}} \boldsymbol{\mathcal{G}}$, respectively. Operator $\circ$ denotes Hadamard (element-wise) product. Given the problem formulation, positiveness of $\boldsymbol{\eta}$ is guaranteed and then can be directly used to perform the linear combination.

4   Results and Discussion

Database: For experiments, the well-known Iris flower open database, extracted from UCI repository [16], is considered. It contains three different types of flowers: Iris Setosa, Iris Versicolor and Iris Virginica, with 50 samples each. Four characteristics were recorded for each sample: width and length of sepal, and width and length of petal. Likewise, there are two overlapping classes and a linearly separable class. This database is used to built and simulate different labels from several experts in order to show the characteristics of the method and its effects. Before carrying out the classification procedures, data matrix is normalized so its maximum value per column be 1.

Methods: As reference methods, we consider the average and the majority vote of the given labeling vectors.

Parameter Settings: To perform our multi-labeler approach, we use a Gaussian kernel whose ij entry are given by $\exp(-0.5\|\boldsymbol{x}_i - \boldsymbol{x}_j\|/\sigma^2)$. Experimentally we set $\sigma = 0.65$.

Performance Measures: To quantify the performance of the considered multi-labeler approaches, conventional measures are used, such as: standard error, statistic mean, margin of error. Cohen's Kappa Index is a measure also used in this work, to evaluate the agreement relation between annotators. It is calculated considering the equal labeled individuals by the experts, where a total agreement equals a Kappa index of 100%, and no agreement at all, a Kappa index of 0%.

Experiments: To test the effectiveness of the proposed method, artificial annotators with different percentages of error in their labels are generated. In order to evaluate the stability of the presented approach, the procedure is iterated 30 times, and five cases are presented, with different induced error rates in the labellers. The labels' noise in labeling vectors is completely random, and the error rates in each annotator was chosen in several quantities and order to try the accuracy of the method. In Table 1, the assigned weights η are presented. These values were used in the experiments below, and are associated to the 'Proposed method' column in Table 2, where the overall results are presented.

Table 1. Weight η values

% η	$y^{(1)}$	$y^{(2)}$	$y^{(3)}$	$y^{(4)}$	$y^{(5)}$	Kappa Index
Experiment 1	25.41 ± 0.41	19.20 ± 0.15	18.52 ± 0.15	18.41 ± 0.19	18.19 ± 0.12	8.1 ± 1.2
Experiment 2	15.65 ± 0.19	21.70 ± 0.27	20.83 ± 0.20	19.46 ± 0.25	22.33 ± 0.39	21.0 ± 1.4
Experiment 3	22.97 ± 0.32	20.52 ± 0.21	19.15 ± 0.13	18.63 ± 0.10	18.71 ± 0.14	3.5 ± 0.8
Experiment 4	20.11 ± 0.18	20.09 ± 0.13	19.86 ± 0.09	19.90 ± 0.15	20.01 ± 0.15	2.0 ± 0.54
Experiment 5	25.12 ± 0.37	20.83 ± 0.40	18.12 ± 0.10	17.91 ± 0.11	18.00 ± 0.12	2.1 ± 0.92

Table 2. Performance results in terms of error percentage $\in$ of wrong classifications.

Experiment	$y^{(1)}$	$y^{(2)}$	$y^{(3)}$	$y^{(4)}$	$y^{(5)}$	Proposed method	Average	Majority vote
1		20	50	60	55	65	**15.59 ± 3.02**	19.03 ± 3.47	26.77 ± 4.88
2		70	20	25	30	20	**4.01 ± 0.82**	4.77 ± 0.87	5.70 ± 1.04
3		30	45	60	75	90	**21.85 ± 4.53**	26.66 ± 4.86	39.03 ± 7.12
4		60	60	60	60	60	**20.74 ± 3.78**	20.88 ± 3.81	31.25 ± 5.70
5		20	40	60	80	100	**12.16 ± 3.16**	18.66 ± 4.17	23.83 ± 5.32

The Kappa Index calculated and presented in Table 1 shows the agreement between labellers. In experiment 2, as all labellers have a low error rate, it is very likely that they have many choices in common. Thus, Kappa index is higher with respect to the other cases. In the fourth experiment, unlike the second one, although the error rate is the same for all labellers, it is hardly expected that they share items in common, so the Kappa index is very low. When the index presents a high value, it is expected that the standard deviation between the η values for a experiment is low, and the annotators will be probably right, at least most of

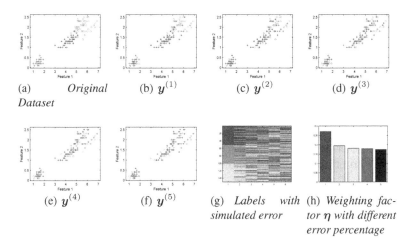

(a)   *Original Dataset*

(b) $\boldsymbol{y}^{(1)}$

(c) $\boldsymbol{y}^{(2)}$

(d) $\boldsymbol{y}^{(3)}$

(e) $\boldsymbol{y}^{(4)}$

(f) $\boldsymbol{y}^{(5)}$

(g) *Labels with simulated error*

(h) *Weighting factor η with different error percentage*

Fig. 1. Generated labellers and data for Experiment No. 5

them, as shown in experiment 2. A low value of this index means that there is much disagreement among labellers, and that does not give much information about the method's accuracy. This index should be taken into account only if it presents high values.

Figure 1 depicts generated data used in the fifth experiment. Figure 1(a) shows the original labels in the three classes. From Fig. 1(b) to (f), the individual corrupted data is shown for the five annotators. The misplaced labels can be noticed based on the color of the classes. In Fig. 1(g), contaminated labels are shown in a clearer way. Figure 1(h) shows the values of η for each annotator, representing the associated weight value. The classification accuracy in terms of percentage of wrong classifications is presented in Table 2. The error rate is decreased for all the cases using the proposed method. As the method assigns different weights to the annotators based on their certainty, the resulting error rate compared to the other methods is lower as the variation of the error in the labellers increases.

Experiment 1: Here, the first annotator $\boldsymbol{y}^{(1)}$ has the lower error rate, unlike the others, exceeding this first one by at least 30%. In this case, it is observed that the weight η, associated to that first annotator is higher, so his opinion will be more relevant in the mixture process. Thus, the performance of the mixture of classifiers will improve in the proposed method, as the worse annotators are not as considered as the first one.

Experiment 2: Unlike last experiment, the first annotator $\boldsymbol{y}^{(1)}$ has now the higher error rate, with a 70%. The other labellers have a maximum of 30%. It is expected that the weight η of $\boldsymbol{y}^{(1)}$ is lower, so, his opinion will be proportionally ignored in the mixture process.

Experiment 3: An ascending error rate values is evaluated in this case, from 30% to 90%, 15 by 15. As a general high error is presented among the annotators, it is expected that the final error is relatively high as well. Even so, the proposed method proves a better response to this noisy data.

Experiment 4: The case of same error rates in all labellers is assessed. The weight η is the same for each annotator, so the improvement in the results is slightly better.

Experiment 5: A similar case to the experiment 3, but in a wider range. The proposed method gives more importance in the mixture of classifiers to those labellers whose error rates are lower, who are $\boldsymbol{y}^{(1)}$ and $\boldsymbol{y}^{(2)}$, so the improvement with respect to other methods is shown.

5   Conclusions and Future Work

Experimentally, we proved that the proposed approach is capable to quantify the confidence of a set of panelists taking into consideration the natural structure of data. Generally, the use of multi-labeler strategy may provide a better design and training of classifiers in comparison with one-labeler approaches. The here proposed method allows to deal with moderate noisy labels, with the capability to penalize labellers, keeping a good performance compared to conventional methods. For future work we are aiming to explore different alternatives that provide optimal conclusions in the mixture of classifiers. Likewise, we are aiming to test other combination weighted methods to improve the classification, and to make a sharper optimization for the kernel function for weights calculation, with a better and more strict penalizing for bad annotators.

References

1. Yan, Y., Fung, G.M., Rosales, R., Dy, J.G.: Active learning from crowds. In: Proceedings of 28th International Conference on Machine Learning (ICML-2011), pp. 1161–1168 (2011)
2. Dekel, O., Gentile, C., Sridharan, K.: Selective sampling and active learning from single and multiple teachers. J. Mach. Learn. Res. **13**(1), 2655–2697 (2012)
3. Dekel, O., Shamir, O.: Good learners for evil teachers. In: ICML, vol. 30 (2009)
4. Donmez, P., Carbonell, J.G., Schneider, J.: Efficiently learning the accuracy of labeling sources for selective sampling. In: Proceedings of 15th ACM SIGKDD International Conference on Knowledge Discovery and Data Mining, pp. 259–268. ACM (2009)
5. Wang, W., Zhou, Z.: Learnability of multi-instance multi-label learning. Chin. Sci. Bull. **57**(19), 2488–2491 (2012)
6. Murillo, S., Peluffo, D.H., Castellanos, G.: Support vector machine-based approach for multi-labelers problems. In: European Symposium on Artificial Neural Networks, Computational Inteligence and Machine Learning (2013)
7. Zhang, Y., Yeung, D.Y.: Multilabel relationship learning. ACM Trans. Knowl. Discov. Data (TKDD) **7**(2), 7 (2013)
8. Cerri, R., de Carvalho, A.C.P., Freitas, A.A.: Adapting non-hierarchical multilabel classification methods for hierarchical multilabel classification. Intell. Data Anal. **15**(6), 861–887 (2011)

9. Murillo-Rendón, S., Peluffo-Ordóñez, D., Arias-Londoño, J.D., Castellanos-Domínguez, C.G.: Multi-labeler analysis for bi-class problems based on soft-margin support vector machines. In: Ferrández Vicente, J.M., Álvarez Sánchez, J.R., Paz López, F., Toledo Moreo, F.J. (eds.) IWINAC 2013. LNCS, vol. 7930, pp. 274–282. Springer, Heidelberg (2013). doi:10.1007/978-3-642-38637-4_28

10. Peluffo-Ordóñez, D.H., Rendón, S.M., Arias-Londoño, J.D., Castellanos-Domínguez, G.: A multi-class extension for multi-labeler support vector machines. In: European Symposium on Artificial Neural Networks (ESANN) (2014)

11. Hsu, C.W., Lin, C.J.: A comparison of methods for multiclass support vector machines. IEEE Trans. Neural Netw. **13**(2), 415–425 (2002)

12. Peluffo-Ordonez, D.H., Aldo Lee, J., Verleysen, M.: Generalized kernel framework for unsupervised spectral methods of dimensionality reduction. In: 2014 IEEE Symposium on Computational Intelligence and Data Mining (CIDM), pp. 171–177. IEEE (2014)

13. Pant, R., Trafalis, T.B.: SVM classification of uncertain data using robust multi-kernel methods. In: Migdalas, A., Karakitsiou, A. (eds.) Optimization, Control, and Applications in the Information Age. PROMS, vol. 130. Springer, Heidelberg (2015). doi:10.1007/978-3-319-18567-5_13

14. Peluffo, D.H., Lee, J.A., Verleysen, M., Rodríguez-Sotelo, J.L., Castellanos-Domínguez, G.: Unsupervised relevance analysis for feature extraction and selection: a distance-based approach for feature relevance. In: International Conference on Pattern Recognition, Applications and Methods - ICPRAM 2014 (2014)

15. Peluffo-Ordóñez, D.H., Castro-Ospina, A.E., Alvarado-Pérez, J.C., Revelo-Fuelagán, E.J.: Multiple kernel learning for spectral dimensionality reduction. In: Pardo, A., Kittler, J. (eds.) CIARP 2015. LNCS, vol. 9423, pp. 626–634. Springer, Heidelberg (2015). doi:10.1007/978-3-319-25751-8_75

16. Lichman, M.: UCI machine learning repository (2013)

Detection of Follicles in Ultrasound Videos of Bovine Ovaries

Alvaro Gómez[1(✉)], Guillermo Carbajal[1], Magdalena Fuentes[1],
and Carolina Viñoles[2]

[1] Facultad de Ingeniería, Universidad de la República, Montevideo, Uruguay
agomez@fing.edu.uy
[2] Instituto Nacional de Investigación Agropecuaria, Tacuarembó, Uruguay

Abstract. Ultrasound imaging is a veterinarian standard procedure for the monitoring of ovarian structures in cattle. Recent studies, suggest that the number of antral follicles can give a cue of the future fertility of a specimen. Therefore, there has been a growing interest in counting the number of antral follicles at early stages in life.

In the most typical procedure, the operator performs a trans-rectal ultrasound scan and counts the follicles on the live video that is seen in the ultrasound machine. This is a challenging task and requires highly trained experts that can reliably detect and count the follicles in a quick sweep of a few seconds.

This work presents the integration of several signal processing techniques to the problem of automatically detecting follicles in ultrasound videos of bovine cattle ovaries. The approach starts from an ultrasound video that traverses the ovary from end to end. Putative follicle regions are detected on each frame with a cascade of boosted classifiers. In order to impose temporal coherence, the detections are tracked across the frames with multiple Kalman filters. The tracks are analyzed to separate follicle detections from other false detections.

The method is tested on a phantom dataset of ovaries in gelatin with dissection ground truth. Results are promising and encourage further extension to in-vivo ultrasound videos.

Keywords: Follicle detection · Cascade classifier · Multitracking

1 Introduction

Recently, there has been an increasing interest in studies concerning antral follicle count (AFC) and its influence on the reproductive performance in cattle, as well as its applications in reproductive biotechnologies [6,10]. AFC is highly variable in different species, but in cattle, there is a high repeatability in the same individual, regardless of race, age, breeding season, lactation or pregnancy conditions [1]. Also, AFC is consistent throughout the estrous cycle of individual cows; therefore, a single routine ultrasound examination is enough to identify

© Springer International Publishing AG 2017
C. Beltrán-Castañón et al. (Eds.): CIARP 2016, LNCS 10125, pp. 352–359, 2017.
DOI: 10.1007/978-3-319-52277-7_43

females with low, intermediate or high AFC. More interestingly, it has been reported that cows with a lower AFC have lower fertility [4, 9].

In order to manually count the antral follicles, veterinarians perform a transrectal ultrasound (TRUS) where they first locate the ovarian region and then scan the ovary while rotating the probe. The procedure must be done in a few seconds and requires highly trained experts in order to reliably detect and count the follicles. Results are accurate when the ovary has few and big follicles, but accuracy drops when the ovary has a large number of follicles and/or follicles are small in size.

Automatically counting the number of antral follicles in the ovary remains an open challenge for state of the art methods [5, 11]. Borders of small follicles are weak and irregular due to the intrinsic characteristics of ultrasound (US) images. The presence of follicle-like structures, such as vessels, makes the problem even harder. However, counting the number of follicles has the advantage that no precise segmentation of the follicles is required. The problem has been addressed before in different ways with 2D and 3D US data but no conclusive results have been achieved in the academy or commercial developments.

A method for fully automated ovary and follicle detection in 3D ultrasound is presented in [2]. The approach proposes a probabilistic framework to estimate the size and location of each individual ovarian follicle by fusing the information from both global and local context.

The commercial product SonoAVC software which is integrated into the Voluson E8 ultrasound machine (GE Medical Systems) performs semiautomatic follicle segmentation and volume measuring on 3D ultrasound data [16].

Methods that detect follicles in 3D ultrasound are probably the most successful ones. Additional information present in 3D ultrasound can be effectively used to discriminate between follicles and other common falsely detected structures [11]. However, veterinarians of our research group, in agreement with observations made in [11], prefer 2D ultrasound scans because they are easier and faster to perform. In addition, 3D US machines are still too expensive for veterinarians in underdeveloped countries. For this reason, our goal is to investigate if it is possible to count the number of follicles using 2D ultrasound videos.

Regarding the detection of follicles in single US images, a research group introduced several fully automated approaches to ovarian follicle detection in single US images based on region-growing [12, 13]. Another group proposed several variants based on a feature extraction and classification scheme [5].

To count all the follicles of the ovary, more than one image is required. One of the possibilities is to segment the images in a frame by frame basis and then group all the detections corresponding to a single follicle as a unique detection. This kind of approach with temporal sequences is presented in [14] and improved in [15] where detected boundaries of the follicles are tracked using a combination of three mutually dependent Kalman filters.

Unfortunately, the lack of publicly-accessible datasets of follicles on ovarian ultrasound images (2D or 3D) prevents from an objective comparison between different methods.

The rest of this paper is organized as follows. Next section introduces the datasets used for developing and evaluating the system. Section 3 presents the signal processing techniques applied to the ultrasound videos in order to detect the follicles. Experiments and results are presented in Sect. 4. Finally, the paper ends with some concluding remarks.

2   Dataset

With the purpose of developing and evaluating the approach, two phantoms were prepared with nine ovaries each immersed in gelatin. The ovaries were collected from the slaughterhouse and conditioned removing tissue debris. Each phantom was built by placing the ovaries at an approximate depth of 1 cm in a box filled with gelatin.

For each ovary three acquisitions with a TRUS probe were performed by an expert. The acquisitions were made by rotating the probe about its axis. The videos were acquired at 30 frames per second and 640 × 480 format with the following criteria: (a) scan from right to left from end to end of the ovary, (b) scan from left to right from end to end of the ovary, (c) scan back and forth from end to end of the ovary while the expert performs a count of the follicles.

Following the US scans, the phantoms were disassembled and the ovaries dissected. For each ovary, all the follicles and corpora lutea were measured. Follicle size ranges from 2 to 20 mm.

Figure 1 shows one of the phantoms and the US acquisition procedure.

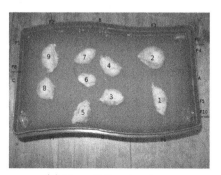

(a) One of the phantoms.   (b) Ultrasound scanning of the phantom

Fig. 1. Phantom with ovaries immersed in gelatin.

3   Signal Processing for Follicle Detection

3.1   Follicle-Like Regions Detection in Each Frame

The first step in the proposed system is to automatically detect in each frame of the video the regions that are likely to be a follicle. Follicles are roughly spherical structures with hard walls and filled with liquid. Echogenicity is high in the walls while the internal fluid is almost anechoic. This gives in US a typical circular pattern brighter in the borders and darker in the middle. A cascade of boosted classifiers [17] based on local binary pattern features is a good and fast alternative to detect this kind of structure. The classifier was trained with a set of follicle regions and negative samples (samples were scaled to 24×24 pixels or equivalently 2.4×2.4 mm for the spatial resolution of the videos). Figure 2a shows an example of follicle like detected regions in a frame.

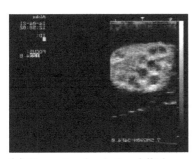

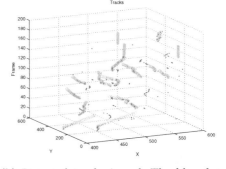

(a) Example of putative follicle regions detected with the cascade classifier in one of the frames of the video.

(b) Detected tracks in red. The blue dots are the centers of the regions detected with the cascade classifier.

Fig. 2. Follicle-like regions are detected with a cascade classifier. In order to impose temporal coherence, the detections are tracked with multiple Kalman filters. (Color figure online)

3.2 Temporal Coherence by Multiple Tracking

The main movement of the probe during an US scan of an ovary is a rotation but there may also be translations. These translations may be involuntary (e.g. trembling of the operator or movement of the animal during the procedure) or due to a necessary panning if the ovary is wider than the size of the US probe. In any case, if these movements are not too brusque, each follicle is normally detected during several frames with the 2D positions of the detections describing a soft movement along the frames. The objective of the tracking step is to group temporal coherent detections into tracks where each track is presumably related to a single follicle.

In order to impose temporal coherence, the detections are tracked with multiple Kalman filters [7] across the video frames with independent constant acceleration models. Detections are assigned to tracks in each frame based on the Hungarian Algorithm [8] using as cost matrix the distance between predicted positions of each track and actual detections (a maximum distance of 1 mm is tolerated in the current implementation). When a detection cannot be assigned to any active track, a new track is created.

Figure 2b shows an example of detected tracks.

3.3   Identification of the Ovarian Region

In the phantoms, the ovary is surrounded by gelatin and that may ease the identification of the ovarian region. In vivo, the ovary is surrounded by other tissues and the boundary of the ovary is usually not easily discernible. Moreover, the surrounding tissues may lead to spurious detections showing follicle-like regions. According to expert veterinarians, the boundary of the ovary is usually hard to identify and the ovarian region is recognized by the grouping of follicles that can be more easily identified in the US video. With this in mind, the approach in this work is to identify the ovarian region as the most important cluster of detected tracks. To cluster the tracks, a probability map is constructed by convolving the track position (weighted by track length) with a Gaussian kernel. The ovary region is detected as the main mode of the probability map using the Mean Shift algorithm [3] considering a typical ovary size. Figure 3 shows the identification of the region of an ovary.

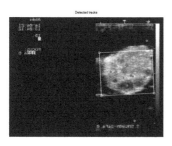
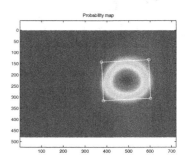

(a) Scatter plot of the xy positions of the tracks weighted by track length on sample frame.

(b) Probability map based on the detected tracks.

Fig. 3. Ovary identification as the main cluster of detected tracks. The white rectangle depicts the main mode.

3.4   Follicle Identification and Measurement

To identify a detected track as a follicle, the track must be active for at least a minimum number of frames. This can allow to differentiate a follicle that is consistently detected across several frames from tracks originated by short term spurious detections of the cascade classifier. With this approach, it is necessary to determine as operating point the best threshold for the minimum number of frames. In this work, the threshold is selected as the one to give the best results in terms of mean square error against the ground truth given by dissection.

Upon detection, the diameter of a follicle can be estimated as the largest detection in the track which corresponds to the frame when the US plane cuts the follicle (roughly a sphere) in its great circle.

4   Experiments and Results

The phantom dataset was used to evaluate the approach. As mentioned in Sect. 2, in the phantom dataset there are two US scans (right to left and left to right) where each ovary is scanned from end to end. In the experiments, a 50 times 6-fold cross validation was performed on the right-to-left sweep to determine the operating point (the threshold for the minimum number of frames to consider a track as a follicle detection). Figure 4 presents the results for the cross validation.

The computed operating point concentrates around a minimum track length of 17 frames (Fig. 4b). Since the videos have 30 frames per second, this means that the tracks must be active for more than half a second to be considered a follicle. Around the operating point, results against dissection can be considered comparable to the expert with a lower correlation factor but better centering (Fig. 4a, c). It is known to experts that the accuracy of follicle count drops when the number of follicles in the ovary is high (the experts tend to underestimate the count). The results on this dataset are consistent with this issue. The expert outperforms the automatic counting in the ovaries with few follicles (eg. less that 18 in this dataset). For the ovaries with more follicles, the automatic approach also underestimates the counts but is in general closer to the ground truth than the expert count.

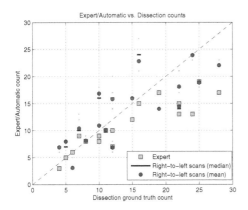

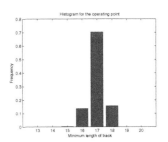

(a) Results of the cross validation on the right-to-left scan

(b) Histogram of the operating points determined by the cross validation

	Expert	Right-to-left scan					Left-to-right scan				
		15	16	17	18	19	15	16	17	18	19
Pearson correlation	0.90	0.82	0.83	0.85	0.85	0.86	0.81	0.78	0.81	0.84	0.85
Max. overestimation	2	11	8	6	5	3	9	8	8	5	3
Min. underestimation	-11	-5	-5	-6	-8	-11	-8	-9	-9	-9	-9
Average of differences	-3.33	2.22	0.78	-0.56	-1.33	-2.78	1.22	-0.33	-1.50	-2.56	-3.44
Median of differences	-2	1.50	0.50	0.00	-0.50	-3.00	1.00	-0.50	-1.50	-2.50	-2.50
Differences std	3.93	4.37	4.18	3.97	3.93	4.05	4.44	4.70	4.37	4.10	4.10

Count results around the operating point (minimum length of track between 15 and 19 frames)

(c) Results for the expert and around the computed operating point for the right-to-left and left-to-right scans

Fig. 4. Results for the 50 times 6-fold cross validation on the right-to-left scan.

5 Concluding Remarks

An automatic algorithm for the problem of detecting follicles in ovarian US videos of bovine cattle ovaries is presented. The proposed approach can work directly on the 2D US videos generated in the typical procedures done by the veterinarians with the only restriction that the US scan must be done in a single sweep from end to end of the ovary. The lightweight processing enables to have the results immediately after the US scan. The method can give also the size of the detected follicles which is an information usually disregarded in the routine follicle count procedure.

Although the phantoms constitute a small dataset, results can be considered promising with count results comparable to an expert in a controlled environment. This encourages future work to robustly extend the approach to in-vivo scans. Future work should include for example the evaluation of other strategies to identify the true follicles from the detections, as well as alternative forms to identify the ovarian region on the video. Also, a method to detect the corpora lutea in the ovary may help to differentiate cavities from real follicles.

References

1. Burns, D.S., Jimenez-Krassel, F., Ireland, J.L.H., Knight, P.G., Ireland, J.J.: Numbers of antral follicles during follicular waves in cattle evidence for high variation among animals, very high repeatability in individuals, and an inverse association with serum follicle-stimulating hormone concentrations. Biol. Reprod. **73**(1), 54–62 (2005)
2. Chen, T., Zhang, W., Good, S., Zhou, K.S., Comaniciu, D.: Automatic ovarian follicle quantification from 3D ultrasound data using global/local context with database guided segmentation. In: 2009 IEEE 12th International Conference on Computer Vision, pp. 795–802. IEEE (2009)
3. Cheng, Y.: Mean shift mode seeking and clustering. IEEE Trans. Pattern Anal. Mach. Intell. **17**(8), 790–799 (1995)
4. Cushman, R.A., Allan, M.F., Kuehn, L.A., Snelling, W.M., Cupp, A.S., Freetly, H.C.: Evaluation of antral follicle count and ovarian morphology in crossbred beef cows: investigation of influence of stage of the estrous cycle, age, and birth weight. J. Anim. Sci. **87**(6), 1971–1980 (2009)
5. Hiremath, P.S., Tegnoor, J.R.: Follicle detection and ovarian classification in digital ultrasound images of ovaries. Advancements and Breakthroughs in Ultrasound Imaging, pp. 167–199. InTechOpen, UK (2013)
6. Ireland, J.J., Smith, G.W., Scheetz, D., Jimenez-Krassel, F., Folger, J.K., Ireland, J.L.H., Mossa, F., Lonergan, P., Evans, A.C.O.: Does size matter in females? An overview of the impact of the high variation in the ovarian reserve on ovarian function and fertility, utility of anti-müllerian hormone as a diagnostic marker for fertility and causes of variation in the ovarian reserve in cattle. Reprod. Ferti. Dev. **23**(1), 1–14 (2010)
7. Kalman, R.M.: A new approach to linear filtering and prediction problems. J. Basic Eng. **82**(1), 35–45 (1960)
8. Kuhn, H.W.: The Hungarian method for the assignment problem. Nav. Res. Logistics Q. **2**(1–2), 83–97 (1955)

9. Martinez, M.F., Sanderson, N., Quirke, L.D., Lawrence, S.B., Juengel, J.L.: Association between antral follicle count and reproductive measures in New Zealand lactating dairy cows maintained in a pasture-based production system. Theriogenology **85**(3), 466–475 (2016)
10. Morotti, F., Barreiros, T.R.R., Machado, F.Z., González, S.M., Marinho, L.S.R., Seneda, M.M.: Is the number of antral follicles an interesting selection criterium for fertility in cattle? Anim. Reprod. **12**(3), 479–486 (2015)
11. Potočnik, B., Cigale, B., Zazula, D.: Computerized detection and recognition of follicles in ovarian ultrasound images: a review. Med. Biol. Eng. Comput. **50**(12), 1201–1212 (2012)
12. Potočnik, B., Zazula, D.: Automated ovarian follicle segmentation using region growing. In: Proceedings of the First International Workshop on Image and Signal Processing and Analysis, IWISPA 2000, pp. 157–162. IEEE (2000)
13. Potočnik, B., Zazula, D.: Automated analysis of a sequence of ovarian ultrasound images. Part I: segmentation of single 2D images. Image Vis. Comput. **20**(3), 217–225 (2002)
14. Potočnik, B., Zazula, D.: Automated analysis of a sequence of ovarian ultrasound images. Part II: prediction-based object recognition from a sequence of images. Image Vis. Comput. **20**(3), 227–235 (2002)
15. Potočnik, B., Zazula, D.: Improved prediction-based ovarian follicle detection from a sequence of ultrasound images. Comput. Methods Programs Biomed. **70**(3), 199–213 (2003)
16. Raine-Fenning, N., Jayaprakasan, K., Clewes, J., Joergner, I., Bonaki, S.D., Chamberlain, S., Devlin, L., Priddle, H., Johnson, I.: SonoAVC: a novel method of automatic volume calculation. Ultrasound Obstet. Gynecol. **31**(6), 691–696 (2008)
17. Viola, P., Jones, M.: Rapid object detection using a boosted cascade of simple features. In: Proceedings of the 2001 IEEE Computer Society Conference on Computer Vision and Pattern Recognition, CVPR 2001, vol. 1, pp. 1–511. IEEE (2001)

Decision Level Fusion for Audio-Visual Speech Recognition in Noisy Conditions

Gonzalo D. Sad, Lucas D. Terissi$^{(\boxtimes)}$, and Juan C. Gómez

Laboratory for System Dynamics and Signal Processing,
Universidad Nacional de Rosario, CIFASIS-CONICET, Rosario, Argentina
{sad,terissi,gomez}@cifasis-conicet.gov.ar

Abstract. This paper proposes a decision level fusion strategy for audio-visual speech recognition in noisy situations. This method aims to enhance the recognition over different noisy conditions by fusing the scores obtained with classifiers trained with different feature sets. In particular, this method is evaluated by considering three modalities, audio, visual and audio-visual, respectively, but it could be employed using as many modalities as needed. The combination of the scores is performed by taking into account the reliability of each modality at different noisy conditions. The performance of the proposed recognition system is evaluated over two isolated word audio-visual databases, a public one and a database compiled by the authors of this paper. The proposed decision level fusion strategy is evaluated by considering different kind of classifier. Experimental results show that a good performance is achieved with the proposed system, leading to improvements in the recognition rates through a wide range of signal-to-noise ratios.

Keywords: Speech recognition · Audio-visual speech · Decision level fusion

1 Introduction

Audio Visual Speech Recognition is a fundamental task in Multimodal Human Computer Interfaces, where the acoustic and visual information (mouth movements, facial gestures, etc.) during speech are taken into account. Several strategies have been proposed in the literature for AVSR [11–13], where improvements of the recognition rates are achieved by fusing audio and visual features related to speech. As expected, these improvements are more notorious when the audio channel is corrupted by noise, which is a usual situation in speech recognition applications. These strategies usually differ in the way the audio and visual information is extracted and combined, and the AV-Model employed to represent the audio-visual information. These approaches are usually classified according to the method employed to combine (or fuse) the audio and visual information, *viz.*, feature level fusion, classifier level fusion and decision level fusion [5].

In feature level fusion (a.k.a. early integration), audio and visual features are combined to form a unique audio-visual feature vector, which is then employed

© Springer International Publishing AG 2017
C. Beltrán-Castañón et al. (Eds.): CIARP 2016, LNCS 10125, pp. 360–367, 2017.
DOI: 10.1007/978-3-319-52277-7_44

for the classification task [2,7]. This strategy is effective when the combined modalities are correlated, since it can exploit the covariations between the audio and video features. This method requires the audio and visual features to be exactly at the same rate and in synchrony, and usually performs a dimensionality reduction stage, in order to avoid large dimensionality of the resulting feature vectors. In the case of classifier level fusion (a.k.a. intermediate integration), the information is combined within the classifier using separate audio and visual streams, in order to generate a composite classifier to process the individual data streams [3,10,11]. This strategy has the advantage of being able to handle possible asynchrony between audio and visual features. In decision level fusion (a.k.a. late integration), independent classifiers are used for each modality and the final decision is computed by the combination of the likelihood scores associated with each classifier [6,9]. This strategy does not require strictly synchronized streams. Different techniques to perform decision level fusion have been proposed. The most commonly used is to combine the matching scores of the individual classifiers with simple rules, such as, *max, min, product, simple sum*, or *weighted sum*.

In this paper, a decision level fusion strategy for audio-visual speech recognition in noisy situations is proposed. The combination of the scores is performed by taking into account the reliability of each modality at different noisy conditions. The performance of the proposed recognition system is evaluated over two audio-visual databases, considering two types of acoustic noise and different classification methods.

The rest of this paper is organized as follows. The description of the proposed system is given in Sect. 2, and the databases used for the experiments are described in Sect. 3. In Sect. 4 experimental results are presented and the performance of the proposed strategy is analyzed. Finally, some concluding remarks are included in Sect. 5.

2 Proposed Approach

Figure 1 shows a schematic representation of the proposed audio-visual speech classification system. The recognition is performed by taking into account three classifiers trained on audio, visual and audio-visual information, hereafter referred as λ_a, λ_v and λ_{av}, respectively. Given an audio-visual observation O_{av} associated with the input word to be recognized, which can be partitioned into acoustic and visual parts, denoted as O_a and O_v, respectively, the probability (or score) vectors $\mathbf{P}(O_a|\lambda^a)$, $\mathbf{P}(O_v|\lambda^v)$ and $\mathbf{P}(O_{av}|\lambda^{av})$ are computed from the audio, visual and audio-visual classifiers, respectively. These vectors are composed by the concatenation of the probabilities associated to each class in the dictionary. Then, the fused probability vector $\mathbf{P}_F(O_{av})$ is defined as

$$\mathbf{P}_F(O_{av}) = c_a * \mathbf{P}(O_a|\lambda^a) + c_v * \mathbf{P}(O_v|\lambda^v) + c_{av} * \mathbf{P}(O_{av}|\lambda^{av}), \qquad (1)$$

where c_a, c_v and c_{av} are the normalized ($c_a + c_v + c_{av} = 1$) reliability coefficients associated to the confidence of the λ^a, λ^v and λ^{av} classifiers, respectively. These

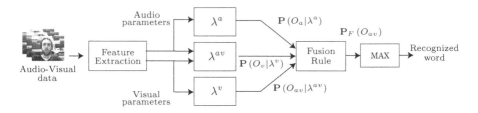

Fig. 1. Schematic representation of the proposed audio-visual speech classification system.

reliability coefficients are computed by taking into account the relative efficiency of each modality, and they are given by the following equations,

$$c_a = \frac{\alpha}{\alpha + \beta + \gamma}, \qquad \alpha = \exp\left(\frac{(s_a - s_v) + (s_a - s_{av})}{s_a}\right),$$

$$c_v = \frac{\beta}{\alpha + \beta + \gamma}, \qquad \beta = \exp\left(\frac{(s_v - s_a) + (s_v - s_{av})}{s_v}\right),$$

$$c_{av} = \frac{\gamma}{\alpha + \beta + \gamma}, \qquad \gamma = \exp\left(\frac{(s_{av} - s_a) + (s_{av} - s_v)}{s_{av}}\right),$$

where s_a, s_v and s_{av} are the confidence factors associated with the audio, visual and audio-visual classifiers, which are computed during the training stage as the recognition rates of the classifiers over a training dataset. Finally, the input data is recognized as the class with the maximum fused probability. In order to employ this recognition scheme at different noisy conditions, the reliability coefficients c_a, c_v and c_{av} are computed for different SNRs.

3   Audio-Visual Databases

The performance of the proposed classification scheme is evaluated over two isolated word audio-visual databases, *viz.*, a database compiled by the authors, hereafter referred as AV-UNR database and the Carnegie Mellon University (AV-CMU) database (now at Cornell University) [1].

(I) *AV-UNR* Database: The AV-UNR database consists of videos of 16 speakers, pronouncing a set of ten words (*up, down, right, left, forward, back, stop, save, open* and *close*) 20 times. The audio features are represented by the first eleven non-DC Mel-Cepstral coefficients, and its associated first and second derivative coefficients. Visual features are represented by three parameters, *viz.*, mouth height, mouth width and area between lips.

(II) *AV-CMU* Database: The AV-CMU database [1] consists of ten speakers, with each of them saying the digits from 0 to 9 ten times. The audio features are represented by the same parameters as in AV-UNR database. To represent the visual information, the weighted least-squares parabolic fitting method proposed in [2] is employed in this paper. Visual features are represented by five parameters, *viz*, the focal parameters of the upper and lower parabolas, mouth's width and height, and the main angle of the bounding rectangle of the mouth.

4   Experimental Results

The proposed decision level fusion strategy for audio-visual speech recognition is tested separately on the databases described in Sect. 3. For the evaluation of the performance of the proposed system over each database, audio-visual features are extracted from videos where the acoustic and visual streams are synchronized. The audio signal is partitioned in frames with the same rate as the video frame rate. The audio parameters at a given frame t is represented by the first eleven non-DC Mel-Cepstral coefficients, and its associated first and second derivative coefficients, computed from this frame. For the case of considering audio-visual information, the audio-visual feature vector at frame t is composed by the concatenation of the acoustic parameters with the visual ones.

To evaluate the recognition rates under noisy acoustic conditions, experiments with additive Gaussian and additive Babble noise, with SNRs ranging from $-10\,\mathrm{dB}$ to $40\,\mathrm{dB}$, were performed. Multispeaker or Babble noise environment is one of the most challenging noise conditions, since the interference is speech from other speakers. This noise is uniquely challenging because of its highly time evolving structure and its similarity to the desired target speech [8]. In this paper, Babble noise samples were extracted from *NOISEX-92* database, compiled by the Digital Signal Processing (DSP) group of the Rice University [4]. In addition, this evaluation is performed by considering three different types of classification methods, Hidden Markov Models, Random Forests and Support Vector Machines, respectively. Thus, the proposed confidence-based fusion rule (CFR) is evaluated in twelve different experiments (2 databases, 2 noise conditions and 3 classification methods).

In order to obtain statistically significant results, a 5-fold cross-validation (CV) is performed at each experiment. In the training stage, the classifiers meta-parameters for each modality are computed with the training set without noise, while the noisy training set is used to compute the reliability coefficients associated to these classifiers at the different noise levels. Then, the noisy Testing set is used to evaluate the performance of the recognition system at each SNR.

The recognition rates obtained for the experiments over the AV-CMU database for different SNRs are depicted in Fig. 2. As it expected, for the six cases, the efficiency of the classifier based on audio-only information deteriorates as the SNR decreases, while the efficiency of the visual-only information classifier remains constant, since it does not depend on the SNR in the acoustic channel. The efficiency of the audio-visual classifier is better than the one for audio classifier, but it also deteriorates at low SNRs. In this figure, are also depicted the performances obtained with the proposed decision level fusion strategy based on the confidence of each modality at different noisy conditions (CFR), and the ones obtained with the *sum* and *product* fusion rules. It can be seen, that the proposed fusion rule leads to improvements in the recognition rates, which are more notorious at low SNR levels. Moreover, the recognition rates obtained for the experiments over the AV-UNR database for different SNRs are depicted in Fig. 3. In this case, it can be also seen that a good performance is achieved with the proposed recognition scheme. On the other hand, the recognition rates

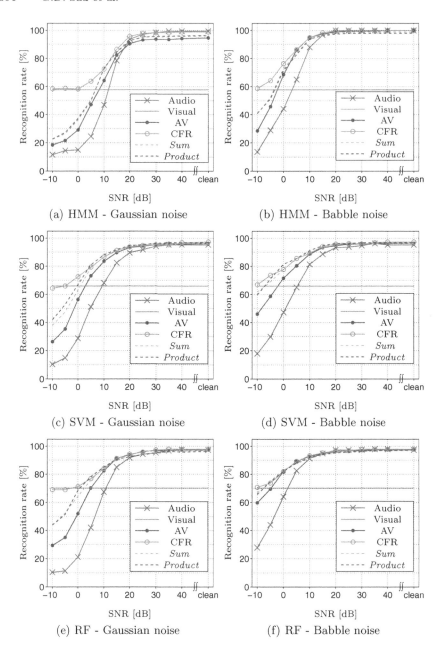

Fig. 2. Recognition rates for different SNRs over the **AV-CMU** database for the cases of considering additive Gaussian noise (left column) and Babble noise (right column), and HMM (first row), SVM (second row) and RF (third row) classifiers. For each case, the recognition rates obtained with audio, visual and audio-visual classifiers are depicted in red, grey and blue solid lines, respectively. The performance obtained with the proposed fusion rule (CFR) is depicted in green solid lines, while the corresponding performances obtained with *sum* and *product* fusion rules are depicted in cyan and magenta dashed lines, respectively. (Color figure online)

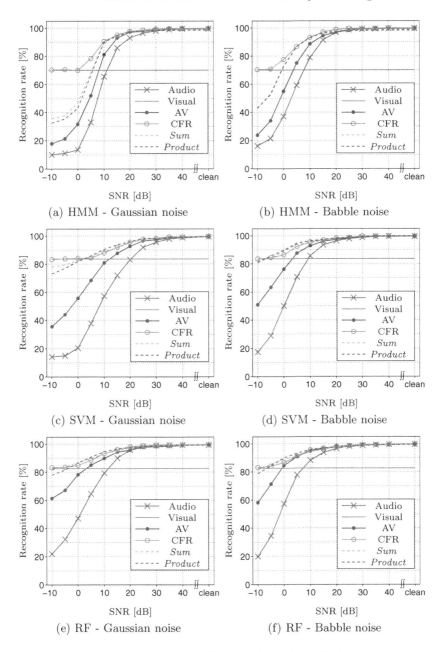

Fig. 3. Recognition rates for different SNRs over the **AV-UNR** database for the cases of considering additive Gaussian noise (left column) and Babble noise (right column), and HMM (first row), SVM (second row) and RF (third row) classifiers. For each case, the recognition rates obtained with audio, visual and audio-visual classifiers are depicted in red, grey and blue solid lines, respectively. The performance obtained with the proposed fusion rule (CFR) is depicted in green solid lines, while the corresponding performances obtained with *sum* and *product* fusion rules are depicted in cyan and magenta dashed lines, respectively. (Color figure online)

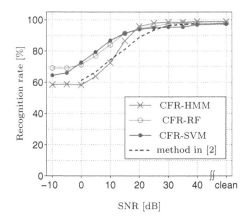

Fig. 4. Comparison between the recognition rates obtained, over the AV-CMU database, with proposed method using HMM, RF, and SVM classifiers, and the recognition rates reported in [2].

obtained with the proposed method over the AV-CMU database for the case of considering Gaussian additive noise and different classifiers are compared with the performance reported in [2], computed over the same database and noisy situation. This comparison is depicted in Fig. 4. It shows that, for the cases of considering RF and SVM classifiers, the proposed method outperforms the one reported in [2], whereas that for case of using HMM classifiers, the improvement is more notorious from 10 dB SNR to clean condition. All these experiments show that the proposed decision level fusion rule performs well for different noisy situations, classification methods and databases, in most cases outperforming the typical *sum* and *product* fusion rules.

5   Conclusions

In this paper, a decision level fusion strategy for audio-visual speech recognition in noisy situations was proposed. In order to enhance the recognition over different noisy conditions, the scores obtained with classifiers trained with different feature sets are fused. The combination of the scores is performed by taking into account the reliability of each modality at different noisy conditions. This method was evaluated by considering three modalities, audio, visual and audio-visual, respectively, but it could be employed using as many modalities as needed. This evaluation was carried out over two isolated word audio-visual databases, a public one and a database compiled by the authors of this paper. The proposed decision level fusion strategy was in addition evaluated by considering HMM, RF and SVM classifier. Experimental results showed that a good performance is achieved with the proposed system, leading to improvements in the recognition rates through a wide range of signal-to-noise ratios, in comparison with other method in the literature.

References

1. Advanced Multimedia Processing Laboratory. Carnegie Mellon University, Pittsburgh. http://chenlab.ece.cornell.edu/projects/AudioVisualSpeechProcessing/
2. Borgström, B., Alwan, A.: A low-complexity parabolic lip contour model with speaker normalization for high-level feature extraction in noise-robust audiovisual speech recognition. IEEE Trans. Syst. Man Cybern. **38**(6), 1273–1280 (2008)
3. Chu, S.M., Huang, T.S.: Audio-visual speech fusion using coupled hidden Markov models. In: Proceedings of IEEE Conference on Computer Vision and Pattern Recognition (2007)
4. Digital Signal Processing (DSP) Group. Rice University: NOISEX-92 Database, Houston
5. Dupont, S., Luettin, J.: Audio-visual speech modeling for continuous speech recognition. IEEE Trans. Multimedia **2**(3), 141–151 (2000)
6. Estellers, V., Gurban, M., Thiran, J.: On dynamic stream weighting for audiovisual speech recognition. IEEE Trans. Audio Speech Lang. Process. **20**(4), 1145–1157 (2012)
7. Kashiwagi, Y., Suzuki, M., Minematsu, N., Hirose, K.: Audio-visual feature integration based on piecewise linear transformation for noise robust automatic speech recognition. In: 2012 IEEE Spoken Language Technology Workshop (SLT), Miami, FL, USA, 2–5 December 2012, pp. 149–152 (2012)
8. Krishnamurthy, N., Hansen, J.: Babble noise: modeling, analysis, and applications. IEEE Trans. Audio Speech Lang. Process. **17**(7), 1394–1407 (2009)
9. Lee, J.S., Park, C.H.: Robust audio-visual speech recognition based on late integration. IEEE Trans. Multimedia **10**(5), 767–779 (2008)
10. Nefian, A.V., Liang, L., Pi, X., Xiaoxiang, L., Mao, C., Murphy, K.: A coupled HMM for audio-visual speech recognition. In: International Conference on Acoustics, Speech and Signal Processing (CASSP02), pp. 2013–2016 (2002)
11. Papandreou, G., Katsamanis, A., Pitsikalis, V., Maragos, P.: Adaptive multimodal fusion by uncertainty compensation with application to audiovisual speech recognition. Trans. Audio Speech Lang. Process. **17**(3), 423–435 (2009)
12. Potamianos, G., Neti, C., Gravier, G., Garg, A., Senior, A.W.: Recent advances in the automatic recognition of audio-visual speech. Proc. IEEE **91**, 1306–1326 (2003)
13. Shivappa, S., Trivedi, M., Rao, B.: Audiovisual information fusion in human computer interfaces and intelligent environments: a survey. Proc. IEEE **98**(10), 1692–1715 (2010)

Improving Nearest Neighbor Based Multi-target Prediction Through Metric Learning

Hector Gonzalez[1], Carlos Morell[2], and Francesc J. Ferri[3(✉)]

[1] Universidad de las Ciencias Informáticas (UCI), La Habana, Cuba
hglez@uci.cu
[2] Universidad Central Marta Abreu (UCLV), Villa Clara, Cuba
cmorellp@uclv.edu.cu
[3] Dept. Informàtica, Universitat de València, Valencia, Spain
Francesc.Ferri@uv.es

Abstract. The purpose of this work is to learn specific distance functions to be applied for multi-target regression problems using nearest neighbors. The idea of preserving the order relation between input and output vectors considering their corresponding distances is used along a maximal margin criterion to formulate a specific metric learning problem. Extensive experiments and the corresponding discussion try to put forward the advantages of the proposed algorithm that can be considered as a generalization of previously proposed approaches. Preliminary results suggest that this line of work can lead to very competitive algorithms with convenient properties.

1 Introduction

Typical problems in pattern recognition and machine learning deal with predictors for a single discrete label or continuous value depending on whether we are dealing with classification or regression. The natural (and most common) extension to formulate the problem of predicting multiple labels/values consists of considering it as an appropriate group of independent predictors. But this approach is prone to obviate correlations among output values which may be of capital importance in many challenging and recent application domains. These methods have been coined with different names as multi-target, multi-variate or multi-response regression [2]. When the different output values are organized using more complex structures as strings or trees we talk about Structured Predictors [1,15]. Among domain applications considered we have ecological modelling [11], gas tank control [8], remote sensing [22] and signal processing [6].

Particular methods for multi-target regression can be categorized either as problem transformation methods (when the original problem is transformed into one or several independent single-output problems), or algorithm adaptation methods (when a particular learning strategy is adapted to deal with multiple

F.J. Ferri—This work has been partially funded by FEDER and Spanish MEC through project TIN2014-59641-C2-1-P.

C. Beltrán-Castañón et al. (Eds.): CIARP 2016, LNCS 10125, pp. 368–376, 2017.
DOI: 10.1007/978-3-319-52277-7_45

interdependent outputs). The latter methods are usually considered as more challenging as an appropriate and interpretable model is obtained usually as a subproduct for the prediction problem [2].

The purpose of this work is to improve previous approaches for muti-target regression by introducing metric learning [12] in the context of nearest neighbor methods [15]. In particular, an input-output homogeneity criterion is introduced to learn a particular distance that consistently leads to improvements according to the empirical validation carried out. In the next section, the proposed methodology is put in the context of distance based muti-target regression while Sect. 3 contains the proposal itself. The empirical section follows with details and results obtained and a final section with conclusions and further work closes the present paper.

2   General Notation and State of the Art

Let $x = [x_1, \ldots, x_p] \in \mathbb{R}^p$, $y = [y_1, \ldots, y_q] \in \mathbb{R}^q$, be two random input and output vectors, respectively. Each training instance is written as $(x^j, y^j) \in \mathbb{R}^p \times \mathbb{R}^q$, and the corresponding multi-target regression problem consists of estimating a unique predictor $h : \mathbb{R}^p \to \mathbb{R}^q$ in such a way that the expected deviation between true and predicted outputs is minimized for all possible inputs.

The most straightforward approach consists of obtaining a univariate predictor for each one of the output variables in an independent way using any of the available methods for single-target prediction [2] which constitutes the simplest of the so-called problem transformation (also known as local) methods that consist of transforming the given multi-target prediction problem into one or more single-target ones [16,21].

The alternative approach to tackle multi-target prediction is through algorithm adaptation (also known as global) methods [2] which consist of adapting any previous strategy to deal directly with multiple targets. Global methods are interesting because they focus on explicitly capturing all interdependencies and internal relationships among targets. According to [2], these methods can be categorized as statistical, support-vector, kernel, trees or rule based, respectively. Apart from these, other strategies can be used. This is the case of one of the best known and used nonparametric methods in classification and estimation: the Nearest Neighbor (NN) family of rules [3]. Using NN for classification and estimation leads to interesting benefits as they behave quite smoothly accross a wide range of applications. These methods are known to approach an optimal behavior regardless of the distance used as the number of samples grows. But nevertheless, distance becomes of capital importance in the finite case.

The K-NN for Structured Predictions (KNN-SP) method [15] has been proposed for different kind of predicition problems and for multi-target regression in particular. Using the size of the neighborhood, K, as a parameter, the KNN-SP method starts by selecting the K nearest neghbors for a given query point according to a fixed distance (usually a weighted version of the Euclidean distance).

The final prediction is constructed as the (weighted) average of the corresponding K target values. These weights are set according to the (Euclidean) distance in the target space [15]. Even though the KNN-SP is very straightforward compared to other approaches, the empirical results show that it is very competitive compared to other methods which constitute the state of the art. Moreover, neighborhood size is the only parameter to tune.

3   Distance Metric Learning for Multi-target Prediction (DMLMTP)

Nearest Neighbor methods have been very widely used, specially for classification. Even though it was introduced very early [18], Distance Metric Learning (DML) has been recently deeply studied as a very convenient way to improve the behavior of distance-based methods [12]. Many powerful methods have been proposed to look for the best distance (in the input space) one may have for a particular problem.

A possible way to improve the results obtained by KNN-SP is by adapting the input space distance to the particular problem according to the final goal in the same way that it has been used for classification.

Many different criteria and approaches have been proposed to learn distances for classification but all of them share the same rationale: a distance is good if it keeps same-class points close and puts points from other classes far away. Many recent approaches implement this rationale as constraints relating pairs or triplets of training points. In the case of pairs, one must select pairs of points that need to be kept close (similar points) or far away (dissimilar points). In the case of triplets, one must select some triplets, (x^i, x^j, x^ℓ), where x^i and x^j are similar and should be kept close, and x^i and x^ℓ are dissimilar and should be taken farther.

In contrast to classification problems, it is far from obvious that similar ideas are to be useful in regression problems without introducing more information about both input and output spaces. In the present work, a first attempt to learn an input distance for multi-target regression is proposed by introducing an homogeneity criterion between input and output spaces using triplets. In particular, we propose to select the same kind of triplets as in classification problems and use a different criterion for similarity. Instead of using labels, similarity between points will be established according to their outputs in such a way that the relative ordering introduced by distances in input and output spaces are preserved.

We formulate an optimization problem to learn an input distance for multi-target regression by following an approach similar to the one in [17] and also in [13, 26]. The goal is to obtain a Mahalanobis-like distance, parametrized by a matrix, W, which maximizes a margin criterion. As usual, this problem is converted into minimizing a regularizer for W (its Frobenius norm) subject to several (soft) constraints using triplets. In our particular case we have

$$\min_{W,\rho,\xi_{ij\ell}} \frac{1}{2}\|W\|_F - \rho + \frac{1}{\nu|\mathcal{T}_K|} \sum_{i,j,\ell \in \mathcal{T}_K} \xi_{ij\ell}$$

$$\text{s.t.} \quad d_W^2(\boldsymbol{x}^i, \boldsymbol{x}^\ell) - d_W^2(\boldsymbol{x}^i, \boldsymbol{x}^j) \geq \rho - \xi_{ij\ell},$$

$$\xi_{ij\ell} \geq 0, \qquad \forall\, i, j, \ell \in \mathcal{T}_K$$

where $d_W^2(\boldsymbol{x}^i, \boldsymbol{x}^j) = (\boldsymbol{x}^i - \boldsymbol{x}^j)^T W (\boldsymbol{x}^i - \boldsymbol{x}^j)$ is the (squared) distance in the input space and the set of triplets is defined as

$$\mathcal{T}_K = \{(\boldsymbol{x}^i, \boldsymbol{x}^j, \boldsymbol{x}^\ell) \,:\, \boldsymbol{x}^j, \boldsymbol{x}^\ell \in \mathcal{N}_K(\boldsymbol{x}^i) \quad \text{and} \quad d(\boldsymbol{y}^i, \boldsymbol{y}^\ell) - d(\boldsymbol{y}^i, \boldsymbol{y}^j) \geq 0\}$$

where $\mathcal{N}_K(\boldsymbol{x})$ is the considered neighborhood around $\boldsymbol{x}$.

Note that the formulation of the optimization problem is the same used for other metric learning and support vector learning approaches and the main change is in the way the particular restrictions have been selected.

In the formulation above, we must introduce an extra constraint to make the matrix W positive semi-definite. This makes the problem considerably more difficult but there are a number of ways in which this can be tackled [13,26]. Nevertheless, in this preliminary work we will simplify the above formulation further. On one hand, we consider only a diagonal matrix, $W = \boldsymbol{w} = [w_1, \ldots, w_p]$, and on the other hand, we will introduce the corresponding restrictions, $w_i \geq 0$, $i = 1, \ldots, q$ into the above optimization. The corresponding dual problem can be written in terms of two new sets of variables as

$$\min_{\alpha_i, \lambda_j} \frac{1}{2}\left(\boldsymbol{\alpha}^T H \boldsymbol{\alpha} + 2\boldsymbol{\alpha}^T \phi\boldsymbol{\lambda} + \boldsymbol{\lambda}^T \boldsymbol{\lambda}\right)$$

$$\text{s.t.} \quad \sum_{i=1}^{|\mathcal{T}_K|} \alpha_i = 1$$

$$0 \leq \alpha_i \leq \frac{1}{\nu|\mathcal{T}_K|} \quad i = 1, \ldots, \mathcal{T}_K$$

$$\lambda_j \succeq 0 \quad j = 1, \ldots, q$$

$\phi \in \mathbb{R}^{|\mathcal{T}_K| \times q}$ is a matrix with a row, $(\boldsymbol{x}^i - \boldsymbol{x}^\ell) \circ (\boldsymbol{x}^i - \boldsymbol{x}^\ell) - (\boldsymbol{x}^i - \boldsymbol{x}^j) \circ (\boldsymbol{x}^i - \boldsymbol{x}^j)$, for each considered triplet where $\circ$ is the Hadamard or entrywise vector product. The kernel matrix is $H = \phi\phi^T$ and the weight vector is obtained as $\boldsymbol{w} = \boldsymbol{\alpha}^T \phi + \boldsymbol{\lambda}$.

An adhoc solver using an adapted SMO approach [10,14] has been implemented specifically for this work. This solver is able to arrive to relatively good results in reasonable times for all databases considered in the empirical work carried out as will be shown in the next section.

4 Experiments

In this section, we describe the experimental setup and discuss the main results of the proposed DMLMTP algorithm. In the first place, we present technical

Table 1. Datasets used in the experimentation and corresponding details. Datasets partitioned in train and test subsets are indicated by the corresponding two sizes in the second column.

Datasets	Instances	Attributes	Targets
Waterquality [5]	1060	16	14
EDM [9]	154	16	2
Solar Flare 1 [23]	323	10	3
Solar Flare 2 [23]	1066	10	3
jura [19]	359	15	3
enb [19]	768	8	2
slump [19]	103	7	3
andro [19]	49	30	6
osales [24]	639	413	12
scpf [25]	1137	23	3
atp1d [19]	201/136	411	6
atp7d [19]	188/108	411	6
rf1 [19]	4108/5017	64	8
rf2 [19]	4108/5017	576	8
OES97 [19]	334	263	16
OES10 [19]	410	298	16

details related to the datasets, parameter setup and implementations. Next, we present comparative results when using the learned distance compared to the Euclidean one when predicting multivariate outputs with the KNN-SP approach over fifteen datasets publicly available for multi target prediction.

In the experiments we distinguish between the number of neighbors used to learn the distance using DMLMTP, K, and the number of neighbors used to obtain the final prediction using the KNN-SP approach, k_p. The value of K should be small to keep the number of triplets small for efficiency reasons. For all experiments reported in this paper, the number of nearest neighbors for training in DMLMTP was set to $K = 2, \ldots, 6$ while the neighborhood sizes for prediction have been taken as odd values from 3 to 35. The final prediction is done computing the average of the target values of these k_p nearest neighbors.

In the experiments, 5-fold cross-validation has been used on each dataset except for 4 of them that have been split into train and test subsets for efficiency and compatibility reasons. Table 1 summarizes the main details [2,21]. The cross validation procedure has been integrated into MULAN software package [20].

As in other similar works, we use the average Relative Root Mean Squared Error (aRRMSE) given a test set, D_{test}, and a predictor, h, which is given as

$$aRRMSE(h; D_{test}) = \frac{1}{q} \sum_{i=1}^{q} \sqrt{\frac{\sum_{(x,y) \in D_{test}} (\hat{y}_i - y_i)^2}{\sum_{(x,y) \in D_{test}} (\overline{y_i} - y_i)^2}} \qquad (1)$$

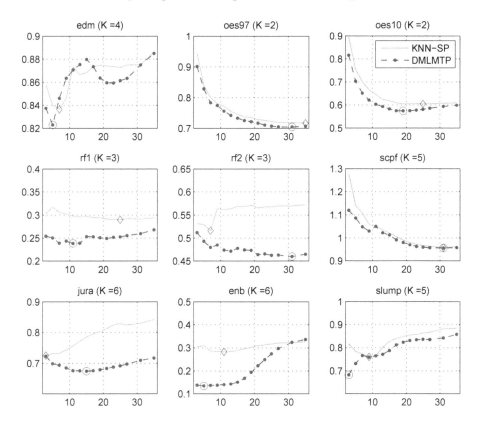

Fig. 1. aRMSE values corresponding to different neighborhood sizes, k_p, using Euclidean (KNN-SP) and learned (DMLMTP) distances. The best neighborhood size used by DMLMTP is indicated along the name of each database.

where $\overline{y}$ is the mean value of the target variable y, and $\hat{y} = h(x)$. We use the Wilcoxon signed rank test and the Friedman procedure with different post-hoc tests to compare algorithms over multiple datasets [4, 7].

Figure 1 contains the aRRMSE versus the neighborhood size, k_p for 9 datasets out of the 16 considered. Only the best neighborhood size used for training, K is shown. Moreover, the best results in the curves are marked with a circle and a diamond, respectively. These best results are shown for all the datasets in Table 2.

Contrary to our expectations, the best performance for DMLMTP over large datasets is obtained with small K. This could be strongly related to the growth in the number of triplets that violate the considered constraints.

The last columns in Table 2 contain the absolute difference between aRRMSE for KNN-SP and DMLMTP, its sign and the average ranking with regard to absolute differences. The DMLMTP method is better with a significance level of 5% according to the Wilcoxon test that leads to a p-value of 6.1035e-5. For all

Table 2. aRRMSE obtained for DMLMTP and KNN-SP algorithms on each dataset along with comparison details.

Dataset	DMLMTP	KNN-SP	abs	sgn	R_i
sf1	0.971	0.973	0.002	+	2.5
sf2	0.976	0.977	0.001	+	1.5
wq	0.947	0.948	0.001	+	1.5
edm	0.823	0.836	0.013	+	6
oes97	0.704	0.716	0.012	+	5
oes10	0.574	0.605	0.031	+	7
atp1d	0.415	0.451	0.036	+	8
atp7d	0.515	0.605	0.090	+	13
rf1	0.238	0.290	0.052	+	10
osales	1.012	1.012	0		-
scpf	0.956	0.958	0.002	+	2.5
jura	0.674	0.725	0.051	+	9
enb	0.134	0.283	0.149	+	15
slump	0.682	0.760	0.078	+	12
andro	0.799	0.931	0.132	+	14
rf2	0.460	0.516	0.056	+	11

datasets, DMLMTP has equal or better performance than (Euclidean) KNN-SP and the difference increases for datasets of higher dimensionality. This situation could be related to the learned input transformation that generates some values equal to zero and ignores some irrelevant attributes. In fact, if we compute the sparsity index of the corresponding transformation vector, as the relative number of zeros with regard to dimensionality, we obtain for our algorithm values below 0.5 except for datasets osales, rf1 and scpf.

5   Concluding Remarks and Further Work

An attempt to improve nearest neighbor based multi-target prediction has been done by introducing an specific distance metric learning algorithm. The mixing of these strategies has lead to very competitive results in the preliminary experimentation carried out. In a wide range of situations and for large variations of the corresponding parameters, the proposal behaves smoothly over the datasets considered paving the way to develop more specialized algorithms. Future work is being planned in several directions. On one hand, different optimization schemes can be adopted both to improve efficiency and performance. On the other hand, different formulations can be adopted by establishing more accurate constraints able to properly capture all kinds of dependencies among input and output vectors in challenging multi output regression problems.

References

1. Bakir, G.H., Hofmann, T., Schölkopf, B., Smola, A.J., Taskar, B., Vishwanathan, S.V.N.: Predicting Structured Data (Neural Information Processing). The MIT Press, Cambridge (2007)
2. Borchani, H., Varando, G., Bielza, C., Larrañaga, P.: A survey on multi-output regression. Wiley Interdisc. Rev. Data Mining Knowl. Discov. **5**(5), 216–233 (2015)
3. Dasarathy, B.V.: Nearest Neighbor (NN) Norms: NN Pattern Classification. IEEE Computer Society, Washington (1990)
4. Demšar, J.: Statistical comparisons of classifiers over multiple data sets. J. Mach. Learn. Res. **7**, 1–30 (2006)
5. Džeroski, S., Demšar, D., Grbović, J.: Predicting chemical parameters of river water quality from bioindicator data. Appl. Intell. **13**(1), 7–17 (2000)
6. Fernández, M.S., de Prado-Cumplido, M., Arenas-García, J., Pérez-Cruz, F.: SVM multiregression for nonlinear channel estimation in multiple-input multiple-output systems. IEEE Trans. Sig. Process. **52**(8), 2298–2307 (2004)
7. García, S., Fernández, A., Luengo, J., Herrera, F.: A study of statistical techniques and performance measures for genetics-based machine learning: accuracy and interpretability. Soft. Comput. **13**(10), 959–977 (2009)
8. Han, Z., Liu, Y., Zhao, J., Wang, W.: Real time prediction for converter gas tank levels based on multi-output least square support vector regressor. Control Eng. Pract. **20**(12), 1400–1409 (2012)
9. Karalič, A., Bratko, I.: First order regression. Mach. Learn. **26**(2–3), 147–176 (1997)
10. Keerthi, S.S., Shevade, S.K., Bhattacharyya, C., Murthy, K.R.K.: Improvements to Platt's SMO algorithm for SVM classifier design. Neural Comput. **13**(3), 637–649 (2001)
11. Kocev, D., Dzeroski, S., White, M.D., Newell, G.R., Griffioen, P.: Using single- and multi-target regression trees and ensembles to model a compound index of vegetation condition. Ecol. Model. **220**(8), 1159–1168 (2009)
12. Kulis, B.: Metric learning: a survey. Found. Trends Mach. Learn. **5**(4), 287–364 (2012)
13. Perez-Suay, A., Ferri, F.J., Arevalillo, M., Albert, J.V.: Comparative evaluation of batch and online distance metric learning approaches based on margin maximization. In: IEEE International Conference on Systems, Man, and Cybernetics, Manchester, SMC 2013, UK, pp. 3511–3515 (2013)
14. Platt, J., et al.: Fast training of support vector machines using sequential minimal optimization. In: Advances in Kernel Methods—Support Vector Learning, vol. 3 (1999)
15. Pugelj, M., Džeroski, S.: Predicting structured outputs k-nearest neighbours method. In: Elomaa, T., Hollmén, J., Mannila, H. (eds.) DS 2011. LNCS (LNAI), vol. 6926, pp. 262–276. Springer, Heidelberg (2011). doi:10.1007/978-3-642-24477-3_22
16. Read, J., Pfahringer, B., Holmes, G., Frank, E.: Classifier chains for multi-label classification. Mach. Learn. **85**, 333–359 (2011)
17. Schultz, M., Joachims, T.: Learning a distance metric from relative comparisons. In: Advances in Neural Information Processing Systems (NIPS), p. 41 (2004)
18. Short, R., Fukunaga, K.: The optimal distance measure for nearest neighbor classification. IEEE Trans. Inf. Theory **27**(5), 622–627 (1981)

19. Spyromitros-Xioufis, E., Tsoumakas, G., Groves, W., Vlahavas, I.: Multi-target regression via input space expansion: treating targets as inputs. Mach. Learn. **104**(1), 55–98 (2016)
20. Tsoumakas, G., Spyromitros-Xioufis, E., Vilcek, J., Vlahavas, I.: Mulan: a Java library for multi-label learning. J. Mach. Learn. Res. **12**, 2411–2414 (2011)
21. Tsoumakas, G., Spyromitros-Xioufis, E., Vrekou, A., Vlahavas, I.: Multi-target regression via random linear target combinations. In: Calders, T., Esposito, F., Hüllermeier, E., Meo, R. (eds.) ECML PKDD 2014. LNCS (LNAI), vol. 8726, pp. 225–240. Springer, Heidelberg (2014). doi:10.1007/978-3-662-44845-8_15
22. Tuia, D., Verrelst, J., Alonso-Chorda, L., Pérez-Cruz, F., Camps-Valls, G.: Multi-output support vector regression for remote sensing biophysical parameter estimation. IEEE Geosci. Remote Sens. Lett. **8**(4), 804–808 (2011)
23. http://archive.ics.uci.edu/ml/datasets/Solar+Flare
24. https://www.kaggle.com/c/online-sales
25. https://www.kaggle.com/c/see-click-predict-fix
26. Wang, F., Zuo, W., Zhang, L., Meng, D., Zhang, D.: A kernel classification framework for metric learning. IEEE Trans. Neural Netw. Learn. Syst. **26**(9), 1950–1962 (2015)

An Approximate Support Vector Machines Solver with Budget Control

Carles R. Riera$^{(\boxtimes)}$ and Oriol Pujol

Department of Applied Mathematics and Analysis,
Universitat de Barcelona, Gran Via, 585 08007 Barcelona, Spain
carles.riera@ub.edu

Abstract. We propose a novel approach to approximately solve online kernel Support Vector Machines (SVM) with the number of support vectors set beforehand. To this aim, we modify the original formulation introducing a new constraint that penalizes the deviation with respect to the desired number of support vectors. The resulting problem is solved using stochastic subgradient methods with block coordinate descent. Comparison with state-of-the-art online methods shows very promising results.

Keywords: SVM · Budget · Coverage · Sparsity · Convex optimization · Online learning

1 Introduction

Kernel Support Vector Machines (SVM) is a very well established supervised machine learning technique. Its training computational complexity depends on the effective number of support vectors. In online settings, the amount of support vectors increases linearly with the size of the dataset. This fact makes online kernel SVM unappealing in front of large scale problems.

Concerning the online capabilities of the SVM, several solvers have been proposed based on the primal formulation of [3], optimized using Stochastic Subgradient Methods (SSM) variations, including Stochastic Gradient Descent (SGD). One of the most widely used is Stochastic Gradient SVM [2]. Pegasos [8] is another well-known solver, which shines due to the use of integer operations that speeds up the training process.

The reduction of the number of support vectors has been an important topic in the machine learning community. Most of the proposals are based on iteratively adding one support vector in a greedy fashion. In SpSVM [6], the newly included support vector is selected from a set of candidates, optimizing its coefficient alone while freezing the other weights. The resulting method is computationally expensive and lacks online capability. SVM^{perf} [4], optimizes the dual formulation using Cutting Planes, while the support vector is chosen solving a preimage problem. This process is also costly and thus unsuitable. The work of [1], LaSVM, extends the original SMO solver [7] with online capabilities.

© Springer International Publishing AG 2017
C. Beltrán-Castañón et al. (Eds.): CIARP 2016, LNCS 10125, pp. 377–384, 2017.
DOI: 10.1007/978-3-319-52277-7_46

The support vector control involves two processes. The first, adds a support at a time and computes its coefficient by using the traditional SMO update step. The second, finds the most violating pair of support vectors and recomputes their coefficients. In the case that one or both coefficients vanish, their associated support vectors are removed. Lastly, we find several works that controls sparsity in the solution by adding an L_1 norm regularizer to the SVM formulation [10], an idea originally applied in the Relevance Support Machine [9]. Other approaches can be found in [5].

However, controlling the exact number of support vectors can be an important matter in many different scenarios. For example, consider a large scale problem, we can control the amount of memory used by the kernel by limiting the number of support vectors. Note that this will also speed up the training process. Moreover, not only this makes the method suitable for ingesting large amounts of data, but also usable in environments with low computational resources. Another scenario where controlling the number of support vectors is important, is in the exploration phase on a machine learning problem. A fast, lower bound on the validation error, can be computed that may help in the hyper-parameter optimization or model selection processes. One way to solve this issue is to set an upper bound on the number of support vectors, called budget, and find an approximate solution. However, the mechanisms to maintain this budget are not trivial and sometimes lead to a large loss of performance.

Our proposal, Elastic Budget SVM (EBSVM) solves the primal kernel SVM formulation [3] with a set of additional constraints for enforcing an upper bound on the number of support vectors using stochastic subgradient methods. Concretely, the magnitude of the coefficients associated to each support vector is required to be above a threshold. This threshold is dynamically optimized during the training process in such a way that minimizes the difference between the current number of support vectors and the required budget. The method does not require additional parameters except for the allowed budget.

The remainder of this article is organized as follows. Section 2 describes the core of the EBSVM algorithm in detail and provide pseudo-code. We validate our proposal with a set of experiments on the UCI database in Sect. 3. Finally, Sect. 4 offers the conclusions.

2   Proposal

Our proposal is founded on the following intuition. We consider the notion of coverage which refers to the effective influence of a support vector on its neighbors. The larger the influence area of a support vector, the smaller the number of support vectors required for modeling/covering all data. In quasi-concave[1] kernel functions of the form $k(x_{sv}, x) = e^{-\gamma d(x_{sv}, x)}$, the coverage is controlled by parameter γ, where $d(\cdot)$ is an arbitrary distance metric. In the most well-known formulation with radial basis functions, $d(\cdot)$ is squared Euclidean distance, and

---

[1] f(x) is called to be quasi-concave when the set $\{x | f(x) = \epsilon\}$ is a convex set.

γ is the inverse of the desired variance $1/\sigma^2$. However, the same effect can be achieved by controlling the magnitude of the solution coefficients α.

The relationship between coverage and coefficients is easily visualized in the level sets of the kernel function. Consider the level set $k(x_{sv}, x) = e^{-\gamma d(x_{sv}, x)} = \epsilon$. The ϵ-coverage can be written as the hyper-volume enclosed by the level set $k(x_{sv}, x) = \epsilon$, i.e. $g(x) = \{x | k(x_{sv}, x) \leq \epsilon\}$, $Vol = \int_{\mathcal{R}} g(x) dx$.

If the kernel function is a quasi-concave function, the volume increases with small values of ϵ and decreases with large values of ϵ. Thus, in the case of a weighted kernel we have $\alpha k(x_{sv}, x)$ with $\alpha > 0$. In that case, $\hat{g}(x) = \{x | \alpha k(x_{sv}, x) \leq \epsilon\}$. This is equivalent to $\hat{g}(x) = \{x | k(x_{sv}, x) \leq \epsilon/\alpha\}$. In that case, it is trivial to see that the greater the α value is the smaller the ϵ value is, and consequently, larger the coverage of the function. Conclusively, we use the magnitude of the coefficients as heuristic to rank the importance of the support vectors.

2.1 Sparsity Constraints

We propose to model the problem adding an additional constraint that depends of a parameter θ, a lower bound θ on the magnitude of α, shown in Eq. 1.

$$
\begin{aligned}
\underset{\alpha}{\text{minimize}} \quad & \frac{1}{2}\lambda\|\alpha\|^2 + \sum_i \xi_i \\
\text{subject to} \quad & y_i \left(\sum_j \alpha_j k(x_i, x_{sv_j}) \right) \geq 1 - \xi_i, \ i = 1, \dots, m \\
& \alpha_i = \begin{cases} \alpha_i & \text{if } |\alpha_i| \geq \theta \\ 0 & \text{otherwise} \end{cases}
\end{aligned}
\tag{1}
$$

The value of θ will allow us to control the amount of support vectors. However, this formulation does not solve the problem of looking for a solution given a budget B of support vectors, i.e. the required number of support vectors. For this task, we define a new function $N = f(\theta)$ that relates the value of theta to the number of support vectors, N, and require the optimization problem to simultaneously optimize the former problem with a new cost function $(f(\theta) - B)^2$ that penalizes the deviation with respect to the desired budget. The problem can be formulated as follows,

$$
\begin{aligned}
\underset{\alpha, \theta}{\text{minimize}} \quad & \frac{1}{2}\lambda\|\alpha\|^2 + \sum_i \xi_i + (B - f(\theta))^2 \\
\text{subject to} \quad & y_i \left(\sum_j \alpha_j k(x_i, x_{sv_j}) \right) \geq 1 - \xi_i, \ i = 1, \dots, m \\
& \alpha_i = \begin{cases} \alpha_i & \text{if } |\alpha_i| \geq \theta \\ 0 & \text{otherwise} \end{cases} \\
& \theta > 0
\end{aligned}
$$

Observe that the optimization of this problem requires solving for θ and α. However, this entails modeling $f(\theta)$, which is the dependency of the number of support vectors with respect to θ. We directly define f as follows, Eq. 2.

$$f(\theta) = \|\alpha\|_0\Big|_\theta.$$
(2)

This involves computing the amount of elements different of zero in the coefficients of the solution.

In order to computationally solve this problem using stochastic subgradient methods, we use a block coordinate descent approach. At each iteration we first solve for α and project the solution into the feasible set, i.e. dropping the coefficients with coefficients below θ. The second step of the iteration is the optimization with respect to θ. The gradient of the loss function with respect to the θ is the following

$$\frac{\partial \mathcal{L}}{\partial \theta} = \frac{\partial \mathcal{L}}{\partial f(\theta)} \frac{\partial f(\theta)}{\partial \theta} = -(B - f(\theta))\frac{\partial f(\theta)}{\partial \theta}$$

The second part of the gradient requieres computing a subgradient of $f(\theta)$. The exact modeling of this term is complex. For this reason we introduce a surrogate of the exact function that follows the expected behavior of the the number of support vectors with respect to the parameter. In particular we can approximate $f(\theta)$ using a linear model as $f(\theta) = b(a - \theta)$. Observe that this capture the notion that the number of support vectors should decrease when θ increases. Thus, the final gradient computation becomes

$$\frac{\partial \mathcal{L}}{\partial \theta} \approx (B - f(\theta))b$$

The update with respect θ results in the formulation of Eq. 3.

$$\theta_{t+1} = \theta_t - \eta'(B - f(\theta))$$
(3)

where η' subsumes the constant term b. Algorithm 1 provides the pseudo-code of our proposal. It starts with an empty set of support vectors. Iteratively, a point is sampled from the dataset uniformly at random. The gradient it is then computed, and if it is non-zero it will be added to the support vector set. The other coefficients are updated using $\eta = \frac{1}{\lambda t}$. Then θ is updated according Eq. 3. Finally the support vectors whose coefficient is below θ are removed. However, we found experimentally that, since we sample one data point at a time, removing more than one is too harsh. Therefore we remove the violating support vector with the smallest coefficient only.

Algorithm 1. EBSVM algorithm.

Require: $D, \lambda, T, \theta, \rho$
    $\mathcal{D}$: A dataset.
    λ: Regularization parameter.
    θ: Minimum magnitude of α desired.
    T: Number of iterations.
    ρ: Steepness of the homotopy function.
    Initialization:
    $\alpha_0 \leftarrow 0$
    $\mathbf{x_{sv}} = \emptyset$, initial set of support vectors,
    for $t = 1, 2, \ldots, T$ **do**
        Choose a pair $(x_i, y_i) \in \mathcal{D}$ uniformly at random
        $\eta := \dfrac{1}{\lambda t}$
        if $y_i \sum_j \alpha_t^j k(x_i, \mathbf{x_{sv}^j}) < 1$ **then**
            $\alpha_{t+1} := \alpha_t - \eta(\lambda K \alpha_t - y_i k(x_i, \mathbf{x_{sv}}))$
            if $x_i \notin \mathbf{x_{sv}}$ **then**
                $\mathbf{x_{sv}} := \mathbf{x_{sv}} \cup x_i$
                $\alpha_t := \alpha_t \cup \eta y_i k(x_i, \mathbf{x_{sv}})$
            end if
        else
            $\alpha_{t+1} := \alpha_t - \eta(\lambda \alpha_t)$
        end if
        $\theta_{t+1} = \theta_t - \eta'(B - \|\alpha_t\|_0)$
        $\hat{x} := \{x_j \in \mathbf{x_{sv}} \,\big|\, |\alpha_j| \leq \theta_t\}$
        $\hat{\alpha} := \{\alpha_j \in \alpha_{t+1} \,\big|\, |\alpha_j| \leq \mu_t \theta\}$
        if $\hat{x} \neq \emptyset$ **then**
            $\alpha_{t+1} := \alpha_{t+1} \setminus \hat{\alpha}$, remove $\hat{\alpha}$ from the vector of coefficients
            $\mathbf{x_{sv}} := \mathbf{x_{sv}} \setminus \hat{x}$, remove $\hat{x}$ from the support vector set
        end if
    end for
Ensure: α_T

---

3 Experiments

In this section, we offer a set of experiments in order to asses its effectivity and performance. We introduce two different experiments comparing our proposal with a set of state-of-the-art SVM solvers. In the first experiment, we set B equal to the number of support vectors of the model showing the best accuracy on each individual problem. With this experiment we show that our approach allows to effectively control the budget without harming generalization. The second experiment sets $B = 10$. This second experiment shows the behavior of the method when the number of support vectors is very small.

3.1 Experimental Setup

Models. The models chosen are the online primal SVM from [3] with L_1 norm, Pegasos and LaSVM. Regarding LaSVM, we have tuned its sparsity parameter τ to achieve the best accuracy.

Performance Measures. We evaluate the classifiers both using the accuracy score to appraise their generalization capabilities, and the number of support vectors for assessing the complexity of the solution.

Datasets. The experiment consists in a comparison conducted over a set of datasets selected from the UCI repository, downloaded from MLData. All the datasets are standardized to $\mu = 0$ and $\sigma = 1$.

Generalization Error Estimation. In order to obtain reliable out-of-sample error estimates we use a nested 3-2 fold cross-validation, where the inner 2 folds are used to optimize the hyperparameters and the 3 outer folds to test its accuracy. Also the number of support vectors is averaged over the outer folds.

The hyperparameters are optimized with a logspace of 13 values from -3 to 9 for the regularization parameter λ and σ and $\{0.01, 0.001, 0.0001\}$ for the τ parameter of LaSVM.

3.2   Discussion

The results are shown in Table 1. Note how our method is certainly bounding the number of support vector to the number of support vectors of the model which exhibits the best accuracy. This offers proof of that our proposal is a effectively capable respect the budget, thus validating it as a budget SVM implementation.

A closer look at the results shows that the difference in terms of accuracy between the proposed model and the best performant one is usually very small.

Table 1. Accuracy and number of support vectors on different datasets.

	LaSVM	OnlineSVM L1	Pegasos	EBSVM	EBSVM 10
australian	0.887(175.33)	0.849(**151.33**)	**0.894**(188.67)	0.877(188.33)	0.854(10.00)
breast-cancer	0.969(76.00)	0.955(**47.33**)	0.968(130.00)	**0.974**(76.00)	0.975(10.33)
breast-cancer-ida	**0.848**(**57.33**)	0.635(87.67)	0.795(79.00)	0.821(58.33)	0.768(9.67)
colon-cancer	**1.000**(19.67)	0.879(20.67)	0.983(20.67)	**1.000**(20.00)	0.777(8.67)
datasets-UCI breast-w	**0.971**(**45.33**)	0.961(100.33)	0.967(89.67)	0.969(46.00)	0.964(10.00)
datasets-UCI heart-statlog	0.870(59.33)	**0.941**(**51.67**)	0.870(84.00)	0.889(52.00)	0.844(9.67)
diabetes	0.813(**140.67**)	0.745(163.33)	**0.823**(222.33)	0.789(223.00)	0.736(10.00)
duke breast-cancer	**1.000**(27.67)	**1.000**(**21.33**)	0.952(28.67)	0.954(22.33)	0.917(11.67)
fourclass	**1.000**(168.67)	**1.000**(262.67)	**1.000**(202.00)	**1.000**(**150.00**)	0.894(10.00)
german.numer	0.798(**187.67**)	0.748(265.67)	0.832(279.67)	**0.855**(279.67)	0.715(10.67)
heart	**0.859**(61.33)	0.833(**59.67**)	**0.859**(85.00)	**0.859**(61.67)	0.837(8.67)
ionosphere	0.969(**58.67**)	0.949(105.00)	**0.986**(99.00)	0.969(93.33)	0.883(10.00)
leukemia	**1.000**(22.00)	**1.000**(**16.00**)	**1.000**(24.00)	0.973(16.33)	0.889(10.33)
liver-disorders	0.794(**80.33**)	0.774(107.33)	**0.820**(112.00)	0.794(111.00)	0.546(8.00)
mg	0.867(**197.67**)	0.851(302.00)	**0.887**(349.67)	0.874(348.67)	0.814(9.33)
mushrooms	**1.000**(2708.00)	0.974(**1317.67**)	0.992(2708.00)	**1.000**(2707.67)	0.922(12.00)
sonar	0.788(**59.67**)	0.443(69.33)	**0.816**(68.00)	0.812(67.67)	0.846(10.33)
splice	**0.988**(**550.00**)	0.801(562.00)	0.946(810.00)	0.861(**550.00**)	0.733(10.00)
svmguide1	**0.974**(265.67)	0.962(338.33)	0.973(684.67)	0.968(**265.67**)	0.953(10.00)
svmguide3	**0.872**(204.00)	0.773(331.67)	0.863(345.33)	0.854(**203.33**)	0.765(12.00)

Table 2. Average rank for each dataset.

	LaSVM	OnlineSVM L1	Pegasos	EBSVM
Rank	1.975	3.55	2.15	2.325

In six out of twenty datasets compared, the proposal achieves the best score. There is a noticeable performance loss in just four of the data sets compared. Table 2 shows the average rank of every model. A model is assigned a rank ranging from 1 for the best to 4 for the worst for each dataset, using the average in case of tie. The ranks are then averaged in order to offer insight of the relative performance among them. The resulting ranks show how our method ranks between the second and the third best on average, notably better than OnlineSVM L1 and no far from Pegasos, thus providing evidence that the loss of performance due the use of a budget is not critical.

Furthermore, if we force our algorithm setting the budget to $B = 10$, the last column of Table 1 shows how our method performs surprisingly well considering the constraints. In some cases like breast-cancer it even outperforms all the other methods. Observe that in at least nine of the data sets, the result achieved is very close to the best performer, while in the rest of the cases the results can be a tight lower bound of the performance achieved without the constraint. It is worth mentioning that using just 10 support vectors effectively reduce the computational complexity and storage of the method by at least an order of magnitude. These results provide confirmatory evidence of the validity of our approach.

4   Conclusion

The control of the number of support vectors in SVM is an important problem in different scenarios such as low computational resources environments, very large scale machine learning, or for preliminary exploratory analysis of a data set, where fast results are needed. In this paper we introduce a novel solver for support vector machines that allows the effective control of the number of support vectors. The results show that the method is competitive with state-of-the-art for online learning and effectively achieves tight lower bounds when enforcing drastic reductions on the amount of support vectors.

Since the method is based on stochastic subgradient methods, this opens the possibility of applying a similar approach to other methods based on SGD and the use of basis functions, such deep neural networks or gradient boosted trees.

Another interesting line of investigation is optimizing θ with respect the training error, allowing to model directly the trade-off between capacity and accuracy. This would enable to traverse the Pareto optimal surface between error and complexity, optimizing the number of support vectors for a given required accuracy.

Acknowledgements. This work has been partially funded by the Spanish MINECO Grant TIN2013-43478-P.

References

1. Bordes, A., Ertekin, E., Weston, J., Bottou, L.: Fast Kernel classifiers with online and active learning. J. Mach. Learn. Res. **6**, 1579–1619 (2005)
2. Bottou, L.: Large-scale machine learning with stochastic gradient descent. In: Lechevallier, Y., Saporta, G. (eds.) Proceedings of COMPSTAT'2010, pp. 177–186. Springer, Heidelberg (2010)
3. Chapelle, O.: Training a support vector machine in the primal. Neural Comput. **19**(5), 1155–1178 (2007)
4. Joachims, T., Yu, C.N.J.: Sparse Kernel SVMs via cutting-plane training. Mach. Learn. **76**(2–3), 179–193 (2009)
5. Jung, H.G., Kim, G.: Support vector number reduction: survey and experimental evaluations. IEEE Trans. Intel. Transp. Syst. **15**(2), 463–476 (2014)
6. Keerthi, S.S., Decoste, D.: Building support vector machines with reduced classifier complexity. J. Mach. Learn. Res. **7**, 1493–1515 (2006)
7. Platt, J.C.: Sequential minimal optimization: a fast algorithm for training support vector machines. Adv. Kernel MethodsSupport Vector Learn. **208**, 1–21 (1998)
8. Shalev-Shwartz, S., Singer, Y., Srebro, N., Cotter, A.: Pegasos: primal estimated sub-gradient solver for SVM. Math. Program. **127**(1), 3–30 (2011)
9. Tipping, M.E.: Sparse Bayesian learning and the relevance vector machine. J. Mach. Learn. Res. **1**, 211–244 (2001)
10. Zhu, J., Rosset, S., Hastie, T., Tibshirani, R.: 1-norm support vector machines. Adv. Neural Inf. Process. Syst. **16**(1), 49–56 (2004)

Multi-biometric Template Protection on Smartphones: An Approach Based on Binarized Statistical Features and Bloom Filters

Martin Stokkenes[1]([envelope]), Raghavendra Ramachandra[1], Kiran B. Raja[1], Morten Sigaard[2], Marta Gomez-Barrero[3], and Christoph Busch[1]

[1] Norwegian University of Science and Technology, Trondheim, Norway
{martin.stokkenes2,raghavendra.ramachandra,kiran.raja,
christoph.busch}@ntnu.no
[2] Denmark Technical University, Kongens Lyngby, Denmark
Mortenkrsi@hotmail.com
[3] Hochschule Darmstadt, Darmstadt, Germany
marta.gomez-barrero@h-da.de

Abstract. Widespread use of biometric systems on smartphones raises the need to evaluate the feasibility of protecting biometric templates stored on such devices to preserve privacy. To this extent, we propose a method for securing multiple biometric templates on smartphones, applying the concepts of Bloom filters along with binarized statistical image features descriptor. The proposed multi-biometric template system is first evaluated on a dataset of 94 subjects captured with Samsung S5 and then tested in a real-life access control scenario. The recognition performance of the protected system based on the facial characteristic and the two periocular regions is observed equally good as the baseline performance of unprotected biometric system. The observed Genuine-Match-Rate (GMR) of 91.61% at a False-Match-Rate (FMR) of 0.01% indicates the robustness and applicability of the proposed system in everyday authentication scenario. The reliability of the system is further tested by engaging disjoint subset of users, who were tasked to use the proposed system in their daily activities for a number of days. Obtained results indicate the robustness of the proposed system to preserve user privacy while not compromising the inherent authentication accuracy without protected templates.

1 Introduction

Biometric recognition involves verifying a claimed identity based on physiological (e.g. face, fingerprint, veins, etc.) and behavioural characteristics (keystroke, mouse dynamics, gait, etc.). Biometrics has become a part of everyone's daily life. The reliability and the accuracy of biometrics based systems, especially in constrained conditions have shown the benefit for applications such as border control, access control, forensics, etc. In spite of broad deployment of biometric

© Springer International Publishing AG 2017
C. Beltrán-Castañón et al. (Eds.): CIARP 2016, LNCS 10125, pp. 385–392, 2017.
DOI: 10.1007/978-3-319-52277-7_47

systems, the accuracy of such systems has remained challenging in unconstrained conditions. Further, the vulnerability of biometric systems being attacked at various levels (sensor, comparison etc.) and the privacy concerns regarding stored biometric templates create a psychological barrier in accepting biometrics in the general sphere of life.

However, access control systems have started to employ smartphone based biometric authentication using face [5], iris [9] and periocular [7] to authenticate the user. Such systems further gained mainstream attention as major smartphone vendors like Apple, Samsung, HTC, etc. provided fingerprint-based authentication that can be used to unlock the smartphone and convenient access control. However, as the ease of presentation attacks on such sensors is known [2], fingerprint recognition should be used in applications including banking and finance transactions only if the transaction volume is limited. For higher transaction volumes face and periocular recognition are a better choice.

As the number of applications using smartphone access control continues to increase, the need to protect the biometric templates when stored on the device has become a task of utmost importance to preserve the privacy of the smartphone user. Even though device manufacturers have claimed to protect the biometric data by keeping the biometric templates in the secured hardware separated from the rest of the device, one cannot rule out the possibility of indirect attacks to retrieve stored biometric templates - for instance fingerprint templates stored in the smartphone [14]. The recent work reported in [14] has exposed the security loop-hole in HTC One Max smartphone where the fingerprint template was accessed from smartphone. Similar attacks carried out on Samsung Galaxy S5 smartphone to avail the fingerprint template further exemplifies the lapse of hardware based security [14]. These incidents illustrate the demand for an efficient biometric template protection scheme to ensure the user's privacy and to avoid misuse of the stolen biometric templates on smartphones.

Recently significant progress has been made in smartphone based biometrics. The majority of the work is devoted to exploring the embedded sensors and designing efficient feature extraction and comparison algorithms of low complexity that are suitable for smartphones [1,3]. However to the best of our knowledge, there is no reported work addressing biometric template protection on smartphones. In general, biometric template protection, which is one of the widely explored areas in biometrics, has resulted in a significant number of algorithms [10]. A very promising template protection approach using Bloom filters was proposed recently [11] and studied for different biometric characteristics such as iris [11] and face [12]. These studies have shown that it is possible to obtain privacy preserving protection which is not degrading the verification performance. Motivated by the lapse of hardware based smartphone security and leveraging the secure nature of Bloom filters for biometric templates, we introduce this template protection scheme for a smartphone based multi-biometric system. Our intention is to create the right balance between the accuracy and privacy, both of which are essential blocks for the success of smartphone based secure access control applications employing biometrics.

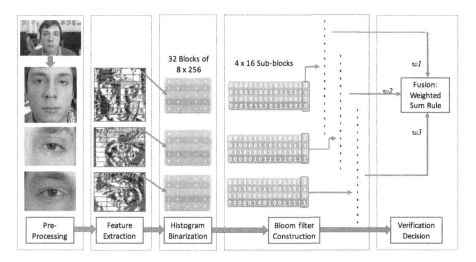

Fig. 1. Block diagram of the proposed multi-biometric template protection scheme for smartphones

Template protection schemes can only be considered secure if they adhere to the principles of Irreversibility and Unlinkability, which are requirements according to the ISO/IEC 24745 standard[1].

Irreversibility - It should not be possible to reconstruct from a protected biometric template the original template or a biometric sample.

Unlinkability - It should be infeasible to link a subject's protected biometric template from one application to another.

The rest of the paper is organized in the following manner: Sect. 2 introduces the proposed approach for multi-biometric template protected system on smartphone. Section 3 provides the details on the experimental set-up, evaluation protocols and the obtained results. Section 4 provides the concluding remarks of the current work and identifies the possible future work.

2   Proposed Multi-biometric Approach

This section provides the details of the proposed approach to enable a multi-biometric template protection system on smartphone. The proposed system includes the extraction of statistical image features from the captured face and periocular images and binarization of the extracted feature vectors followed by the transformation to protected templates using Bloom filters. The choice of Bloom filters is motivated by the non-degraded biometric performance while securing the biometric template as reported in [4].

---

[1] International Organization for Standardization, ISO/IEC 24745:2011, 'Information technology – Security techniques – Biometric information protection', 2011.

Figure 1 shows the block diagram of the proposed biometric template protection scheme. The proposed system consists of three principal steps: (1) Segmentation of face and periocular region (left and right) from the captured image. (2) Generalized feature extraction and binarization scheme to extract the discriminant binary features from all segmented image regions. (3) Template protection using Bloom filters.

2.1   Face and Periocular Segmentation

In this work, we have carried out the face and periocular segmentation based on the Haar cascade classifier [13]. We have implemented two different classifiers as proposed in [8] to segment the face and periocular region (left and right) in real-time. Figure 1 shows an example and the results for the segmented face and periocular regions. The segmented face images are resized to 64×80 pixels and the periocular image to a size of 120×88 pixels that will be used for feature extraction. The parameters used for the resize step are selected based on the testing database (see Sect. 3) by considering both accuracy and the need for low complexity computations on the smartphone.

2.2   Generalized Feature Extraction and Binarization

The selection of a suitable feature extraction technique plays an important role to achieve both biometric template protection and also the expected verification performance on the smartphone. The Bloom filter, as many other template protection approaches, requires binary feature vectors as input. The main challenge is to use a low computation based binary feature extraction scheme that can provide good verification accuracy on both face and periocular biometrics. In this work we explore the binarized statistical image features (BSIF) [6] that can provide binary features by performing convolution with a number of filters that are learned on natural images. These statistically independent filters are optimized by exploring natural image statistics via unsupervised learning using Independent Component Analysis (ICA). Depending upon the various patch sizes and the number of ICA basis functions, one can learn a number of BSIF filters with various sizes (adjusted to the patch size) and bits (number of basis selected from ICA). In this work, we propose a feature extraction scheme based on eight different filter sizes such as: 3×3, 5×5, 7×7, 9×9, 11×11, 13×13, 15×15 and 17×17 and each filter is of 8 bit in length. The choices of the BSIF filter size and length is based on empirical trials on a testing dataset yielding higher verification accuracy (see Sect. 3).

Given a face (or periocular) image captured using the smartphone, we first divide the entire image into 32 different blocks such that each block size is of 8×20 pixels in the case of the face image and 15×22 pixels in the case of a periocular image. Then, for each block, we extract the BSIF features by convolving eight different BSIF filters each of 8-bit length. Finally the response of each filter with 8-bit size is encoded to a histogram of dimension 1×256. Since we have eight different filters, the feature size for each block is 8×256. We repeat this process for each block and concatenate the result to a final BSIF feature vector set of dimension 65536. The extracted histogram features h is binarized as follows:

$$h(i,j) = \begin{cases} 1 & \text{if } h(i,j) > 0 \\ 0 & \text{if } h(i,j) = 0 \end{cases} \tag{1}$$

We finally reorganize the binarized features to have a matrix of dimension 4×8 rows and 8×256 columns that can be used to generate a protected template using Bloom filter [4][2].

2.3   Biometric Template Protection: Bloom Filters

In this work we have employed the Bloom filter by considering its irreversibility property in generating protected biometric templates [4]. The Bloom filter parameters are defined as follows: $nBits$ denotes the number of bits used to address a location in a Bloom filter. $nWords$ denotes the number of inputs to each Bloom filter. The length of each Bloom filter is defined by 2^{nBits}. Based on recommendations in [4] we use $nBits = 4$, which makes the length of each Bloom filter 16-bits. The number of inputs to one Bloom filter is restricted to $nWords \leq 2^{nBits}$. In our case we set $nWords = 16$.

Input to each Bloom filter is derived in the following manner: As described in the previous section, each template consists of 32 blocks of 8×256 binary elements, $x_{ij} \in 0,1$. To address a location in a Bloom filter of length 16, we further divide each of the 32 blocks into 4×16 sub-blocks where each column (containing 4 binary values) is used as a codeword denoting an index in a Bloom filter. The value in the location indicated by the index is then set to 1. This approach will generate 1024 Bloom filters of length 2^{nBits}, achieving a final template of 2 kB.

To compare two templates we obtain a dissimilarity score using the distance metric as shown in Eq. 2 where $|b|$ is the hamming weight or number of bits set to one in a Bloom filter, and $HD(b_i, b_j)$ is the Hamming Distance between two Bloom filters or number of disagreeing pairs of bits [4].

$$DS(b_i, b_j) = \frac{HD(b_i, b_j)}{|b_i| + |b_j|} \tag{2}$$

The unprotected and protected templates for face and periocular (left and right) regions are processed independently to obtain three separate dissimilarity scores. The final decision result is obtained by taking the majority vote on the decisions from each of the three scores.

3   Experimental Setup and Results

In this section we present the results of the proposed biometric template protection scheme for face and periocular biometrics on a smartphone. First, we evaluate the proposed system on a newly collected multi-biometric database using

---

[2] The reorganization of the features will ease the implementation of the Bloom filter protection technique.

Table 1. Quantitative results of proposed scheme on testing database

Verification performance				
Characteristic	Without Bloom filter		With Bloom filter	
	GMR@FMR = 0.01%	EER%	GMR@FMR = 0.01%	EER%
Face	90.05	1.67	82.63	2.90
Left periocular	83.32	3.20	72.22	5.39
Right periocular	83.78	4.53	68.03	5.48
Fused	95.95	1.12	91.61	1.80

the Samsung Galaxy S5 smartphone constituting of 94 subjects. Table 1 shows the results from evaluation on the testing database in terms of Genuine Match Rate (GMR) for a fixed value of False Match Rate (FMR) and Equal Error Rate (EER). We observe that there is some performance degradation after Bloom filter is applied. However, with fusion of the different biometric characteristics we obtain a GMR of 91.61% when FMR = 0.01%, indicating robustness and applicability of the proposed template protection scheme on smartphones for multi-biometrics. Then, a real-time evaluation, described in the next section, is carried out by providing the template protected biometric system for access control to 22 subjects on a Samsung Galaxy S5.

3.1   Real-Time Evaluation of the Proposed System

In this experiment the smartphone application was distributed on a Samsung S5 phone to 22 unique subjects along with the instruction to use the proposed system in order gain access to a secure application using multi-biometric (face and periocular) recognition. The threshold for genuine accept rate was set on the basis of threshold obtained on testing set at an FAR of 0.1%. The subjects were instructed to enrol into the system with 10 images and they were further instructed to test the application independently in various lighting conditions for a number of times.

On average 10 genuine attempts were recorded for the entire subset of users. The users were also asked to attempt to login as impostors to some other subjects than themselves to gauge the robustness of the system against zero-effort impostor attempts. These subjects repeatedly tried to gain impostor access for a period of 4 days. Experiments were carefully designed and the users were asked to attack target subjects that had the same ethnical background, for instance, an American subject was asked to pose as another compatriot to account for the robust impostor attacks.

Figure 2a shows the number of successful attempts when the system was not operating with protected templates and Fig. 2b shows the impostor attempts. The section indicated in the green color signifies the successful genuine attempts while the sections in red color indicate the rejection of the attempts for both genuine subjects and impostors. It can be observed from Fig. 2c and 2d that the

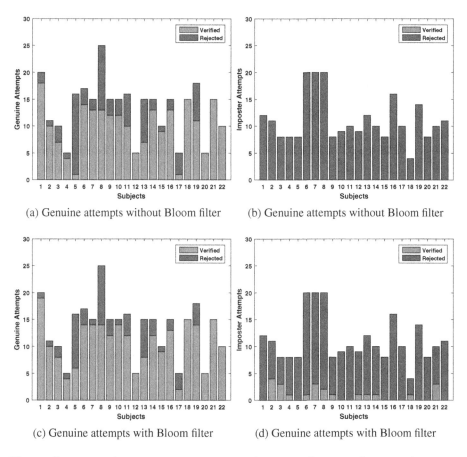

(a) Genuine attempts without Bloom filter   (b) Genuine attempts without Bloom filter

(c) Genuine attempts with Bloom filter   (d) Genuine attempts with Bloom filter

Fig. 2. Genuine and impostor attempts in real-time application of proposed system. Green color indicated number of successful attempts and red color indicates the number of failed attempts to login to the system. (Color figure online)

system with the proposed template protection scheme works under operational conditions as good as the system without template protection. Very equivalent performance in both cases indicates the robust nature of the proposed template protection scheme while not compromising the verification accuracy significantly.

Based on the extensive analysis conducted from the above experiments, it can be observed that the proposed template protection system for multi-biometric based smartphone application is promising with an acceptable degradation of verification accuracy.

4 Conclusion

Preserving privacy of the smartphone user's biometric templates is of utmost importance as the biometric characteristics cannot be renewed in case of compromise arising out of attacks on the biometric references. Wide spread use of smartphones for biometrics based authentication further justifies the need for protecting the templates on smartphones. In this work we have proposed a novel system using the strengths of Bloom filter and generalizability of BSIF on both face and periocular characteristics to devise a template protected biometric system for smartphones. Extensive experiments conducted on the database of 94 subjects indicate high verification accuracy even with protected templates. The GMR of 91.61% justifies the applicability of the proposed system for a real-life authentication scenario. Further, evaluation using disjoint set of subjects to test the reliability of system indicates the robust nature of proposed system against zero impostor attempts even with same ethnical origin.

References

1. Barra, S., Casanova, A., Narducci, F., Ricciardi, S.: Ubiquitous iris recognition by means of mobile devices. Pattern Recogn. Lett. **57**, 66–73 (2015)
2. Cao, K., Jain, A.: Hacking mobile phones using 2D printed fingerprints. Michigan State University, MSU, Technical report (2016)
3. De Marsico, M., Galdi, C., Nappi, M., Riccio, D.: Firme: face and iris recognition for mobile engagement. Image Vis. Comput. **32**(12), 1161–1172 (2014)
4. Gomez-Barrero, M., Rathgeb, C., Galbally, J., Busch, C., Fierrez, J.: Unlinkable and irreversible biometric template protection based on bloom filters. Inf. Sci. **370**, 18–32 (2016). http://www.sciencedirect.com/science/article/pii/S0020025516304753
5. Ijiri, Y., Sakuragi, M., Lao, S.: Security management for mobile devices by face recognition. In: 7th International Conference on Mobile Data Management, MDM 2006, p. 49, May 2006
6. Kannala, J., Rahtu, E.: BSIF: Binarized statistical image features. In: 21st International Conference on Pattern Recognition (ICPR), pp. 1363–1366 (2012)
7. Raja, K.B., Raghavendra, R., Stokkenes, M., Busch, C.: Smartphone authentication system using periocular biometrics. In: 2014 International Conference of the Biometrics Special Interest Group (BIOSIG), pp. 1–8, September 2014
8. Raja, K.B., Raghavendra, R., Stokkenes, M., Busch, C.: Multi-modal authentication system for smartphones using face, iris and periocular. In: Proceedings of 2015 International Conference on Biometrics, ICB 2015, pp. 143–150 (2015)
9. Marsico, M.D., Galdi, C., Nappi, M., Riccio, D.: Firme: face and iris recognition for mobile engagement. Image Vis. Comput. **32**(12), 1161–1172 (2014)
10. Nandakumar, K., Jain, A.K.: Biometric template protection: bridging the performance gap between theory and practice. IEEE Sig. Process. Mag. **32**(5), 88–100 (2015)
11. Rathgeb, C., Breitinger, F., Busch, C.: Alignment-free cancelable iris biometric templates based on adaptive bloom filters. In: Proceedings - 2013 International Conference on Biometrics, ICB 2013 (2013)
12. Rathgeb, C., Gomez-Barrero, M., Busch, C., Galbally, J., Fierrez, J.: Towards cancelable multi-biometrics based on bloom filters: a case study on feature level fusion of face and iris. In: 2015 International Workshop on Biometrics and Forensics (IWBF), pp. 1–6, March 2015
13. Viola, P., Jones, M.: Robust real-time face detection. Int. J. Comput. Vis. **57**(2), 137–154 (2004)
14. Zhang, Y., Chen, Z., Xue, H., Wei, T.: Fingerprints on mobile devices: abusing and eaking. In: Black Hat Conference (2015)

Computing Arithmetic Operations on Sequences of Handwritten Digits

Andrés Pérez[1], Angélica Quevedo[2], and Juan C. Caicedo[2(✉)]

[1] Universidad Nacional de Colombia, Bogotá, Colombia
afperesm@unal.edu.co
[2] Fundación Universitaria Konrad Lorenz, Bogotá, Colombia
{angelicad.quevedoc,juanc.caicedor}@konradlorenz.edu.co

Abstract. This paper studies the problem of sequential visual processing to solve arithmetic operations using handwritten digits. We feed a sequence of digits with an arithmetic operator to a trained system, and then ask for the resulting symbolic answer. All digits and operators in the input sequence are images, while the output is a real number rounded up. The proposed architecture is a hybrid recurrent-convolutional network with a regression module that is trainable end-to-end. The experimental results show that the proposed architecture is able to add or subtract sequences of up to five elements with high accuracy, and that long sequences require long training times.

Keywords: Digit operations · Arithmetics · Recurrent networks · Convolutional networks · Regression

1 Introduction

Our capacity to recognize numbers and the ability to do arithmetics with them is remarkable. We use it everyday to calculate bills, check available cash, estimate times and dates, and many other activities. Although machines are more precise and faster than humans to operate numbers, we can recognize them visually in a variety of forms and apply abstract concepts over them. An example of this is the ability to perform arithmetic operations involving the visual interpretation of symbol sequences.

In this work, we study the problem of sequential processing of visual signals. In particular, we study the problem of computing arithmetic operations on sequences of handwritten digits, as a proxy task to understand the difficulty of sequential-visual analysis tasks. At the core of these problems is the ability to recognize a set of objects, analyze their spatial arrangement, and apply some reasoning to make a decision.

We propose and evaluate a deep learning architecture to solve arithmetic operations from visual information. The input to this architecture is a sequence of images with handwritten digits and handwritten symbols of addition or subtraction. The output is a single decimal number representing the result of the

© Springer International Publishing AG 2017
C. Beltrán-Castañón et al. (Eds.): CIARP 2016, LNCS 10125, pp. 393–400, 2017.
DOI: 10.1007/978-3-319-52277-7_48

depicted operation. We evaluate the proposed architecture using sequences of different lengths to assess its robustness for long-term recognition. Our results indicate that the architecture can predict the correct output for sequences of length 3 and 5, but longer sequences are harder to train in a reasonable amount of time.

2   Previous Work

Image classification and object recognition are vision problems that have experienced unprecedented progress during the last few years thanks to the resurgence of convolutional networks [7]. Originally proposed during the late 80s [8], convolutional networks are a differentiable model that makes multiple non-linear transformations to an image to extract relevant features for a particular task.

Recurrent Neural Networks (RNNs) are a family of models that have shown excellent results for modeling sequential data. They are being actively investigated for language modeling [9], and also incorporated into vision systems to predict sentences [6], segment images [12], and analyze video [1]. One of the most interesting properties of RNN models such as the Long-Short Term Memory (LSTM) [3] is that they can learn to remember long-term dependencies of sequential data. In this way, a single observation made at time t may change the whole interpretation of the input sequence or signal. In our work, we evaluate the capacity of LSTM models to understand the order of digits in a sequence and produce correct interpretations associated to the corresponding arithmetic operation.

Digit classification is one of the most widely studied problems in the machine learning community [11], other related studies include image generation [2], visual attention models [10], and spatial transformations of objects [5]. In this frame the closest study our work is the one proposed in [4] where the authors use a deep neural network (DNN) to process two input images, each showing a 7-digit number, in order to produce an image displaying the number result of an arithmetic operation. Unlike their model, our proposal works on sequences of varying length composed by images of handwritten digits and arithmetic symbols rather than images of fixed length numbers made of digits written electronically in a standard font. However both models are incomparable since their experimental conditions (input, output and objective function) are fundamentally different.

3   Handwritten Arithmetics with Deep Learning

3.1   Problem Description

Assume two participants involved in the task of solving an arithmetic operation. The first participant writes the operation by hand using a pen, while the second participant observes from left to right each character of the operation one at a time, and computes the result. The second participant has to remember important information of the sequence to interpret the numbers and the operation correctly in order to compute the result. In our setup the first participant is a computer program that generates arithmetic operations and feeds them as sequences to the second participant which is the proposed algorithm.

An easy way to model this problem with high accuracy would be to use a handwritten digit classifier and a module engineered to keep the prediction results and solve the operation deterministically. However, we want to study the properties of a system that has to *learn* the process of remembering, interpreting the input, and associating the correct output. The design of such system is presented in the following subsections.

3.2   Deep Learning Architecture

We propose a deep learning architecture with three main components: a convolutional network, a recurrent network, and a regressor. The convolutional network takes as input images of handwritten digits and symbols and extracts from them visual features. The recurrent network (RNN) reads these visual features in order to recognize the most important information and to keep a compact representation of the entire sequence. Finally, the regressor reads and interprets the last representation of the sequence to produce a real number as result. Figure 1 depicts the proposed architecture.

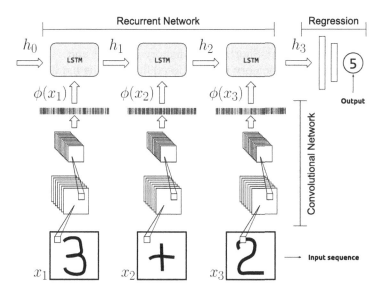

Fig. 1. Architecture for visual arithmetics. The inputs x_i are images of handwritten digits and symbols, which are processed independently by a convolutional network. The feature representation $\phi(x_i)$ of each digit is processed by an LSTM that encodes a latent state h_i sequentially. The last state of the recurrent network is read by the regression network to produce an output.

Convolutional Network. The convnet is a function that builds a feature embedding for an image. We use a three-layer convnet that takes inputs of 28×28 pixels with a single gray-scale channel. The first two are convolutional layers that transform the input into a feature map that applies the ReLU non-linearity ($R(x) = max(0, x)$) and a 2×2 max-pooling operation. The first layer has 32 filters of 5×5 pixels and the second layer has 64 filters also of 5×5 pixels. At last there is a fully connected layer that embeds the output into a 1.024 feature vector.

Recurrent Network. An RNN models a dynamic system whose inputs vary with time and can take a sequence $X = \{\phi(x_1), \phi(x_2), \ldots, \phi(x_T)\}$ of arbitrary length T. The goal of an RNN is to progressively encode information of the sequence in a vector representation called *hidden state*. Each element of X is processed sequentially, and the output is the new hidden state h_t that depends on the previous state h_{t-1} and the current observation $\phi(x_t)$. A simple RNN function can be modeled as

$$h_t(\phi(x_t), h_{t-1}) = \tanh\left(W^r \phi(x_t) + U h_{t-1}\right), \tag{1}$$

where W^r and U are linear transformations of the inputs, and tanh is an element-wise nonlinear activation function. Simple RNN models are difficult to train when input sequences are long because the gradient tends to vanish through time. However, alternative formulations of RNN functions have been proposed, which have trainable memory mechanisms for dealing with longer sequences. In particular, we use the Long-Short Term Memory (LSTM) unit.

An LSTM unit has a memory vector c in addition to the hidden state h, and it implements *gates* that allow reading, writing and resetting information into the memory vector. The memory and hidden state vectors are formulated as

$$c_t(\phi(x_t), c_{t-1}) = f \odot c_{t-1} + i \odot g \qquad h_t(\phi(x_t), h_{t-1}) = o \odot \tanh(c_t), \tag{2}$$

where c_t is the memory content at time t, i and f are functions that control writing and resetting the memory content, g is the transformed input, and o is a function that controls the output to the hidden state. Notice that the writing (i), resetting (f), and reading (o) gates have element-wise multiplicative interaction ($\odot$) with the information in the memory vector.

We adopt the LSTM architecture as the recurrent network in our model, using a hidden state $h \in \mathbb{R}^{512}$, and input vectors $\phi(x) \in \mathbb{R}^{1024}$ produced by the convnet. The LSTM architecture is useful in the problem of processing sequences of handwritten digits, because the memory structure allows to preserve information of the order in which digits appear in the sequence, and also the position of the arithmetic operator. Getting this information right in the final representation is crucial to produce a correct interpretation of the operation.

Regression Network. The final component of our architecture is a network that reads the last state of the recurrent network (h_T) and computes the result of the arithmetic operation. We use a two-layer, fully connected network to transform h_T into a single real number. This network is supervised with respect to the true result of an arithmetic operation using the regression loss function

$$\mathcal{L}(y, \hat{y}) = \begin{cases} |y - \hat{y}| - 0.5 & \text{if } |y - \hat{y}| > 1 \\ 0.5(y - \hat{y})^2 & \text{otherwise} \end{cases}, \tag{3}$$

where y is the ground truth result. This loss function is also known as smooth-$L1$ loss, and has very stable gradients for regression problems of arbitrary output. It has constant gradient if the absolute value of the error is greater than 1, and linear gradient when the error is smaller than 1. It works well in our arithmetic operations problem, because large sequences usually result in outputs with big numbers.

4 Experiments and Results

We evaluated the proposed architecture with three setups considering only additions, only substractions and a mix considering both. In all cases, we conducted experiments with sequences involving two operands and one operator. Both operands are natural numbers with the same number of digits. We considered operands with one, two and three digits each.

4.1 Dataset and Sequences

To generate sequences for feeding the proposed architecture we use the MNIST handwritten digit database, which is composed by 55.000 training examples and 10.000 testing examples. This dataset was extended with a set of arithmetic symbol images of which 144 are for training and 36 for testing. Each dataset example is a 28×28 pixels gray scale image.

We generate sequences of arithmetic operations as follows, the operator is chosen from the set of symbols depending on the model that we are training, and the operands are constructed digit by digit. At each position of the sequence, we sample a random digit from the MNIST database with uniform probability. Corresponding ground truth is also calculated. Both training and testing sequences were generated following this procedure, but using separate sets of data.

4.2 Training

The CNN was pre-trained to classify each of the 10 digits with a softmax classification layer. This layer was connected to the last fully connected layer of the proposed architecture with 1.024 features. The softmax classifier was removed, and the weights of the network were used as initialization values. This architecture was trained end-to-end using the Adam optimization algorithm with a learning rate of 0,0001. Training was run until achieving a training accuracy of at least 95%. We used mini-batches of 64 sequences, each sequence with 3 to 7 images. The hyper-parameters of the network were cross-validated until a stable configuration was found.

Table 1 presents an overview of the training results for the considered models. Notice how longer sequences need longer training times to reach a useful solution. This is explained by the large amount of examples that an operation can have as we add more digits to the operands. The combination of digits leads to an exponential growth on the number of potential operations to consider, making harder the problem of training the model for long sequences.

Table 1. Training configuration of the three trained models

Sequence length	Addition		Subtraction		Add+Sub	
	Training iterations	Training time	Training iterations	Training time	Training iterations	Training time
3	50,000	1 h 36 min	30,000	57 min	30,000	57 min
5	60,000	3 h 7 min	30,000	1 h 33 min	30,000	1 h 33 min
7	150,040	10 h 48 min	150,190	10 h 51 min	150,910	10 h 52 min

Figure 2 shows the performance of the addition and subtraction model for the three considered lengths of sequences, a similar behavior was described for the only addition and only subtraction models. First thing to notice is that mini-batch error is inversely proportional with the training accuracy as the number of iterations grows. It can also be observed that for longer sequences, training error is higher during the whole training session. This is due to the range of the regression output, as well as the difficulty of training longer sequences. Interestingly, subtraction reaches the best performance in less iterations than addition, and similarly happens with the model that computes both addition and subtraction.

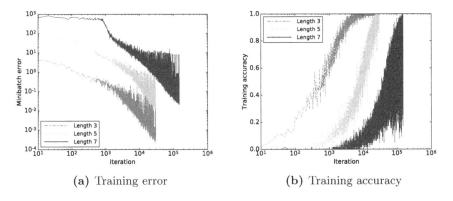

(a) Training error   (b) Training accuracy

Fig. 2. Effect of varying sequence length during training for the addition and subtraction model.

4.3 Testing

In general, all tested models are relatively successful as shown in Table 2. Accuracy was computed by rounding up the predictions and comparing the percent of matches with the ground truth. For sequences of length 3 we reach accuracy above 97%, more than 90% for sequences of length 5, and between 75% and 85% for sequences of length 7. Longer sequences are more difficult to interpret possibly due to the lack of robustness of the regression network.

Table 2. Test results on the held-out dataset for the three models.

	Addition		Subtraction		Add+Sub	
Sequence length	Testing accuracy	Testing time	Testing accuracy	Testing time	Testing accuracy	Testing time
3	97.9%	0.0090	99.7%	0.0076	99.90%	0.0057
5	90.0%	0.0569	95.8%	0.0690	94.20%	0.104484
7	76.8%	8.636843	86.1%	0.0611	85.70%	0.540550

Among all three models, subtraction exhibits the best accuracy. The testing accuracy of the addition-subtraction model is slightly lower than the testing accuracy for subtraction only. This can be attributed to the introduction of addition operations, whose symbol seems to be harder to identify than the subtraction symbol.

Table 3 depicts some sample test sequences. The failure cases for short sequences are typically due to a few decimals from the ground truth. However for longer sequences the error can be off by a large margin. Although our goal was never to achieve 100% accuracy on the output prediction, but rather to study the problem of visual sequences analysis in the context of RNNs, the proposed system was still able to achieve a good performance, motivating further studies in this direction in problems where needs to be taken into account long-term dependencies in visual sequential data.

Table 3. Success and failure examples of test sequences.

Input	Prediction	Rounded	Ground truth	Result
2+7	9.05	9	9	Success
97+88	184.69	185	185	Success
502+924	1426.26	1426	1426	Success
8+0	8.71	9	8	Failure
70+45	116.10	116	115	Failure
701+294	106.77	107	955	Failure

5   Conclusions and Future Work

This work presented and evaluated an architecture to compute arithmetic operations from sequences of hand-written digits. The model is very successful at predicting the correct output for sequences with operands of one or two digits each. However, the model becomes harder to train for longer sequences, reaching a moderately successful result for operands with three digits. The proposed architecture has also shown the effectiveness of RNN models for visual sequence analysis. In particular LSTM capabilities to memorize and summarize sequential information. In our future work, we aim to explore more efficient training strategies to optimize the exploration of combinations of operations of certain length. We also plan to experiment with other operators in arbitrary positions of the sequence.

References

1. Donahue, J., Anne Hendricks, L., Guadarrama, S., Rohrbach, M., Venugopalan, S., Saenko, K., Darrell, T.: Long-term recurrent convolutional networks for visual recognition and description. In: Proceedings of the IEEE Conference on Computer Vision and Pattern Recognition, pp. 2625–2634 (2015)
2. Gregor, K., Danihelka, I., Graves, A., Wierstra, D.: Draw: a recurrent neural network for image generation. arXiv preprint arXiv:1502.04623 (2015)
3. Hochreiter, S., Schmidhuber, J.: Long short-term memory. Neural Comput. **9**(8), 1735–1780 (1997)
4. Hoshen, Y., Peleg, S.: Visual learning of arithmetic operations. arXiv preprint arXiv:1506.02264 (2015)
5. Jaderberg, M., Simonyan, K., Zisserman, A., et al.: Spatial transformer networks. In: Advances in Neural Information Processing Systems, pp. 2008–2016 (2015)
6. Karpathy, A., Fei-Fei, L.: Deep visual-semantic alignments for generating image descriptions. In: Proceedings of the IEEE Conference on Computer Vision and Pattern Recognition, pp. 3128–3137 (2015)
7. Krizhevsky, A., Sutskever, I., Hinton, G.E.: Imagenet classification with deep convolutional neural networks. In: Advances in Neural Information Processing Systems, pp. 1097–1105 (2012)
8. LeCun, Y., Boser, B., Denker, J.S., Henderson, D., Howard, R.E., Hubbard, W., Jackel, L.D.: Backpropagation applied to handwritten zip code recognition. Neural Comput. **1**(4), 541–551 (1989)
9. Mikolov, T., Karafiát, M., Burget, L., Cernocký, J., Khudanpur, S.: Recurrent neural network based language model. In: INTERSPEECH, vol. 2, p. 3 (2010)
10. Mnih, V., Heess, N., Graves, A., et al.: Recurrent models of visual attention. In: Advances in Neural Information Processing Systems, pp. 2204–2212 (2014)
11. Sermanet, P., Chintala, S., LeCun, Y.: Convolutional neural networks applied to house numbers digit classification. In: 2012 21st International Conference on Pattern Recognition (ICPR), pp. 3288–3291. IEEE (2012)
12. Zheng, S., Jayasumana, S., Romera-Paredes, B., Vineet, V., Su, Z., Du, D., Huang, C., Torr, P.H.: Conditional random fields as recurrent neural networks. In: Proceedings of the IEEE International Conference on Computer Vision, pp. 1529–1537 (2015)

Automatic Classification of Non-informative Frames in Colonoscopy Videos Using Texture Analysis

Cristian Ballesteros[1], Maria Trujillo[1(✉)], Claudia Mazo[1], Deisy Chaves[1], and Jesus Hoyos[2]

[1] Multimedia and Computer Vision Group, Universidad del Valle, Cali, Colombia
{cristian.ballesteros,maria.trujillo,
claudia.mazo,deisy.chaves}@correounivalle.edu.co
[2] Hospital Universitario del Valle Evaristo Garcia ESE, Cali, Colombia

Abstract. Colonoscopy is the most recommended test for preventing/detecting colorectal cancer. Nowadays, digital videos can be recorded during colonoscopy procedures in order to develop diagnostic support tools. Once video-frames are annotated, machine learning algorithms have been commonly used in the classification of normal-vs-abnormal frames. However, automatic analysis of colonoscopy videos becomes a challenging problem since segments of a video annotated as abnormal, such as cancer or polypos, may contain blurry, sharp and bright frames. In this paper, a method based on texture analysis, using Local Binary Patterns on the frequency domain, is presented. The method aims to automatically classify colonoscopy video frames into either *informative* or *non-informative*. The proposed method is evaluated using videos annotated by gastroenterologists for training a support vector machines classifier. Experimental evaluation shown values of accuracy over 97%.

Keywords: Colonoscopy videos · Discrete Fourier Transform · Local Binary Pattern · Support Vector Machines

1 Introduction

Automatic detection of polyps and cancer, in colonoscopy videos, is frequently based on machine learning algorithms using supervised learning [3,12,13]. Machine learning algorithms build models based on training sets and those training sets are created with annotations manually made by gastroenterologists. Those annotations are done by segmenting colonic video files into shots representing endoscopic findings (lesions). For instance, a file annotation of an observed lesion (cancer) is 10:14:30 (time begin) and 10:16:27 (time end). It is highly possible that shots annotated as lesion contain frames with blur, low contrast, noise and/or brightness—called *non-informative*.

In fact, machine learning is an inverse and ill-posed problem, there is a set of assumptions that a learning algorithm makes about the true function that it is trying to learn a model off. In general, the training of a classifier should be done including a sufficiently big number of frames representing most of the

© Springer International Publishing AG 2017
C. Beltrán-Castañón et al. (Eds.): CIARP 2016, LNCS 10125, pp. 401–408, 2017.
DOI: 10.1007/978-3-319-52277-7_49

possible frame configurations. Otherwise the classifier is obviously bad trained and does not provide the expected result. In our case, annotations—of cancer or polyps—contain *non-informative* frames that may affect the learning of a model and, consequently, produce a classifier with low accuracy. Thus, a preprocessing is needed before annotating colonoscopy videos and building machine learning models for the classification of normal-vs-abnormal frames.

Research has been conducted on identifying *non-informative* frames in colonoscopy videos. Methods based on edge detection and brightness segmentation in order to remove *non-informative* frames from colonoscopy videos are presented in [14,19]. Other techniques based on color transformations [10], lumen detection [6,9], video tracking framework [11], global features [8,20], and texture analysis [14] were proposed to address this problem.

Results reported in [14] indicate that precision, sensitivity, specificity, and accuracy for the edge-based and the clustering-based classification techniques are greater than 90% and 95%, respectively, using the specular reflection detection technique. However, the comparison done by Rungseekajee and Phongsuphap in [18] between their proposed approach and the edge-based classification in [14] yields lower values for the edge-based classification presented in [14]. In [18] precision 90%, sensitivity 61%, specificity 50%, and accuracy 60% are reported.

In [2], we proposed a method based on edge detection using the hypothesis that *non-informative* frames usually do not contain many edges. Experimental evaluation showed values of accuracy and precision over 95%. In this paper, we explore the use of texture analysis to automatically classify *informative* and *non-informative* frames, in colonoscopy videos. Local Binary Pattern (LBP) operator [15] is used as texture descriptor and it is calculated on the frequency domain, and the Support Vector Machines (SVM) [5] is used for building a classifier. The proposed method aims to preprocess data sets before being used in machine learning algorithms by eliminating *non-informative* frames. Experimental evaluation shown values of accuracy over 97%.

2   Automatic Classification of Non-Infomative Frames

For the sake of completeness, a definition of *non-informative* and *informative* frames is presented in order to clarify the meaning in the application domain. The former corresponds to a frame out-of-focus, containing bubbles and/or light reflection artifacts due to wall contact and/or light reflections on water used to clean the colon wall and/or motion blur. The latter corresponds to a frame with well-defined content and spread over whole frame.

A description of the proposed automatically identification of *non-informative* frames is as follows. Given a video sequence, each frame is converted into gray-level scale and gray-level frame is transformed into the frequency domain using the Discrete Fourier Transform (DFT). Then, the LBP operator is applied at each pixel and a histogram is built to represent the content per frame. Finally, a classifier is created using the linear-SVM algorithm.

2.1 Frequency Domain Transform

The results shown in Fig. 1 indicate that the *informative* frame contain more low frequencies than the higher ones. The *non-informative* frame contains components of all frequencies, with smaller magnitudes for higher frequencies. The *non-informative* frame also contains a dominating direction in the Fourier image, passing vertically through the centre. These originate from the regular patterns in the original frame. We can observe that the frequency domain provides discriminant information about the frame content.

Each frame is converted into gray-scale and then transformed into the Fourier domain using the following equation [1].

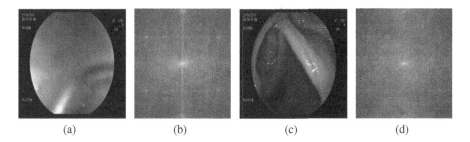

(a)   (b)   (c)   (d)

Fig. 1. Illustration of the frequency domain using a *non-informative* and an *informative* frames. (a) spatial domain of a *non-informative* frame, (b) frequencies spectrum of (a), (c) spatial domain of an *informative* frame and (d) frequencies spectrum of (c).

$$F(u,v) = \frac{1}{MN} \sum_{x=0}^{M-1} \sum_{y=0}^{N-1} f(x,y) \exp[-i2\pi(\frac{ux}{M} + \frac{uy}{N})], \tag{1}$$

where $M \times N$ is the frame dimension, $f(x,y)$ is a value at (x,y) position in the spatial domain and the exponential term is the basis function corresponding to each point $F(u,v)$ in the Fourier space. The equation can be interpreted as the value of each point $F(u,v)$ is obtained by multiplying the spatial domain with the corresponding base function and summing the result.

2.2 Texture Analysis

Initially, the texture analysis on the frequency spectrum was conducted using Haraclik features such as Angular Second Moment, Contrast, Correlation, Dissimilarity, Entropy, Energy and Uniformity [17], as it was done in [14]. However, the obtained results in experiments were not discriminant enough for the classification using SVM.

The Local Binary Pattern (LBP) is a common texture descriptor that contains several advantages, such as invariance and low computational cost [15]. In our approach, the LBP works on the frequency spectrum using a 3 × 3 kernel, where pixels around at the central pixel are thresholded with the value at the

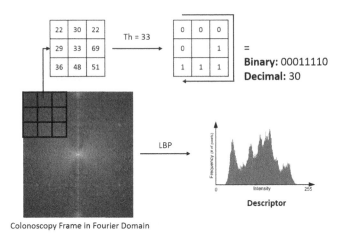

Fig. 2. Illustration of the LBP descriptor calculation on the frequency domain.

central pixel. A binary string is obtained as result of the LBP thresholding, it is converted into a decimal number and used to replace the central pixel value. Finally, a histogram of the LBP decimal numbers is calculated. The obtained histograms are used as frame content representations to classify colonoscopy frames. Figure 2 illustrates the LBP calculation.

2.3   Classification Using Support Vector Machines

The original algorithm for SVMs was proposed by Vapnik and Lerner in 1963 [21]. The algorithm solves a classification problem for linearly separable data. The algorithm finds a separating hyperplane that has the maximum distance from the closest input points. The hyperplane, if exists, is called *maximum-margin hyperplane*. The decision rule only depends on the dot product of the training vector and the unknown point.

We chose Support Vector Machines (SVMs) for building the classifier based on the problem, *i.e.* a binary classification, and the available training data, *i.e.* limited training dataset size. The training dataset are annotated by gastroenterologists and they usually have not spare time for this task.

3   Experimental Evaluation

The performance of the proposed method is evaluated in this section. Data set, experimental settings and evaluation criteria are presented. Tests were conducted using a Laptop with Windows 8 Pro X64, Intel (R) Core (TM) i5 @ 2.60 GHz and 4,00 GB RAM and the implementation was done using the C++ programming language. The classification of colonoscopy frames is performed using the LIBSVM Library [4] and its implementation requires the following parameters: a feature matrix that is constructed with the histograms of LBP descriptors; a column vector of labels that contain a class value. We used the class value 1 for *informative* frames and −1 for *non-informative* ones. And the cost value C; a low cost value, such as $c = 1$, has a minor sensibility in the classification errors than a big one.

Several videos of complete colonoscopic procedures were recorded at the Hospital Universitario del Valle and three videos were selected to present evaluation results. Those videos have length of 6, 12 and 14 min, respectively, frame resolution of 636×480 and they were recorded at 10 fps, using MP4 format and H264 compression. Frames are extracted—using the FFmpeg Multimedia Framework [7]—from video sequences. A total of 600 frames—taken 100 *informative* and 100 *non-informative* per video—were used. The selected frames were manually annotated by a gastroenterologist.

A set of metrics commonly used to evaluate the performance of a binary classification is employed [16]. The metrics are based on the confusion matrix and correspond to *Sensitivity, Specificity, Accuracy, Precision* and *F-Measure*.

3.1 Evaluation Conditions

The proposed method was evaluated using three evaluation conditions. The evaluation conditions are illustrated in the Fig. 3 and described as follows.

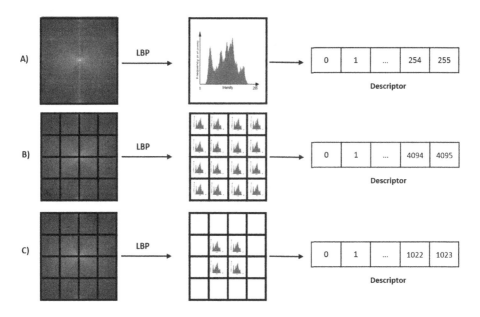

Fig. 3. Illustration of evaluation conditions of the proposed method.

(A) The LBP is calculated using the frequency domain image. The obtained descriptor is an array of 256 length.

(B) The frequency domain image is divided into 4 × 4 blocks and the LBP is calculated on each block. This yields 16 histograms that are concatenated. The obtained descriptor is an array of 4096 length.
(C) The frequency domain image is divided into 4×4 blocks and the 2×2 central blocks are used in the calculation of the LBP. This yields 4 histograms that are concatenated. The obtained descriptor is an array of 1024 length.

3.2 Results and Discussion

The SVM classification model is calculated using 80% of the data set—400 frames—as a training set and the remaining frames as test set—200 frames.

Table 1. Confusion matrices calculated under the three evaluation conditions

Evaluation condition	TI	FI	TN	FN
A	89	8	92	11
B	97	62	38	3
C	93	3	97	7

Table 1 contains the confusion matrices calculated under the three evaluation conditions and presents the number of frames correctly using TI for symbolising True Informative and TN for symbolising True Non-Informative, and the number of frames incorrectly classified using FI for symbolising False Informative and FN for symbolising False Non-Informative. The evaluation condition C yields the lowest false informative whilst the evaluation condition B yields the lowest false non-informative.

Having in mind that performance measures are calculated for assessing the capability of correctly classifying non-informative frames, Table 2 contains the obtained performance measure values using the evaluation conditions. The highest values of accuracy and precision were obtained under the evaluation condition C and the lowest values were obtained in the evaluation condition B.

Obtained results may be interpreted looking at Fig. 3 where frequency magnitudes get smaller for higher frequencies, that are located in the external blocks. In this way, LBP calculated using the four central blocks encodes the frequency domain.

Table 2. Performance metric values

Evaluation condition	Sensitivity (%)	Specificity (%)	Accuracy (%)	Precision (%)	F-measure (%)
A	89	92	90	92	90
B	97	38	67	61	75
C	93	97	95	97	95

Table 3 contains the results using the edge-based approach presented in [2]. Our proposed approach, using LBP computed at the central blocks in the frequency domain, outperforms the edge-based approaches proposed by us in [2] using the same training image set.

Table 3. Performance metrics reported in [2]

Algorithm based on	Sensitivity (%)	Specificity (%)	Accuracy (%)	Precision (%)
Edges	92	99	96	99

4   Final Remarks

In this paper, a method based on texture analysis was proposed to classify colonoscopy frames into two categories: *informative* and *non-informative*. The method uses the LBP descriptor as texture feature that are calculated on the frequency domain and the SVM as classifier.

The proposed classification method is able to correctly detect frames without relevant information, that should not be used for training machine learning algorithms in the classification of normal-vs-abnormal frames.

Moreover, the proposed method may be used to significantly reduce duration of videos—if frames classified as *non-informative* are deleted—before being analysed by gastroenterologists.

Metrics used to evaluate the performance of the proposed method shown that the accuracy and the precision are over 95% when the central blocks in the frequency domain are used to calculate the LBP descriptor and in general the proposed method outperforms the proposed approach in [2].

References

1. Acharya, T., Ray, A.K.: Image Processing: Principles and Applications. Wiley Interscience, Reading (2005)
2. Ballesteros, C., Trujillo, M., Mazo, C.: Automatic classification of non-informative frames in colonoscopy videos. In: 6th Latin American Conference on Networked and Electronic Media. LACNEM, facultad de Minas, Universidad Nacional sede Medellín (2015)
3. Bernal, J., Gil, D., Sánchez, C., Sánchez, F.J.: Discarding non informative regions for efficient colonoscopy image analysis. In: Luo, X., Reichl, T., Mirota, D., Soper, T. (eds.) CARE 2014. LNCS, vol. 8899, pp. 1–10. Springer, Heidelberg (2014). doi:10.1007/978-3-319-13410-9_1
4. Chang, C.C., Lin, C.J.: LIBSVM: a library for support vector machines. ACM Trans. Intell. Syst. Technol. **2**, 27:1–27:27 (2011). Software: http://www.csie.ntu.edu.tw/~cjlin/libsvm
5. Cortes, C., Vapnik, V.: Support-vector networks. Mach. Learn. **20**(3), 273–297 (1995)

6. Fanand, Y., Meng, M.Q., Li, B.: A novel method for informative frame selection in wireless capsule endoscopy video. In: 33rd Annual International Conference of the IEEE EMBS, Boston, Massachusetts, USA, pp. 4864–4867 (2011)

7. FFMPEG: Ffmpeg, multimedia framework (2015). https://www.ffmpeg.org/. Accessed 06 Nov 2015

8. Grega, M., Leszczuk, M., Duplaga, M., Fraczek, R.: Algorithms for automatic recognition of non-informative frames in video recordings of bronchoscopic procedures. Inf. Technol. Biomed. **69**, 535–545 (2010)

9. Hwang, S., Lee, J., Cao, Y., Liu, D., Tavanapong, W., Wong, J., Oh, J., de Groen, P.: Automatic measurement of quality metrics for colonoscopy videos. In: International Collegiate Programming Contest, Singapore, pp. 912–921 (2005)

10. Khunand, P., Zhuo, Z., Yang, L., Liyuan, L., Jiang, L.: Feature Selection and Classification for Wireless Capsule Endoscopic Frames. Institute for Infocomm Research, Singapore (2009)

11. Liuand, J., Subramanian, K.R., Yoo, T.S.: A robust method to track colonoscopy videos with non-informative images. Int. J. Comput. Assist. Radiol. Surg. **08**(4), 575–592 (2013)

12. Manivannan, S., Wang, R., Trucco, E., Hood, A.: Automatic normal-abnormal video frame classification for colonoscopy. In: 2013 IEEE 10th International Symposium on Biomedical Imaging (ISBI), pp. 644–647 (2013)

13. Manivannan, S., Wang, R., Trujillo, M.P., Hoyos, J.A., Trucco, E.: Video-specific SVMs for colonoscopy image classification. In: Luo, X., Reichl, T., Mirota, D., Soper, T. (eds.) CARE 2014. LNCS, vol. 8899, pp. 11–21. Springer, Heidelberg (2014). doi:10.1007/978-3-319-13410-9_2

14. Oh, J., Hwang, S., Lee, J., Tavanapong, W., Wong, J., de Groen, P.C.: Informative frame classification for endoscopy video. Med. Image Anal. **11**(1), 110–127 (2007)

15. Matti, P., Abdenour, H., Zhao, G., Ahonen, T.: Computer Vision Using Local Binary Patterns. Springer, Dordrecht (2011)

16. Powers, D.M.W.: Evaluation: from precision, recall and F-measure to ROC, informedness, markedness and correlation. J. Mach. Learn. Technol. **02**(1), 37–63 (2011)

17. Puetz, A., Olsen, R.: Haralick texture feature expanded into the spectral domain. In: Processing of SPIE, p. 6233 (2006)

18. Rangseekajee, N., Phongsuphap, S.: Endoscopy video frame classification using edge-based information analysis. In: Computing in Cardiology, pp. 549–552 (2011)

19. Rungseekajee, N., Lohvithee, M., Nilkhamhang, I.: Informative frame classification method for real-time analysis of colonoscopy video. In: 6th International Conference ECTI-CON 2009, vol. 02, no. 1, pp. 1076–1079 (2009)

20. Seguil, S., Drozdzal, M., Vilarino, F., Malagelada, C., Azpiroz, F., Radeva, P., Vitria, J.: Categorization and segmentation of intestinal content frames for wireless capsule endoscopy. IEEE Trans. Inf. Technol. Biomed. **16**(6), 2341–2352 (2006). New York City, USA

21. Vapnik, V., Lerner, A.: Pattern recognition using generalized portrait method. Autom. Remote Control **24**, 774–780 (1963)

Efficient Training Over Long Short-Term Memory Networks for Wind Speed Forecasting

Erick López[1(⊠)], Carlos Valle[1], Héctor Allende[1,3], and Esteban Gil[2]

[1] Departamento de Informática,
Universidad Técnica Federico Santa María, Valparaíso, Chile
{elopez,cvalle,hallende}@inf.utfsm.cl
[2] Departamento de Ingeniería Eléctrica,
Universidad Técnica Federico Santa María, Valparaíso, Chile
esteban.gil@usm.cl
[3] Facultad de Ingeniería y Ciencias, Universidad Adolfo Ibañez, Santiago, Chile

Abstract. Due to its variability, the development of wind power entails several difficulties, including wind speed forecasting. The Long Short-Term Memory (LSTM) is a particular type of recurrent network that can be used to work with sequential data, and previous works showed good empirical results. However, its training algorithm is expensive in terms of computation time. This paper proposes an efficient algorithm to train LSTM, decreasing computation time while maintaining good performance. The proposal is organized in two stages: (i) first to improve the weights output layer; (ii) next, update all weights using the original algorithm with one epoch. We used the proposed method to forecast wind speeds from 1 to 24 h ahead. Results demonstrated that our algorithm outperforms the original training algorithm, improving the efficiency and achieving better or comparable performance in terms of MAPE, MAE and RMSE.

Keywords: Wind speed forecasting · Recurrent neural networks · Long Short-Term Memory · Multivariate time series

1 Introduction

In order to integrate wind into an electric grid it is necessary to estimate how much energy will be generated in the next few hours. Nevertheless, this task is highly complex because wind power depends on wind speed, which has a random nature. It is worth to mention that the prediction errors may increase the operating costs of the electric system, as system operators would need to use peaking generators to compensate for an unexpected interruption of the resource, as well as reducing the reliability of the system [12]. Several models have been proposed in the literature to address the problem of forecasting wind power or wind speed [7]. Among the different alternatives, machine learning models have gained popularity for achieving good results with a smaller number of restrictions in comparison with statistical models [11]. In particular, recurrent neural networks (RNN)

© Springer International Publishing AG 2017
C. Beltrán-Castañón et al. (Eds.): CIARP 2016, LNCS 10125, pp. 409–416, 2017.
DOI: 10.1007/978-3-319-52277-7_50

[15] have become popular because they propose a network architecture that can process temporal sequences naturally, relating events from different time periods (with memory). However, gradient descent type methods present the vanishing gradient problem [1], which makes difficult the task of relating current events with events of the distant past, i.e., it may hurt long-term memory. An alternative to tackle this problem are Long Short-Term Memory (LSTM) networks [4], a new network architecture that changes the traditional artificial neuron (perceptron) for a memory block formed by gates and cells memories that control the flow of information. This model was compared in [9] to predict the wind speed from 1 to 24 steps ahead. Empirical results showed that LSTM are competitive in terms of accuracy against two neural networks methods. However, its training algorithm demands high computation time due to the complexity of its architecture. In this paper we propose an efficient alternative method to train LSTM; the proposed method divides the training process in two stages: First stage uses ridge regression in order to improve the weights initialization. Next, LSTM is trained to update the weights in a online fashion. The proposal will be evaluated using standard metrics [10] for the wind speed forecasting, of three geographical points of Chile. In these areas it is necessary to provide accurate forecasting in less than one hour. We consider wind speed, wind direction, ambient temperature and relative humidity as input features of the Multivariate time series. Here is the outline of the rest of the paper. In Sect. 2 we describe the LSTM model. Section 3 describes the proposed approach for training the LSTM. Next, we describe the experimental setting on which we tested the method for different data sources in Sect. 4. Finally, the last section is devoted to conclusions and future work.

2   Long Short-Term Memory

Long Short-Term Memory (LSTM) [4] is a class of recurrent network which replaces the traditional neuron in the hidden layer (perceptron) by a memory block. This block is composed of one or more memory cells and three gates for controlling the information flow passing through the blocks, by using sigmoid activation functions with range $[0, 1]$. Each memory cell is a self-connected unit called "Constant Error Carousel" (CEC), whose activation is the state of the cell (see Fig. 1).

All outputs on each memory block (gates, cells and block) are connected with each input of all blocks, i.e., full-connectivity among hidden units. Let $net_{in_j}(t)$, $net_{\varphi_j}(t)$ and $net_{out_j}(t)$ be the weighted sum of the inputs for the *input, forget* and *output* gates, described in Eqs. (1), (3) and (7), respectively, where j indexes memory blocks. Let $y^{in_j}(t)$, $y^{\varphi_j}(t)$ and $y^{out_j}(t)$ be the output on the activation functions ($f_{in_j}(.)$, $f_{\varphi_j}(.)$, $f_{out_j}(.)$, logistic functions with range $[0, 1]$) for each gate. Let $net_{c_j^v}(t)$ be the input for the vth CEC associated to the block j and $s_{c_j^v}(t)$ its state at time t. Let $y^{c_j^v}$ be the output of the vth memory cell of the jth block and S_j is the number of cells of the block j. Then, the information flow (*forward* pass) following the next sequence:

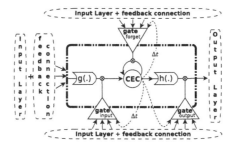

Fig. 1. LSTM architecture with 1 block and 1 cell.

$$net_{in_j}(t) = \sum_m w_{in_j m} \cdot y^m(t-1) + \sum_{v=1}^{S_j} w_{in_j c_j^v} \cdot s_{c_j^v}(t-1), \tag{1}$$

$$y^{in_j}(t) = f_{in_j}(net_{in_j}(t)), \tag{2}$$

$$net_{\varphi_j}(t) = \sum_m w_{\varphi_j m} \cdot y^m(t-1) + \sum_{v=1}^{S_j} w_{\varphi_j c_j^v} \cdot s_{c_j^v}(t-1), \tag{3}$$

$$y^{\varphi_j}(t) = f_{\varphi_j}(net_{\varphi_j}(t)), \tag{4}$$

$$net_{c_j^v}(t) = \sum_m w_{c_j^v m} \cdot y^m(t-1), \tag{5}$$

$$s_{c_j^v}(t) = y^{\varphi_j}(t) \cdot s_{c_j^v}(t-1) + y^{in_j}(t) \cdot g(net_{c_j^v}(t)), \tag{6}$$

$$net_{out_j}(t) = \sum_m w_{out_j m} \cdot y^m(t-1) + \sum_{v=1}^{S_j} w_{out_j c_j^v} \cdot s_{c_j^v}(t), \tag{7}$$

$$y^{out_j}(t) = f_{out_j}(net_{out_j}(t)), \tag{8}$$

$$y^{c_j^v}(t) = y^{out_j}(t) \cdot h(s_{c_j^v}(t)). \tag{9}$$

where w_{rm} is the weight from the unit m to the neuron r; $y^m(t-1)$ is the mth input of the respective unit at time $t-1$; $g(.)$ and $h(.)$ are hyperbolic tangent activation functions with range $[-2,2]$ and $[-1,1]$ respectively. For a more comprehensive study of this technique, please refer to [3].

The CEC solves the problem of vanishing (or explosion) gradient [1], since the local error back flow remains constant within the CEC (without growing or decreasing), while a new instance or external signal error does not arrive. Its training model is based on a modification of the algorithm *BackPropagation Through Time* [13] and a version of the *Real-Time Recurrent Learning* [14]. The main parameters are the block number, the number of cells of each block, and the number of input and output neurons as well. For the training process the following hyperparameters need to be defined: the activation functions, the number of iterations and the learning rate $\alpha \in [0,1]$. This technique has shown accurate results in classification and forecasting problems. However, this method is computationally expensive. Therefore, its architecture is not scalable [8].

3   An Efficient Training for LSTM

LSTM architecture involves high computation time during the training process to find the optimal weights. And the computational training cost considerably increases when either the number of blocks or the number of cells increases.

To address the above-mentioned problem, we propose a new training method that reduces the computational cost, while maintaining its level of performance. The classical LSTM randomly initializes the weights. However, this point may be far from optimal and subsequently, the training algorithms may take more epochs to converge. To improve this particular drawback, we propose a fast method to find a better starting point in the hypothesis space by evaluating a number of instances, and using these output signals to perform a ridge regression to obtain the output layer weights. Finally, we train the LSTM in an online form.

Algorithm 1 describes our training method. It considers a network of three layers (input-hidden-output), where the hidden layer is composed by memory blocks and the output layer by simple perceptron units. Moreover, T is the length of the training series, n_{in} is the number of units of the input layer (the number of lags), n_h is the number of units in the hidden layer, n_o is the number of units of the output layer. Let Y matrix $T \times n_o$ with the current outputs associated with each input vector.

Algorithm 1. Efficient Training Method for LSTM

Input: A set of instances $\{(x_1(t), \ldots, x_{n_{in}}(t))\}_{t=1,\ldots,T}$
1. Randomly initialize the weights $w_{in_j m}, w_{\phi_j m}, w_{out_j m}, w_{in_j c_j^v}, w_{\phi_j c_j^v}, w_{out_j c_j^v}, w_{c_j^v m}$
   from a uniform distribution with range $[-0.5; 0.5]$ $\forall m, \forall j = 1, \ldots, n_h; v = 1, \ldots, S_j$.
2. Set matrix S as

$$
S = \begin{bmatrix}
x_1(1) & \ldots & x_{n_{in}}(1) & y^{c_1^1}(1) & \ldots & y^{c_1^v}(1) & y^{c_2^1}(1) & \ldots & y^{c_j^v}(1) & \ldots & y^{c_{n_h}^{S_j}}(1) \\
x_1(2) & \ldots & x_{n_{in}}(2) & y^{c_1^1}(2) & \ldots & y^{c_1^v}(2) & y^{c_2^1}(2) & \ldots & y^{c_j^v}(2) & \ldots & y^{c_{n_h}^{S_j}}(2) \\
\vdots & \ddots & \vdots & \vdots & \ddots & \vdots & \vdots & \ddots & \vdots & \ddots & \vdots \\
x_1(T) & \ldots & x_{n_{in}}(T) & y^{c_1^1}(T) & \ldots & y^{c_1^v}(T) & y^{c_2^1}(T) & \ldots & y^{c_j^v}(T) & \ldots & y^{c_{n_h}^{S_j}}(T)
\end{bmatrix}.
$$

3. $W_{out} \leftarrow (S' \times S + \lambda \times I)^{-1} \times S' \times Y$

4. **for** $t = 1$ **to** T **do**
5.   Old weights of LSTM are updated making one epoch with the LSTM model using the instance $\mathbf{x}(t)$.
6. **end for**
Output: All LSTM's weights.

The first stage (steps 1 to 3) of the algorithm finds a good starting point for the LSTM. In step 1 all network weights are initialized from a uniform distribution with range $[-0.5; 0.5]$. Next, the matrix S, containing all memory cells outputs, is computed. Here each row of this matrix corresponds to the outputs of the units directly connected to the output layer given an input vector $\mathbf{x}(t) = (x_1(t), \ldots, x_{n_{in}}(t))$ as is described in step 2. Thus, the target estimations can be written as:

$$\hat{Y} = S \cdot W_{out},$$

where W_{out} is a $(n_{in} + \sum_{j=1}^{n_h} S_j) \times n_o$ matrix containing the output layer weights. Then, W_{out} can be estimated by rigde regresion, as shown in step 3, where S' is the transpose of matrix S and I is the identity matrix. In the second stage, the LSTM network is trained with a set of instances by using incremental learning, i.e., the weights are updated after receiving a new instance. Note that this approach is similar to the way that extreme learning machines (ELM) [5] adjust the weights of the output layer. This approaches is well-known because its interpolation and universal approximation capabilities [6]. In contrast, here we use this fast method just to find a reliable starting point for the network.

4 Experiments and Results

In order to assess our proposal, we use three data sets from different geographic points of Chile: Data 1, code b08, (22.54°S, 69.08°W); Data 2, code b21, (22.92°S, 69.04°W); and Data 3, code d02, (25.1°S, 69.96°W). These data are provided by the Department of Geophysics of the Faculty of Physical and Mathematical Sciences of the University of Chile, commissioned by the Ministry of Energy of the Republic of Chile[1].

We worked with the hourly time series, with no missing values. The attributes considered for the study are: wind speed at 10 meters height (m/s), wind direction at 10 m height (degrees), temperature at 5 m (°C), and relative humidity at 5 m height. The series starting at 00:00 on December 1, 2013 to 23:00 on March 31, 2015. Each feature is scaled to $[-1, 1]$ using the min-max function.

To evaluate the model accuracy, each available series is divided in $R = 10$ subsets using a 4 months sliding window approach with a shift of 500 points (20 days approximately), as depicted in Fig. 2.

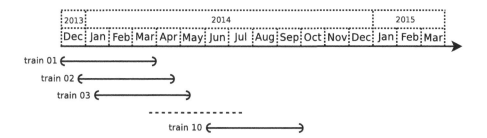

Fig. 2. Sliding window approach to evaluate model accuracy.

---

[1] http://walker.dgf.uchile.cl/Mediciones/.

Then to measure the accuracy of the model, we consider the average over the subsets, computing three standard metrics [10] based on $e_r(T + h|T) = y_r(T + h) - \hat{y}_r(T + h|T)$:

$$\text{MAE}(h) = \frac{1}{R} \sum_{r=1}^{R} |e_r(T + h|T)| \tag{10}$$

$$\text{MAPE}(h) = \frac{1}{R} \sum_{r=1}^{R} \left| \frac{e_r(T + h|T)}{y_r(T + h)} \right| \tag{11}$$

$$\text{RMSE}(h) = \sqrt{\frac{1}{R} \sum_{r=1}^{R} (e_r(T + h|T))^2}. \tag{12}$$

Here $y_r(T + h)$ is the unnormalized target at time $T + h$. T is the index of the last point, and h is the number of steps ahead; $\hat{y}_r(T + h|T)$ is the target estimation of the model at the time $T + h$. The forecasting of several steps ahead was made with the multi-stage approach prediction [2]. Table 1 shows the parameters tuned.

Table 1. Parameters of LSTM for tuning.

number of blocks	: $n_h \in \{1, 2, 3, 4\}$
number of cells	: $S_j \in \{1, 2, 3, 4\}$
learning rate	: $\alpha \in \{10^{-4}, 10^{-3}, 10^{-2}, 10^{-1}\}$
number of lags	: $n_{in} \in \{20, 25, 30, 35\}$
maximum of iterations	: 100

The results show that the proposed method achieves a better overall computational time for 10 runs, based in the best model that minimizes MSE for each data set. Table 2 shows the training time of the original algorithm and of our proposal (columns two and three respectively). And the remainder columns exhibit the parameters selected to train the models.

Table 2. Overall time for 10 runs (in minutes) and the selected values of the parameters for each model.

			Parameters tuned							
	Models		(original)				(proposal)			
Source	(original)	(proposal)	n_h	S_j	α	lag	n_h	S_j	α	lag
Data 1	204.57	2.78	1	2	10^{-4}	30	3	1	10^{-4}	30
Data 2	394.73	3.18	3	2	10^{-4}	25	2	4	10^{-4}	35
Data 3	60.21	4.35	2	3	10^{-4}	20	3	4	10^{-4}	20

Figure 3 shows that the proposed algorithm achieves a better or comparable performance using MAPE and RMSE. The results for MAE are omitted
because they show similar behavior that MAPE, but different scale. There is a
important insights from these experiments, we observe that the proposed method
outperformed the original model by several steps, especially for the MAPE when
forecasting several steps ahead.

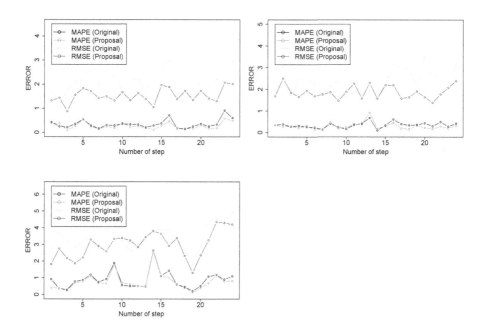

Fig. 3. Data 1 (top-left), Data 2 (top-right), Data 3 (botton-left)

5   Conclusions and Future Work

This work presents an efficient training algorithm for LSTM networks. We
observed that our proposed training method outperforms the original algorithm
reducing by 98%, 99% and 92% the computational time for each data set. One
can also notice that although our proposal uses a greater number of blocks or
cells or lags, it is most efficient. Results suggests that our proposal, besides being
efficient, in general achieved a better performance when forecasting several steps
ahead. As future work, we would like to research how to increase the forecasting accuracy and evaluating our algorithm against other models derived from
LSTM. Another interesting issue is to explore the performance of our proposal
in large datasets.

Acknowledgments. This work was supported in part by Research Project DGIP-UTFSM (Chile) 116.24.2 and in part by Basal Project FB 0821 and Basal Project FB0008.

References

1. Bengio, Y., Simard, P., Frasconi, P.: Learning long-term dependencies with gradient descent is difficult. IEEE Trans. Neural Netw. **5**(2), 157–166 (1994)
2. Cheng, H., Tan, P.-N., Gao, J., Scripps, J.: Multistep-ahead time series prediction. In: Ng, W.-K., Kitsuregawa, M., Li, J., Chang, K. (eds.) PAKDD 2006. LNCS (LNAI), vol. 3918, pp. 765–774. Springer, Heidelberg (2006). doi:10.1007/11731139_89
3. Gers, F.: Long short-term memory in recurrent neural networks. Ph.D. thesis, École Polytechnique Fédérale de Laussanne (2001)
4. Hochreiter, S., Schmidhuber, J.: Long short-term memory. Neural Comput. **9**(8), 1735–1780 (1997)
5. Huang, G., Huang, G.-B., Song, S., You, K.: Trends in extreme learning machines: a review. Neural Netw. **61**, 32–48 (2015)
6. Huang, G.-B., Wang, D.H., Lan, Y.: Extreme learning machines: a survey. Int. J. Mach. Learn. Cybern. **2**(2), 107–122 (2011)
7. Jung, J., Broadwater, R.P.: Current status and future advances for wind speed and power forecasting. Renew. Sustain. Energy Rev. **31**, 762–777 (2014)
8. Krause, B., Lu, L., Murray, I., Renals, S.: On the efficiency of recurrent neural network optimization algorithms. In: NIPS Optimization for Machine Learning Workshop (2015)
9. López, E., Valle, C., Allende, H.: Recurrent networks for wind speed forecasting. In: IET Seminar Digest (ed.) International Conference on Pattern Recognition Systems, ICPRS 2016, vol. 2016 (2016)
10. Madsen, H., Pinson, P., Kariniotakis, G., Nielsen, H.A., Nielsen, T.S.: Standardizing the performance evaluation of short-term wind prediction models. Wind Eng. **29**(6), 475–489 (2005)
11. Perera, K.S., Aung, Z., Woon, W.L.: Machine learning techniques for supporting renewable energy generation and integration: a survey. In: Woon, W.L., Aung, Z., Madnick, S. (eds.) DARE 2014. LNCS (LNAI), vol. 8817, pp. 81–96. Springer, Heidelberg (2014). doi:10.1007/978-3-319-13290-7_7
12. Sideratos, G., Hatziargyriou, N.D.: An advanced statistical method for wind power forecasting. IEEE Trans. Power Syst. **22**(1), 258–265 (2007)
13. Werbos, P.J.: Backpropagation through time: what it does and how to do it. Proc. IEEE **78**(10), 1550–1560 (1990)
14. Williams, R.J., Zipser, D.: A learning algorithm for continually running fully recurrent neural networks. Neural Comput. **1**(2), 270–280 (1989)
15. Zimmermann, H.-G., Tietz, C., Grothmann, R.: Forecasting with recurrent neural networks: 12 tricks. In: Montavon, G., Orr, G.B., Müller, K.-R. (eds.) NN: Tricks of the Trade. LNCS, vol. 7700, pp. 687–707. Springer, Heidelberg (2012). doi:10.1007/978-3-642-35289-8_37

Classifying Estimated Stereo Correspondences Based on Delaunay Triangulation

Cristina Bustos, Elizabeth Vargas, and Maria Trujillo$^{(\boxtimes)}$

Multimedia and Computer Vision Group, Universidad del Valle, Cali, Colombia
{cristina.bustos,elizabeth.vargas,maria.trujillo}@correounivalle.edu.co

Abstract. The stereo vision problem gives rise to two sub-problems: the correspondence and the stereo calibration. Once the sub-problems have been solved, depth is estimated by triangulation. Depth estimation, using stereo images, is based on disparity – the relative displacement between corresponding points – and the camera geometry. Small errors in disparity may produce large errors in depth estimates due to the ill-posedness nature of the stereo vision problem. Moreover, if the camera geometry is unknown, it is estimated using corresponding points. A solution of stereo vision problem may be improved by identifying corresponding points which are inaccurately estimated. In this paper, the classification of a set of estimated corresponding points uses Delaunay triangulation by restricting it to a given subset of estimated corresponding points. Delaunay edges among estimated corresponding points are used to build undirected graphs. Classification criteria based on adjacencies are defined in order to decide whether or not a corresponding point is correctly estimated. Experimental evaluation, using ground truth image sets for quantitative analysis, shown values of specificity around 70% while sensitivity up to 96%.

Keywords: Corresponding points · Delaunay triangulation · 3D-reconstruction

1 Introduction

Stereo vision refers to the ability to infer 3D structure, such as depth, of a scene using two or more images taken from different viewpoints. The correspondence problem has to be solved in order to infer the 3D structure of a scene. The correspondence problem consists in determining which point on the left image corresponds to which point on the right one [20]. Depth estimation has application in different sectors of industry, such as: autonomous robots [9], self-driving cars [7], intelligent cars [26], immersive 3D video conferencing [8], 3D-TV [13] and urban planning [14], among others.

The stereo correspondence problem is not solved yet, since it is an inverse and ill-posed problem. Different techniques have been employed in order to find corresponding points given two images [2,4,18]. When the correspondence problem is addressed, corresponding points may be inaccurately estimated: falsely matched

© Springer International Publishing AG 2017
C. Beltrán-Castañón et al. (Eds.): CIARP 2016, LNCS 10125, pp. 417–425, 2017.
DOI: 10.1007/978-3-319-52277-7_51

or badly located. Small inaccuracies, in estimated corresponding points, may have a large impact on depth estimation [20], due to the ill-posedness nature of the stereo vision problem. The ill-posedness nature gives rise to a set of constraints and rules in order to keep only disparities of reliable points by rejecting points that are mismatched. However, a solution to the inverse problem does not always exist. Thus, there is a room for proposals on identifying corresponding points which are inaccurately estimated. A set of correctly estimated corresponding points may be used as anchors for estimating dense disparity maps along with the camera geometry.

Delaunay triangulation has been used in feature correspondence [19], region corresponding [22] and block matching [23]. In [10], it is presented a method for establishing dense correspondence maps between two images, in a video sequence or in a stereo pair, by combining feature matching with Delaunay triangulation. This approach has advantages, such as: it allows estimating large displacements and subsequently taking into account disparity discontinuities. However, the results depend on the presence of texture. The method works well in images with strong textures, where a large number of small triangles are built. In untextured areas, the results are not accurate due to homogeneous regions. In [27], it is presented an improved SUSAN (Smallest Univalue Segment Assimilating Nucleus) corner detection algorithm and also an effective corners correspondence algorithm based on Delaunay Triangulation. This algorithm is robust to scaling, rotation and translation with short baseline, since interior angles and local structural similarity are robust to noise and motion transform. Dense disparity maps are calculated using Delaunay triangulation in [24], where image pairs are triangulated and triangles are classified into matched and unmatched triangles using a matching criterion: matched triangle and unmatched one, and disparity values in two types of triangle region are solved, respectively.

A generative probabilistic model to binocular stereo for fast matching is proposed in [6]. A Bayesian approach that uses a 2D mesh via Delaunay triangulation for stereo matching is presented, which is able to compute accurate disparity maps of high resolution images at frame rates close to real time without requiring global optimisation. The algorithm reduces the search space and can be parallelised. In [3], it is presented a method for stabilising the computation of stereo correspondences using Delaunay triangulation. The input images are partitioned into small and localised regions. Adjacent triangles are merged into larger polygonal patches and the planarity assumption is verified in order to achieve the planarity test robustly. Point correspondences are established by planar homographies, and are used as control points in the final matching process using dynamic programming. The approach was tested on three different types of data, including medical data. The results showed that it can speed up the stereo matching process. In [25], it is proposed a Delaunay-based stereo matching method, which relies on Canny edge detector to extract key-points. Using those points, Delaunay triangulation is performed and an initial disparity estimation is calculated, based on the histogram of each triangular area. This matching is further refined by considering that depth values of adjacent triangles should have similar depth values.

An approach for classifying corresponding points based on the Delaunay triangulation is presented in [21]. However, we did not consider that corresponding points are constrained by image content. Thus, a triangulation may be built using points belonging to different objects. In this paper, given a set of estimated corresponding points, a classification approach using Delaunay triangulation is presented. Delaunay edges are used to build undirected graphs, points belonging to the right image built one graph and points belonging to the left image built other graph. Classification criteria based on adjacencies are defined in order to decide whether or not a corresponding point is correctly estimated. Without loss of generality, the propose approach is validated using ground truth image sets.

2   Corresponding Points Classification

Once corresponding points are estimated, some of them may be inaccurately estimated. Thus, the classification of corresponding points consists in deciding whether or not a matching point is reliable. Estimated corresponding points are mapped into two graphs, points belonging to the right image are mapped into one graph and points belonging into the left image are mapped into the other graph. Thus, corresponding points are classified as correctly estimated if and only if the two graphs are isomorphic. Additional conditions are imposed in order to deal with the inverse and ill-posed characteristics of this classification.

2.1   Fundamentals

Suppose that $G = (V, E)$ is an undirected graph where $|V| = n$. The adjacency matrix A, with respect to an ordered set of vertices, is the $n \times n$ zero-one matrix with one at the $(i, j)th$ entry when v_i and v_j are adjacent and zero at the $(i, j)th$ entry when v_i and v_j are not adjacent. A bijective function $\phi : V \Leftrightarrow V$ is a graph isomorphism if:

$$v_i, v_j \in E \Leftrightarrow \phi(v_i), \phi(v_j) \in E'. \tag{1}$$

Moreover, a bijective function ϕ preserves the adjacency between vertices [5].

A Delaunay triangulation is a triangulation of the vertex set with the property that the circumcircle of every triangle is empty, that is, there is no point from the vertex set in its interior.

2.2   Classification Based on Delaunay Triangulation

The proposed classification approach is presented, without lost of generality, under the assumption of the stereo images are rectified. Given a set of estimated corresponding points, two Delaunay triangulation are build, one with points from the right image and the other one with points from the left image. After the triangulation, points from each image are labelled with the same number in order to identify the correspondence. Points are vertices and Delaunay edges are edges in an undirected graph.

In the proposed approach, the bijective function ϕ is represented by estimated corresponding points and a set of adjacent vertices may have cardinality larger than 3. We will use the term correctly estimated to indicate that the points represent the same 3D point in both images. The classification of estimated corresponding points is based on the following restrictions:

Restriction 1 (Adjacency). *Given two matching points, they are correctly estimated if and only if the set of adjacent vertices in the right image is equal to the set of adjacent vertices in the left image.*

Restriction 2 (Cardinality). *Given two matching points and two sets of adjacent vertices of cardinalities n and m with n < m, the matching points are correctly estimated if and only if the set of vertices with low-cardinality is a proper subset of the set of vertices with high-cardinality.*

Bearing in mind that corresponding points are representing as disparity vectors and disparity is inversely proportional to depth, corresponding points are constrained by image content. In this way, object features may provide information to the Delaunay triangulation indicating a way in which feature points are related.

Restriction 3 (Object boundaries). *Given two matching points and two sets of adjacent vertices of cardinalities n and m with n < m, also edges represent object boundaries, matching points must be triangulated in groups belonging to the same objected. The matching points are correctly estimated if and only if they belong to the same 3D object.*

3   Experimental Evaluation

The proposed approach is validated using existing feature detection and matching schemes. An initial set of key-points is obtained using algorithms from the state of the art. In particular, the Scale Invariant Feature Transform (SIFT) [12], the Features from Accelerated Segment Test (FAST) [15], the Speeded Up Robust Features (SURF) [1] and Good Features to Track (GFTT) [17]. The matching SIFT key-points algorithm is provided in [11]. The FAST, SURF and GFTT key-points are matched using Sum of Squared Differences (SSD). The obtained matching points in the matching stage are classified using the proposed approach based on the Delaunay triangulation and the restrictions.

In order to enforce Restriction 3, a pre-processing step is conducted to obtain a binary image where the main contours could be identified. In this step, binarisation is obtained by applying a threshold using the Otsu algorithm. A closing and a opening morphological operations are applied to the resulting binary image, independently. Contours are obtained by subtracting resultant images after opening and closing operations. Every key-point is assigned to one contour, in order to triangulate points belonging to the same object by finding whether the key-point is inside or outside or on the contour. This process is independently conducted using key-points from the left and the right images.

The proposed approach is quantitatively evaluated using a "gold standard". The Middlebury Stereo Page (http://vision.middlebury.edu/stereo/) provides stereo images and each dataset includes the true disparity map [16]. In total, 8 sets of images are selected: Art, Books, Cones, Dolls, Laundry, Moevius, Reindeer and Teddy. The metrics employed to evaluate the proposed approach performance are Sensitivity, Specificity, Positive Predictive Value (PPV) and Negative Predictive Value (NPV). The ground-truth information is used for validating an identified correct or incorrect estimated corresponding point. Initial estimated corresponding points are compared with the ground-truth using the absolute error equation. The error at the (i,j) position is calculated using the following formula:

$$error_{i,j} = |DT_{i,j} - d_{i,j}|, \tag{2}$$

where $DT_{i,j}$ represents the ground-truth disparity value at (i,j) and $d_{i,j}$ is the estimated disparity value at (i,j). An estimated corresponding point is labelled as estimate incorrectly with an error value bigger than 1, otherwise is considered a estimate correctly. In this way, the performance metrics are built based on estimate incorrectly and estimate correctly.

Tables 1, 2, 3 and 4 contain the performance metrics obtained using SIFT, FAST, SURF and GFTT key-points. In general, the sensitivity decreased for most datasets while the specificity increased. The reason, is that there are some edges missing in comparison with the unconstrained triangulation in [21], and some of them were not supposed to be there, decreasing the number of false positives (which increases the specificity) and some had to be there, increasing the number of false negatives (which decreases the sensitivity).

Table 1. Performance metrics of the proposed classification using SIFT keypoints.

Dataset	Sensitivity (%)	Specificity (%)	PPV (%)	NPV (%)
Art	75,76	55,23	24,51	92,23
Books	68,63	70,86	25,55	93,94
Cones	42,62	71,67	18,44	89,26
Dolls	58,26	60,54	27,57	84,91
Laundry	64,21	54,76	44,53	73,02
Moebius	46,43	62,21	20,80	84,46
Reindeer	49,00	40,91	79,03	15,00
Teddy	64,41	55,76	24,20	87,72

In general, key-points calculated using SURF and GFTT were not useful for the proposed classification. Using SIFT and FAST, while the specificity value incremented for most datasets (for books there was no change), the sensitivity decremented for some of them, and for others keep equal.

Algorithms' run-time is used as a measure of the complexity. Table 5 shows run-times of the proposed approach using keypoints calculated with the SIFT,

Table 2. Performance metrics of the proposed classification using FAST keypoints.

Dataset	Sensitivity (%)	Specificity (%)	PPV (%)	NPV (%)
Art	63,89	31,52	42,20	52,73
Books	93,29	23,38	54,09	78,26
Cones	71,79	51,05	37,58	81,51
Dolls	67,06	62,05	43,51	81,21
Laundry	95,88	39,39	75,61	82,98
Moebius	68,25	49,33	36,13	78,72
Reindeer	77,59	35,66	32,85	79,69
Teddy	78,86	36,79	44,29	73,20

Table 3. Performance metrics of the proposed classification using SURF keypoints.

Dataset	Sensitivity (%)	Specificity (%)	PPV (%)	NPV (%)
Art	63,16	50,41	37,11	74,70
Books	61,61	51,60	27,49	81,86
Cones	43,88	61,12	17,62	85,18
Dolls	60,87	59,71	38,89	78,37
Laundry	67,67	53,37	48,13	72,08
Moebius	52,63	57,92	22,90	83,73
Reindeer	52,24	63,13	34,65	77,93
Teddy	62,50	56,04	29,57	83,50

Table 4. Performance metrics of the proposed classification using GFTT keypoints.

Dataset	Sensitivity (%)	Specificity (%)	PPV (%)	NPV (%)
Art	60,98	61,25	28,74	85,96
Books	65,00	50,69	10,79	94,04
Cones	45,95	64,68	9,94	93,38
Dolls	61,90	53,80	13,68	92,27
Laundry	73,48	57,29	61,86	69,62
Moebius	82,72	56,59	33,17	92,63
Reindeer	65,71	45,84	10,22	93,44
Teddy	67,74	49,75	17,43	90,78

Table 5. Run-time using keypoints calculated with the SIFT, the FAST, the SURF and the GFTT (in seconds).

Dataset	SIFT	SURF	FAST	GFTT
Art	0.43	0.37	0.25	0.31
Books	0.36	0.43	0.23	0.32
Cones	0.46	0.49	0.24	0.32
Dolls	0.52	0.45	0.27	0.35
Laundry	0.43	0.40	0.25	0.32
Moebius	0.39	0.33	0.26	0.33
Reindeer	0.33	0.37	0.38	0.33
Teddy	0.37	0.34	0.24	0.43

the FAST, the SURF and the GFTT. In general, the FAST took fewer seconds than the others except for the Reindeer dataset.

4 Conclusions

In this paper, we presented an approach for classifying estimated corresponding points based on the constrained Delaunay triangulation and considering that corresponding points are constrained by image content. The proposed approach uses graph isomorphism to formulate three restrictions: adjacency, cardinality and object boundaries, in order to classify estimated corresponding points as correctly estimated if and only if the two graphs are isomorphic.

Regarding the keypoints, the SURF and the GFTT algorithms did not yield key-points useful for the proposed approach. The proposed approach exhibits the best performance with the FAST corners, where an initial set of corresponding points contains more than 50% estimate incorrectly. However, the proposed approach produced error propagations since a corresponding point incorrectly estimated affects directly the neighbour in the graphs.

In comparison to the unconstrained Delaunay triangulation presented in [21], the use of constrains contribute to the increment of specificity, given that not all the matching points are triangulated together, reducing the number of false positives. However, it also decrements the sensitivity, because for some points that are "estimate incorrectly", there are missing edges, generating a false negative.

References

1. Baya, H., Essa, A., Tuytelaarsb, T., Goo, L.V.: Speeded-up robust features (SURF). Comput. Vis. Image Underst. **110**(3), 346–359 (2008)
2. Bleyer, M., Gelautz, M.: Graph-cut-based stereo matching using image segmentation with symmetrical treatment of occlusions. Sig. Process. Image Commun. **22**, 127–143 (2007)

3. Chen, C.-I., Sargent, D., Tsai, C.-M., Wang, Y.-F., Koppel, D.: Stabilizing stereo correspondence computation using Delaunay triangulation and planar homography. In: Bebis, G., et al. (eds.) ISVC 2008. LNCS, vol. 5358, pp. 836–845. Springer, Heidelberg (2008). doi:10.1007/978-3-540-89639-5_80
4. Cho, N.: Stereo matching using multi-directional dynamic programming and edge orientation. In: Korea-Japan Joint Workshop on Frontiers of Computer, vol. 1, pp. 233–236 (2007)
5. Gallardo, P.: Notas de matematica discreta (2002). http://www.uam.es/personal_pdi/ciencias/gallardo/capitulo8a.pdf
6. Geiger, A., Roser, M., Urtasun, R.: Efficient large-scale stereo matching. In: Kimmel, R., Klette, R., Sugimoto, A. (eds.) ACCV 2010. LNCS, vol. 6492, pp. 25–38. Springer, Heidelberg (2011). doi:10.1007/978-3-642-19315-6_3
7. Google: German scientists unveil self-driving car (2012). http://www.google.com/hostednews/afp/article/ALeqM5izCRPtPhsc2k6kCY2_-YOeUs7r_w?docId=CNG.c668be9320e3376a10d767deb0b0649f.a91
8. Ho, Y.S., Jang, W.S.: Gaze correction using 3D video processing for videoconferencing. In: IEEE China Summit and International Conference on Signal and Information Processing (ChinaSIP), pp. 496–499 (2015)
9. Jet Propulsion Laboratory, California Institute of Technology: Mars exploration rovers (2012). http://marsrovers.jpl.nasa.gov/home/index.html
10. Kardouchi, M., Konrad, J., Vazquez, C.: Estimation of large-amplitude motion and disparity fields: application to intermediate view reconstruction. Process. SPIE (2001). http://proceedings.spiedigitallibrary.org/data/Conferences/SPIEP/34800/340_1.pdf
11. Lowe, D.: Demo software: Sift keypoint detector (2005). http://www.cs.ubc.ca/~lowe/keypoints/siftDemoV4.zip
12. Lowe, D.: Distinctive image features from scale-invariant keypoints. Int. J. Comput. Vis. **60**, 1–28 (2004)
13. Oh, J., Sohn, K.: A depth-aware character generator for 3DTV. IEEE Trans. Broadcast. **58**(4), 523–532 (2012)
14. Romanoni, A., Matteucci, M.: Incremental reconstruction of urban environments by edge-points Delaunay triangulation. In: IEEE/RSJ International Conference on Intelligent Robots and Systems (IROS), pp. 4473–4479 (2015)
15. Rosten, E., Drummond, T.: Machine learning for high-speed corner detection. In: Leonardis, A., Bischof, H., Pinz, A. (eds.) ECCV 2006. LNCS, vol. 3951, pp. 430–443. Springer, Heidelberg (2006). doi:10.1007/11744023_34
16. Scharstein, D.: Middlebury stereo datasets (2011). http://vision.middlebury.edu/stereo/data/
17. Shi, J., Tomasi, C.: Good features to track. In: 1994 IEEE Computer Society Conference on Computer Vision and Pattern Recognition, Proceedings of CVPR 1994, pp. 593–600. IEEE (1994)
18. Sun, J., Zheng, N., Shum, H.: Stereo matching using belief propagation. Pattern Anal. Mach. Intell. **25**(7), 787–800 (2003)
19. Tan, H., Zhou, F., Zhang, W., Wang, Y.: Automatic correspondence approach for feature points in camera calibration. J. Optoelectr. Laser **22**(5), 736–739 (2011)
20. Trucco, E., Verri, A.: Introductory Techniques for 3-D Computer Vision. Prentice Hall, Englewood Cliffs (1998)
21. Vargas, E., Trujillo, M.: A corresponding points classification approach by Delaunay triangulation. In: De La Fuente Rubio, E. (ed.) Proceedings of 4th Latin American Conference on Networked and Electronic Media, pp. 6–12. Escuela de Informática, Universidad Nacional Andrés Bello, Santiago de Chile (2012)

22. Wang, J., Wang, L., Chan, K.L., Constable, M.: A linear programming based method for joint object region matching and labeling. In: Lee, K.M., Matsushita, Y., Rehg, J.M., Hu, Z. (eds.) ACCV 2012. LNCS, vol. 7725, pp. 66–78. Springer, Heidelberg (2013). doi:10.1007/978-3-642-37444-9_6
23. Zhang, H., Chen, X.: Effective scene matching for intelligent video surveillance. In: International Conference on Graphic and Image Processing (2013)
24. Zhang, S., Qu, X., Ma, S., Yang, Z., Kong, L.: A dense stereo matching algorithm based on triangulation. J. Comput. Inf. Syst. **8**, 283–292 (2012)
25. Zhang, X.H., Li, G., Li, C.L., Zhang, H., Zhao, J., Hou, Z.X.: Stereo matching algorithm based on 2D Delaunay triangulation. Math. Probl. Eng. **2015**, 1–14 (2015)
26. Zhao, H.: Motion planning for intelligent cars following roads based on feasible neighborhood. In: IEEE International Conference on Control Science and Systems Engineering (CCSSE), pp. 27–31 (2014)
27. Zhou, D., Li, G.: Effective corner matching based on Delaunay triangulation. In: Robotics and Automation Proceedings, pp. 2730–2735 (2004)

Data Fusion from Multiple Stations for Estimation of PM2.5 in Specific Geographical Location

Miguel A. Becerra[1,2]([✉]), Marcela Bedoya Sánchez[1],
Jacobo García Carvajal[1], Jaime A. Guzmán Luna[2],
Diego H. Peluffo-Ordóñez[3,4], and Catalina Tobón[5]

[1] GEA Research Group, Institución Universitaria Salazar y Herrera,
Medellín, Colombia
migb2b@gmail.com, marcelabedoyita@gmail.com,
jacobogarcia04@gmail.com
[2] SINTELWEB Research Group,
Universidad Nacional de Colombia, Medellín, Colombia
Jaime.guzman@unal.edu.co
[3] Facultad de Ingeniería en Ciencias Aplicadas-FICA
from Universidad Técnica del Norte, Ibarra, Ecuador
dhpeluffo@unal.edu.co
[4] Department of Electronics, Universidad de Nariño, Pasto, Colombia
[5] Universidad de Medellín, Medellín, Colombia
ctobon@udem.edu.co

Abstract. Nowadays, an important decrease in the quality of the air has been observed, due to the presence of contamination levels that can change the natural composition of the air. This fact represents a problem not only for the environment, but also for the public health. Consequently, this paper presents a comparison among approaches based on Adaptive Neural Fuzzy Inference System (ANFIS) and Support Vector Regression (SVR) for the estimation level of PM2.5 (Particle Material 2.5) in specific geographic locations based on nearby stations. The systems were validated using an environmental database that belongs to air quality network of Valle de Aburrá (AMVA) of Medellin Colombia, which has the registration of 5 meteorological variables and 2 pollutants that are from 3 nearby measurement stations. Therefore, this project analyses the relevance of the characteristics obtained in every single station to estimate the levels of PM2.5 in the target station, using four different selectors based on Rough Set Feature Selection (RSFS) algorithms. Additionally, five systems to estimate the PM2.5 were compared: three based on ANFIS, and two based on SVR to obtain an aim and an efficient mechanism to estimate the levels of PM2.5 in specific geographic locations fusing data obtained from the near monitoring stations.

Keywords: ANFIS · PM2.5 estimation · Support Vector Regression

© Springer International Publishing AG 2017
C. Beltrán-Castañón et al. (Eds.): CIARP 2016, LNCS 10125, pp. 426–433, 2017.
DOI: 10.1007/978-3-319-52277-7_52

1 Introduction

The World Health Organization (WHO) has studied the harmful effects of the polluted air on the human health, showing the need of monitoring different types of pollutants, including pollutants like PM2.5 particles in the cities of developed and developing countries [1]. It is because there are evidences that these particles have a high association with cancer, heart diseases, lung diseases, and low tract respiratory infections, increasing the morbidity and mortality of the population in a considerable way. Therefore, to measure or estimate the PM2.5 in order to determine the values of the air quality, everywhere in an effective way, is very important in order to establish preventive measures and reduce the risk on the health of the population.

There are three types of models, which are widely used, for predicting the quality of the air: Probabilistic, autoregressive and hybrids [2]. Probabilistic models use different techniques to assess the relationship between the air quality and the meteorological factors [3]. These are usually cheap computational models, such as Hidden Markov Models (HMM), which allow to do the accuracy statistic prediction with relatively less detailed data in relation to the prognostic in many areas [2–4]. However, these techniques have problems when the sampling data are limited, and the model may not work as expected which limits their ability of predicting future events. The nonlinear relationship between PM2.5 and meteorological data pose a problem of nonlinear regression between predictors, so models of artificial neural networks (ANN) are used, which have the ability to detect underlying nonlinear relationships between the responses and the predictors. These can be trained using some algorithms and require much less informatics resources [4–8]. The Autoregressive models have been widely used to predict the air quality, but with a variable precision due to their capacity of application to nonlinear processes and their dependence of the quality of the input data [9, 10]. The support vector machines (SVM) are designed to solve nonlinear classification problems; but because of its ability to generalize, it has shown its application in regression problems and time series forecasting. The SVM makes use of a kernel function, attributing its great capacity for generalization, even when the training set is small, i.e., where both the generalization and the training process of the machine do not necessarily depend on the attributed number; allowing an excellent performance in high dimensional problems [11]. Referring to hybrid models as ANFIS [12], they have better performance, because they have the capacity to predict the maximum PM2.5 concentrations, that are considered a critical factor in the prediction system of air pollution [8]; and they also give adequate solutions to nonlinear problems, ambiguity, and randomness of the data. Other reported studies for the prediction of meteorological variables and contaminants have, in most of them, limitations in predictions respect to time and generality. Also, they do not focus on the estimation of variables in specific geographic locations and cities with a limited number of measurement stations and/or limited variables measurements, as is the case in Medellín city [13–19], which is the principal contribution of this work.

In this study, a comparison among three ANFIS approaches, and two SVM approaches to estimate the PM2.5 was applied together with a relevant analysis based on Rough Set Feature Selection. This aimed at reducing the number of features obtained from the fused measures of near monitoring stations, in order to provide an objective, and an accurate mechanism for more reliable estimation of the PM2.5 concentration on a specific geographical location. Thus, this was obtained from meteorological data and air quality variables in order to give the community an adequate information about the air quality to make decisions on the environmental exposure.

Fig. 1. Geographical location of the Medellín stations

2   Materials and Methods

2.1   Database

The air quality network of "*Valle de Aburrá*" (AMVA) has 22 measuring places of air quality and meteorology that are fixed. These places are distributed in their different municipalities and the jurisdiction of the Metropolitan area. Three stations were selected taken the completeness and their locations as the criteria, as shown in Fig. 1. They are described as follows: (*i*) "Urbana Museo de Antioquia" (MED-MAINT) station, with the following coordinates: an altitude of 1488 [MASL], a latitude of 6°15′ 08.48″ North, and a length of 75°34′07.37″ West. (*ii*) "Fondo Urbano Universidad Nacional de Colombia Sede Medellín – Núcleo el Volador" (MED-UNNV) station, with the following coordinates: an altitude of 1506 [MASL], a latitude of 6°15′34.39″ North, and a length of 75°34′32.46″ West. (*iii*) "Suburbana UNE – Casa Yalta, Loma Los Balsos" (MED-UNEP) station, with the following coordinates: an altitude of 1848 [MASL], a latitude of 6°11′11.07″ North, and a length of 75°33′26.55″ West. These stations have measures of particles smaller than 2.5 micrometers (PM2.5), automatic monitors of Ozone (O3), and meteorological stations that generate data of temperature, humidity, wind speed, wind direction, and radiation.

2.2   Theoretical Background

Adaptive Neural Fuzzy Inference System (ANFIS). This method needs Fuzzy Logic to change the given inputs into wanted outputs through highly interconnected Neural

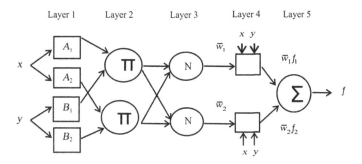

Fig. 2. ANFIS architecture

Network processing elements and information connections, which are weighted to locate and assign inputs into an output. To ease the use of this technique, two fuzzy IF-THEN rules based on a first order Sugeno model [20] are implemented. *Rule*(1) : *IF x is A_1 AND y is B_1, THEN $f_1 = p_1x + q_1y + r_1$* and *Rule*(2) : *IF x is A_2ANDy is B_2, THEN $f_2 = p_2x + q_2y + r_2$*. Where, x and y are inputs, A_i and B_i are fuzzy sets, f_i are outputs within the fuzzy region specified by the fuzzy rule, and p_i, q_i, r_i are design parameters which are adjusted during the training process. The layer 1 is a mapping input variable to corresponding fuzzy membership $O_i = \mu A_i(x)$. In the layer 2, Π is an AND operator to fuzzify the inputs i.e. $w_i = \mu A_i(x) * \mu A_2(y) i = 1, 2, 3, \ldots, N$. Layer 3 has N-nodes that indicate normalization to the firing strengths from the previous layer. In the fourth layer, the nodes i are adaptive and the output of each node is the product of the normalized firing strength with a first order polynomial (for a first order Sugeno model) $O_i = w_i f_i = w_i(px + qy + r) i = 1, 2, 3, \ldots, N$. Finally, the overall output f of the model is given by one single fixed i.e. $f = \sum_i \bar{w} f_i$ (Fig. 2).

Support Vector Machines. The support vector machines (SVMs) were proposed by [21], to solve a nonlinear problem, an input variable that corresponds to predictor variable is non-linearly mapped into a high-dimension feature space [22], and the SVR function is formulated as $y = \omega\phi(x) + b$. Where ϕ (x) is called the feature, which is nonlinear mapped from the input space x. The coefficients w and b are estimated by minimizing:

$$R(c) = c\frac{1}{N}\sum_{i=1}^{N}L_\varepsilon(d_i, y_i) + \frac{1}{2}\|\omega\|^2$$

$$L_\varepsilon(d, y) = \begin{cases} |d - y| - \varepsilon & |d - y| \geq \varepsilon \\ 0 & Others \end{cases}$$

Where both c and ε are prescribed parameters. The first term $L_\varepsilon(d, y)$ is called the ε - intensive loss function. Here, $k(\tilde{x}, \bar{x})$ is called the kernel function. The value of the kernel is equal to the inner product of two vectors $\tilde{x}$ and $\bar{x}$ in the feature space ϕ ($\tilde{x}$) and

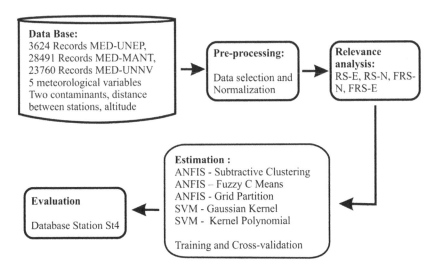

Fig. 3. Proposed procedure

$\phi(\bar{x})$, such that $k(\tilde{x}, \bar{x}) = \phi(\tilde{x}) * \phi(\bar{x})$. Any function satisfying Mercer's condition can be used as the kernel function.

Proposed Procedure. Figure 3 shows the proposed procedure to estimate the PM2.5 in specific geographical location. First, we eliminate the samples with outliers, and then the database was normalized between 0 and 1. Then, the data of the selected stations were fused. Next, the relevance analysis was carried out using Rough Set- Neighbor (RS-N), Rough Set - Entropy (RS-E), Fuzzy Rough Set - Neighbor (FRS-N), and Fuzzy Rough Set – Entropy (FRS-E). These techniques are widely explained in [23]. Each of the parameters (inclusion and neighbor) of the algorithms was adjusted heuristic, taking values between 0.05 and 1 with increments of 0.05. Next, the results obtained were computed according to common relevance of the data given by the four selection methods, obtaining a reduct, which was used for training 5 regression approaches. Three system based on ANFIS were trained; each used different algorithms (Grid Partition-GP, Fuzzy C-means-FC [21] and Subtractive Clustering-SC [24, 25]) for establishing the initial FIS. Their parameters were adjusted in order to maximize accuracy. The same way, two SVR systems with two different kernels (polynomial kernel -KP and Gaussian kernel-KG) were trained, and their parameters C and gamma were adjusted for maximizing the accuracy. Finally, all approaches were validated using a 30-fold cross-validation procedure (70/30 split).

3   Results and Discussion

Figure 4 shows the results obtained of the relevance analysis carried out applying 4 selection algorithms (RS-N, FRS–N, RS-E and FRS–E) obtaining as relevant variables: wind speed, wind direction, and humidity to estimate the PM2.5.

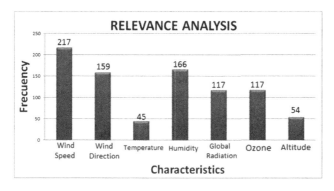

Fig. 4. Global relevance analysis

Table 1 shows the accuracy of the different techniques on the estimation of PM2.5 concentration in each station. The notation ANFIS-SC, ANFIS-GP, ANFIS-FC, SVR-KP, SVR-KG means Adaptive neural fuzzy inference system- subtractive clustering, Adaptive neural fuzzy inference system-Grid Partition, Adaptive neural fuzzy inference system-Fuzzy C means, Support vector machines-Kernel Polynomial, Support vector machines-Kernel Gaussian, respectively. The best results were obtained with ANFIS-FC and SVR-KP 93.4% for each. However, the difference among the systems can be considered as minimal 3%. The worst results were obtain by SVR-KG 87.5%.

Table 1. PM2.5 estimation -ANFIS and SVR

	Accuracy			
Technique	St-1 (%)	St- 2 (%)	St- 3 (%)	Mean
ANFIS-SC	94.0±0.4	92.3±0.3	93.4±0.3	93.23
ANFIS-GP	92.1±0.8	89.3±1.7	91.5±0.8	90.97
ANFIS-FC	94.4±0.4	92.2±0.4	93.6±0.3	93.4
SVR-KP	94.0±0.2	92.1±0.3	94.1±0.2	93.4
SVR-KG	87.2±0.5	88.3±0.5	87.1±0.6	87.5

4   Conclusions

In this study, meteorological and pollutant variables obtained from three nearby measurement stations were fused, and a relevance analysis using techniques based on RS for selecting was carried out. This allowed to obtain objectives, and accurate mechanisms to estimate the PM2.5. Besides, five regression systems were compared, where the results were similar. The best results were obtained using the three ANFIS and one SVM systems, which have demonstrated the capability of the variables selected together with data fusion of nearby stations to estimate the PM2.5 in specific geographical location, which demonstrated the effectiveness of our proposed procedure. However, the adjustment of parameters of the regression systems can be optimized using metaheuristic algortithms in order to obtain major results in terms of accuracy. In addition, this study should be extended using other database from other geographical locations in order to increase its generality.

Acknowledgments. This work was supported by the research project identified with code 267 at the "Institución Universitaria Salazar y Herrera" of Medellin, Colombia, CALAIRE Laboratory of "Universidad Nacional of Colombia", and the Area Metropolitana de Medellín, who supplied the database.

References

1. OMS | Calidad del aire (exterior) y salud, *WHO*. http://www.who.int/mediacentre/factsheets/fs313/es/. Accessed 24 Oct 2015
2. Dong, M., Yang, D., Kuang, Y., He, D., Erdal, S., Kenski, D.: PM2.5 concentration prediction using hidden semi-Markov model-based times series data mining. Expert Syst. Appl. **36**(5), 9046–9055 (2009)
3. Sun, W., Zhang, H., Palazoglu, A., Singh, A., Zhang, W., Liu, S.: Prediction of 24-hour-average PM2.5 concentrations using a hidden Markov model with different emission distributions in Northern California. Sci. Total Environ. **443**, 93–103 (2013)
4. Mishra, D., Goyal, P., Upadhyay, A.: Artificial intelligence based approach to forecast PM2.5 during haze episodes: a case study of Delhi, India. Atmos. Environ. **102**, 239–248 (2015)
5. Perez, P., Gramsch, E.: Forecasting hourly PM2.5 in Santiago de Chile with emphasis on night episodes. Atmos. Environ. Part A **124**, 22–27 (2016)
6. Zhou, Q., Jiang, H., Wang, J., Zhou, J.: A hybrid model for PM2.5 forecasting based on ensemble empirical mode decomposition and a general regression neural network. Sci. Total Environ. **496**, 264–274 (2014)
7. Antanasijević, D.Z., Pocajt, V.V., Povrenović, D.S., Ristić, M.Đ., Perić-Grujić, A.A.: PM10 emission forecasting using artificial neural networks and genetic algorithm input variable optimization. Sci. Total Environ. **443**, 511–519 (2013)
8. Feng, X., Li, Q., Zhu, Y., Hou, J., Jin, L., Wang, J.: Artificial neural networks forecasting of PM2.5 pollution using air mass trajectory based geographic model and wavelet transformation. Atmos. Environ. **107**, 118–128 (2015)
9. Kumar, A., Goyal, P.: Forecasting of daily air quality index in Delhi. Sci. Total Environ. **409**(24), 5517–5523 (2011)
10. Qin, S., Liu, F., Wang, J., Sun, B.: Analysis and forecasting of the particulate matter (PM) concentration levels over four major cities of China using hybrid models. Atmos. Environ. **98**, 665–675 (2014)
11. Velásquez, J.D., Olaya, Y., Franco, C.J.: Time series prediction using support vector machines. Ingeniare, 64–75 (2010)
12. Popoola, O., Munda, J., Mpanda, A., Popoola, A.P.I.: Comparative analysis and assessment of ANFIS-based domestic lighting profile modelling. Energy Build. **107**, 294–306 (2015)
13. Klaić, Z.B., Hrust, L.: Neural network forecasting of air pollutants hourly concentrations using optimised temporal averages of meteorological variables and pollutant concentrations. Atmos. Environ. **43**(35), 5588–5596 (2009)

14. Gardner, M.W., Dorling, S.R.: Neural network modelling and prediction of hourly NOx and NO2 concentrations in urban air in London. Atmos. Environ. **33**(5), 709–719 (1999)
15. Yildirim, Y., Bayramoglu, M.: Adaptive neuro-fuzzy based modelling for prediction of air pollution daily levels in city of Zonguldak. Chemosphere **63**(9), 1575–1582 (2006)
16. Hoshyaripour, G., Noori, R.: Uncertainty analysis of developed ANN and ANFIS models in prediction of carbon monoxide daily concentration. Atmos. Environ. **44**(4), 476–482 (2010)
17. Assareh, E., Behrang, M.A.: The potential of different artificial neural network (ANN) techniques in daily global solar radiation modeling based on meteorological data. Sol. Energy **84**(8), 1468–1480 (2010)
18. Pai, T.-Y., Hanaki, K., Su, H.-C., Yu, L.-F.: A 24-h forecast of oxidant concentration in Tokyo using neural network and fuzzy learning approach. CLEAN – Soil Air. Water **41**(8), 729–736 (2013)
19. Polat, K.: A novel data preprocessing method to estimate the air pollution (SO2): neighbor-based feature scaling (NBFS). Neural Comput. Appl. **21**(8), 1–8 (2001)
20. Jang, J.S.R.: ANFIS: adaptive-network-based fuzzy inference system. IEEE Trans. Syst. Man Cybern. **23**(3), 665–685 (1993)
21. Vapnik, V.N.: The Nature of Statistical Learning Theory. Springer New York, New York (2000)
22. Deo, R.C., Wen, X., Qi, F.: A wavelet-coupled support vector machine model for forecasting global incident solar radiation using limited meteorological dataset. Appl. Energy **168**, 568–593 (2016)
23. Orrego, D.A., Becerra, M.A., Delgado-Trejos, E.: Dimensionality reduction based on fuzzy rough sets oriented to ischemia detection. In: 2012 Annual International Conference of the IEEE Engineering in Medicine and Biology Society, pp. 5282–5285 (2012)
24. Chiu, S.L.: Fuzzy model identification based on cluster estimation. J. Intell. Fuzzy Syst. **2**(3), 267–278 (1994)
25. Lohani, A.K., Goel, N.K., Bhatia, K.K.S.: Improving real time flood forecasting using fuzzy inference system. J. Hydrol. **509**, 25–41 (2014)

Multivariate Functional Network Connectivity for Disorders of Consciousness

Jorge Rudas[1]([✉]), Darwin Martínez[2,3], Athena Demertzi[4,5], Carol Di Perri[5], Lizette Heine[5], Luaba Tshibanda[5], Andrea Soddu[6], Steven Laureys[5], and Francisco Gómez[7]

[1] Department of Biotechnology,
Universidad Nacional de Colombia, Bogotá, Colombia
jrudascas@gmail.com
[2] Department of Computer Science,
Universidad Nacional de Colombia, Bogotá, Colombia
[3] Department of Computer Science, Universidad Central, Bogotá, Colombia
[4] Institut du Cerveau et de la Moelle épinière,
Hôpital de la Pitié-Salpêtrière, Paris, France
[5] Coma Science Group, GIGA Research Center, University of Liège, Liège, Belgium
[6] Department of Physics and Astronomy, Western University, London, Canada
[7] Department of Mathematics, Universidad Nacional de Colombia, Bogotá, Colombia

Abstract. Recent evidence suggests that healthy brain is organized on large-scale spatially distant brain regions, which are temporally synchronized. These regions are known as resting state networks (RSNs). The level of interaction among these functional entities has been studied in the so called functional network connectivity (FNC). FNC aims to quantify the level of interaction between pairs of RSNs, which commonly emerge at similar spatial scale. Nevertheless, the human brain is a complex functional structure which is partitioned into functional regions that emerge at multiple spatial scales. In this work, we propose a novel multivariate FNC strategy to study interactions among communities of RSNs, these communities may emerge at different spatial scales. For this, first a community or hyperedge detection strategy was used to conform groups of RSNs with a similar behavior. Following, a distance correlation measurement was employed to quantify the level of interaction between these communities. The proposed strategy was evaluated in the characterization of patients with disorders of consciousness, a highly challenging problem in the clinical setting. The results suggest that the proposed strategy may improve the capacity of characterization of these brain altered conditions.

Keywords: Community · Disorders of consciousness · Hyperedge · Multivariate functional network connectivity · Resting state networks

1 Introduction

After severe brain injury, some patients may fall in coma. After coma, some of them may evolve to severely altered states of consciousness, such as, Vegetative

© Springer International Publishing AG 2017
C. Beltrán-Castañón et al. (Eds.): CIARP 2016, LNCS 10125, pp. 434–442, 2017.
DOI: 10.1007/978-3-319-52277-7_53

State/Unresponsive Wakefulness Syndrome (VS/UWS), in which patients open their eyes but remain unresponsive to external stimuli [1], or Minimally Conscious State (MCS), in which patients exhibit signs of fluctuating yet reproducible remnants of non-reflex behavior [2]. These conditions are collectively known as Disorders of Consciousness (DOCs).

In recent years, a novel neuroimaging protocol called resting state fMRI (functional Magnetic Resonance Imaging) has been used to study brain activity in pathological/pharmacological altered brain conditions. In this protocol, the subject rests in a magnetic resonator few minutes without being exposed to any stimuli, while their hemodynamic brain activity is recorded. Using this protocol, a sets of spatial regions with common functional behaviors have been identified. The different set of regions have been called Resting State Network (RSN) [3]. At least ten of these entities have been consistently identified in healthy subjects (default mode network (DMN), executive control network left (ECL), executive control network right (ECR), saliency, sensorimotor, auditory, cerebellum and three visual networks medial, lateral and occipital). These networks have been linked to cognitive/sensorial high level processes [3]. Recently, the interactions among these RSNs have been studied by using the so called functional network connectivity (FNC). FNC have been used to study brain dynamic during altered brain conditions, including, pharmacological alterations, pathological conditions and brain reconfiguration in response to external task modulations, among others [4–6]. Particularly, for DOC conditions it has been suggested that a dysfunctional time-sustained hyper-connectivity between executive control left (ECL) - executive control right (ECR) and visual medial - salience networks may be linked to the severe states of loss of consciousness [7]. In addition, the level of modularity between RSNs has been shown to be reduced in patients with DOC in comparison with healthy subjects [8]. Nevertheless, despite of recent advances in the description of these brain dynamics for patients in DOC conditions, a complete characterization of these neurological alterations remains as a major clinical challenge [9].

FNC aims to assess the level of interactions between spatially remote neurophysiological events. These interactions are typically studied for brain regions of similar spatial scale [10]. However, the human brain is a complex functional structure that may be partitioned into regions that will emerge at multiple spatial scales [11]. For instance, brain activity may result from the interaction of large scale regions, such as, lobes or large functional systems (visual system, default mode network, etc.) and small scale regions (Broca's area, hypothalamus and thalamus). These interactions across different spatial scales remains poorly studied [11]. The understanding of this phenomenon can potentially improve our knowledge about the DOC conditions [8].

Recently, the concept of community has been proposed to capture the segregation and integration phenomena, two major features that have been described for the brain activity in resting state [12]. Segregation refers to the organization of the brain dynamic in groups of specialized behavior, and integration is related to the existence of high levels of integration among some of these regions [12].

These two concepts have been used to group brain regions resulting in the concept of brain communities [12]. On the other hand, an extension of traditional graph approach also has been used in order to study the groups emerging of fMRI data, particularly, it has been extended the edge formulation to hyperedge, which the relation among the nodes into graph may be associated to more two nodes [13,14]. These hyperedges then, are similar to community definition, because, are groups of brain entities based an particular association measure. Recent evidence suggests that segregation and integration properties may be altered in DOC condition [8]. In this work, we propose a novel strategy to quantify the functional interaction between RSN brain communities or hyperedge. The proposed strategy accounts for linear and non-linear interactions between these entities, and considers the possibility of having communities/hyperedge emerging at different spatial scales. The proposed strategy was evaluated in the characterization of patients with DOC.

2   Materials and Methods

Figure 1 illustrates the proposed method. First, the fMRI data was preprocessed to account for the anatomical variability of the subjects (Sect. 2.2). Following, the individual fMRI resting state signal was decomposed into functional spatio-temporal components by using Independent Component Analysis (ICA). Later, an automatic matching procedure was applied to identify the set of functional components that corresponded to a set of predefined RSNs templates (Sect. 2.2). Then, a multi-objective optimization strategy was used to find a

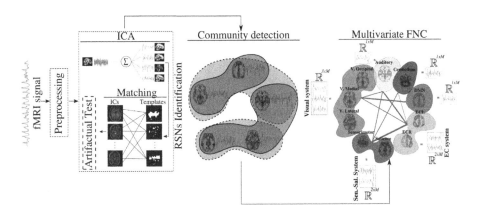

Fig. 1. Multivariate FNC approach. Firstly, the fMRI data were preprocessed to remove anatomical variability. Following, the fMRI signal was decomposed into functional components by using Independent Component Analysis (ICA). Later, a RSN identification procedure was applied on the resulting components to identify a set of RSNs. Subsequently, a multi-objective optimization method was used to find a set of RSN communities/hyperedge. Finally, the interaction among these groups was computed by using distance correlation.

set of communities/hyperedge among these RSNs (Sect. 2.3). Finally, the interactions between these groups were quantified by using a distance correlation measurement (Sect. 2.4).

2.1 Participants and Data Acquisitions

Data from 76 subjects were used for this study: 27 healthy controls (14 women, mean age 47 ± 16 years), 24 patients in minimally conscious state and 25 with vegetative state/unresponsive wakefulness syndrome (20 women, mean age 50 $\pm$ 18 years). All patients were clinically examined using the French version of the Coma Recovery Scale Revised (CRS-R) [15]. For each subject, fMRI resting data were acquired in a 3T scanner (Siemens medical Solution in Erlangen, Germany). Three hundred fMRI volumes multislice $T2^*$-weighted functional images were captured (32 slices; voxel size: $3 \times 3 \times 3\,mm^3$; matrix size 64; repetition time $= 2000\,ms$; echo time $= 30\,ms$; flip angle $= 78$; field of view $= 192 \times 192\,mm^2$). A structural T1 image was also acquired for anatomical reference. More details of the subjects demography can be found in [16]. These data are used exclusively for research of our collaborators.

2.2 fMRI Preprocessing

fMRI data was processed using SPM8[1]. Preprocessing included: realignment, coregistration of functional onto structural data, segmentation of structural data, normalization into MNI space and spatial smoothing with a Gaussian kernel of 8 mm. Large head motions were corrected using ArtRepair[2].

RSNs Identification. The first step for the RSN identification was the fMRI signal decomposition into sources of neuronal/physiological origin. For this task, we used ICA, which aims to decompose the signal into a set of statistically independent components (ICs) of brain activity. In traditionally ICA, one considers the mixture as linear and the sources as statistically mutually independent and non-Gaussian [17]. In the fMRI data, the spatial dimension is much greater than temporal one, then, we used spatial ICA (sICA), which decompose the signal into maximally independent spatial maps [18]. In sICA each spatial map (source) have an associated time course, which corresponds to the common dynamic exhibit by this component. The RSNs time-courses obtained with sICA were subsequently used for all the FNC computations. For the sICA decomposition 30 components were used, this selection was performed based on previous work that have shown that this number of components is enough to characterize the different RSNs both for healthy controls and patients with DOC [16]. After the sICA decomposition, the different RSNs were identified at individual level. A machine learning based labeling method was applied to discriminate between IC of "neuronal" or "artifactual" origin [16].

---

[1] http://www.fil.ion.ucl.ac.uk/spm.
[2] http://cibsr.stanford.edu/tools/ArtRepair/ArtRepair.htm.

2.3 Community Detection

A community in *graph* theory refers to a groups of nodes which share common properties within a graph [12]. This definition naturally assumes the existence of an underlying graph representation, i.e., the existence of a binary relation which characterizes the interaction between pairs of RSNs. This assumption may oversimplify the brain dynamic [8]. In order to account for this limitation, the notion of community can be reformulated to be not dependent of the graph representation. For our case, a community was defined as the set of nodes that *share* some commons feature or properties, similarly to hyperedge definition into hyper graph theory [8]. By using this formulation, the RSN community detection problem can be reformulated as a clustering problem. For this case, a segregation-modularity measurement was used to quantify the level of integration for groups of RSNs. This measure guided the clustering process. More particularly, two measures were used to quantify the level of segregation of a specific partition of the RSNs set [8]. The first measure, aims to quantify the level of information of each community by using a similar idea to the Kapur criterion, i.e., maximize the sum of entropy for all possible communities, high values for this measure can be expected when each community contains highly informative RNSs [19]. The second measure, considers the combination between both inter-community and intra-community variances by using the Otsu criterion [20]. These two measures were used in a multi-objective optimization approach to find the optimal partition that maximizes the brain segregation for the set of RSNs [8]. This process was applied to the data and a community model was obtained for the DOC subjects [8]. This model was composed by six communities: auditory, cerebellum, DMN, executive control system (ECN left and right), sensorimotor-salience and visual system (visual occipital, medial and lateral).

2.4 Multivariate FNC Method

After the RSN community construction, each one is composed by one or more RSNs. Therefore, each community can be described by one or more time-courses. In order to quantify the level of interaction between different communities a distance correlation (DC) measure was used [21]. DC aims to quantify the level of non-linear interaction between two time series of arbitrary dimension [21]. In this work, this ability was exploited to study the interactions between RSN communities at different spatial scales. DC measures the dependencies between two random variables X and Y with finite moments in arbitrary dimension [21]. For defining DC, let $(X, Y) = \{(X_k, Y_k)|k = 1, 2, \ldots, n\}$ an observed random sample of the joint distribution of random vectors X in $\mathbb{R}^p$ and Y in $\mathbb{R}^q$. Using these samples a transformed distance matrix A can be defined as follows:

$$a_{kl} = \|X_k - X_l\|, \quad \bar{a}_{k.} = \frac{1}{n}\sum_{l=1}^{n} a_{kl}, \quad \bar{a}_{.l} = \frac{1}{n}\sum_{k=1}^{n} a_{kl}, \quad \bar{a}_{..} = \frac{1}{n^2}\sum_{k,l=1}^{n} a_{kl}$$

$A_{kl} = a_{kl} - \bar{a}_{k\cdot} - \bar{a}_{\cdot l} + \bar{a}_{\cdot\cdot}$, where $k, l = 1, 2, \ldots, n$. Similarly, B is defined to characterize distances between samples for Y. An empirical distance between X and Y can be defined by

$$V_n^2(X,Y) = \frac{1}{n^2}\sum_{k,l=1}^{n} A_{kl}B_{kl} \quad \rightarrow \quad R_n(X,Y) = \begin{array}{cc} \frac{V_n^2(X,Y)}{\sqrt{V_n^2(X)V_n^2(Y)}} & V_n^2(X)V_n^2(Y) > 0 \\ 0 & V_n^2(X)V_n^2(Y) = 0 \end{array}$$

where $V_n^2(X) = V_n^2(X,X)$. Note that A and B can be computed independently of p and q, and both contain information about the distance between sample elements in X and Y. $V_n^2(X,Y)$ is a measure of the distance between the probability distribution of the joint distribution and the product of the marginal distributions, i.e., $V_n^2(X,Y)$ quantifies $\|f_{X,Y} - f_X f_Y\|$, with f_X and f_Y the characteristic function of X and Y, respectively, and $f_{X,Y}$ the joint characteristic function [21]. The DC corresponds to a normalized version of $V_n^2(X,Y)$, which takes values between 0 and 1, with zero corresponding to statistical independence between X and Y, and 1 total dependency.

To quantify the level of interactions among communities, previously the DC measure was obtained. For this, let $r = 10$, the number of RSNs after applying the RSN identification process (see Sect. 2.2). Each community can be represented by the time-courses of their RSNs, i.e., $c_i = \{t_1, t_2, \ldots, t_{k_i}\}$ with k_i the number of RSN that conform the ith-community and $C = \{c_1, c_2, \ldots c_p\}$, $p \leq r$ the set of emerging communities on fMRI signal for a particular subject, as shown in Fig. 2A. The time-courses in the set c_i can be used to build a multivariate random variable Q_i by concatenating the set of RSN time courses in arbitrary order, with $Q_i \in \mathbb{R}^{k_i \times M}$, where M is the size of the time course. The interaction between pairs of communities was computed by using distance correlation. A total of $p!/(2!(p-2)!)$ possible pairs were assessed. The interaction between pairs of communities was defined as the distance correlation between the community c_i and the community c_j, i.e., $R(Q_i, Q_j)$, see Fig. 2B. This multivariate FNC approach was calculated at individual level using the communities previously found. A total of 15 interactions were studied.

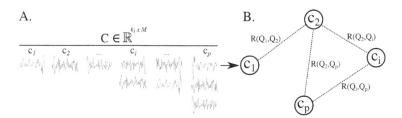

Fig. 2. Graph construction. Each community c_i is composed of k_i time-courses and they represent the nodes in the graph representation. The edges in the graph of communities, are the distance correlation measure R between each possible relationship among communities.

2.5   Group Analysis

To quantify the capacity of the proposed method to characterize patients with DOC a group analysis was performed. In particular, the discrimination capacity between different diagnostic categories was evaluated in two settings. Firstly, in the characterization of differences between MCS *vs* VS/UWS subjects. Secondly, in the discrimination of healthy subjects and patients with DOC (MCS and VS/UWS). A Student's t-test was used to assess significant values of interaction at the group level ($p < 0.05$). Differences among healthy, VS/UWS and MCS subjects, were assessed by using a two sample t-test ($p < 0.0033$), these computations were Bonferroni corrected to account multiple comparisons [7,22].

3   Results

Figure 3 shows the average of the multivariate FNC among communities for the three studied groups: healthy (left), MCS (middle) and VS/UWS (right) subjects. As observed, there is no a reconfiguration of the connectivity through the groups and the level of strength for some interactions changed in DOC conditions in contrast to healthy subjects, for instance, there is a reduction of the average functional connectivity between Visual System and Salience-Sensorimotor communities in VS/UWS patients in contrast to healthy subjects. In addition, there is an increase of the average functional connectivity between the auditory and the executive control communities in MCS patients in contrast to healthy subjects. Significant differences between VS/UWS and MCS patients were found for the relation between the visual system and cerebellum (relation marked with ** in Fig. 3). In addition, significant differences were found among healthy subjects and VS/UWS patients for the visual system - cerebellum and cerebellum - auditory connections (relation marked with *** in Fig. 3). These results suggest

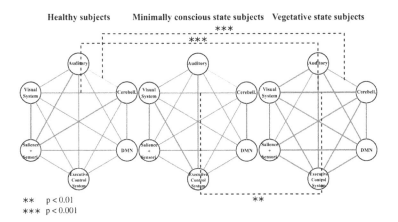

Fig. 3. Multivariate functional connectivity at group level. The line thickness represents the connectivity strength between two different communities.

that an increase of FNC among different communities maybe linked to loss of consciousness, this is consistent with our previous results [7,22].

4   Conclusions

In this work, we proposed a novel strategy to study of brain dynamic in resting state by using a multivariate FNC approach. For this, a community or hyper-edge detection process was used to determinate the RSNs with similar behavior. Recently, similar approaches has been proposed [13,14], however, have not been applied in the study of disorders of consciousness. On the other hand, in this work distance correlation measure was used to quantify the amount of interaction between different communities or hyperedge, probably into different spatial scale. Recently, analogous approach was proposed to redefine the functional network connectivity, using distance correlation to estimate the level of interaction based in all the signals from two regions of interest without the need to average over voxels first [23]. Finally, the proposed strategy was evaluated in patients with DOC. The results suggest that the new strategy improves the capacity of characterization of these patients. In particular, the interaction among visual system and cerebellum can be a potential biomarker for these conditions.

Acknowledgment. This work was supported by the James McDonnell Foundation, the European Space Agency, Mind Science Foundation, the French Speaking Community Concerted Research Action, the Belgian inter university attraction pole, the Public Utility Foundation "Université Européenne du Travail", "Fondazione Europea di Ricerca Biomedica", the University Hospital of Liége and agreement Gobernación del Magdalena - COLCIENCIAS (Instituto Colombiano para el Desarrollo de la Ciencia y la Tecnologa).

References

1. Laureys, S., et al.: BMC Med. **8**(1), 1–4 (2010)
2. Giacino, J., et al.: Nat. Rev. Neurol. **10**, 99–114 (2014)
3. Damoiseaux, J., et al.: Proc. Nat. Acad. Sci. **103**(37) (2006)
4. Liu, X., et al.: Neuropsychiatr. Dis. Treat **11**, 2573–2581 (2015)
5. He, H., et al.: J. Affect. Disord. **190**, 483–493 (2016)
6. Dae-Jin, K., et al.: NeuroImage Part A **124**, 550–556 (2016)
7. Rudas, J., Guaje, J., et al.: Conf. Proc. IEEE Eng. Med. Biol. Soc. (2014)
8. Rudas, J., et al.: Process. SPIE **9681**, 96 810U–96 810U–9 (2015)
9. Hannawi, Y., et al.: Neurology **84**(12), 1272–1280 (2015)
10. Kolchinsky, A., et al.: Front. Neuroinform. **8**, 66 (2014)
11. Fujita, A., et al.: arXiv e-prints, December 2015
12. Fortunato, S.: CoRR, vol. abs/0906.0612 (2009). http://arxiv.org/abs/0906.0612
13. Davison, E.N., et al.: PLoS Comput. Biol. **11**(1), e1004029 (2015)
14. Jie, B., et al.: Med. Image Anal. **32**, 84–100 (2016)
15. Schnakers, C., et al.: Brain Inj. **20**(10), 786–792 (2008)
16. Demertzi, A., Gómez, F., et al.: Cortex **52**, 35–46 (2014)
17. Cecile, B., et al.: J. Stat. Softw. **44**(9) (2011)

18. McKeown, M.: Hum. Brain Mapp. **6**, 160–188 (1998)
19. Kapur, J., et al.: Comput. Vis. Graph. Image Process. **29**(3), 273–285 (1985)
20. Otsu, N.: IEEE Trans. Syst. Man Cybern. **9**(1), 62–66 (1979)
21. Dae-Jin, K., et al.: Ann. Stat. **35**(6), 2769–2794 (2007)
22. Rudas, J., et al.: Process. SPIE **9287**, 92 870P–92 870P–6 (2015)
23. L. G. et al.: NeuroImage **135**, 16–31 (2016)

Non-parametric Source Reconstruction via Kernel Temporal Enhancement for EEG Data

C. Torres-Valencia$^{(\boxtimes)}$, J. Hernandez-Muriel, W. Gonzalez-Vanegas,
A. Alvarez-Meza, A. Orozco, and M. Alvarez

Engineering Faculty, Automatics Research Group,
Universidad Tecnológica de Pereira, Pereira, Colombia
`cristian.torres@utp.edu.co`

Abstract. Source reconstruction from EEG data is a well know problem in the neuroscience field and affine areas. There are a variety of applications that could be derived form an adequate source reconstruction of the cerebral activity. In recent years, non-parametric methods have been proposed in order to improve the reconstruction results obtained from the original Low Resolution Tomography (LORETA) like approaches. Nevertheless, there is room for improvement since EEG data could be processed to enhance the reconstruction process via some temporal and spatial transformations. In this work we propose the use of a Kernel-based temporal enhancement (kTE) of the EEG data for a preprocessing stage that improves the results of source reconstruction into the non-parametric framework. Three metrics of source error localization named as Dipole Localization Error (DLE), Euclidean Distance (ED) and Dipole dispersion (DD) are computed for comparing the performance of swLORETA in different scenarios. Results shows an evident improvement in the reconstruction of brain source from the proposed kTE in comparison to the state of art non-parametric approaches.

Keywords: LORETA · Kernel · EEG data · Temporal enhancement · swLORETA

1 Introduction

The correct understanding of brain activity has been attracting significant relevance in the development of systems of assistance in the neuroscience field. In the last decade several works have been developed in order to improve the mapping and accurate detection of neural activity inside the brain in different scenarios [1,4,8]. The brain activity has been modeled by currents, more precisely by current dipoles that fit the neuronal potentials being generated within the brain cortex. Electroencephalography (EEG) has proved to be an effective method for capturing the electric brain activity. The EEG data has the property of a high temporal resolution but low spatial resolution [1,4,8].

Authors are with the Automatic Research Group from the Engineering Faculty of the "Universidad Tecnológica de Pereira".

© Springer International Publishing AG 2017
C. Beltrán-Castañón et al. (Eds.): CIARP 2016, LNCS 10125, pp. 443–450, 2017.
DOI: 10.1007/978-3-319-52277-7_54

Since there are few orders of difference between the number of possible brain sources with the number of electrodes, the problem of source reconstruction becomes ill-posed [1]. Due the ill-posed nature of the problem, some spatial and temporal priors should be included in order to find a unique solution. Methods that are used to solve the inverse problem are categorized in two approaches [6]. The first approach known as parametric methods, assume few dipoles inside the brain with unknown position and orientation, then, a non-linear problem is solved in order to estimate these parameters. The second approach of non-parametric methods, assume several dipoles within the brain volume, with fixed positions and orientations. A linear problem is solved in order to find the amplitudes and orientations of these assumed sources from the EEG data [4]. There is still a wide open field in which the EEG data could be effectively used for brain source reconstruction. Since the Minimum norm Estimate (MNE) and Low Resolution Tomography (LORETA) were proposed, some variations to these methods have been used for addressing this problem [5,10]. From the non-parametric methods, a Bayesian framework could be used to deduce most of the methods based on regularization approaches of the inverse problem. In [8], a technical note for solving the source reconstruction problem from M/EEG data within a Bayesian framework is presented [8].

The objective when source reconstruction from EEG data is addressed is to reduce the localization error. The original LORETA and MNE approaches show a considerable high error localization of the sources [1]. Some variations of LORETA have tried to address the low resolution with some spatial and temporal transforms of the data and the prior covariance definition. Standardized LORETA (sLORETA) and standardized weighted LORETA (swLORETA) methods give lower error localization of the sources. Standardized LORETA (sLORETA) [11] proposes the use of a initialization of the current density estimate, given by MMN solution and then standardized values of the current density are inferred [11]. The sLORETA algorithm proves to reduce the localization dipole error results compared to LORETA. The swLORETA proposed in [9], uses a Single Value Decomposition (SVD) of the leadfield matrix in order to improve the spatial resolution. Some unsolved issues of this approach are related to the capability of the SVD linear relationships to modeling the noise adequately.

In this work, we propose the use of kernel functions in order to enhance and model the temporal relationships in the EEG data. This enhancement could derive in a considerable improvement in the source reconstruction results within the LORETA framework. A Gaussian kernel is applied over the EEG data, projecting it to another dimension in which a selection of the most relevant temporal modes within the data could be performed. This reduced data then is the input for the sLORETA and swLORETA methods to obtain the results of the source reconstruction. Different metrics of source localization are computed to compare the proposed method against the classic LORETA framework. The Dipole localization error (DLE), the euclidean distance (ED) and the dipole dispersion (DD) shows a considerable improvement of the proposed method in the source reconstruction problem.

2   Materials and Methods

Description of the data used to develop this work is included in this section. A theoretical background of LORETA as the non-parametric method selected for source reconstruction is also presented. Finally the description of the metrics employed and the experimental setup are depicted. The Statistical Parametric Mapping (SPM) [3] is used in the 12.1 version, and is combined into Matlab 2014a. The data for source reconstruction analysis corresponds to generated synthetic EEG data. The SPM implementation for LORETA is used in order to compare and validate the results obtained from the proposed methodology.

2.1   Inverse Problem

The source reconstruction problem from M/EEG data could be formulated as follows:

$$\Phi = \mathbf{KJ}, \tag{1}$$

where $\Phi \in \mathbb{R}^{N \times s}$ is the set of observations from the N electrodes in s time samples, $\hat{J} \in \mathbb{R}^{N \times M}$ is known as the current density in M possible sources inside the brain and $K \in \mathbb{R}^{N \times M}$ is a gain matrix known as the Leadfield matrix [10]. For solving the inverse problem, the forward problem should be solved first in order to establish the relationships between the neural activity and how it maps into the set of M/EEG sensors [7].

2.2   LORETA

The solution of the inverse problem proposed in [10] sacrifices spatial resolution by assuming smoothness in the solution space. The functional that has to be minimized is presented in Eq. (2);

$$\min_{j} \left\| BW\hat{J} \right\|^{2}, \tag{2}$$

where W is the diagonal matrix with elements $w_{ii} = \|K_i\|$ and $B_{3M \times 3M}$ is the Laplacian operator that ensures smoothness in the solution [10]. The solution for the estimated current density is obtained following Eq. (3);

$$\hat{J} = T\Phi, \tag{3}$$

with $T = (WB^{T}BW)^{-1}K^{T}(K(WB^{T}BW)^{-1}K^{T})^{\dagger}$ and $\dagger$ denotes the Moore-Penrose inverse operator. Furthermore, swLORETA proposes a variation of the sLORETA method which includes a initialization of the current density from the Minimum Norm solution (MN) $\hat{J}_{MN}$ and the use of a regularization parameter α in the functional to be minimized, see Eq. (4). The swLORETA algorithm incorporates a SVD over the leadfield matrix in order to compensate the variations of the sensor sensitivity to currents at different depths [2,9]. The SVD over $\mathbf{K}$

is performed on the three columns corresponding to each dipole source location l in the three primary axes as $K_l = U_i \Sigma_i V_i^T$. From these singular values, the current source density covariance matrix is constructed as Eq. (5).

$$F = \|\Phi - KJ\|^2 + \alpha \|J\|^2 \tag{4}$$

$$\left[S_j^{1/2} \right] = S^{-1/2} \otimes I_{3 \times 3}, \tag{5}$$

with $I_{3 \times 3}$ is the identity matrix, $\otimes$ is the Kronecker product and $S \in \mathbb{R}^{M \times M}$ a diagonal matrix with S_l elements being the maximum sensitivity at voxel l [2]. Furthermore, works like [1,8] have proposed the use of a pre-processing stage. This pre-processing is related to the computing of the discrete cosine transform (DCT) over the EEG data. The DCT allows a filtering in the frequency domain before computing the SVD over the data. This approach of the DCT computing is performed within the kTE approach.

2.3   Kernel Temporal Enhancement

In order to improve the swLORETA method, we propose the use of a Gaussian kernel function to perform a temporal mapping of the data obtained from the sensors Φ (and after the DCT is computed).

$$\hat{\Phi} = exp\left(-\frac{\|\Phi^T - \Phi\|^2}{2\sigma^2} \right), \tag{6}$$

obtaining a transformed input space $\hat{\Phi}$. From this transformed input space the first N_r eigenvalues with higher magnitude are selected to form the new input space from the corresponding eigenvectors as $\hat{V} \in \mathbb{R}^{s \times N_r}$. The kernel parameter σ works as a noise filter, depending on a selected value, the transformation to the new input space will consider more or less noise. The threshold for the selection of the number N_r of temporal modes is set as the 95% of the normalized magnitude of the sum of all the eigenvalues. Then using the mapped input space, the covariance matrix $S \in \mathbb{R}^{s \times s}$ is computed as a tensorial product between the linear and a Gaussian kernel relationships in the projected space, with $\mathbb{K}_{\hat{\Phi}_{cc'}} = exp(- \left\| \hat{\Phi}_c V - \hat{\Phi}'_c V \right\| / 2\sigma_{\hat{\Phi}}^2)$.

$$\mathbf{S} = \hat{\Phi} V (\hat{\Phi} V)^T \otimes \mathbb{K}_{\Phi}. \tag{7}$$

2.4   Reconstruction Error Metrics

For the validation of the results, some metrics that measure the correct source reconstruction are implemented. The Dipole Localization Error (DLE) measures the difference between the position of the simulated dipole $\hat{V}_o$ against the position of the estimated dipole $\hat{V}_e$, see Eq. (8). Another metric that measures the error in the source reconstruction is the Euclidean distance (ED) between the simulated and estimated dipole locations, see Eq. (9). Finally, The percent relative error is computed for the variance of the potential in the 1% source points surrounding the simulated dipole location. This gives us a measure of the dipole dispersion (DD), see Eq. (10).

$$DLE = \|\hat{V}_o - \hat{V}_e\| \tag{8}$$

$$E_p = \sqrt{\left(V_{o_x} - V_{r_x}\right)^2 + \left(V_{o_y} - V_{r_y}\right)^2 + \left(V_{o_z} - V_{r_z}\right)^2} \tag{9}$$

$$DD = \frac{\left|Var(E(\hat{V}_o)_{0.1}) - Var(E(\hat{V}_o)_{0.1})\right|}{Var(E(\hat{V}_o)_{0.1})} * 100\% \tag{10}$$

2.5 Experimental Setup

A set of $M = 8196$ dipoles correspond to the discretization of the head volume and the sources that are simulated could take any of this M possible locations. The input data correspond of synthetic EEG signals with $N = 128$ electrodes and $s = 250$ time samples. The EEG signals are simulated with random locations of the source at different depths across the head volume. An analysis of different values of the SNR in the EEG data is performed. Signals with levels of 3dB, 5dB, 7dB, 10dB and 20dB are generated. From the LORETA framework, four scenarios of the method and pre-processing stages are analyzed. First, an swLORETA with spatial-SVD of the leadfield, temporal-SVD and linear covariance matrix calculation of the current density, with tag $LORLL$. Second, swLORETA with spatial-SVD of the leadfield, temporal-SVD and the Gaussian kernel covariance computation of the current density with tag $LORLG$. Finally, third and four scenario corresponde to swLORETA+kTE with spatial-SVD of the leadfield and kTE enhancement (with the DCT included) with linear covariance of the current density for $LORGL$ and Gaussian kernel covariance of the current density for $LORGG$. For the kTE the kernel parameter σ in Eq. (6) is selected as factors of 1 times the value of the median of the kernelized input space. Finally, 20 EEG signals are generated for each SNR scenario in order to asses the statistical significance of each method following the error metrics computation.

3 Results

Figure 1 shows the reconstruction of the brain source for one random dipole. As can be seen form the figure the different combinations of preprocessing shows similar graphically results, due to this quantification of the dipole localization error is needed. Table 1 presents the DLE related to the different experiments for the three proposed SNR levels.

In Table 1, it can be seen that the methods involving kTE, show less DLE compared with the other strategies for different SNR, with 7.13 being the lowest

Table 1. Quantified results of source reconstruction, Dipole Localization Error (DLE), norm l1 distance error (ED) and Percent relative error/dipole dispersion (DD)

Method	Signal to noise ratio				
	3dB	5dB	7dB	10dB	20dB
Dipole Localization Error (DLE)					
LORLL	10.45 ± 7.72	9.01 ± 8.15	8.49 ± 8.35	8.15 ± 8.57	7.15 ± 8.92
LORLG	10.45 ± 7.72	9.01 ± 8.15	8.49 ± 8.35	8.15 ± 8.57	7.15 ± 8.92
LORGL	8.52 ± 9.15	8.52 ± 9.15	8.52 ± 9.15	7.13 ± 9.16	8.26 ± 8.47
LORGG	8.52 ± 9.15	8.52 ± 9.15	8.52 ± 9.15	7.13 ± 9.16	8.26 ± 8.47
Norm l1 distance (ED) error					
LORLL	18.10 ± 13.37	15.69 ± 14.11	14.71 ± 14.47	14.11 ± 14.85	12.38 ± 15.44
LORLG	18.10 ± 13.37	15.69 ± 14.11	14.71 ± 14.47	14.11 ± 14.85	12.38 ± 15.44
LORGL	14.76 ± 15.86	14.76 ± 15.86	14.76 ± 15.86	12.35 ± 15.87	14.30 ± 14.68
LORGG	14.76 ± 15.86	14.76 ± 15.86	14.76 ± 15.86	12.35 ± 15.87	14.30 ± 14.68
Percent relative error (DD)					
LORLL	292.79 ± 240.44	281.13 ± 233.68	275.72 ± 222.77	259.05 ± 179.14	202.28 ± 134.26
LORLG	292.79 ± 240.44	281.13 ± 233.68	275.72 ± 222.77	259.05 ± 179.14	202.28 ± 134.26
LORGL	104.12 ± 147.27	79.91 ± 94.62	93.96 ± 128.61	85.66 ± 111.35	70.82 ± 68.83
LORGG	104.12 ± 147.27	79.91 ± 94.62	93.96 ± 128.61	85.66 ± 111.35	70.82 ± 68.83

mean value for 10dB SNR. Also, it can be seen how for the lower values of SNR the strategies of *LORGL* and *LORGG* obtained lower mean value of the DLE in comparison with the other two strategies. From the second metric of error from Eq. (9), it can be seen a similar behavior compared to the DLE. For the lower values of SNR the kTE method proves to achieve lower mean errors across the 10 repetitions of the experiment. The lower mean level is 12.35 obtained for the 10dB signals. The last metric computed in this analysis is the percent relative error between the variance of the 1% points (from the whole head model) surrounding the dipole location in the original data, see Eq. (10).

The results presented in Table 1 for the DD metric, show that in the analysis of this metric, the proposed method of kTE improves the accuracy in the detection of the brain source. It can be seen that for all the SNR levels, the kTE method proves to reduce considerably the error in the source dispersion in comparison to the methods without kTE. In this particular metric the lower result is presented for the SNR of 20dB with a mean percent relative error of 70.82%. Some of the results in Table 1 are presented as box diagrams in Fig. 1(d), (e) and (f). A graphical analysis of these figures allows to compare the behavior of the three metrics for different experiments. While the DLE and ED metrics shows an improved mean value for the kTE methods against the classical methods, the DD shows an important improvement in the cases were kTE was used.

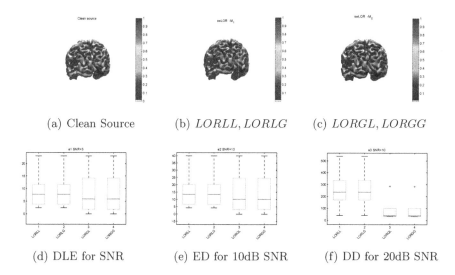

(a) Clean Source   (b) *LORLL, LORLG*   (c) *LORGL, LORGG*

(d) DLE for SNR   (e) ED for 10dB SNR   (f) DD for 20dB SNR

Fig. 1. Source reconstruction from EEG data for a simulated dipole. (a) presents the original source, (b) presents the results for reconstruction using the methods without kTE and figure (d) shows the reconstruction when kTE is used as was proposed in the experimental setup, (d, e and f) are Box diagram for the DLE, ED and DD of some source reconstruction experiments

4   Discussion and Conclusions

An analysis of the results presented in the Sect. 3, allows us to determine that an improvement on the source recognition within the LORETA framework could be performed. In this case, the results when the kTE+LORETA method was employed shows less error on the localization than the classical LORETA methods. As higher as the influence of noise is, the kTE proves to model and remove the noise interference of the signals in the process of source reconstruction. This shows that the kernel function maps adequately the original data within a space where there are less influence of noise. The three metrics proposed to compare the methods shows lower mean levels of DLE, ED and DD when the kTE is employed. Even there is no statistical difference in DLE and ED metrics, since the standard deviation of the results is higher in the kTE case, the DD metric shows high difference for all the SNR levels. The results presented by the DD metric show a considerable improvement of the low resolution exhibited by LORETA. The results presented in this work allow us to conclude that an improvement with a pre-processing stage for EEG source reconstruction is possible. The mapping of the data using a kernel function as the kTE propose, allows to filter the noise of the data and improves the source localization results. Some improvements could be studied for this method, as the process of selection of the kernel parameters and the temporal modes to be included from the kernelized input space.

Acknowledgment. The authors would like to thank Universidad Tecnológica de Pereira, Singleclick SAS and Instituto de Epilepsia y Parkinson del eje cafetero NEU-ROCENTRO S.A. Author C.A.T-V. was funded by the program "Formación de alto nivel para la ciencia, la tecnología y la innovación - Doctorado Nacional - Convoctoria 647 de 2014" of COLCIENCIAS. Authors would like to thank the project "Diseño y desarrollo de una plataforma basada en cloud computing para la prestacin de servicios de apoyo al diagnstico clnico: Aplicacin en el rea de las neurociencias" funded by NEUROCLOUD. Code: FP44842-584-2015.

References

1. Belardinelli, P., Ortiz, E., Barnes, G., Noppeney, U., Preissl, H.: Source reconstruction accuracy of MEG and EEG Bayesian inversion approaches. PLoS ONE **7**(12), 1–16 (2012)
2. Boughariou, J., Jallouli, N., Zouch, W., Slima, M.B., Hamida, A.B.: Spatial resolution improvement of EEG source reconstruction using swLORETA. IEEE Trans. NanoBiosci. **14**(7), 734–739 (2015)
3. Dale, A.M., Liu, A.K., Fischl, B.R., Buckner, R.L., Belliveau, J.W., Lewine, J.D., Halgren, E.: Dynamic statistical parametric mapping: combining FMRI and MEG for high-resolution imaging of cortical activity. Neuron **26**(1), 55–67 (2000)
4. Grech, R., Cassar, T., Muscat, J., Camilleri, K.P., Fabri, S.G., Zervakis, M., Xanthopoulos, P., Sakkalis, V., Vanrumste, B.: Review on solving the inverse problem in EEG source analysis. J. NeuroEng. Rehabil. **5**(1), 1–33 (2008)
5. Hämäläinen, M.S., Ilmoniemi, R.J.: Interpreting magnetic fields of the brain: minimum norm estimates. Med. Biol. Eng. Comput. **32**(1), 35–42 (1984)
6. Hansen, P.C.: Rank-Deficient and Discrete Ill-Posed Problems: Numerical Aspects of Linear Inversion. Society for Industrial and Applied Mathematics, Philadelphia (1998)
7. Haufe, S.: An extendable simulation framework for benchmarking EEG-based brain connectivity estimation methodologies. In: 2015 37th Annual International Conference of the IEEE Engineering in Medicine and Biology Society (EMBC), pp. 7562–7565, August 2015
8. López, J.D., Litvak, V., Espinosa, J.J., Friston, K., Barnes, G.R.: Algorithmic procedures for Bayesian MEG/EEG source reconstruction in SPM. NeuroImage **84**, 476–487 (2014)
9. Palmero-Soler, E., Dolan, K., Hadamschek, V., Tass, P.A.: swLORETA: a novel approach to robust source localization and synchronization tomography. Phys. Med. Biol. **52**(7), 1783–1800 (2007)
10. Pascual-Marqui, R.D., Michel, C.M., Lehmann, D.: Low resolution electromagnetic tomography: a new method for localizing electrical activity in the brain. Int. J. Psychophysiol. **18**(1), 49–65 (1994)
11. Pascual-Marqui, R.D., et al.: Standardized low-resolution brain electromagnetic tomography (sLORETA): technical details. Methods Find. Exp. Clin. Pharmacol. **24**(Suppl D), 5–12 (2002)

Tsallis Entropy Extraction
for Mammographic Region Classification

Rafaela Alcântara$^{(\boxtimes)}$, Perfilino Ferreira Junior, and Aline Ramos

Computer Science Departament, Mathematics Institute,
Federal University of Bahia, Salvador, Bahia, Brazil
rafa.alcantara23@gmail.com, perfeuge@gmail.com, shaysantanar@gmail.com

Abstract. Breast cancer is the second disease responsible for women's death in the world. To reduce the number of cases, screening mammography is used to detect this disease. To improve exam accuracy results, computer-aided systems (CAD) have been developed to analyze the mammography and provide statistics based on image features extracted. This paper presents a novel approach for a computer-aided detection system (CADe) based on Tsallis entropy extraction from quantized gray level co-occurrence matrix (GLCM) from mass images. A comparison study is presented based on a feature extraction scheme using weigthed Haralick features. The best result accuracy rate was 91.3% from Tsallis entropy based on GLCM matrix using 24 feature measures.

Keywords: Tsallis entropy · Mammography · Classification · SVM · Haralick features

1 Introduction

According to [1], breast cancer mortality rate has been increasing among the years. In 2012, 521.907 million women in the world died from this disease and the prediction for 2015 was around 560.407 million deaths. Breast cancer is considered a heterogeneous disease and can be caused by many facts including genetic mutation on genes BRCA1 and BRCA2, responsible for 5% to 10% of breast cancer cases.

Screening mammography aids the radiologists to find anomalies on initial stage although, in some cases, diagnosis can be difficult by breast density and a false-positive diagnosis can be provided, resulting on an unnecessary biopsy. Furthermore, about 10% to 30% of results provided by radiologists analysis are failed [2].

To solve this problem, computer-aided detection and diagnosis (CADe and CADx, respectively) have been developed to assist radiologists and doctors on a final result for the patient, avoiding false-positive results and unnecessary biopsy. The main propose is to reduce the mammography review using a computer instead of a second human evaluation and to improve the accuracy based on different methods extracted from this type of image [2].

© Springer International Publishing AG 2017
C. Beltrán-Castañón et al. (Eds.): CIARP 2016, LNCS 10125, pp. 451–458, 2017.
DOI: 10.1007/978-3-319-52277-7_55

This paper presents a new CADe system for mammography image classification into mass and non-mass tissue. Its main contribution is the mass classification based on Tsallis entropy from quantized version of gray-level co-occurrence matrix (GLCM) of mammograms.

2   Related Works

Among the years, many researchers presented different textural extraction methods to improve breast tissues analysis. In [3] a CADe system for mammogram images based on feature extraction from a co-occurrence matrix (GLCM) and a gray-level run-length matrix (GLRLM) was developed. The main proposal was a comparative study over some classifiers options and the best accuracy rate from [3] was 83.8%.

In [4] another work involving breast cancer detection from GLCM matrices are presented. From these matrices, four statistical features [5] were extracted. In addition, three other shape features were used to compose the feature vector. In this work, two classifiers were used: K-means and Support Vector Machine. The best accuracy rate was 93.11%.

Tsallis entropy is frequently used for region segmentation. The q-index value variation provides a set of threshold values that can be used to binary segmentation as shown in [6].

3   Methodology

This section presents the methodology used to compose all experiments in this paper.

3.1   Gray-Level Co-occurrence Matrix (GLCM)

According to [5], a digital image can be expressed as a cartesian product between the spatial domains L_x and L_y composing the resolution cells, as follows:

$$f : L_x \times L_y \to G. \tag{1}$$

where $L_x = \{1, 2, 3, \ldots, N_x\}$, $L_y = \{1, 2, 3, \ldots, N_y\}$ and G represents the gray-levels of each cell.

Based on these concepts, [5] proposed a GLCM to extract texture information from an image f. Such matrices are calculated from a specific direction θ, where $\theta \in \{0°, 45°, 90°, 135°\}$ and a specific distance d, where $d \in \{1, 2, 3, 4\}$. From these matrices, fourteen statistical features were calculated to compose the vector of texture information for image classification.

In this work, two GLCM-based strategies were used for feature extraction: the first one with Tsallis entropy and the second one with a weighted Haralick features [5]. Both cases according to a multilevel decomposition scheme as follows.

3.2   Multilevel Decomposition

Gray-level quantization allows to reduce the misclassification of breast lesions. Hence the noise-induced effects are decreased. Normalized co-occurrence matrices represents a relationship between the gray-levels of two pixels and it can be seen as gray-level images. Here, an uniform quantization scheme were used in which the gray-levels are quantized into separated bins. In this case, no consideration is made about the gray-levels distribution of image. A quantized version of GLCM image f can be expressed as

$$g = \lfloor \alpha_1 f + \alpha_2 \rfloor. \tag{2}$$

where g is the quantized GLCM image, $\lfloor \bullet \rfloor$ is the floor function, α_1 and α_2 are coefficients defined as

$$\begin{cases} \alpha_1 = \frac{l}{f_{max} - f_{min}} \\ \\ \alpha_2 = -\alpha_1 f_{min} \end{cases} \tag{3}$$

where l is the desired number of quantization levels, f_{min} and f_{max} are the minimum and maximum gray-levels values within image f, respectively. Notice that the quantized image g is in the range $[0, l]$.

3.3   Weighted Haralick Features

In effect to compare the aforementioned feature vectors performance another test was made. Six Haralick features were choosed-four out of 14 most used and two others. Fixing θ, for each quantization level l and corresponding distance d are computed

$$F_{i,d} = \frac{f_i^{(1)}(d).\ 3 + f_i^{(2)}(d).\ 4 + \cdots + f_i^{(6)}(d).\ 8}{3 + 4 + \cdots + 8} = \frac{1}{33}.\sum_{l=3}^{8} f_i^{(l-2)}(d).\ l\ . \tag{4}$$

where $f_i^{(l-2)}(d)$ is the $i-th$ Haralick feature for the quantization level l ($l = \{3, 4, 5, 6, 7, 8\}$), and a distance d ($d = 1, 2, 3, 4$), with $i = \{1, 2, 3, 4, 5, 6\}$. In this case, f_1, f_2, f_3, f_4, f_5 and f_6 are correlation, ASM, entropy, energy, sum of entropy and sum of variance, respectively [5]. Value of quantization level works as a weight to the corresponding calculated feature from GLCM.

3.4   Tsallis Entropy

From normalized GLCM, as described above, a set of features were extracted using concepts of Tsallis entropy. Such measure generalizes the Boltzmann-Gibbs entropy from thermodynamics concepts. Equation (5) represents Tsallis entropy as follows

$$S = \frac{1 - \sum_{i=0}^{k}(p_i)^q}{q - 1}. \tag{5}$$

where p_i corresponds to the probability of a gray level in a range of size k from a normalized GLCM matrix. The q parameter represents a real value that varies according to the application.

3.5   Classification

After the feature set creation, the next step was the classification of mammography images. In this paper the concepts of Support Vector Machine (SVM) were used through an auxiliary library called libSVM [7]. In order to provide a better result for classification, the radial basis function (RBF) kernel was chosen. Equation (6) represents this kernel.

$$K(x, y) = e^{-\gamma \|x - y\|^2}. \tag{6}$$

where γ is a positive number which could be choosen by the user. In our experiments a best value of γ and a classifying error parameter, C, were provided by a python script called **grid.py**, implemented on [7].

4   Database Acquisition

For this work, 594 images (297 mass and 297 non-mass ROIs) were randomly chosen from Digital Database Screening Mammography (DDSM) [8]. Such database was a collaborative work and contains 2620 cases, divided by normal, benign and cancer cases.

4.1   ROI Extraction

For ROI extraction step, an approach based on chain code values provided from DDSM database text files was developed to minimize muscular region and mammography background interference.

Using chain code coordinates for nodule boundary, four values were calculated: maximum and minimum height, and maximum and minimum width. Thus, a central point coordinate (x_0, y_0) was calculated to compose bounding box center as shown in Fig. 1(a).

According to [9], the two most efficient bounding box size for mammography classification are 32×32 and 64×64. In this paper, all ROIs were extracted with a 64×64 size bounding box from the central pixel coordinate (x_0, y_0) (see Fig. 1(b)).

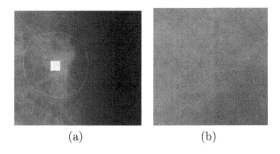

| (a) | (b) |

Fig. 1. (a) ROI selection (b) ROI after extraction

5   Experiments

For all experiments a fixed θ was used, where $\theta \in \{0°, 45°, 90°, 135°\}$ and the distance d varied between four others values ($d \in \{1, 2, 3, 4\}$). These values were selected from experiments described in [5]. For each ROI, six quantized new images ($2^3, 2^4, 2^5, 2^6, 2^7, 2^8$) were created. From these six new images, GLCM matrix with the fixed θ was calculated for each four distances d. Each GLCM matrix was normalized and Tsallis entropy was extracted using Eq. (2). As a comparative study, weighted Haralick features were extracted from normalized GLCM to compose another texture information vector.

Five values were used to measure classification method performance: accuracy (ACC), sensibility (SENS), specificity (SPEC), positive predict value (PPV) and negative predict value (NPV).

5.1   Experiment I: Fixed Direction θ and $q \in [0.1, 1.9]$

For the first step of this experiment, 19 q-index values were selected on the range $[0.1, 1.9]$ with an increase rate equal to 0.1 for each value. Figure 2 presents results from Tsallis entropy feature extraction steps.

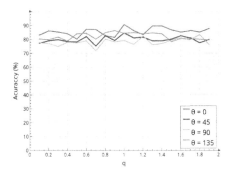

Fig. 2. Accuracy rates with fixed direction θ and $q \in [0.1, 1.9]$

It is possible to conclude that for the most q-index values, the best results were provided for $\theta = 0°$.

Table 1 presents three best results from Tsallis entropy feature extraction. The largest rate of sensibility was achieved in last case (85.43%) and the corresponding specificity was 94.33%. This indicates a better capability, for the pair of pointed parameters, to identify the non-mass group. The largest value of accuracy was 89.88% for two couple of parameters $(q, \theta) = (1.3, 0°)$ and $(q, \theta) = (1.4, 0°)$.

Table 1. Performance of mass/non-mass classification for $q \in [0.1, 1.9]$

q	θ	C	γ	SPEC (%)	SENS (%)	ACC (%)	PPV (%)	NPV (%)
1.3	0°	2048	0.03125	95.55	84.21	89.88	94.98	85.82
1.4	0°	32768	0.0078125	94.33	85.43	89.88	93.78	86.62
1.9	135°	512	0.0078125	89.47	86.23	87.85	89.12	86.67

The second step consists on weighted Haralick features average extraction as a comparative study on mammography classification field. Table 2 provides three best accuracy values for this strategy.

Table 2. Performance of mass/non-mass classification for weighted Haralick features

θ	C	γ	SPEC (%)	SENS (%)	ACC (%)	PPV (%)	NPV (%)
0°	32768	0.00195312	85.83	80.97	83.40	85.11	81.85
45°	128	0.5	68.02	72.87	70.45	69.50	71.49
90°	512	0.125	81.78	83.00	82.39	82.00	82.79
135°	8192	0.03125	74.09	66.00	70.04	71.81	68.54

The largest accuracy rate was 83.40% and the attained sensibility for this case was 80.97%.

5.2 Experiment II: Two Refined Ranges: [1.31, 1.39] and [1.41, 1.49]

Based on the Experiment I results, two new q-index ranges were refined: [1.31, 1.39] and [1.41, 1.49] with an increase rate equal 0.01 for each value. Figure 3(a) and (b) shows the accuracy rate results obtained for each q-index on these ranges. Again, it is possible to conclude that $\theta = 0°$ provide the best results.

The best accuracy were 91.3%, for the couple of parameters $(q, \theta) = (1.31, 0°)$, in the first range and 89.68%, for $(q, \theta) = (1.49, 0°)$, in the second one. Table 3 shows the three best results for these cases.

Finally, Table 4 provides a comparative board with some related works and the approach developed in this paper.

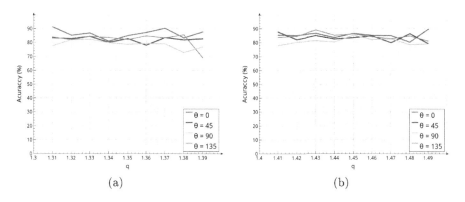

Fig. 3. (a) Accuracy rates with fixed direction θ and $q \in [1.31, 1.39]$ (b) Accuracy rates with fixed direction θ and $q \in [1.41, 1.49]$.

Table 3. Performance of mass/non-mass classification for $q \in [1.31, 1.39] \cup [1.41, 1.49]$

q	θ	C	γ	SPEC (%)	SENS (%)	ACC (%)	PPV (%)	NPV (%)
1.31	0°	32768	0.0078125	93.52	89.06	91.3	93.22	89.53
1.39	0°	32768	0.00195312	90.69	84.62	87.65	90.09	85.5
1.49	0°	2048	0.0078125	96.76	82.59	89.68	96.23	84.75

Table 4. Related works comparative

Title	Method	Classifier	Database	ACC (%)
Mavroforakis, M.E (2002) [3]	GLCM/GLRLM	Neural network	130	85.4%
Martins, L.O. et al. (2009) [4]	GLCM	K-Means/SVM	1177	93.11%
Mata, B.N.B.U. (2010) [10]	GLCM	Naive Bayes	11	82.4%
Our approach	GLCM/SVD	SVM	594	91.3%

6   Conclusions and Future Works

This paper provides an exhaustive amount of tests using the concepts of GLCM, Tsallis entropy and Haralick features. The main propose was to analyze and select the best value of q-index that provides highest accuracy rate for mass and non-mass classification on mammography exam. Entropic and the weighted Haralick features strategies were compared. In this way, all the experiments took into account the same number of features. Results lead to the conclusion that Tsallis entropy is a promising measure, once that the size of feature vector used does not need be long.

As a future work the authors intend to investigate how automatically to select the q-index value. For the choose range [0.1, 1.9] in this paper, the best value of q-index was 1.31, providing an accuracy rate of 91.3%.

References

1. Globocan Cancer Fact Sheets: Breast Cancer. http://globocan.iarc.fr/Default. aspx. Accessed 03 July 2015
2. Jalalian, A., Mashohor, S.B., Mahmud, H.R., Saripan, M.I.B., Ramli, A.R.B., Karasfi, B.: Computer-aided detection/diagnosis of breast cancer in mammography and ultrasound: a review. Clin. Imaging **37**(3), 420–426 (2013)
3. Mavroforakis, M., Georgiou, H., Cavouras, D., Dimitropoulos, N., Theodoridis, S.: Mammographic mass classification using textural features and descriptive diagnostic data. In: 2002 14th International Conference on Digital Signal Processing, DSP 2002, vol. 1, pp. 461–464 (2002)
4. Martins, L., Junior, G.B., Silva, A.C., de Paiva, A.C., Gattass, M.: Detection of masses in digital mammograms using k-means and support vector machine. ELCVIA: Electron. Lett. Comput. Vis. Image Anal. **8**(2), 39–50 (2009)
5. Haralick, R., Shanmugam, K., Dinstein, I.: Textural features for image classification. IEEE Trans. Syst. Man Cybern. **6**, 610–621 (1973)
6. Mohanalin, B., Kalra, P.K., Kumar, N.: A novel automatic microcalcification detection technique using Tsallis entropy a type II fuzzy index. Comput. Math. Appl. **60**(8), 2426–2432 (2010)
7. Chang, C.-C., Lin, C.-J.: A library for support vector machines. ACM Trans. Intell. Syst. Technol. **2**, 27:1–27:27 (2011)
8. Heath, M., Bowyer, K., Kopans, D., Kegelmeyer, P., Moore, R., Chang, K., Munishkumaran, S.: Current status of the digital database for screening mammography. In: Karssemeijer, N., Thijssen, M., Hendriks, J., van Erning, L. (eds.) Digital Mammography, vol. 13, pp. 457–460. Springer, Heidelberg (1998)
9. Garcia-Manso, A., Garcia-Orellana, C.J., Gonzalez-Velasco, H., Gallardo-Caballero, R., Macias Macias, M.: Consistent performance measurement of a system to detect masses in mammograms based on blind feature extraction. BioMed. Eng. Online **12**, 2–18 (2013)
10. Mata, B., Meenaksh, M.: A novel approach for automatic detection of abnormalities in mammograms. In: 2011 IEEE Recent Advances in Intelligent Computational Systems (RAICS), pp. 831–836 (2011)

Edge Detection Robust to Intensity Inhomogeneity: A 7T MRI Case Study

Fábio A.M. Cappabianco[1,4]([⊠]), Lucas Santana Lellis[1,4], Paulo Miranda[2,4], Jaime S. Ide[3,4], and Lilianne R. Mujica-Parodi[3,4]

[1] Inst. de Ciência e Tecnologia, Univ. Federal de S. Paulo, S. José dos Campos, Brazil
fcappabianco@gmail.com
[2] Dept. of Computer Science, Univ. of São Paulo (USP), São Paulo, Brazil
[3] Dept. of Biomedical Engineering, Stony Brook Univ. School of Medicine, Stony Brook, NY 11794, USA
[4] Dept. of Radiology, A. A. Martinos Center for Biomedical Imaging, Massachusetts General Hospital, Charlestown, MA 02129, USA

Abstract. Edge detection is a fundamental operation for computer vision and image processing applications. As of 1986, John Canny proposed a methodology that became known due to its simplicity, small number of parameters, and high accuracy. The method was designed to optimally detect, locate, and trace single edges over each local gradient maximum. Since then, a number of works were proposed but none of these improvements were capable of dealing with non-uniform intensity, which are notably present in ultra high field magnetic resonance imaging (MRI). In this paper, we evaluate the effects of inhomogeneity correction over automatic edge detection methods over 7T MRI. Importantly, we propose a non-supervised edge detection method which improves the accuracy of state of the art in 28.0% as detecting head and brain edges.

Keywords: Biomedical imaging · Edge detection · Inhomogeneity · MRI

1 Introduction

Edge detection has been of ultimate importance for numerous applications in the context of computer vision and image processing such as feature extraction [1], and noise detection [2]. As it is the primary step for several other computational procedures, its is widely applied to medical imaging [3,4] and to other areas.

Since 1986, the methodology proposed by Canny [5] became widely used because of its simplicity, small number of parameters, and due its features such as good detection, good localization, and edge singleness. Afterwards, a number of works proposed improvements specially to parameter selection [6–10] and relevant edge recovery [11–13]. Nevertheless, no previous works enhanced Canny methodology to address Intensity non-uniformity in a satisfactory way.

F.A.M. Cappabianco—Thanks to CNPq (486988/2013-9) and Fapesp for funding.

C. Beltrán-Castañón et al. (Eds.): CIARP 2016, LNCS 10125, pp. 459–466, 2017.
DOI: 10.1007/978-3-319-52277-7_56

Intensity non-uniformity may be described as a low-frequency noise, as it appears in the literature [14]. In this context, image inhomogeneity is described as a multiplicative low frequency noise showed in Fig. 1(a) and in Eq. 1, where J is the ideal image, I is the input image, B is the inhomogeneity, N is a high-frequency noise, and operator $*$ is the Hadamard product.

$$I = J * B + N, \tag{1}$$

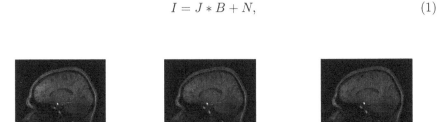

(a)   (b)   (c)

Fig. 1. Example 7 Tesla T1-weighted magnetic resonance image of the head. (a) The input image with very high inhomogeneity effect. (b),(c) Inhomogeneity effect partially corrected using N3 and N4 methodologies, respectively.

There are quite a few methodologies designed to correct the inhomogeneity effect of MRI without inserting artifacts into the image [15,18]. Still, we are obviously not interested in methods that require edge detection or image segmentation as a pre-processing, since accurate edge detection is our final goal. Therefore, only two of them are well suited for our experiments. The first is the Non-parametric Nonuniform intensity Normalization [16] (N3) that was proposed in 1998 and estimates the inhomogeneity effects employing B-splines with a regular grid of control points in a interactive process. The second is denominated N4, and was proposed in 2010 by Tustison et al. [17] as an improvement to N3. It uses a more robust B-spline and a modified optimization strategy. Figure 1(b) shows the partial correction of the inhomogeneity effect by applying N3 over the input image (a).

In this paper, we make an evaluation of the effects of inhomogeneity correction prior to edge detection in 7T magnetic resonance imaging (MRI) of the brain. 7T images are particularly susceptible to intensity non-uniformity due to the strong bias field, and have attracted the attention of recent development of inhomogeneity correction methods [18]. We also propose a novel methodology to handle inhomogeneity while detecting edges in a non-supervised fashion.

The remaining of this paper is organized as follows: Sect. 2 presents related works; Sect. 3 describes the proposed methodology; Sects. 4 and 5 contain the experiments and the conclusions, respectively.

2   Related Works

Canny operator was one of the first methods introduced with mathematically formulated goals. The goals consist of detecting edge free from false borders, preserve all important edges, keep the minimum distance between the detected and actual borders, and set a single occurrence for each of them.

The basic Canny operator implementation follows the steps below:

(1) Smooth the image applying a convolution with a Gaussian kernel removing image noise and irrelevant details. This step requires the definition of the Gaussian standard deviation σ.
(2) Compute the directional gradient of the pixels. A convolution with Sobel kernel may be used here.
(3) Suppress pixels whose gradient are not maximal in the gradient direction, to achieve edge singleness.
(4) Apply a hysteresis threshold, eliminating the weaker edges corresponding to less important details. This step requires the definition of higher (τ_h) and lower (τ_l) threshold parameters values.

The main parameters of Canny algorithm are σ, τ_h, and τ_l. These parameters influence the amount of noise and details to be suppressed from the image.

Two kinds of strategy were designed to improve the original Canny method: automatic parameter selection; and recovery of mistakenly suppressed edges.

2.1   Parameter Selection

In [10], the authors adopted the automatic Otsu thresholding for τ_h with $\sigma = 2.0$.

Huo et al. [9] proposed a probabilistic model to compute τ_l. The hypothesis is that lower intensity edges correspond to the texture that must be avoided.

Hui et al. [8] proposed the use of a sliding window to compute image signal to noise level and define σ after it.

Instead of computing the τ_h and τ_l to find the edges in a so called "instability zone", [7] proposed to find the most appropriate thresholds in a large range, based on edge intensity and connectivity.

More recently, Rong et al. [6] proposed the use of adaptive thresholding methods, depending on the quantity of edges contained in the input image. It also introduced the concept of gravitational field intensity to replace image gradient preserving more relevant borders during non-maximum suppression.

2.2   Edge Refining

In [13] the authors computed the "major edges" by the traditional maximum suppression step, and detected "minor edges", i.e. maximal intensity points in other directions than their gradient direction. Afterwards, only minor edges that are connected to major edges are preserved.

In the same fashion, Worthington [12] proposed a method that uses local curvature consistency to adaptively control the smoothness of gradient estimates, based on their neighborhood curvature, reducing the lost of relevant pixels.

Finally, Wang and Jin [11] tried to recover more suppressed edge pixels by including diagonal direction information both in gradient computation and in the non-maximum suppression step.

3   Proposed Methodology

The proposed methodology detects edges while dealing with non-uniformity effects in a non-supervised framework based on Canny algorithm. As inhomogeneity is a regional scale effect, the idea behind the methodology is to apply the thresholding process in multiple different scales and partitions of the image, so that important low contrast edges are detected. Also, it may be interesting to apply a filter, removing the components that are not connected to the edges found in higher scales. Algorithm 1 describes the proposed procedure:

Algorithm 1. Bias Free Edge Detection
Inputs: Image Img_0, canny parameters σ, τ_{fl}, τ_{fh}, and subscales div_{max}
Outputs: Edge map $Edge$

1: $Img_s \leftarrow Gaussian(Img_0, \sigma)$.
2: $Grad \leftarrow Gradient(Img_s)$.
3: $Supr \leftarrow Suppression(Grad)$.
4: Compute τ_l, τ_h from τ_{fl}, τ_{fh}.
5: $Edge \leftarrow Hysteresis(Supr, \tau_l, \tau_h)$
6: **for** $div \leftarrow 2$ to div_{max} **do**
7:    $dims \leftarrow Dimensions(Img_s)/div$
8:    **for** each partition $Part$ of $Supr$ with dimensions $dims$ **do**
9:       Compute τ_l, τ_h from τ_{fl}, τ_{fh}.
10:       $Edge \leftarrow Edge + Hysteresis(Part, \tau_l, \tau_h)$.
11:       Remove the edges that are not connected to the image delineation. (Optional).
12:    **end for**
13: **end for**
14: Return $Edge$.

Input parameters are: the input image Img_0, the standard deviation of the Gaussian filter (σ); fractions ($\tau_{fl}, \tau_{fh} \in [0.0, 1.0]$), used to compute hysteresis thresholds τ_l, τ_h; and the number times (div_{max}) we divide the input image in smaller sections. The output is a binary image ($Edge$) with the detected edges.

In line 1, we smooth Img_0 applying a Gaussian filter with standard deviation σ and generate Img_s. Then, in line 2, we compute gradient magnitude and direction ($Grad$) from Img_s. The non-maximum edges are suppressed in line 3 resulting in $Supr$. Hysteresis with parameters τ_l and τ_h is applied to $Supr$ in line 5, generating the first set of edges $Edge$. The calculation of τ_l and τ_h may be a simple multiplication of τ_{fl} and τ_{fh} by the maximum gradient value, or a more sophisticated method as proposed by [7].

The external loop in lines 6–13 computes the size *dims* of image partition with successively smaller dimensions. Then, in the internal loop (lines 8–12) *Supr* is partitioned into $Supr_p$ accordingly. For each partition, in line 10, a new hysteresis is applied, extending the current set of edges *Edge*. Note that τ_h and τ_l computed in line 9 might be different for each partition and for each scale, since they are based on local gradient values. Optionally, the new edges that are not connected to the previous ones are discarded in line 11. The external loop is repeated until the maximum number of subscales div_{max} of the image domain is reached. Finally, *Edge* is returned in line 14.

The key idea behind the algorithm is that instead of getting lower thresholds to detect all important edges influenced by inhomogeneity effect, it might be more insightful to employ distinct thresholds for each image section. In this way, there will be no spurious details in brighter regions.

The optional step in line 11 may be interesting for applications in which structures are strongly connected.

4   Experiments

In this paper, we implemented the unsupervised method proposed in [7] (MEDINA) and compared it with the addition of the non-uniformity treatment proposed in this paper (NOVEL). Notice that any improvement described in Sect. 2 could also be added to the framework proposed in this paper.

We used Baddeley's measure [19] as comparison metric, given by Eq. 2:

$$\Delta_w^p(A, B) = \left[\frac{1}{|X|} \sum_{x \in X} |w(d(x, A)) - w(d(x, B))|^p \right]^{1/p} \tag{2}$$

where A and B are the processed and ground-truth edge pixel sets, respectively. X is the set of all pixel in the image domain, $d(x, A)$ is the distance of pixel x to the closest border pixel in A, $w(t)$ is a continuous and concave function on interval $[0, \infty[$, $|X|$ is the size of set X, and p is a scalar greater than 1. For an uncut distance function we chose $w(t) = \min(t, c)$, with $c = \sqrt{M^2 + N^2}$, where M and N are the number of columns and lines in the image, and $p = 2$.

For our database, we used 6 ultra high field of 7 Tesla (7T) T1-weighted magnetic resonance (MR) images of the human head. These are state-of-the-art images with higher resolution obtained at cost of field inhomogeneity. MR data was collected in a Siemens Magnetom at Martinos Center for Biomedical Imaging, Massachusetts General Hospital Charlestown, MA.

The experiments consisted in detecting edges belonging to the head and to the brain, without taking too many details. This procedure is extremely important to estimate the pose of the brain by means of the head position, and to strip the brain in the image [20]. It is very important to avoid detecting too many edges of small structures since they may confuse skull-stripping algorithms. Evidently, it is also fundamental to keep the most relevant and long borders.

In a initial qualitative analysis we manually set parameters to the best set of values to detect external brain and head edges in 3D images. Figure 2 shows the improvement achieved by our methodology (Inhom) with the optional step (line 11 of Algorithm 1) with 3 scales over 7 Tesla MR images. We executed traditional Canny algorithm over the input image, and over corrected images by N3, and N4. It is clear that it is impossible to detect the lower edges around the head and neck and at the same time avoid marking several edges from brain gyri and sulci without using the proposed methodology (Compare the first two columns in Fig. 2 to the last column).

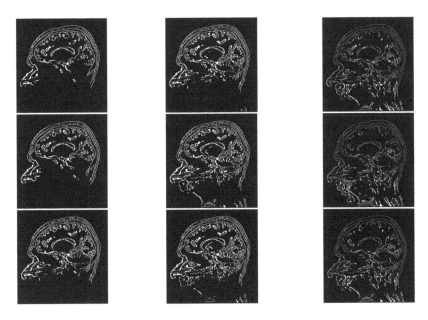

Fig. 2. Qualitative experimental results with best parameters tuned manually. Input images showed in Fig. 1(a)(b)(c) generated results in the first, second, and third lines, respectively. First column: Canny algorithm with $\tau_l = 0.6$, $\tau_h = 0.95$, $\sigma = 1.0$. The cerebellum was not delineated, but borders of small structures already appear inside the brain. Second column: Canny with $\tau_l = 0.45$, $\tau_h = 0.9$, $\sigma = 1.0$. The neck was not delineated, but almost all internal borders already appear in the brain and part of the cerebellum. Third column: Proposed approach with $\tau_l = 0.45$, $\tau_h = 0.9$, $\sigma = 1.0$, $div_{max} = 3$. The main edges were correctly delineated, with a few borders inside the brain. Dark gray, light gray, and white pixels were found in the first, second, and third scale iteration, respectively.

In our qualitative evaluation of 7T MR images, we run NOVEL with 2 scales and MEDINA with $\sigma \in \{0.6, 1.2, 1.8\}$ against manually segmented brain and head contours by a specialist. We used 18 test cases, being two sagittal slices extracted from each 3D image. Table 1 shows the accuracy of the methods according to Baddeley's measure. The best results were achieved with $\sigma = 0.6$, in which NOVEL improved MEDINA in 28.0% in average. All comparisons between corresponding NOVEL and MEDINA are significant according to paired t-test, achieving $p < 0.0062$.

Table 1. Mean and standard deviation of Baddeley's measure between estimated edges and the ground truth in 7T dataset. Testes executed over input images, and corrected images by N3 and N4. The lower is the value, the more accurate is the method.

σ	Input		N3		N4	
	MEDINA	NOVEL	MEDINA	NOVEL	MEDINA	NOVEL
0.6	0.088 ± 0.020	0.073 ± 0.012	0.097 ± 0.028	0.077 ± 0.024	0.128 ± 0.080	0.091 ± 0.032
1.2	0.121 ± 0.035	0.084 ± 0.017	0.187 ± 0.128	0.111 ± 0.055	0.194 ± 0.144	0.111 ± 0.046
1.8	0.482 ± 0.389	0.185 ± 0.100	0.528 ± 0.354	0.240 ± 0.081	0.566 ± 0.362	0.258 ± 0.085

Contrarily to what happened to MEDINA, our method improved the accuracy with higher σ values. Higher σ values blur the smaller details of the image, but NOVEL was able to get stronger contours in darker areas due to its local analysis. Also, executing MEDINA over N3, or N4 corrected images did not improve the results, as someone may expect. This happened because the automatic parameter selection tend to give higher priority to the borders in brighter regions (i.e. gyri and sulci) to the detriment of the darker edges that seems to be more relevant to human eyes (i.e. inferior border of the brain and cerebellum, neck and chin), even though they were considerably enhanced by N3 and N4. Therefore, correcting inhomogeneity is not sufficient for a more precise and clean head and brain edge detection.

5   Conclusion

We showed in this paper that current inhomogeneity correction algorithms are not sufficient for head and brain edge detection in 7T MRI images using the original framework proposed by John Canny. It became clear that both Canny's and Medina's methods lack a better treatment for images corrupted by non-uniform low-frequency noise. We also proposed a non-supervised method that solved this problem efficiently, detecting important local edges weakened by shades or inhomogeneity in MRI. Future works include an evaluation of the proposed method with respect to non-maximum suppression edge refinement.

References

1. Stehling, R.O., Nascimento, M.A., Falcão, A.X.: A compact and efficient image retrieval approach based on border/interior pixel classification. In: Proceedings of 11th International Conference on Information and Knowledge Management, pp. 102–109 (2002)
2. Palma, C.A., Cappabianco, F.A.M., Ide, J.S., Miranda, P.A.V.: Anisotropic diffusion filtering operation and limitations-magnetic resonance imaging evaluation. World Congr. Int. Fed. Autom. Control **47**, 3887–3892 (2014)

3. Gang, X., Zhou, Y, Zhou, H., Zheng, Y.: Ultrasound image edge detection based on a novel multiplicative gradient and Canny operator. Ultrasonic Imaging 1–13 (2014)
4. Salman, N.H., Ghafour, B.M., Hadi, G.M.: Medical image segmentation based on edge detection techniques. Adv. Image Video Process. 3(2), 1 (2015)
5. Canny, J.: A computational approach to edge detection. IEEE Trans. Pattern Anal. Mach. Intell. 8(6), 679–698 (1986)
6. Rong, W., Li, Z., Zhang, W., Sun, L.: An improved Canny edge detection algorithm. In: 2014 IEEE International Conference on Mechatronics and Automation (ICMA), pp. 577–582. IEEE (2014)
7. Medina-Carnicer, R., Muñoz-Salinas, R., Yeguas-Bolivar, E., Diaz-Mas, L.: A novel method to look for the hysteresis thresholds for the Canny edge detector. Pattern Recogn. 44(6), 1201–1211 (2011)
8. Hui, P., Ruifang, Z., Shanmei, L., Youxian, W., Lanlan, W.: Edge detection of growing citrus based on self-adaptive Canny operator. In: 2011 International Conference on Computer Distributed Control and Intelligent Environmental Monitoring (CDCIEM), pp. 342–345. IEEE (2011)
9. Huo, Y.-K., Wei, G., Zhang, Y.-D., Wu, L.-N.: An adaptive threshold for the canny operator of edge detection. In: 2010 International Conference on Image Analysis and Signal Processing (IASP), pp. 371–374. IEEE (2010)
10. Fang, M., Yue, G., Yu, Q.: The study on an application of Otsu method in canny operator. In: International Symposium on Information Processing (ISIP), pp. 109–112. Citeseer (2009)
11. Xun, W., Jin, J.-Q.: An edge detection algorithm based on improved Canny operator. In: 2007 Seventh International Conference on Intelligent Systems Design and Applications (ISDA 2007), pp. 623–628. IEEE (2007)
12. Worthington, P.L.: Enhanced Canny edge detection using curvature consistency. In: 2002 Proceedings of 16th International Conference on Pattern Recognition, vol. 1, pp. 596–599. IEEE (2002)
13. Ding, L., Goshtasby, A.: On the Canny edge detector. Pattern Recogn. 34(3), 721–725 (2001)
14. Cappabianco, F.A.M., de Miranda, P.A.V., Ide, J.S., Yasuda, C.L., Falcao, A.X.: Unraveling the compromise between skull stripping and inhomogeneity correction in 3T MR images. In: 2012 25th SIBGRAPI Conference on Graphics, Patterns and Images (SIBGRAPI), pp. 1–8. IEEE, August 2012
15. Vovk, U., Pernus, F., Likar, B.: A review of methods for correction of intensity inhomogeneity in MRI. IEEE Trans. Med. Imaging 26(3), 405–421 (2007)
16. Sled, J., Zijdenbos, A., Evans, A.: A nonparametric method for automatic correction of intensity nonuniformity in MRI data. IEEE Trans. Med Imaging 17(1), 87–97 (1998)
17. Tustison, N.J., Avants, B.B., Cook, P.A., Zheng, Y., Egan, A., Yushkevich, P.A., Gee, J.C.: N4ITK: improved N3 bias correction. IEEE Trans. Med. Imaging 29(6), 1310–1320 (2010)
18. Ganzetti, M., Wenderoth, N., Mantini, D.: Intensity inhomogeneity correction of structural MR images: a data-driven approach to define input algorithm parameters. Front. Neuroinform. 10 (2016)
19. Baddeley, A.J.: An error metric for binary images. Robust Comput. Vis. 5978 (1992)
20. Miranda, P.A.V., Cappabianco, F.A.M., Ide, J.S.: A case analysis of the impact of prior center of gravity estimation over skull-stripping algorithms in MR images. In: IEEE International Conference on Image Processing, pp. 675–679. IEEE (2013)

Fine-Tuning Based Deep Convolutional Networks for Lepidopterous Genus Recognition

Juan A. Carvajal[1], Dennis G. Romero[1(✉)], and Angel D. Sappa[1,2]

[1] Escuela Superior Politécnica del Litoral, ESPOL,
Facultad de Ingeniería en Electricidad y Computación,
Campus Gustavo Galindo Km 30.5 Vía Perimetral,
P.O. Box 09-01-5863, Guayaquil, Ecuador
dgromero@espol.edu.ec
[2] Computer Vision Center, Edifici O, Campus UAB,
Bellaterra, 08193 Barcelona, Spain

Abstract. This paper describes an image classification approach oriented to identify specimens of lepidopterous insects at Ecuadorian ecological reserves. This work seeks to contribute to studies in the area of biology about genus of butterflies and also to facilitate the registration of unrecognized specimens. The proposed approach is based on the fine-tuning of three widely used pre-trained Convolutional Neural Networks (CNNs). This strategy is intended to overcome the reduced number of labeled images. Experimental results with a dataset labeled by expert biologists is presented, reaching a recognition accuracy above 92%.

1 Introduction

The ever increasing computational capabilities of mobile devices together with the advances in multimedia technology has opened novel possibilities that go beyond communication between users. One of this novel possibilities lies on the image acquisition and recognition, from anywhere at any time. In this sense, different tools have been developed to contribute with biodiversity categorization, driven by recognition techniques based on features obtained from images, for a variety of applications: registration of known animal species; identification of unrecognized genus; or academic tools for learning about animal species.

In biology, for example, students learn to identify insect species as a part of their formal instruction. For this task, it is necessary to consider tools that provide summarized information in order to be presented to students, which represents a challenge for teachers who must summarize large amounts of information about class, family, genus and details that distinguish between them. A suitable method for automatic recognition of insect species would facilitate the registration of biodiversity, by enabling a wider use of these records and the knowledge we have about the species, also by reinforcing the inclusion of different areas of research such as life sciences, computer science and engineering.

The current work is a continuation of previous studies about automatic identification of lepidopterous insects, which discussed an image feature extraction

© Springer International Publishing AG 2017
C. Beltrán-Castañón et al. (Eds.): CIARP 2016, LNCS 10125, pp. 467–475, 2017.
DOI: 10.1007/978-3-319-52277-7_57

methodology based on the butterfly shape. The current approach goes beyond classical pattern recognition strategies by using convolutional neural networks (CNNs). More specifically, a fine-tuning scheme of pre-trained CNNs is considered. The manuscript is organized as follows. Related works are presented in Sect. 2. Then, the proposed approach, which is based on the usage of different CNN's architectures, is introduced in Sect. 3. Experimental results are provided in Sect. 4. Finally, conclusions are given in Sect. 5.

2 Related Work

The problem addressed in this paper is related to biology. This discipline covers a specific approach oriented to the study and description of living beings, either as individual organisms or species. These species have in some cases significant differences between them, facilitating their identification, although this is not always easy when categorized by genus.

The study presented in [1] describes a system for visual identification of plants, by using pictures taken by the user, returning additional images of some specie along with a description thereof, using a mobile device. Some other studies have been carried out identifying automatically patterns through traditional classification algorithms. In a recently published paper [2], three classifiers were selected (k-NN, MLP, SVM) and evaluated considering metrics such as simplicity, popularity and efficiency. Different tests were conducted by varying the training set, considering in all the cases 75% for training and 25% for evaluation, from 2 to 8 classes, the validation method used was Leave-One-Out. The best result was obtained with SVM reaching a 75.5% of accuracy. The mentioned previous work considered 8 classes from the same dataset used to evaluate the current work, but now considering 15 classes.

Over the past few years, deep CNNs have revolutionized large-scale image recognition and classification. Virtually all of today's high achieving algorithms and architectures for image classification and recognition make use of deep CNN architectures in some way [3,4]. In large part, these advances have been made possible by large public image repositories and the use of high performance GPUs. CNNs are a set of layers in which each layer performs a non-linear transformation function learned from a labeled set of data. The most important type of layers are convolutional and fully connected layers. Convolutional layers can be thought as a bank of filters that are convoluted to produce a feature map as an output. When relating CNNs to multilayer perceptrons (i.e., classical neural networks) these banks of filters can be seen as shared weights for all of the neurons in a layer. Fully connected layers, while also applying convolution, connect every single neuron of a layer, thus creating a large number of weights, and having an n-dimensional vector as an output [5]. Other widely used layers are pooling layers, which produce a sub-sampling of the input, and ReLu layers, which apply the rectifier activation function to add non-linearity to the network. For training purposes, a layer called dropout can be inserted between fully connected layers to avoid over-fitting [6]. These dropout layers set to zero the output of a neuron randomly, preventing the dependence of that neuron to particular other neurons.

The dataset to be used for training and validation is an important element, as in any pattern recognition approach. The current work is specifically oriented to a particular genus of lepidopterous (butterflies) in their last stage of life cycle. The process of collecting samples of these genus takes time and effort before being subjected to studies, in order to determine the most relevant characteristics. Experts in this area have generated over time, knowledge bases on families, subfamilies, tribes and genre of butterflies with different relevant information about them. Because of this, biology students now have a lot of information about Lepidopterous, however, this amount of information turns hard to be used as learning resources in a more efficiently and effective way.

There are different datasets to be used in studies of pattern recognition; the most widely used is *ImageNet* [7], which has a considerable quantity of images in different categories; in this dataset there is a category called "Butterfly" that has 2115 images of butterflies classified from four main families: nymphalid, danaid, pierid, lycaenid. Figure 1 shows butterflies obtained from the ImageNet dataset.

Fig. 1. Images of *butterflies* category taken from *ImageNet* dataset

Leeds [8] is another dataset; it contains images and textual descriptions for ten categories (species) of butterflies. The image dataset comprises 832 images in total, with the distribution ranging from 55 to 100 images per category. Images were collected from Google Images by querying with the scientific (Latin) name of the species, for example "Danaus plexippus", and manually filtered for those depicting the butterfly of interest.

One of the richest place in lepidoptera specimens is the *"Sangay National Park"*[1], which has been listed by UNESCO as a World Heritage Site since 1983. Several samples of lepidoptera specimens have been collected by expert biologist in this reserve, some of them not included in *ImageNet* or *Leeds datasets*. The dataset provided by Sangay National Park has 2799 images of butterflies, classified in 32 genus (Table 1). Figure 2 shows an image per category; the inter-class similarity of some of them makes the problem quite challenging.

3   Proposed Approach

This section presents the different approaches proposed to perform the classification of 15 lepidoptera species. All these approaches are based on the usage of deep convolutional neural networks (CNNs). Even though a deep CNN can

---

[1] Ecological reserve in Ecuador: http://www.sangay.eu/index.php?lang=en.

Table 1. Name and number of images per butterfly's genus in *Sangay National Park* *dataset* (in bold the 15 selected classes, those that have a larger number of images)

Genus	Images	Genus	Images	Genus	Images
Adelpha	**120**	Catonephele	28	Eunicini but Eunica	38
Anaeini but Memphis	65	Cealenorrhini	27	**Euptychia et al.**	**197**
Ancyluris et al.	40	Coloburini	16	**Eurybie Teratophtalma**	**127**
Anteros	39	Corades	58	**Euselasia**	**156**
Anthoptini	68	**Dalla**	**208**	Euselasiini but Euselasia	10
Astraptes	**122**	Dynamini	21	**Forsterinaria**	**93**
Callicore et al.	30	**Emesis**	**86**	Haeterinae	120
Calpodini	**89**	Epargyreus	31	**Helicopini others**	**83**
Carcharodini	**160**	Eretris	37	**Hesperiini**	**236**
Catagrammidi	**80**	**Erynnini**	**111**	Hesperiini incertae sedis	33
Eudaminae	**242**	Eunica	28		

achieve very high accuracy, one of its drawbacks is the need of having a relatively large dataset to train a CNN from scratch. Solutions to this problem have been proposed, such as using the output of a layer (before the last one) from a pre-trained CNN network as a feature extractor. These features can be then used as inputs to train a classical classifier, for instance a SVM [9]. Another way to tackle the limitation related with the size of the dataset consists of taking a network already trained and *adapt* it for the current work. In the adaptation process, which is referred to in the literature as *transfer learning*, the output layer of the pre-trained CNN network is randomly initialized; then the network is trained again but now with the dataset of the new classification problem [10]. Intuitively, the transfer learning is taking advantage of the filters originally conformed by the network, instead of starting from random initialization as would be the case when training from scratch. Moreover, the filter learned at the early layers are usually very generic (edge detectors, blob detectors) than can generalize to a variety of problems.

Having in mind the two possibilities mentioned above, and being aware of the limitation of our dataset (we are working with real dataset taken by expert biologists that cannot be that easily extended), we have decided to use pre-trained networks. The selected networks have been trained on the *Imagenet* dataset for the Imagenet Large Scale Visual Recognition Challenges. Imagenet classes include various types of animals, insects and more importantly different species of Lepidopterous, assuring that some filters have learned to identify different species of butterfly. Filters along the network that have learned this, generalize well to our classification problem. In conclusion, the fine tuning of the filters, specially the ones in the last fully connected layer, which were previously randomly initialized, takes advantage of whatever filters the pre-trained network had already learned. When training it with the new dataset, it learns to differentiate the different species of Lepidopterous in the new dataset.

a) Adelpha   b) Astraptes   c) Calpodini   d) Carcharodini   e) Catagrammidi

f) Dalla   g) Emesis   h) Erynnini   i) Eudaminae   j) Euptychia

k) Eurybie T.   l) Euselasia   m) Forsterinaria   n) Helicopini   o) Hesperiini

Fig. 2. Illustration of butterflies categorized by genus, for space limitation just one image per category is depicted (note the similarity between classes: (c), (d) and (e)).

In this section we present the three pre-trained networks selected to be fine tuned with our dataset. These networks have been trained to perform classification on Imagenet ILSVRC. Originally the networks have an output of 1000 classes, which correspond to the Imagenet classes. They were modified to output the 15 classes from our dataset. A brief description of them is given below.

AlexNet. This architecture has been introduced in [3] and inspires much of the work that is being done today on deep CNNs. It has eight layers with learned parameters, five convolutional and three fully connected. In its original implementation it was trained on two GPUs. It was the winning model of the ILSVRC-2012 challenge. An illustration of this CNN is depicted in Fig. 3.

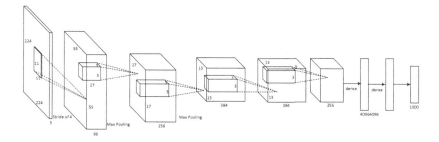

Fig. 3. Illustration of the architecture of the AlexNet CNN (see [3] for details about the configuration of each layer).

VGG-F. This network has been introduced in [11] and is very similar in architecture to Alexnet [3], but it is considerably faster when training and evaluating.

Because Alexnet was trained on two GPUs, it used sparse connections between convolutional layers. VGG-F, on the other hand, has dense connectivity between these layers [11]. Originally trained on ILSVRC-2012 challenge dataset.

VGG-Very Deep Convolutional Networks. This network has been presented in [12]. This work proposes two deep networks with 16 and 19 layers with learned parameters respectively, with the last three being fully connected. Besides the increased depth of the networks, they also introduce very small convolution filters. Initially submitted to ImageNet ILSVRC Challenge 2014.

4   Experimental Results

This section presents results obtained with the fine tuning of the three pre-trained network presented above; the lepidoptera dataset from *Sangay National Park* introduced in Sect. 2 has been considered. Our training and evaluation of the networks was done with the MatConvNet toolbox for Matlab. This toolbox is efficient and friendly for computer visions research with CNNs [13]. The usage of GPUs has also become widespread when training CNNs, allowing for faster computation and deeper networks. In our case, when training networks as deep as a VGG-Very Deep Convolutional Network the usage of GPU is completely necessary. While this is not mandatory for the other two pre-trained networks, it certainly speeds up the computation. We used the NVIDIA GeForce GTX950 GPU.

Regarding the data, we apply data augmentation so that the network sees more training and test images. Specifically, we mirror a random number of the images presented to the network on each epoch. When mirroring there is no change done to the species of the lepidoptera. Even though it is not as good as having more independent samples, data augmentation improves performance [11]. 25 percent of the images in each category were used for testing, while the other 75 percent were used for training purposes. All images were normalized and re-sized when given as input to each network. All networks originally accept color images of size 224×224, but in the particularly case of the AlexNet network, because an Matconvnet implementation of the network that was imported from Caffe was used, the color images had to be re-sized to 227×227.

Figure 4 depicts the percentage of error versus the number of epochs for the four networks with the 15 classes presented in Table 1. In the case of the less deep networks it can be apreciated that less epochs are needed for the training and test error to converge compared to the VGG-Very Deep Networks. Furthermore, in the deeper networks, it can be seen that there is more over fitting and the validation error oscillates more. We can attribute this behavior to the much larger number of parameters which are learned from the training set in the deeper networks. While this contributes to a very small training error, it makes it harder and longer for the test error to converge. Quantitative results are presented in Table 2, note that in all the cases the recognition ratio is higher than 92%, which is more than acceptable and considerably better that previous recently presented works [2].

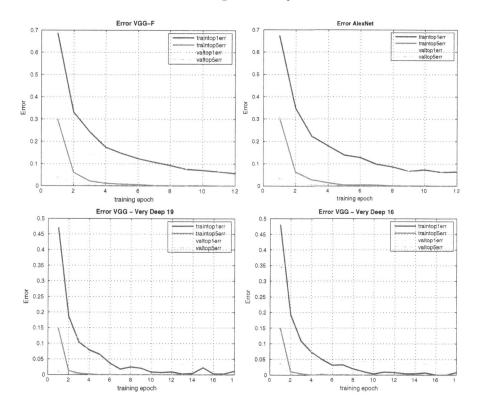

Fig. 4. Percentage of error vs number of epochs for each of the fine tuned networks (top1 error percentage for the train and test sets are shown in full blue and dotted yellow, respectively). (Color figure online)

Table 2. Results obtained with 15 classes from Sangay National Park dataset. Error values are percentages relative to the total number of images in the dataset.

Pre-trained CNN	Training Error	Validation Error
AlexNet	6.30%	6.81%
VGG-F	5.59%	7.17%
VGG - Very Deep 16	0.8%	2.97%
VGG - Very Deep 19	1.17%	3.84%

5 Conclusions

This paper tackles the challenging problem of lepidopterous insects recognition by transfer learning from three pre-trained CNNs. This strategy is proposed to overcome the lack of large data set to be used during the training stage. Fortunately, the networks have been trained with ImageNet dataset, which contains a category called "Butterfly" with more than 2000 images. Although the lepidoptera specimens considered in the current work are not included in ImageNet, we assume the pre-trained networks already have some capabilities to discriminate them. Experimental results shows that, even thought the size of training set from *Sangay National Park* dataset is quite reduced, an acceptable performance is reached (note that in all the cases more than 92% of recognition ratio has been reached). Future work will be focused on evaluating different data augmentation strategies to enlarge the training set.

Acknowledgments. This work has been supported by the Escuela Superior Politécnica del Litoral (ESPOL) within the research project "Reconocimiento de Patrones: Casos de estudio en la Agricultura y Acuicultura", M1-D1-2015. Third author has been partially supported by the Spanish Government under Project TIN2014-56919-C3-2-R. Authors would like to thanks Jean-Claude Petit for providing us the labeled dataset used in this work.

References

1. Belhumeur, P.N., Chen, D., Feiner, S., Jacobs, D.W., Kress, W.J., Ling, H., Lopez, I., Ramamoorthi, R., Sheorey, S., White, S., Zhang, L.: Searching the World's Herbaria: a system for visual identification of plant species. In: Forsyth, D., Torr, P., Zisserman, A. (eds.) ECCV 2008. LNCS, vol. 5305, pp. 116–129. Springer, Heidelberg (2008). doi:10.1007/978-3-540-88693-8_9
2. Cosquillo, Y., Romero, D.: Reconocimiento automático de especímenes lepidópteros en dispositivos de bajo poder computacional. Revista Tecnológica-ESPOL 28 (2015)
3. Krizhevsky, A., Sutskever, I., Hinton, G.E.: Imagenet classification with deep convolutional neural networks. In: Advances in neural information processing systems, pp. 1097–1105(2012)
4. Zeiler, M.D., Fergus, R.: Visualizing and understanding convolutional networks. In: Fleet, D., Pajdla, T., Schiele, B., Tuytelaars, T. (eds.) ECCV 2014. LNCS, vol. 8689, pp. 818–833. Springer, Heidelberg (2014). doi:10.1007/978-3-319-10590-1_53
5. Reyes, A.K., Caicedo, J.C., Camargo, J.E.: Fine-tuning deep convolutional networks for plant recognition. In: Working Notes of CLEF 2015 Conference (2015)
6. Hinton, G.E., Srivastava, N., Krizhevsky, A., Sutskever, I., Salakhutdinov, R.R.: Improving neural networks by preventing co-adaptation of feature detectors. arXiv preprint arXiv:1207.0580 (2012)
7. Deng, J., Dong, W., Socher, R., Li, L.J., Li, K., Fei-Fei, L.: Imagenet: a large-scale hierarchical image database. In: IEEE Conference on Computer Vision and Pattern Recognition, CVPR 2009, pp. 248–255. IEEE (2009)
8. Wang, J., Markert, K., Everingham, M.: Learning models for object recognition from natural language descriptions. In: Proceedings of the British Machine Vision Conference (2009)
9. Razavian, A., Azizpour, H., Sullivan, J., Carlsson, S.: CNN features off-the-shelf: an astounding baseline for recognition. In: Proceedings of the IEEE Conference on Computer Vision and Pattern Recognition Workshops, pp. 806–813 (2014)
10. Bengio, Y.: Deep learning of representations for unsupervised and transfer learning. Unsupervised Transf. Learn. Challenges Mach. Learn. **7**, 19 (2012)

11. Chatfield, K., Simonyan, K., Vedaldi, A., Zisserman, A.: Return of the devil in the details: delving deep into convolutional nets. In: British Machine Vision Conference (2014)
12. Simonyan, K., Zisserman, A.: Very deep convolutional networks for large-scale image recognition. arXiv preprint arXiv:1409.1556 (2014)
13. Vedaldi, A., Lenc, K.: Matconvnet: convolutional neural networks for matlab. In: Proceedings of the 23rd Annual ACM Conference on Multimedia Conference. ACM 689–692 (2015)

Selection of Statistically Representative Subset from a Large Data Set

Javier Tejada[1]($\boxtimes$), Mikhail Alexandrov[2,3], Gabriella Skitalinskaya[4], and Dmitry Stefanovskiy[3]

[1] Catholic University of San Pablo, Arequipa, Peru
jtejada@itgrupo.net
[2] Autonomous University of Barcelona, Barcelona, Spain
MAlexandrov@mail.ru
[3] Russian Presidential Academy of National Economy and Public Administration, Moscow, Russia
dstefanovskiy@gmail.com
[4] Moscow Institute of Physics and Technology, Moscow, Russia
gabriellasky@icloud.com

Abstract. Selecting a representative subset of objects is one of the effective ways for processing large data sets. It concerns both automatic time-consuming algorithms and manual study of object properties by experts. 'Representativity' is considered here in a narrow sense as the equality of the statistical distributions of objects parameters for the subset and for the whole set. We propose a simple method for the selection of such a subset based on testing complex statistical hypotheses including an artificial hypothesis to avoid ambiguity. We demonstrate its functionality on two data sets, where one is related to the companies of mobile communication in Russia and the other – to the intercity autobuses communication in Peru.

Keywords: Sampling · Representative objects · Big data · Statistics · Complex hypothesis

1 Introduction

1.1 Related Works

The topic of selecting a representative subset of objects from a large data set is becoming very popular in the era of big data. One could find many publications on this topic related to specific machine learning problems and specific domains. This topic in literature is often named 'sampling'. The well-known report [6] presents variants of sampling. The following two principal types of random sampling are domain independent: *case-based sampling* and *group-based sampling*.

Work done under partial support of FINCyT (Peru).

C. Beltrán-Castañón et al. (Eds.): CIARP 2016, LNCS 10125, pp. 476–483, 2017.
DOI: 10.1007/978-3-319-52277-7_58

The typical problem of case-based sampling big data consists in searching the minimal number of objects to evaluate the number of case-similar objects in the whole data set. Algorithms of case-based sampling consist in the calculation of proximity between objects and a given case (pattern), evaluation of distribution of case-similar objects inside the whole data collection, and application of statistics formulas taking into account the number of objects in a subset and in the whole object set. The survey of case-based sampling techniques is presented in [2].

Group-based sampling is the most popular type of sampling. The typical problem of group-based sampling big data consists in searching the minimal number of objects, whose structure is close (similar) to the structure of the whole data set. One of the first algorithms of group-based sampling is described in [3]. Here the number of group centers is fixed and only these centers are considered as the representative objects. The search algorithm determines the location of these centers taking into account the density of the objects distribution in a given space of parameters. The same approach was used in our work [5]. Here the whole object set was grouped with the method MajorClust [7]. Unlike the previous case, this method determines the number of clusters automatically, and this circumstance makes this method one of the most effective for cluster analysis. Each cluster is presented by 3 objects (semantic descriptors), reflecting relation to objects inside this cluster and outside: this cluster. These descriptors from all of the clusters are considered as the representative subset. Finally we present a recently published paper [9]. Here, the algorithm k-means forms clusters using n objects and finds m objects near each center. Besides the algorithm randomly selects dxk other objects. The values k, m and d are supposed to be assigned by a user, where $m \ll n$ and $d \ll n$. These $(m + d)xk$ objects is the so-called T-representative subset if the quality of clustering this subset exceeds a given threshold T. Otherwise m and d are increased.

The main disadvantage of group-based sampling unlike case-based sampling is the necessity to calculate pairwise distances that needs time and memory when working with big data. Indeed, for $N = 10^3$ objects, we have $\sim 10^6/2$ calculations, for $N = 10^6$ objects, we have already $\sim 10^{12}/2$ calculations, etc.

Sampling presented in the paper aims to select the minimal number of objects, whose parameters distributions should be close (similar) to those in the whole data set. With such a sample it would be possible easy to find given cases and easy to reveal data structure. This problem reduces to the well-known testing hypothesis about uniformity of two data sets [4]. The corresponding algorithm forms an increasing subset of objects until the taken hypothesis of uniformity will be accepted for all the objects parameters. However such a simple algorithm doesn't take into account both independence of data and ambiguity related to the acceptance of hypothesis. The proposed method tries to avoid these disadvantages.

1.2   Problem Setting

According the goal formulated above we need to form a representative subset of objects under the following conditions:

(1) The subset should reflect the parameters distributions of the whole data set;
(2) The subset should not depend on objects location in the data set;
(3) The subset should reduce ambiguity related to acceptance of statistical hypothesis;
(4) The subset should have minimum volume.

The problems solution consists in testing several statistical hypotheses covering the requirements (1), (2) and (3). The step-by-step procedure with an increasing number of objects allows to satisfy the requirement (4). The proposed method is tested on two data sets related to the companies of mobile communication in Russia and to the intercity autobuses communication in Peru.

The other sections of the paper are organized as follows. Section 2 describes the complex hypotheses. Section 3 presents the steps of the algorithms. Section 4 demonstrates the results of the experiments. Section 5 concludes the paper.

2   Testing Complex Hypotheses

2.1   Testing Hypothesis in Hard and Soft Modes

The procedure of testing the similarity of two sets of random values is well known in statistics and we only remind it [4].

Let us have two sets of objects each presented by one parameter. If both sets come from a normal distribution with an unknown mean and dispersion, then we form the so-called z-statistics and compare it with the critical value $z_{crit}(\alpha)$, where a is the accepted level of the statistical type 1 error. If the law of distribution is unknown then one should use any non-parametric test. In our paper, we apply the Kolmogorov-Smirnov test, where the tabulated α-dependent critical value is used.

Let we have two sets of objects presented by n parameters each. We can consider two modes: hard mode and soft mode. The hard mode considers two multi-dimensional sets as similar ones if all n parameters of the 1-st set have the same distribution as the corresponding parameters of the 2-nd set. The similarity of each parameter can be tested using z-statistics or the Kolmogorov-Smirnov test mentioned above. The soft mode considers two multi-dimensional sets as the similar ones if at least one parameter of the 1-st set has the same distribution as the corresponding parameter of the 2-nd set.

2.2   Null-Hypothesis and Artificial Contrary Hypothesis

Classical statistics says that the acceptance of any null hypothesis means only the absence of a significant contradiction between a given hypothesis and existing data. But it doesn't mean the hypothesis is correct [4].

In our practice such a situation occurred for the first time almost 15 years ago, when objects were classified by using the statistical measure of proximity between objects and given classes [1]. This measure consisted in testing the null-hypothesis and it proved that objects often belonged simultaneously to several classes.

In our case, the experiments show the following: with limited volume of data any hypothesis concerning data distribution even the most unlikely one can be accepted.

Then with increasing the volume of data the discrepancies between the data and the hypothesis become more and more apparent and finally the hypothesis may be rejected. Therefore, at the initial stage of selection with a very small number of objects there is a high probability to accept any hypothesis, which leads to the statistical type II error. To reduce this effect we propose to use a contrary artificial hypothesis: when this hypothesis is accepted then the acceptance of the basic null-hypothesis should be reevaluated.

The proposed artificial hypothesis concerns the uniform distribution of parameters in the subset under consideration. Such a solution can be supported by the assumption that such distribution is rarely used for practically useful applications and excluded from consideration in algorithms with object selection. It is necessary emphasize that the uniform distribution may be not the best one to be fit for our case. But at least it serves as a filter for the type II errors for the null-hypothesis. To test the hypothesis about the uniform distribution we use the Kolmogorov-Smirnov test.

3   Method

3.1   Steps of Algorithm

We build the algorithm according the problem setting described in the Sect. 1.2. The algorithm includes two stages: stage 1 (preprocessing) and stage 2 (processing). Preliminary we fix the total number of objects N, the initial number of representative objects M, the step for increasing the number of representative objects m, the number of parameters to be considered k. We fix also the probability of the type I error for the main hypothesis α, and the probability of the type I error for the artificial hypothesis β. In each step we use 3 sets of objects: the increasing representative subset, an additional random subset of the same volume, and the original object set.

Stage 1: preprocessing

(1)   Calculation of statistics for the whole data collection, such as the mean, dispersion or empirical function of distribution for the formulae from Sect. 2.1.
(2)   The data collection is randomly intermixed to avoid any influence of the initial object distribution.

Stage 2: processing

(1)   Selection of the initial representative subset of M objects. It can be the first objects from the whole object set.
(2)   Testing the complex hypothesis of similarity between the representative subset and the whole object set (the hypothesis-1). Here the formulae from Sect. 2 are used. The null-hypothesis is accepted in hard mode.
(3)   Testing the hypothesis of similarity between the representative subset and the additional random subset (the hypothesis-2). We use here the formulae from Sect. 2. The volume of both subsets must be equal. The null-hypothesis is accepted in the hard mode.

(4) Testing the artificial hypothesis. For this, the concordance between the representative subset and the uniform distribution is checked. We use here the formulae from Sect. 2. The volume of the subset and the theoretical sample must be equal. The null-hypothesis is accepted in the soft mode.

(5) If the hypotheses about similarity between the representative subset, the additional subset and the whole data set are accepted and the artificial hypothesis is rejected then the process interrupts: the current representative subset can be considered as the final solution. Otherwise a new portion of objects are added to the current representative subset $M = M + m$ and the algorithm returns to the step (2).

3.2   Evaluation of Results

We use the approach proposed in [9]. This approach can be presented in the following simplified form. Let D_n and D_v be the whole data set and a sample from this data set containing n and v objects respectively, here $v < n$. Let P, $Q(P, D_v)$ and T be the problem, the quality of problem solution with D_v and the threshold of quality respectively.

<u>Definition 1.</u> The value v is the T-Critical Sampling Size if (a) for any $k \geq v$ the quality $Q(P, D_k) \geq T$, (b) there is $k < v$ that $Q(P, D_k) < T$.

<u>Definition 2.</u> The value $Q_s = v/n$ is the Sample Quality of D_n, where v is the Critical Sampling Size of D_n.

So, to determine practically the T-Critical Sampling Size for a given problem P one needs to complete the following: to fix T and repeat several experiments with different subsets D_k for different k. Then one needs to select cases $Q \geq T$ and take the maximal value k in these cases. We used this method in our experiments.

4   Experiments

4.1   Selecting Companies of Mobile Communication (Russia)

The Russian market of the companies of mobile communication includes about 600 companies (data of 2015). The problem consists in the selection of a representative subset from the list of companies in order to approximately evaluate its attraction for investments. Such problem was considered by master students from the Russian Presidential Academy of National Economy and Public Administration. This research was supported by the Academy.

For the analysis the following 4 parameters of company activity were taken: Profitability of Production (*PP*), Return on Assets (*RA*), Return on Equity (*RE*), Coefficient of Financial Stability (*FS*). The normalized data of the first 3 companies are presented in Table 1.

In the experiment, we varied: (1) the probability of the type I error α for the hypothesis-1 and for the hypothesis-2, and (b) the probability of type I error β for the artificial hypothesis. The initial subset included 5 objects. The step was also equal 5 objects. We completed 10 experiments for each combination (α, β) and fixed the

Table 1. Characteristics of companies (first 3 of 600).

PP	RA	RE	FS
−0.32	−0.45	0.23	0.68
0.27	0.13	−0.37	−0.01
−0.38	−0.52	0.98	−0.09
......	...	...	...

maximal number of the selected objects for these combinations. Therefore we revealed the so-called (α, β)-critical sampling sizes in terms of Sect. 3.2. The results are presented in the Table 2. The sign '−' means the artificial hypothesis was not used. The most significant combination here is (0.05, 0.01) for which the sample quality equals $Q_s = 55/600 \sim 0.1$.

Table 2. The critical sampling sizes (companies).

$\alpha \backslash \beta$	-	0.01	0.05	0.10	0.50
0.05	5	**55**	55	15	10
0.10	5	65	65	15	10
0.50	260	260	260	260	260

We evaluated the quality of sampling on the basis of results of clustering. For this we considered the mentioned (0.05, 0.01)-combination and completed the following procedures:

(1) Clustering the whole data set, whose clusters were considered as classes in cluster F-measure (see below);
(2) Clustering 55 random objects taken from the whole data set, this procedure was repeated 10 times;
(3) Comparison of both results using cluster F-measure introduced in [8]. This measure reflects the correspondence between the classes and clusters revealed on steps (1) and (2). Theoretically the best and the worst values of cluster F-measure are equal 1 and 0 respectively.

In this experiment we used k-means method with $k = 3$. The value k was assigned by experts. The resulting F-measure proved to belong the interval [0.45, 0.65].

4.2 Selecting Modes of Movements in Intercity Autobus Communication (Peru)

Daily there are approximately 6000 intercity bus trips in Peru. They connect about 100 destinations (data of 2015). All movements of autobuses are registered by GPS in special databases. The frequency of registration is 1 min. We introduce the notion 'object of movement'. It is the distance, which is covered by a vehicle in 10 min. Therefore, 10 sequential records form one object of movement. We collected about 10000 records reflecting 24 bus trips in Peru and collocated them sequentially one bus trip after the other one. These records can be considered as 10000/10 = 1000 objects of movement. The problem consists in the selection of a representative subset of this object set for data mining the Peruvian intercity autobuses communication. The work was completed in the Computer Science Department of the Catholic University of San Pablo in the framework of the project with a Peruvian company responsible for monitoring the intercity autobus communication.

For the analysis we used the following 3 parameters for each object: average velocity (AV km/h), variation of velocity (VV), and deviation of direction (DD degrees). The normalized data of the first 3 objects on the route Lima-Arequipa is presented in Table 3.

Table 3. Characteristics of objects of movement (first 3 of 1000).

AV	VV	DD
−0.55	0.13	0.34
0.92	−0.30	−0.49
0.89	−0.95	2.42
...	...	...

We repeated the experiment concerning various combinations (α, β) with objects of movement. The results are presented in the Table 4. The most significant combination here is also (0.05, 0.01), for which the sample quality equals Qs = 20/1000 = 0.02.

Table 4. The critical sampling sizes (objects of movement).

$\alpha \backslash \beta$	-	0.01	0.05	0.10	0.50
0.05	5	*20*	15	15	10
0.10	5	25	15	15	10
0.50	445	445	345	345	345

We evaluated the quality of sampling on the basis of results of clustering as we did for the companies. In this procedure we used 20 and 1000 objects. The number of clusters equaled 4 that corresponded to the best value of Dunn-index [8]. The resulting F-measure proved to belong to the interval [0.85, 1.00].

5   Conclusions

In the paper, we propose the method for selecting representative subset of objects from a large object set, which provides:

- forecast of parameters distributions in the whole object set on the basis of their distributions in the subset;
- independence of the selection on the objects collocation in the object set;
- increased accuracy by means of disambiguation related to acceptance of null-hypotheses.

These advantages are reached on the basis of using complex statistical hypotheses. The method has lineal complexity and can be used with any distributions unknown in advance. The method was checked on two real data sets and demonstrated promised results.

In future we intend to consider the following applications of the method: selection of representative subsets of universities and colleges in given countries, selection of representative subsets of students in given universities and colleges, other applications concerning sciences and education.

References

1. Alexandrov, M., Gelbukh, A., Lozovoi, G.: Chi-square classifier for document categorization. In: Gelbukh, A. (ed.) CICLing 2001. LNCS, vol. 2004, pp. 457–459. Springer, Heidelberg (2001). doi:10.1007/3-540-44686-9_45
2. Au, T., Chin, M.-L.I., Ma, G.: Mining rare events data by sampling and boosting: a case study. In: Prasad, S.K., Vin, H.M., Sahni, S., Jaiswal, M.P., Thipakorn, B. (eds.) ICISTM 2010. CCIS, vol. 54, pp. 373–379. Springer, Heidelberg (2010). doi:10.1007/978-3-642-12035-0_38
3. Chaudhur, D., Murthy, C., Chaudhur, B.: Finding a subset of representative points in a data set. IEEE Trans. Syst. Man Cybern. **24**(9), 1416–1424 (1994)
4. Cramer, H.: Mathematical Methods of Statistics. Princeton Landmark in Mathematics. Princeton University Press, Princeton (2016)
5. Gelbukh, A., Alexandrov, M., Bourek, A., Makagonov, P.: Selection of representative documents for clusters in a document collection. In: Natural Language Processing and Information Systems, GI-Edition, LNI, Germany, vol. 29, pp. 120–126 (2003)
6. National Research Council: Frontiers in Massive Data Analysis, Report of NRC, National Academies Press, USA (2013)
7. Stein, B., Niggemann, O.: On the nature of structure and its identification. In: Widmayer, P., Neyer, G., Eidenbenz, S. (eds.) WG 1999. LNCS, vol. 1665, pp. 122–134. Springer, Heidelberg (1999). doi:10.1007/3-540-46784-X_13
8. Stein, B., Meyer zu Eissen, S., Wilssbrock, F.: On cluster validity and the information need of users. In: Proceedings of 3rd IASTED International Conference on AI and Applications (AIA-2003), Acta Press, pp. 216–221 (2003)
9. Sung, A., Ribeiro, B., Liu, Q.: Sampling and evaluating the big data for knowledge discovery. In: Proceedings of International Conference on Internet of Things and Big Data (IoTBD 2016), Science and Technology Publications, pp. 378–382 (2016)

Non-local Exposure Fusion

Cristian Ocampo-Blandon$^{(\boxtimes)}$ and Yann Gousseau

LTCI, CNRS, Telecom ParisTech, Universite Paris-Saclay, 75013 Paris, France
{ocampobl,yann.gousseau}@telecom-paristech.fr

Abstract. Exposure fusion is an efficient method to obtain a well exposed and detailed image from a scene with high dynamic range. However, this method fails when there is camera shake and/or object motions. In this work, we tackle this issue by replacing the pixel-based fusion by a fusion between pixels having similar neighborhood (patches) in images with different exposure settings. In order to achieve this, we compare patches in the luminance domain. We show through several experiments that this procedure yield comparable or better results than the state of the art, at a reasonable computing time.

Keywords: Exposure fusion · Image fusion · Patches · Non-local methods · High dynamic range

1 Introduction and Previous Works

Natural scenes often exhibit luminance ranges that are beyond the capacity of most imaging sensors. In such cases, a single standard photograph cannot capture details in dark regions without producing saturated values in the brightest regions. A classical way to create a high dynamic range (HDR) image faithfully representing the luminance of a scene is to combine photographs acquired with different exposure times [8]. Once the response function of the camera is known (or when working with linear RAW images) and when the scene is static, this combination task is essentially a statistical problem [4]. In order to visualize such an image on a classical display, contrast is then compressed by using a so-called *tone mapping* operator [10,11]. An alternative to this two-step procedure (HDR image creation followed by tone mapping) was proposed in [15] and is called *exposure fusion*. The main idea is to bypass the HDR creation step and to directly create a low dynamic range image (typically made of 8 bits per color channel) by fusing the input images corresponding to different acquisition times. Specific weights are used to ensure that the final result is contrasted enough, has vivid colors and avoid over and under-exposure. Using classical ideas from computer graphics [16], the images are fused in a multi-scale framework, enabling one to blend images seamlessly.

This procedure is very efficient and has yielded numerous softwares and plug-ins. It has also triggered an abundant research literature proposing variants on the original method: involving image decomposition into several components at

C. Beltrán-Castañón et al. (Eds.): CIARP 2016, LNCS 10125, pp. 484–492, 2017.
DOI: 10.1007/978-3-319-52277-7_59

the patch level [13], formulating the problem as a MAP estimation [21] or involving the bilateral filter [20]. Nevertheless, such approaches have a common drawback: they fail when there is either camera shake or moving objects in the scene. Indeed, they all perform the fusion at pixel level and assume that images are registered and that objects are still. Of course, this is a serious limitation in practice, and several works have tackled this issue. Anti-ghosting algorithms [5,17,22] propose to perform image alignment and possibly to explicitly detect moving objects to prevent using them in the fusion. Recently, the methods from [12,19] proposed to solve the problem by an iterative optimization procedure relying on correspondences between patches. In particular, the authors of [12] propose to create, from a reference image and a set of differently exposed images, a set of images that are all aligned and radiometrically coherent with the reference, and whose content is automatically modified (in the case of moving objects) to match the reference. This is achieved thanks to contrast prescription and patch-based content comparison between the images. As illustrated in [12], classical exposure fusion can then be applied to the aligned set to yield a satisfactory final image, even in the case of camera shake and object motion.

In this paper, we propose an exposure fusion method that also deals both with camera shake and object motion. The basic idea is very simple and illustrated by Fig. 1: instead of fusing values taken by pixels at the same spatial position, we fuse values of pixels having similar neighborhood (thereafter called a patch) in the different images. The method is therefore in the spirit of non-local restoration methods such as the non-local means [7], and also share similarities with the non-local method introduced in [3] for the creation of HDR images. Such a simple approach has the ability to deal with camera motion (although a previous global image registration may be useful to accelerate the process or to enhance the similarity between patches) and with moving objects. By construction, the method also boils down to the original exposure fusion algorithm in the absence of camera and object motions. The rest of the paper is organized as follows. In Sect. 2, the original exposure fusion algorithm is first briefly recalled and a detailed presentation of the proposed method is provided. In Sect. 3, we show experimental results and comparisons with the recent state-of-the-art algorithm from [12].

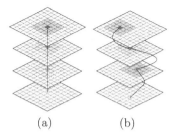

(a)   (b)

Fig. 1. Instead of fusing values at the same position (a), we use a non-local fusion (b) with values from different positions, to account for object motions and camera shake.

2   Non-local Exposure Fusion

2.1   Classical Exposure Fusion

The method introduced in [15] proposes to fuse a series of images $I_1, \ldots, I_N$ acquired with different exposure settings (the knowledge of these is not needed). For each pixel x and index i a weight $W_i(x)$ is defined by taking into account the quality of contrast, color saturation and well-exposedness at this pixel. The idea is then to fuse the values $I_i(x)$ according to the weights. The most straight-forward approach would be to define the resulting image R as

$$R(x) = \sum_{i=1}^{N} W_i(x) I_i(x), \tag{1}$$

but this yields incoherences in flat regions and visible seams at slow transitions. In order to achieve seamless fusion, the blending is performed in a multi-scale framework. For each level l of a Laplacian pyramid [16], the Laplacian pyramid of the resulting image R is computed as

$$\mathcal{L}_l(R) = \sum_{i=1}^{N} \mathcal{G}_l(W_i) \mathcal{L}_l(I_i),$$

where $\mathcal{L}_l(I_i)$ is the Laplacian pyramid at level l of image i and $\mathcal{G}_l(W_i)$ is the Gaussian pyramid at level l of the weight map W_i. The final image R is then reconstructed from its Laplacian pyramid. As explained in the introduction, this (otherwise very efficient) method fails when there is camera shake or object motions.

2.2   Non-local Fusion

In order to deal with dynamic scenes we propose, for each pixel x, to replace the fusion of values at this pixel with the fusion of values at well chosen pixels $f_1(x), \ldots, f_N(x)$ in the series of images. In the static case these pixels would simply be all equal to x, and in the general case, they will be defined in order to compensate motions, as we will now see. In a way similar to the HDR creation method from [3], we define $f_i(x)$ to be the pixel having its neighborhood in I_i the closest to the one of x in I_{ref} (where $ref \in (1, \ldots, N)$) a reference image to be defined later. More precisely, we define $P_i(x)$, a patch at pixel x, to be the collection of values in I_i in a $(2s+1) \times (2s+1)$ square window centered at x. We assume that we have a distance d between patches (to be defined in the next section) and then define the pixel most similar to x in I_i to be

$$f_i(x) = \arg \min_{y \in x + \mathcal{N}} \left(d(P_{ref}(x), P_i(y)) \right), \tag{2}$$

where $\mathcal{N}$ is a search window (bigger than the patch). In the (unlikely) case where the minimum is achieved for several values, one is chosen at random. If the minimum is achieved for only one value, we have in particular that $f_{ref}(x) = x$.

Let us now assume for a moment that we perform the fusion at a single scale (that is, as in Formula (1)). Then the resulting image could be defined as

$$R(x) = \sum_{i=1}^{N} W_i(f_i(x)) I_i(f_i(x)).$$

Of course, as in the classical exposure fusion, defining the resulting image R this way would yield incoherences and seams. Therefore, the non-local fusion is defined, for each level l of a multi-scale pyramid, as:

$$\mathcal{L}_l(R) = \sum_{i=1}^{N} \mathcal{G}_l(\tilde{W}_i) \mathcal{L}_l(\tilde{I}_i), \tag{3}$$

where $\tilde{I}_i$ is defined as $\tilde{I}_i(x) = I_i(f_i(x))$. The weights $\tilde{W}_i$ are obtained by multiplying the original weights w_i^1 from [15] (accounting for well-exposedness, color saturation and contrast) at position $f_i(x)$ by a weight asserting the similarity between x and $f_i(x)$. Similarly to what is done in the NL-means algorithm [7] we define this similarity weight to be

$$w_i^2(x) = \exp\left(-\frac{d(P_{ref}(x), P_i(f_i(x)))^2}{\sigma^2}\right), \tag{4}$$

where σ is a parameter. The resulting weights are $\tilde{W}_i(x) = w_i^1(f_i(x)) \times w_i^2(x)$.

In short, for a given pixel, the fusion is performed by using the most similar pixels in each image of the sequence and by modulating the original weights of the fusion by a similarity score.

2.3 Computing Pixel Similarity

Here, we detail how to compute the distance between patches. For this, images acquired with different exposure settings (e.g. different exposure times) should be comparable. In this paper, we assume that we have access to the linear RAW images of the captured scene. This is not a real limitation in the common case of embedded processing. When these RAW images are not available, other scenarios are possible (such as matching all image histograms to the one of the reference image, in a way similar to [12]), but these will not be investigated in this paper.

In addition to the processed images $I_1, \ldots, I_N$ (typically JPEG images after demosaicing, white balance and gamma transform), we consider the corresponding RAW (linear) images $R_1, \ldots, R_N$. We assume that these have been acquired with exposure times $t_1, \ldots, t_N$. The luminance images can then be computed as $L_i = gt_i^{-1}(R_i - O)$, where O is the offset of the camera and g the gain (see e.g. [2]). Without loss of generality, we will assume that $g = 1$ and that the offset is known. The comparison of pixels is then carried out in the luminance domain, so that the distance d between two patches $P_i(x)$ and $P_j(y)$ is defined as the Sum of Squared Differences (SSD) between the values of the patches in the respective luminance images L_i and L_j, that is,

$$d(P_i(x), P_j(y)) = \sum_{k=-s}^{s} (L_i(x+k) - L_j(y+k))^2. \tag{5}$$

If the camera offset is unknown, or in order to achieve results that are more robust to spatial variability of the gain (photo response non-uniformity, PRNU) or imprecision in the exposure time, the L^2 distance can be replaced by a similarity measure that is invariant to affine transforms of the luminance. Although this aspect will not be developed here, we have observed that the affine invariant distance proposed in [9] is very robust to errors in the luminance conversion.

2.4   Algorithm Details

First, the images are aligned through a global SIFT-based homography estimation and bicubic interpolation. Although not necessary, we observed that this step yields more reliable patch comparisons. A reference image is picked as in [12] by choosing the image having the least number of saturated pixels. Then, saturated regions in the reference are removed (see below). Second, and for each pixel in the reference, its best match in image i is chosen using Formula (2) and the similarity measure defined by (5). In order to accelerate the algorithm, the best match is found using the PatchMatch algorithm [6], yielding, for each patch of the reference image, an approximate nearest neighbor (ANN) in each image L_i. The PatchMatch algorithm is run with its default parameters, which in particular implies that the search window $\mathcal{N}$ is the entire image.

The corresponding value of the final image is obtained by fusing the values of the best matches using Formula (3). In this formula, the weights $\tilde{W}_i$ are obtained by multiplying the original weight of [15] by the similarity weight defined by Formula (4). The original weights from [15] depend on three exponents that we all choose to be equal to 1. The parameter σ of the similarity measure (4) is chosen as the quantile of level 5% of the distribution of the distances between patches (separately for each couple L_{ref}, L_i). We empirically found this choice to provide a reasonable compromise in order to simultaneously obtain a result similar to the original algorithm from [15] and efficiently discard moving objects.

(a)   (b)

Fig. 2. Image fusion: (a) non-local exposure fusion (NLEF), (b) exposure fusion [15].

(a) Our Method - NLEF, (69 (b) Fusion with [12], (143 sec).
sec).

(c) Our Method - NLEF, (84 sec).   (d) Fusion with [12], (116 sec).

(e) Our Method - NLEF, (184 sec).   (f) Fusion with [12], (394 sec).

Fig. 3. Sets of bracketed exposure images and their fusion. (Third image set is courtesy of EMPA Media Technology.)

A last point that we did not explain is what to do with saturated pixels. We could have chosen a strategy similar to the one of [12]: use only the non-saturated pixels when comparing patches. However, this fails as soon as the saturated region is wider than the patch, which is very common in practice. In order to get more robust results, we use the same strategy as in [3]. For each connected region Ω of saturated pixels, we detect the region from the other images that is the most similar to Ω on its immediate neighborhood (a ring outside Ω). This region is then blended into the reference using Poisson editing [18].

3   Experiments

In all experiments (see Fig. 3), we use a search window equal to the full image. The patch size is set to $s = 1$ (that is 3×3 patches).

First, as a sanity check, we compare the result of our method, NLEF (non-local exposure fusion), with that of the original exposure fusion [15] on a static scene with a moving object, from the database [1]. As we can see in Fig. 2, the results are very similar on static parts (everything except one object on the table), but no ghost is produced with our method for the moving object.

Then, we provide three comparisons between our method and the recent method from [12], using the code kindly provided by the authors. Since exposure fusion does not aim at producing a real HDR image (a faithful representation of the luminance values), we only provide visual comparisons. A full user study would be the best way to evaluate the quality of the results [14] but is beyond the scope of this paper. Both methods rely on PatchMatch to get patch correspondences, and we use the same parameters for this algorithm in all experiments and for the two approaches (full search over the complete image and patch size of $s = 1$). The examples corresponding to Figs. 3(a) and (c), are acquired with a Canon 400D and Canon 7D camera, respectively, for which the offsets are 256 and 2046.

In Figs. 3(a) and (e), one can see that our method produces more accurate colors. However, the color errors from [12] could probably be solved by taking into account the luminance computed from linear images. Second, one observes that both methods produce accurate details and no ghosts, with occasional artifacts at different positions. However, our approach is both faster and much simpler, and in particular involves no complex optimization procedure.

4   Conclusions

In this work, we have introduced a non-local exposure fusion yielding good results in the presence of camera shake and objects' motions. There are several ways this work could be continued. First, the saturated parts are handled using a Poisson editing procedure, but this step could probably be integrated in the PatchMatch algorithm, at the propagation step. Second, the method may produce noisy results, in particular when the reference image is noisy. We plan to tackle this issue by using more patches for the fusion, in a way similar to the NL-means algorithm. Last, similar non-local approaches could be studied in different fusion schemes, for instance to perform focus stacking.

Acknowledgments. The authors thank C. Aguerrebere for her help and advice. This work is supported by Colciencias under the grant "Programa doctoral de Becas" call 529 and the French Research Agency (ANR) under project MIRIAM.

References

1. HDR imaging website: International University of Sarajevo (2016). http://projects. ius.edu.ba/computergraphics/hdr/
2. Aguerrebere, C., Delon, J., Gousseau, Y., Musé, P.: Study of the camera acquisition process and statistical modeling of the sensor raw data. Prep. HAL
3. Aguerrebere, C., Delon, J., Gousseau, Y., Muse, P.: Simultaneous HDR image reconstruction and denoising for dynamic scenes. In: 2013 IEEE International Conference on ICCP, pp. 1–11. IEEE (2013)
4. Aguerrebere, C., Delon, J., Gousseau, Y., Musé, P.: Best algorithms for HDR image generation. A study of performance bounds. SIAM J. Imaging Sci. **7**(1), 1–34 (2014)
5. An, J., Lee, S.H., Kuk, J.G., Cho, N.I.: A multi-exposure image fusion algorithm without ghost effect. In: 2011 IEEE International Conference on Acoustics, Speech and Signal Processing (ICASSP), pp. 1565–1568. IEEE (2011)
6. Barnes, C., Shechtman, E., Finkelstein, A., Goldman, D.: Patchmatch: a randomized correspondence algorithm for structural image editing. ACM Trans. Graph. (TOG) **28**(3), 24 (2009)
7. Buades, A., Coll, B., Morel, J.-M.: A non-local algorithm for image denoising. In: IEEE Conference on Computer Society CPVR, vol. 2, pp. 60–65. IEEE (2005)
8. Debevec, P.E., Malik, J.: Recovering high dynamic range radiance maps from photographs. In: ACM SIGGRAPH 2008 Classes, p. 31. ACM (2008)
9. Delon, J., Desolneux, A.: Stabilization of flicker like effects in image sequences through local contrast correction. SIAM J. Imaging Sci. **3**, 703–734 (2010)
10. Durand, F., Dorsey, J.: Fast bilateral filtering for the display of high-dynamic-range images. ACM Trans. Graph. (TOG) **21**(3), 257–266 (2002)
11. Fattal, R., Lischinski, D., Werman, M.: Gradient domain high dynamic range compression. ACM TOG **21**, 249–256 (2002). ACM
12. Hu, J., Gallo, O., Pulli, K., Sun, X.: HDR deghosting: how to deal with saturation? In: 2013 IEEE Conference on CVPR, pp. 1163–1170. IEEE (2013)
13. Ma, K., Wang, Z.: Multi-exposure image fusion: a patch-wise approach. In: IEEE International Conference on Image Processing (ICIP) (2015)
14. Ma, K., Zeng, K., Wang, Z.: Perceptual quality assessment for multi-exposure image fusion. IEEE Trans. Image Process. **24**(11), 3345–3356 (2015)
15. Mertens, T., Kautz, J., Van Reeth, F.: Exposure fusion. In: 2007 15th Pacific Conference on Computer Graphics and Applications, pp. 382–390 (2007)
16. Ogden, J.M., Adelson, E.H., Bergen, J.R., Burt, P.J.: Pyramid-based computer graphics. RCA Eng. **30**(5), 4–15 (1985)
17. Pece, F., Kautz, J.: Bitmap movement detection: HDR for dynamic scenes. In: 2010 Conference on Visual Media Production (CVMP), pp. 1–8. IEEE (2010)

18. Pérez, P., Gangnet, M., Blake, A.: Poisson image editing. ACM Trans. Graph. (TOG) **22**, 313–318 (2003). ACM
19. Qin, X., Shen, J., Mao, X., Li, X., Jia, Y.: Robust match fusion using optimization. IEEE Trans. Cybern. **45**, 1549–1560 (2015)
20. Raman, S., Chaudhuri, S.: Bilateral filter based compositing for variable exposure photography. In: Proceedings of Eurographics (2009)
21. Song, M., Tao, D., Chen, C., Bu, J., Luo, J., Zhang, C.: Probabilistic exposure fusion. IEEE Trans. Image Process. **21**(1), 341–357 (2012)
22. Zheng, J., Li, Z., Zhu, Z., Wu, S., Rahardja, S.: Hybrid patching for a sequence of differently exposed images with moving objects. IEEE Trans. Image Process. **22**(12), 5190–5201 (2013)

Analysis of the Geometry and Electric Properties of Brain Tissue in Simulation Models for Deep Brain Stimulation

Hernán Darío Vargas Cardona$^{(\boxtimes)}$, Álvaro A. Orozco,
and Mauricio A. Álvarez

Department of Electric Engineering,
Universidad Tecnológica de Pereira, Pereira, Colombia
{hernan.vargas,aaog,malvarez}@utp.edu.co

Abstract. Deep Brain Stimulation (DBS) of Subthalamic Nucleus (STN) has proved to be the most effective treatment for Parkinson's disease. DBS modulates neural activity with electric fields. However, the mechanisms regulating the therapeutic effects of DBS are not clear, in fact there is not a full knowledge about the voltage distribution generated in the brain by the stimulating electrodes. Knowledge of voltage distribution is useful to find the optimal parameters of stimulation, that allow the neurosurgeons to get the best clinical outcomes and minimal side effects. In this paper, we analyze the geometry and electric characteristics in DBS models with a Colombian population study. We characterized the electric conductivity of the brain using diffusion Magnetic Resonance imaging dMRI and we define three types of geometries to be modeled in DBS simulations. Finally we estimate the voltage propagation in brain tissue generated by DBS using the finite element method.

1 Introduction

Deep Brain Stimulation (DBS) of Subthalamic Nucleus (STN) has become in the best treatment for Parkinson's disease [1]. DBS consists of the implantation of a neurostimulator in one of the movement control centers (thalamus, subthalamic nucleus, globus palidus), which transmits an electric current to the surrounding neurons. The fundamental purpose of DBS is to modulate neural activity with electric fields. But, despite the clinical success of this procedure, there is not a full knowledge about the therapeutic mechanisms of DBS [2], limiting opportunities to improve treatment efficacy and simplify selection of stimulation parameters [3]. In addition, there are many questions related to the consequences that DBS generates in the nervous system, difficulting to determine the amount of brain tissue which presents excitation or electrical response to stimulation [2].

Recently the scientific community has developed methodological tools that link scientific analysis and both electrical and anatomical models of DBS applied in humans [4]. Models based on finite element method to determine the voltage distribution generated by DBS, show that therapeutic benefit can be achieved

© Springer International Publishing AG 2017
C. Beltrán-Castañón et al. (Eds.): CIARP 2016, LNCS 10125, pp. 493–501, 2017.
DOI: 10.1007/978-3-319-52277-7_60

by direct stimulation in a wide range of the subthalamic region [2]. It is known that several factors influence the behavior of the electric field, and therefore the neural response of patients to therapeutic DBS. In addition to stimulation parameters, the electric potential distribution depends on size and shape of the electrode [5], and the postoperative cell growth around the implant resulting in an encapsulation layer [6].

In most studies, researchers have employed conductor homogeneous domains with square, rectangular and spherical shapes. The main problem with these models is that the optimal geometry and dimensions for the volume conductor are not defined. For example in [7] was quantified the role that geometry dimensions and boundary conditions play in the model. They found that for any variation of lengths in a box or ground positioning, the results are different for fundamental electrical quantities such as potential, field and neural activation function. Similarly, electrical properties of brain tissue affect the simulation results [8]. In [9] has been proved the influence of anisotropic and heterogeneous brain tissue in results of DBS applied in a human head model with finite elements. This brain model includes the gray matter, white matter and cerebrospinal fluid with their corresponding electrical properties and anisotropy values, through the inclusion of diffusion tensors. The authors of [10] showed that the tissue layers in the head (white matter, gray matter, cerebrospinal fluid, skull, muscle, stimulating device) and electrical properties of brain tissue are critical in results. These studies show a gap in the modeling of the DBS and most of them have been performed in North American and European populations. Therefore, little is known about the electrical behavior during DBS in patients from South America. For this reason, it is necessary to consider all the critical aspects of the DBS model applied to South American population. In this regard must be considered an optimal geometry, a heterogeneous brain tissue, the encapsulation tissue around the electrode, some layers of tissue inside the head and the anisotropic electrical behavior of neurons.

The aim of this paper is to show an objective comparison among three kind of volumes which represent the geometry, layers and structure of the head, likewise, the influence of electrical properties of anisotropic tissue and ground positioning. Another worth contribution of this work is the analysis and evaluation of the anisotropy level in relevant brain tissue (i.e. Thalamus-Thal, Subthalamic Nucleus-STN and Substantia Nigra reticulata-SNr) in Colombian patients. For this purpose, we use diffusion Magnetic Resonance Imaging (dMRI) obtained from five patients with Parkinson's Disease located in the west-central region of Colombia.

2 Materials and Methods

2.1 Simulation Models for DBS and Geometries

The phenomenon of electrical brain activity is described by the Laplace equation [7,8]. Here, the voltage propagation is a spatial function depending of tensorial tissue parameters:

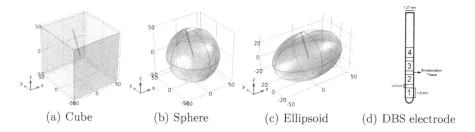

(a) Cube   (b) Sphere   (c) Ellipsoid   (d) DBS electrode

Fig. 1. Geometric domains and DBS electrode.

$$\nabla \cdot \sigma \nabla V_e = 0, \tag{1}$$

where V_e is the electrical potential, σ is a conductivity tensor calculated from dMRI and ∇ is the gradient operator given by $\frac{\partial}{\partial x}, \frac{\partial}{\partial y}, \frac{\partial}{\partial z}$. Equation 1 is solved by finite element approximation, because it is not possible to find the analytical solution. We employ the software Comsol Multiphysics 4.3.

We propose three solution domains that represent the skull cavity. First, a $10 \times 10 \times 10\ cm^3$ cubic domain (see Fig. 1a), second a spherical 3D domain with 10 cm of radius (see Fig. 1b), and third a ellipsoid domain (see Fig. 1c). Also, we establish Dirichlet boundary conditions for all models and we consider the encapsulation tissue ($\sigma_t = 0.042\ S/m$) of 0.1 mm thickness generated by the DBS electrode. The medium is anisotropic and heterogeneous. To address this formulation, we apply conductivity tensors estimated from dMRI obtained from 2 patients with Parkinson's disease located in the west-central region of Colombia.

The stimulation electrode is the *Medtronic DBS lead model* 3389 (radius 0.635 mm)[1]. This device have four platinum electrode contacts (length 1.5 mm, spaced 0.5 mm), with constant conductivity $\sigma_e = 8.6 \times 10^6\ S/m$. We consider an unipolar stimulus located in the Subthlamic Nucleus (STN) (see Fig. 1d). The waveform is a train of squared pulses with amplitude -1 V and frequencies between 100–150 Hz.

2.2   Brain Conductivity Through Diffusion Magnetic Resonance Imaging (dMRI)

The diffusion Magnetic Resonance Imaging (dMRI) studies the diffusion of water particles in the human brain. The direction of the diffusion is related to local micro-structure of the brain tissue and it is usually aligned through oriented pathways like the connective fibers. Diffusion can be described by a 3×3 tensor matrix proportional to the covariance of a Gaussian distribution.

---

[1] Medtronic DBS Implant Manual 2006: lead kit for deep brain stimulation.

$$D = \begin{bmatrix} D_{xx} & D_{xy} & D_{xz} \\ D_{yx} & D_{yy} & D_{yz} \\ D_{zx} & D_{zy} & D_{zz} \end{bmatrix} \tag{2}$$

For water, the diffusion tensor (DT) is symmetric, so that $D_{ij} = D_{ji}$ to $i, j = x, y, z$. The water DT D will be completely defined by six elements: $D_{xx}, D_{yy}, D_{zz}, D_{xy}, D_{xz}, D_{yz}$. The diffusion tensor for each voxel of the dMRI is calculated using the Stejskal-Tanner formulation [11]:

$$S_k = S_0 e^{-b\hat{\mathbf{g}}_\mathbf{k} D \hat{\mathbf{g}}_\mathbf{k}^\top}, \tag{3}$$

where, S_k is the diffusion weighted image (DWI), S_0 is the reference image, $\hat{\mathbf{g}}_\mathbf{k}$ is the gradient vector and b is the diffusion coefficient. It is necessary at least 7 DWIs for each slice ($K = 0, 1, ..., 7$). Usually, DTs are estimated using re-weighted least squares. However, there are robust methods for DT estimation. In this work, we use the RESTORE algorithm [12] for solving the DTs.

Diffusion tensor D can be transformed to a conductivity tensor σ using the scalar transformation proposed in [13]:

$$\sigma = k \frac{S \times sec}{mm^3} D, \tag{4}$$

where $k = 0.844 \pm 0.0545$, S: Siemens, sec: seconds, mm^3: cubic millimeters.

For evaluating anisotropy level in brain tissue, first we locate the Thal, STN and SNr through Atlas registration. Then, we calculate the fractional anisotropy (FA) from the diffusion tensors using the following relation:

$$FA = \frac{\sqrt{(\lambda_1 - \lambda_2)^2 + (\lambda_2 - \lambda_3)^2 + (\lambda_3 - \lambda_1)^2}}{\sqrt{2(\lambda_1^2 + \lambda_2^2 + \lambda_3^2)}}, \tag{5}$$

where, λ_1, λ_2 and λ_3 are the eigenvalues of the each diffusion tensor D. The FA measures the fraction of the total magnitude of the tensor that is attributable to the anisotropic diffusion. For example a value of FA equal to 0 represents a perfect sphere, while a FA of 1 is an ideal linear diffusion. It is known that well-defined tracts have a value of FA greater than 0.20 and that very few regions have a higher FA to 0.9 [14].

3 Experimental Results

3.1 Anisotropy Level of Brain Structures for Each Patient

We evaluate the anisotropy level of relevant brain structures for each patient. The Table 1 shows the average fractional anisotropy (FA) for the Thalamus (Thal), Subthalamic Nucleus (STN) and Substantia Nigra reticulata (SNr).

Table 1. Average results of fractional anisotropy (FA) for Thal, STN and SNr.

	Thal	STN	SNr
Patient 1	0.342 ± 0.051	0.324 ± 0.012	0.435 ± 0.046
Patient 2	0.358 ± 0.083	0.316 ± 0.025	0.452 ± 0.039
Patient 3	0.383 ± 0.043	0.368 ± 0.023	0.397 ± 0.026
Patient 4	0.326 ± 0.019	0.392 ± 0.031	0.411 ± 0.044
Patient 5	0.354 ± 0.065	0.403 ± 0.036	0.409 ± 0.026

3.2   Simulations of DBS Models

We evaluate simulations for two patients (A and B). We test the cubic, spherical and ellipsoidal geometries and we analyze the electrical propagation for each domain. For example, we observe if there are significant modifications in potential distribution respect to a type of geometry, ground positioning and tissue properties.

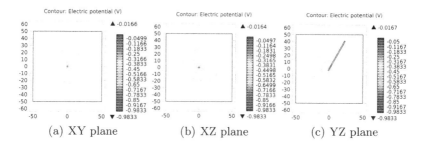

(a) XY plane   (b) XZ plane   (c) YZ plane

Fig. 2. Simulation of a STN-DBS for a cubic domain (patient A)

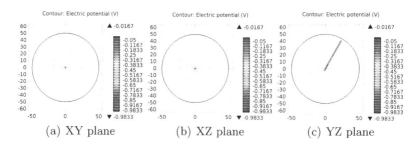

(a) XY plane   (b) XZ plane   (c) YZ plane

Fig. 3. Simulation of a STN-DBS for a spherical domain (patient A)

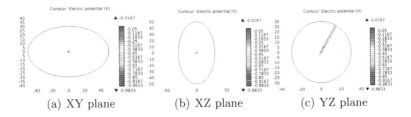

Fig. 4. Simulation of a STN-DBS for a ellipsoidal domain (patient A)

Results for Patient A: Figures 2, 3 and 4 show the electric potential generated by DBS in cubic, spherical and ellipsoidal domains for patient A.

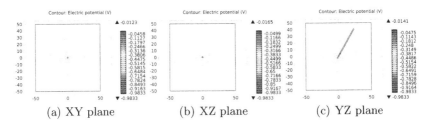

Fig. 5. Simulation of a STN-DBS for a cubic domain (patient B)

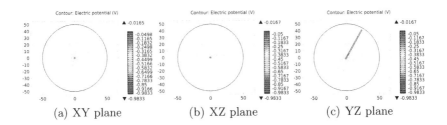

Fig. 6. Simulation of a STN-DBS for a spherical domain (patient B)

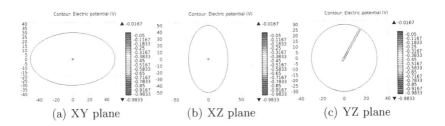

Fig. 7. Simulation of a STN-DBS for a ellipsoidal domain (patient B)

Results for Patient B: Figures 5, 6 and 7 show results for patient B.

3.3   Discussion

We synthesize the highlights of results in the following points:

- The FA values are very important because they allow us to establish if the diffusion in Thalamus (Thal), Subthalamic Nucleus (STN) and Substantia Nigra reticulata (SNr) can be considered as anisotropic or not. The threshold value established for considering anisotropy is ≥ 0.3. FA results in Table 1 show that Thal, STN and SNr are highly anisotropic in five Colombian patients. This is a clear trend of anisotropy and heterogeneity in the tissues of relevant structures in the DBS. According to this, we can infer that a simulation model that considers the medium as isotropic and homogeneous has low accuracy and it is not consistent with the real behavior of the brain tissue from a patient with Parkinson's disease.
- Diffusion tensors provide quantitative information about water diffusion in brain tissue, allowing more accurate conductivity models in comparison to previous approaches of brain electrical conductivity. They also offer information of directionality and anisotropy. In the other hand, homogeneous and isotropic models consider the brain tissue as a single conductor volume with a scalar value (i.e. 0.3 S/m). This is a wrong approximation because the brain tissue is divided in several classes: gray matter, white matter, cerebrospinal fluid, skull, among others. Each tissue type has unique electrical properties.
- All simulations performed in our experiments show that voltage propagation generated by DBS does not depend of geometry. We can observe from all figures that shape, quantity and direction of propagation did not change when we use a cube, a sphere or an ellipsoid in the same patient. In contrast, in previous approaches [7,8] reported for DBS models with homogeneous and isotropic domains, was found a strong dependence with the selected geometry. According to this analysis, we can establish that geometry is not a key factor when the medium is anisotropic and heterogeneous.
- We also analyze the ground positioning in all DBS models. We tested ground poles in different sides of each geometry and we found a strong dependence with the ground placement in our models. We obtained different voltage propagation for several ground locations. This demonstrates that DBS simulation models are highly sensible to the ground placement.
- If we compare results for patients A and B, we can see a different voltage propagation. In heterogeneous models the tissue properties have a fundamental role in electrical magnitudes generated by DBS. Following this notion, the propagation of the electric potential generated by a DBS electrode is related to the stimulation parameters (amplitude, frequency, wide pulse, etc) and the electrical properties of the brain tissue. However, we found that another factors inside the simulation models as ground position are very critical. The inclusion of anisotropy with diffusion tensors allow the development of detailed models for specific patients analysis. Similarly, it is possible to realize more reliable

simulations according to the real electrical behavior of neurons affected by DBS. This point has a remarkable relevance, because the therapeutic outcomes of DBS in Parkinson's disease patients depends strongly of an accurate stimulation in the STN and surrounding structures.

4 Conclusions and Future Work

In this paper, we presented an analysis of the role of geometry and electric properties of brain tissue in simulation models for DBS. We found that type of geometry is not relevant when the medium is anisotropic and heterogeneous. However, the tissue properties and ground location are key in voltage propagation. The electric properties of brain tissue are different in all analyzed patient. For this reason, a homogeneous and isotropic model might be a wrong approach to simulate the DBS. As future work, we propose models with real brain shapes reconstructed from MRI. For example: 3D brain reconstructions to perform patient specific analysis of voltage propagation and to correlate it with therapeutic outcomes.

Acknowledgments. H.D. Vargas is funded by Colciencias under the program: *formación de alto nivel para la ciencia, la tecnología y la innovación - Convocatoria 617 de 2013*. This research is developed under the project financed by Colciencias with code 111065740687.

References

1. Limousin, P., Krack, P., Pollak, P., Benazzouz, A., Ardouin, C., Hoffmann, D., Benabid, A.: Electrical stimulation of the subthalamic nucleus in advanced Parkinson's disease. N. Engl. J. Med. **339**, 1105–1111 (1998)
2. Maks, C., Butson, C., Walter, B., Vitek, J., McIntyre, C.: Deep brain stimulation activation volumes and their association with neurophysiological mapping and therapeutic outcomes. J. Neurol. Neurosurg. Psychiatry **80**, 659–666 (2009)
3. Johnson, M., Miocinovic, S., McIntyre, C., Vitek, J.: Mechanisms and targets of deep brain stimulation in movement disorders. Neurotherapeutics **5**, 294–308 (2008)
4. Butson, C., Maks, C., Walter, B., Vitek, J., McIntyre, C.: Patient-specific analysis of the volume of tissue activated during deep brain stimulation. Neuroimage **34**, 661–670 (2007)
5. Gimsa, U., Schreiber, U., Habel, B., Flehr, J., Van Rienen, U., Gimsa, J.: Matching geometry and stimulation parameters of electrodes for deep brain stimulation experiments: numerical considerations. J. Neurosci. Methods **150**, 212–227 (2006)
6. Yousif, N., Richard, B., Liu, X.: The influence of reactivity of the electrode-brain interface on the crossing electric current in therapeutic deep brain stimulation. Neuroscience **156**, 597–606 (2009)
7. Liberti, M., Apollonio, F., Paffi, A., Parazzini, M., Maggio, F., Novellino, T., Ravazzani, P., D'Inzeo, G.: Fundamental electrical quantities in deep brain stimulation: influence of domain dimensions and boundary conditions. In: Proceedings of Conference on IEEE Engineering Medicine Biology Society, pp. 6668–6671 (2007)

8. Grant, P., Lowery, M.: Electric field distribution in a finite-volume head model of deep brain stimulation. Med. Eng. Phys. **31**, 1095–1103 (2009)

9. Schmidt, C., Van Rienen, U.: Modeling the field distribution in deep brain stimulation: the influence of anisotropy of brain tissue. IEEE Trans. Biomed. Eng. **59**, 1583–1592 (2012)

10. Walckiers, G., Fuchs, B., Thiran, J., Mosig, J., Pollo, C.: Influence of the implanted pulse generator as reference electrode in finite element model of monopolar deep brain stimulation. J. Neurosci. Methods **186**, 90–99 (2010)

11. Tanner, J., Stejskal, E.: Spin diffusion measurements: spin echoes in the presence of a time-dependent field gradient. journal of Chemical. J. Chem. Physiol. **42**, 288–292 (1995)

12. Lin-Chin, C., Jones, D., Pierpaoli, C.: RESTORE: robust estimation of tensors by outlier rejection. Magn. Reson. Med. **53**, 1088–1095 (2005)

13. Tuch, D., Weeden, V., Dale, A.: Conductivity tensor mapping of the human brain using diffusion tensor MRI. Proc. Natl. Acad. Sci. **98**, 11697–11701 (2001)

14. Basser, P.: Inferring microstructural features and the physiological state of tissues from diffusion weighted images. NMR Biomed. **8**, 333–344 (1995)

Spatial Resolution Enhancement in Ultrasound Images from Multiple Annotators Knowledge

Julián Gil González[✉], Mauricio A. Álvarez, and Álvaro A. Orozco

Faculty of Engineering, Universidad Tecnológica de Pereira, Pereira, Colombia 660003
{jugil,malvarez,aaog}@utp.edu.co

Abstract. Enhancement of spatial resolution for medical images improves clinical procedures such as diagnosis of different diseases, image registration, and tissue segmentation. Although different methods have been proposed in the literature to tackle this problem, each of them comes with their own strengths and their own weaknesses. In this work, we present a novel approach for the enhancement of spatial resolution in ultrasound images that aims at improving resolution enhancement by combining different interpolation methods. The methodology is based on learning from multiple annotators, also known as learning from crowds, a recent development in supervised learning to incorporate the diverse levels of knowledge that different experts can have on a prediction problem, in order to leverage the prediction performance in a single model. In particular, we consider each pixel intensity value in each new high resolution image as a corrupted version of a gold standard. Each of the single interpolation algorithms acts as an expert that provides a level of intensity for a particular pixel. We then use a regression scheme for multiple annotators based on Gaussian Processes with the aim of computing an estimate of the actual image from the noisy annotations given by the interpolation algorithms. We compare our approach against two super resolution schemes based on Gaussian process regression. This comparison is performed using the mean square error (MSE) for the interpolation validation and the Dice coefficient (DC) for the morphological validation. Results obtained show that our approach is a promising methodology for enhancing spatial resolution in ultrasound images.

1 Introduction

In contrast to other types of medical imaging (e.g. Magnetic Resonance Imaging, MRI, diffusion Magnetic Resonance Imaging, dMRI, and x-rays computed tomography), ultrasound imaging (UI) offers several advantages such as real-time imaging, no radiation, and small movable devices [1]. However, even the best ultrasound images, have a lower resolution compared with the corresponding MRI and x-ray computed tomography [2]. The loss of spatial resolution in ultrasound images is mainly caused by two types of limitations: intrinsic and extrinsic. Intrinsic resolution limitation is related to the characteristics of the interaction between a tissue and the ultrasound wave, such as attenuation

© Springer International Publishing AG 2017
C. Beltrán-Castañón et al. (Eds.): CIARP 2016, LNCS 10125, pp. 502–510, 2017.
DOI: 10.1007/978-3-319-52277-7_61

and scattering. Extrinsic resolution limitation is given by spontaneous and non-spontaneous movements of the analyzed tissue or the patient (e.g. respiration and heart breathing) [3].

The lost of spatial resolution in medical imaging is a critical issue for clinical analysis since it complicates the accomplishment of different procedures such as diagnosis of diseases, segmentation (tissue, nerves and bone) and needle guidance in peripheral nerve blocking (PNB) procedures [4]. For this reason, several approaches have been developed to enhance the spatial resolution in ultrasound images from the post-processing of low-resolution (LR) images. For instance, in [1] the ultrasonic images are modeled as a convolution of a point spread function of the image process, and the tissue function. Spatial resolution is increased by deconvolving the observed radio frequency image by the point spread function using homomorphic filtering. However, this method requires great computational effort. Another approach is proposed in [5]. Here, the spatial resolution enhancement is considered as an ill-posed inverse problem where the regularization is performed using anisotropic diffusion. Nevertheless, its accuracy is affected in presence of high levels of speckle noise. Recently, a new approach based on patch learning have been proposed. For instance, in [6,7], a patch-based Gaussian Processes Regression (GPR) for enhancing the spatial resolution was introduced. Nevertheless, due to the natural smooth behavior of Gaussian Processes (GPs), the method tends to generate a blurring effect over the edges in the High Resolution (HR) images. This drawback was minimized for MRI in [6] by adding a post-processing step, where a specific 2D filter is used in order to highlight the edges.

As we have pointed above, different approaches have been proposed to deal with the issue of low spatial resolution in medical images. However, each of them approaches exhibits their strengths and their weakness. In this sense, the HR resolution images generated with these approaches correspond to noisy versions of the actual (and unknown) HR image. This problem can be minimized using supervised learning with multiple annotators systems to estimate the ground true (actual HR image) from multiple noisy images.

Learning with multiple annotators is an emergent area in the context of supervised learning. Its aim is to deal with supervised problems, where the gold standard is not available, and in contrast, we just have access to an amount of noisy annotations provided by multiple annotators [8,9]. Recently, many approaches have been proposed to address different problems such as classification [8], regression [9], and sequence labeling [10], under the framework of multiple annotators. In the area of machine learning with multiple annotators, we can recognize two main goals using the training data: First, to estimate the ground-truth from the multiple annotations. Second, to build a supervised scheme (e.g. a classifier or a regressor). In this work we focus in the first goal, in order to estimate a HR image from multiple HR images (possibly noisy). The idea of our methodology is to use different interpolations methods (e.g. Bilinear interpolation, Bicubic interpolation and Lanczos interpolation) to generate HR images. Then, we consider each pixel intensity value in each of these HR images as a corrupted version of the pixel intensity value in the corresponding hidden and true HR image (considered as the gold standard). Finally, we use the regression scheme for multiple annotators based on Gaussian Processes, proposed in [9], with the aim of computing an estimate of the actual HR image from the noisy HR images given by the interpolation algorithms. We compare our approach against two super-resolution schemes based on Gaussian process regression. One of these schemes uses the pixel intensities in the nearest neighbors as the features [7]. The second scheme uses the position of each pixel as features [6]. Validation is carried out based on the Mean Square Error (MSE) metric and the Dice coefficient (DC) for the morphological validation (nerve segmentation).

2   Materials and Methods

2.1   Dataset

The dataset used in this work was collected by the Universidad Tecnológica de Pereira, Pereira, Colombia and the Santa Mónica Hospital, Dosquebradas, Colombia. The dataset comprises recordings of ultrasound images from patients who underwent regional anesthesia using the Peripheral nerve blocking procedure. The dataset in conformed by 31 ultrasound images from the ulnar nerve (18 images) and median nerve (13 images). Each ultrasound image was collected using a Sonosite Nano-Maxx device (The resolution of each image is 360×370 pixels). Each image in the dataset was labeled by an expert in anesthesiology to indicate the location of the nerve structures. Figure 1 shows the types of images in the dataset.

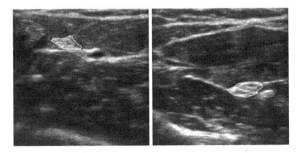

Fig. 1. Images belonging to the dataset. In the left, the ulnar nerve is shown and the median nerve is shown on the right. Each image has been labeled by an expert in anesthesiology to locate the nerve structures.

2.2   Spatial Resolution Enhancement Based on Gaussian Processes Regression with Multiple Annotators

We follow the regression model with multiple annotators proposed by [9]. We assume that there are R HR images, generated by R different interpolation methods. In this sense, the training set comprises $\mathscr{D} = \left\{ \mathbf{x}_i, y_i^1, ...y_i^R \right\}_{i=1}^N$, where x_i is the feature vector (in this case we consider as features the pixel coordinates x and y) and y_i^j is the intensity value of the pixel i in the HR image generated by the $j-$th interpolation algorithm. So, the model assumes that each annotation is generated following $y_i^j = f_i + \epsilon^j$, where f_i is the unknown ground-truth (in this case the intensity value pixel in the actual but hidden HR image) and $\epsilon^j = \mathscr{N}(0, \sigma_j^2)$ is the distribution associated with the noise.

Assuming that each annotator (in this case each interpolation method) labels the observation $\mathbf{x}_i$ independently and assuming independence between the annotators, the likelihood is given as

$$p(\mathbf{y}|\mathbf{f}) = \prod_j \prod_i \mathscr{N}(y_i^j|f_i, \sigma_j^2),$$

where $\mathbf{y} = \left\{ y_i^1, ...y_i^R \right\}_{i=1}^N$. Assuming a Gaussian process prior over $\mathbf{f}$ such as $p(\mathbf{f}) = \mathscr{N}(\mathbf{0}, \mathbf{K})$, where the covariance function is computed using a specific kernel function, it can be shown that for a new observation $f(\mathbf{x}_*)$, the posterior distribution is given by

$$p(f(\mathbf{x}_*)|\mathbf{y}) = \mathscr{N}(f(\mathbf{x}_*)|\bar{f}(\mathbf{x}_*), k(f(\mathbf{x}_*), f(\mathbf{x}'_*))), \tag{1}$$

where, $\bar{f}(\mathbf{x}_*) = k(\mathbf{x}_*, \mathbf{X}) \left(\mathbf{K} + \hat{\Sigma} \right)^{-1} \hat{\mathbf{y}}$,

$k(f(\mathbf{x}_*), f(\mathbf{x}'_*)) = k(\mathbf{x}_*, \mathbf{x}'_*) - k(\mathbf{x}_*, \mathbf{X}) \left(\mathbf{K} + \hat{\Sigma} \right)^{-1} k(\mathbf{X}, \mathbf{x}_*)$ and

$$\frac{1}{\hat{\sigma}_i^2} = \sum_j \frac{1}{\hat{\sigma}_j^2}, \quad \hat{y}_i = \hat{\sigma}_i^2 \sum_j \frac{y_i^j}{\hat{\sigma}_j^2}, \quad \hat{\Sigma} = \mathrm{diag}\left(\hat{\sigma}_1^2, ...\hat{\sigma}_N^2 \right).$$

However, in this work we are not interested in making a prediction for new samples $\mathbf{x}_*$, but in computing a gold standard estimate from the multiple annotations. In this sense, we consider as a new instances the whole samples in the training data $\mathbf{X}$ and use the expression (1) with the aim of computing a probabilistic estimate for the gold standard (i.e. the actual intensity value for each pixel in the HR image) The unspecified parameters can be estimated by minimizing the negative log of the evidence (see [9]).

2.3   Procedure

GPRMA for Resolution Enhancement. For developing our methodology we generate five HR images from five different interpolation methods: Nearest Neighborhood interpolation (NN) [11], Bilinear interpolation (Bil) [12], Bicubic interpolation (Bic) [12], Lanczos interpolation (Lan) [11] and a methodology where the files in the images are interpolated using linear interpolation and the rows are estimated using nearest interpolation (LN) Then, each pixel intensity value in each of these HR images is considered as a corrupted version of the pixel intensity value in the corresponding hidden and true HR image (considered as the gold standard). Finally, we perform a patch-based Gaussian process regression with multiple annotators (GPRMA) for computing an estimated of the actual HR image from the annotations (i.e. the HR images generated using the interpolation methods described above). We use 10×10 patches with no overlap. The patch size was chosen by cross-validation. A GPRMA is trained in each patch, considering as features the relative position x and y with respect to the beginning of the patch, and considering as annotations, the pixel intensity values in each HR image. We use the well-known radial basis function (RBF) kernel for all the experiments. Parameter estimation is performed by minimizing the log of the evidence (see [9]) using gradient descend.

Experimental Setup. In order to validate the methodology proposed in this paper, first, we down-sample each image in the dataset described in Sect. 2.1. Hence, we obtain 31 180×175 LR images. Next, we use our methodology based on GPRMA for building the 360×370 pixels HR images. We compare our approach with two super-resolution schemes based on Gaussian process regression. One of these schemes uses the position of each pixel as features [6]. The other scheme uses the nearest neighbors as the features [7]. We compare all the HR images with the respective Gold Standard (i.e. the images in the initial dataset). The performance of all methods is measured in terms of the mean squared error (MSE) obtained with each HR image with respect to the Ground truth. Finally, we morphologically validate the methodologies by segmenting nerve structures. This segmentation process is carried out using an active shape model [13]. This model is initialized using a methodology based on Graph Cuts [14]. The performance for the morphological validation is measured in terms of the Dice Coefficient [15].

3   Results and Discussions

First, we perform a direct comparison between the up-sampled ultrasound images and the gold standard (i.e. the images in the dataset). As we pointed in Sect. 2.3 this comparison is carry out in terms of MSE (specifically, the average MSE computed for the whole images in the dataset). Table 1, shows the average MSE results for the interpolations methods used as annotations in the GPRMA scheme. Moreover, this table reports the average MSE for the methodology proposed in [6] (GPR-1), the methodology proposed in [7] (GPR-2) and our methodology (GPRMA). In Fig. 2, we show the graphical error with the up-sampled images and the gold standard. We choose as gold standard one of the images in the dataset corresponding to a ulnar nerve. Sub-figures (a)-(e) show the absolute error for the interpolation methods used as annotations for the regression scheme with multiple annotators. Sub-figures (f), (g) and (h), show the absolute error for the methodologies proposed in [6,7] and the approach proposed in this paper.

From Table 1 it is possible to note that our methodology for the spatial resolution enhancement outperforms the approaches based on Gaussian process

Table 1. Average MSE between the up-sampled images and the ground truth.

Method	Mean squared error (MSE)
	$\mu \pm \sigma$
NN	61.7090 ± 12.9012
Bil	36.5130 ± 7.4665
Bic	34.9892 ± 7.1502
Lan	35.3365 ± 7.2193
LN	35.5640 ± 6.9246
GPR-1 [6]	798.4395 ± 193.1096
GPR-2 [7]	38.8651 ± 7.9923
GPRMA	$\mathbf{27.0990 \pm 6.1855}$

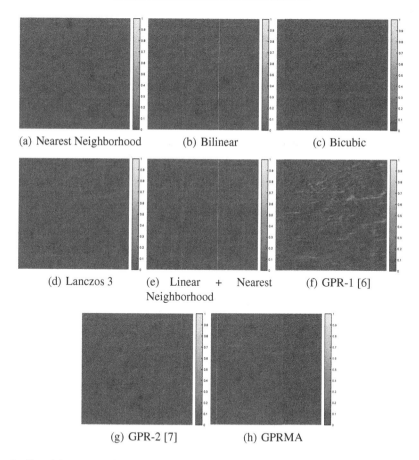

(a) Nearest Neighborhood   (b) Bilinear   (c) Bicubic

(d) Lanczos 3   (e) Linear + Nearest Neighborhood   (f) GPR-1 [6]

(g) GPR-2 [7]   (h) GPRMA

Fig. 2. Graphics errors for the spatial resolution enhancement in ultrasound images. Sub-figures (a)–(e), correspond to the error images for the interpolations methods considered as annotations. Similarly, in (f), (g) and (h), we show respectively the graphical error for the methods proposed in [6,7] and our methodology

regression (GPR-1 and GPR-2) in terms of the MSE. However, according to the Sub-figure (h), it is possible to observe that our methodology interpolates efficiently all the intensity of pixels in the HR image, excepts for those located at the edges, were a blur effect is observed. The above is due to the natural smooth behavior of Gaussian processes. Furthermore, we note that the method GPR-1 has a low performance in ultrasound images. Taking into account that the approach GPR-1 was originally developed for the resolution enhancement in Magnetic resonance images, which are affected by a different type of noise to the present in ultrasound images (speckle noise). The low performance obtained by GPR-1 is explained by its sensitivity to the speckle noise present in ultrasound images.

Then, we validate the HR images morphologically to define which method is appropriate or not. The morphological validation is performed by segmenting nerves structures in the image. The segmentation of nerves is a key issue in anesthesiology since it is necessary to perform peripheral nerve blocking (which is used for regional anesthesia and for pain management). Table 2, shows the results for morphological validation measured in terms of DC between the gold standard (manual segmentation by a specialist) and the segmentation obtained with the HR images and the original images in the dataset.

Table 2. Morphological validation in terms of the Dice coefficient.

Method	Dice coefficient (DC)
	$\mu \pm \sigma$
Original images	0.5702 ± 0.1277
NN	0.5583 ± 0.1246
Bil	0.5567 ± 0.1256
Bic	**0.5651 ± 0.1280**
Lan	0.5642 ± 0.1270
LN	0.5629 ± 0.1235
GPR-1 [6]	0.5157 ± 0.1487
GPR-2 [7]	0.5568 ± 0.1256
GPRMA	0.5578 ± 0.1244

From Table 2, it is possible to note that there are not significant differences between the performance of our methodology (GPRMA), the interpolations methods and the methodology proposed in [7] (GPR-2). Moreover, we can note that the methodology proposed in [6] has the lowest performance. On the other hand, if we take into account the whole validations schemes proposed in this work, it is possible to determine that the proposed methodology outperforms the approaches considered in this work.

4   Conclusion

In this paper, we introduced a methodology for spatial resolution enhancement in ultrasound images based on a novel area in supervised learning known as learning from multiple annotators. Results achieved with the proposed methodology outperform to the GPR methods proposed in [6,7] for both validation schemes: interpolation validation and morphological validation. Hence, it is possible to determine that GPRMA is a promising methodology for enhancing spatial resolution in ultrasound images. Future work can be oriented in developing a post-processing step with the aim of highlight the edges in the HR image. Moreover, it is possible to extend the GPRMA model used in order to model the dependence between the annotator expertise and the samples in the input space.

Acknowledgement. Julián Gil González is funded by Colciencias under the program: *Jóvenes Investigadores e Innovadores - Convocatoria 645 de 2014*. This work was developed under the project: "Stochastic modeling of medical imaging for spatial resolution enhancement and improvement of clinical analysis", with financial support from the Universidad Tecnológica de Pereira.

References

1. Jiřík, R., Taxt, T.: High-resolution ultrasonic imaging using two-dimensional homomorphic filtering. IEEE Trans. Ultrason. Ferroelectr. Freq. Control **53**(8), 1440–1448 (2006)
2. Taxt, T., Jiřík, R.: Superresolution of ultrasound images using the first and second harmonic signal. IEEE Trans. Ultrason. Ferroelectr. Freq. Control **51**(2), 163–175 (2004)
3. Kouame, D., Ploquin, M.: Super-resolution in medical imaging: an illustrative approach through ultrasound. In: 2009 IEEE International Symposium on Biomedical Imaging: From Nano to Macro, ISBI 2009, pp. 249–252. IEEE (2009)
4. Trinh, D.-H., Luong, M., Dibos, F., Rocchisani, J.-M., Pham, C.-D., Nguyen, T.Q.: Novel example-based method for super-resolution and denoising of medical images. IEEE Trans. Image Process. **23**(4), 1882–1895 (2014)
5. Dai, Y., Wang, B., Liu, D.C.: A fast and robust super resolution method for intima reconstruction in medical ultrasound. In: 2009 3rd International Conference on Bioinformatics and Biomedical Engineering, ICBBE 2009, pp. 1–4. IEEE (2009)
6. Cardona, H.D.V., López-Lopera, A.F., Orozco, Á.A., Álvarez, M.A., Tamames, J.A.H., Malpica, N.: Gaussian processes for slice-based super-resolution MR images. In: Bebis, G., et al. (eds.) ISVC 2015. LNCS, vol. 9475, pp. 692–701. Springer, Heidelberg (2015). doi:10.1007/978-3-319-27863-6_65
7. He, H., Siu, W.-C.: Single image super-resolution using gaussian process regression. In: 2011 IEEE Conference on Computer Vision and Pattern Recognition (CVPR), pp. 449–456. IEEE (2011)
8. Raykar, V.C., Yu, S., Zhao, L.H., Valadez, G.H., Florin, C., Bogoni, L., Moy, L.: Learning from crowds. J. Mach. Learn. Res. **11**, 1297–1322 (2010)

9. Groot, P., Birlutiu, A., Heskes, T.: Learning from multiple annotators with Gaussian processes. In: Honkela, T., Duch, W., Girolami, M., Kaski, S. (eds.) ICANN 2011. LNCS, vol. 6792, pp. 159–164. Springer, Heidelberg (2011). doi:10. 1007/978-3-642-21738-8_21

10. Rodrigues, F., Pereira, F., Ribeiro, B.: Sequence labeling with multiple annotators. Mach. Learn. **95**(2), 165–181 (2014)

11. Acharya, T., Tsai, P.-S.: Computational foundations of image interpolation algorithms. ACM Ubiquity **8**(42), 1–17 (2007)

12. Szeliski, R.: Computer Vision: Algorithms and Applications. Springer Science and Business Media, Heidelberg (2010)

13. Chan, T.F., Vese, L.A.: Active contours without edges. IEEE Trans. Image Process. **10**(2), 266–277 (2001)

14. Boykov, Y., Funka-Lea, G.: Graph cuts and efficient ND image segmentation. Int. J. Comput. Vis. **70**(2), 109–131 (2006)

15. Sampat, M.P., Wang, Z., Gupta, S., Bovik, A.C., Markey, M.K.: Complex wavelet structural similarity: a new image similarity index. IEEE Trans. Image Process. **18**(11), 2385–2401 (2009)

How Deep Can We Rely on Emotion Recognition

Ana Laranjeira[1], Xavier Frazão[2], André Pimentel[2], and Bernardete Ribeiro[1(✉)]

[1] CISUC - Department of Informatics Engineering,
University of Coimbra, Coimbra, Portugal
afolgado@student.dei.uc.pt, bribeiro@dei.uc.pt
[2] EyESee Solutions, Lisboa, Portugal

Abstract. The emerging success of digital social media has had an impact on several fields ranging from science to economy and business. Therefore, there is an invested interest in emotion detection and recognition technology from facial expressions, in order to increase their market competitiveness. This area still presents many challenges, namely the difficulty in achieving real-time facial recognition. Herein we tackle this problem by crossing methods targeting both static images and active images. In this work, we explore the recent technological breakthroughs in deep learning and develop a system based on automatic recognition of human face expressions using Convolutional Neural Networks (CNN). We use the Cohn-Kanade Extended (CKP) dataset for testing our proposed CNN model along with an augmented version, which demonstrated effectiveness in seven basic expressions. In order to enhance the quality of the results instead of the overlapping method for building the augmented dataset we propose random perturbations from a wide set including: skew, translation, scale, and horizontal flip. Moreover, we built a real-time video framework using our model (a version of LeNet-5) which is fed with frames detected with Viola-Jones face tracker that reproduce the CKP dataset. The results are promising.

Keywords: Emotion recognition · Convolutional neural networks

1 Introduction

Facial Expressions are the source of Human Emotion Recognition and a deep understanding on these emotional responses gives a major advantage to any dependent field. The information retrieved from emotion detection and recognition technology will leverage a wide range of applications such as lie detections, pain assessment, surveillance, healthcare, consumer electronics, law enforcement and many others dependent on Human Computer Interfaces. The challenge in the field is to develop a way to integrate a large spectrum of expressions intrinsically related to the capacity of humans to express feelings. However, there are six basic (or prototypic) expressions that can be proven consensual within any culture, namely *Anger, Disgust, Fear, Happiness, Sadness, Surprise* [4]. These have being used by the majority of Facial Expression Recognition (FER) systems.

© Springer International Publishing AG 2017
C. Beltrán-Castañón et al. (Eds.): CIARP 2016, LNCS 10125, pp. 511–520, 2017.
DOI: 10.1007/978-3-319-52277-7_62

A classic FER system is structured according to three stages, starting with facial detection, followed by the feature extraction and the final stage, expression recognition. The first stage is the most explored and can be considered as an exhausted topic. A few surveys have identified and categorized many proposed improvements and extensions [1]. Viola and Jones made a boost in facial detection [16] and their classifier is still widely used despite the recent appearance of more effective cases [6]. The CIFE dataset survey [12] includes a descriptive table of the algorithms used for the above stages, and concludes that feature-based approaches perform better. Most of the drawbacks come from the dataset construction, performed under a controlled environment (positions, lightning, no occlusions), therefore jeopardizing the robustness of the system in presence of invariance. Despite these constraints, in the past few years with hardware technological innovations and the consequent development of deep learning hierarchical models it was possible to extract complex data interactions from large scale datasets and build advanced solutions on demand. Within this context, an important and possible solution has been revisited - **Convolutional Neural Networks**, leading to significant results in image base classification, as demonstrated in the annual contest *ImageNet Large Scale Visual Recognition Challenge* (ILSVRC). There are some works that stand out for their performance in this large image classification problem, the *AlexNet* [10] from 2012 and a more recent, *GoogLeNet* [15] inspired in a deeper concept - "Network inside Network".

In this paper, we propose and develop a Convolutional Neural Network for emotion facial recognition. Our proposed model was trained with seven expressions, introducing also the expression of *Contempt* to the prototypic set, based on its presence in a widely explored dataset, the **Cohn-Kanade Extended** [13] and therefore considered as a reliable source. In order to enhance the quality of the results we have augmented the dataset based on random perturbations from a wide set including: skew, translation, scale, and horizontal flip. We specified several heuristics for the model configuration and ran the experiments in a stand-alone version of *Caffe* [9], supported by an instance from *Amazon EC2* which allows access to GPU advanced hardware. We are aware that in the literature the methods are dependent on the type of dataset, falling into static-images or dynamic images sequences. In the case of sequence-based approaches, the temporal information could be a concern due to real time constraints. In this work, we look at how the static methods can be applied to sequence events without compromising the reliability and even improving speed of recognition. We put forward, a real-time video framework composed by Viola-Jones *OpenCV*[1] face tracker. This framework shows promising preliminary results with our deep approach and matches the question we try to answer: How deep can we rely on emotion recognition? Future work should fully answer this question.

This paper is organized as follows. In the next section we present the related work highlighting inspiring studies in the field of emotion recognition. In Sect. 3 our proposed model is described from both static image data and sequence based data for real-time application. In Sect. 4 the datasets, the experimental setup,

---

[1] http://opencv.org/.

the technicalities of the approach and the evaluation metrics of our model classifier are described. In Sect. 5 the results are discussed. Finally, in Sect. 6 the conclusions are presented with the future work.

2   Related Work

An annually scientific contest *Emotion Recognition in the Wild Challenge* (EmotiW) is defining the state of the art focusing on affective sensing with uncontrolled environments[2]. From the latest year contest (EmotiW 2015), two versions of dealing with expressions can be selected: one uses static images from Static Facial Expressions in the Wild (SFEW); and the other uses an acted point of view resorting to an Acted Facial Expressions in the Wild (AFEW) dataset. Among the static-image approaches the project [17] proposes a 3-way detection of the face, with a hierarchical selection from the Join Cascade Detection and Alignment (JDA), Deep CNN-Based (DCNN) and MoT along with a simple network (11 layers). These detectors are processed in a multiple network framework in order to enhance the performance. It also includes a pre-processing phase to improve accuracy, which might be considered a drawback in the classification response. An approach resorting to video is strongly dependent on spatio-temporal issues. In [3] a Synchrony Autoencoder (SAE) was introduced for local motion feature extraction together with assembling an hybrid network, CNN-RNN.

3   Proposed Emotion Recognition Model

Our proposed model was constructed with LeNet-5 as baseline and has been progressively developed. Its current stage is depicted in Fig. 1.

The classic LeNet-5 architecture [11], is a combination of seven layers represented by a convolutional layer with 16 feature maps with a 5×5 kernel size, followed by a sub-sampling by half. This sub-sampling (pooling layer) and the next convolutional layer (16 feature maps) are not connected in order to break symmetry. The network includes also another pooling stage, maintaining the 16 feature maps and the kernel to 5×5. As opposed to our version of the LeNet-5, it includes another convolutional full sized (kernel 1×1), mapping 120 units before the last full connected layers with 84 and 10 neurons, respectively (since LeNet is addressed to digit recognition).

Our model is composed by an initial convolutional set with a 5×5 kernel size and 20 feature maps plus a shared bias ending up with 520 parameters. The next layer or Pooling Layer performs a downsampling with a maximum value of a 2×2 kernel size. This process is repeated except for the pooling stride which changed to 2 and the convolution process expecting 50 instead of 20 feature maps, augmenting the parameters to 1300. We used in our work Krizhevsky [10] alternative model neurons output. Instead of the standard functions $f(x) = tanh(x)$

---

[2] https://cs.anu.edu.au/few/emotiw2015.html.

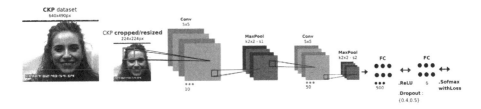

Fig. 1. Flowchart of the different stages of our CNN model, adapted from the classic LeNet-5 - Lenet Ov (our version)

or $f(x) = (1 - e^{-x})^{-1}$, we use a faster version $f(x) = max(0, x)$ designated as *Rectified Linear Units* (ReLUs). The convolutional neural networks with ReLU proved to be 6 times faster than an equivalent network using saturating neurons, reaching 25% of training error rate. Taking this into account, the full connected layer that follows, containing 500 filter numbers, is connected with an ReLU. So far, the structure is similar to Lenet, however we introduced a *Dropout* between the full connected layers, reducing between 0.4 and 0.5 percent of their connectivity by dropping randomly some units or neurons which do not contribute to the forward pass. This procedure will overcome overfitting. Dropout prevents co-adaption showing a significant improvement by 10% of accuracy, namely in the ILSVRC 2012 validation and test sets [14]. Finally, the last full connected layer is responsible for shrinking the feature maps to our class problem - 7 (expressions). The weights presented to the net follow the *xavier* type except for the first convolutional layer which is set between *gaussian, unitball, xavier*.

4   Experimental Design

The **Cohn-Kanade Extended** dataset [13] (CKP) set is labeled between 0–7 corresponding to *Neutral, Anger, Contempt, Disgust, Fear, Happiness, Sadness, Surprise* and contains 593 sequences across 123 subjects with posed and non-posed (candid) expressions captured into a $640x \times 490$ px or 640×480 px frame, depending on the channel. The class Contempt despite being excluded from the range of the six basic expressions, was used mainly because it was reported to be found above 75% both in Western and non-Western cultures [5]. On the other hand, the neutral face was not considered in the training stage since it is hardly present in video-based classification. Only images with labels and in the peak of expression (apex state) were considered (1631 images), cropped as depicted in Fig. 1. The CKP dataset was split into 70% for training; the remaining were taken for the validation and test phases. In order to feed properly the network, the **CKP** set of images were **augmented** with random perturbations, based on the expressive results [17] from an experiment over a lower resolution dataset (**FER dataset** [8]). The perturbation set, skew, translation, scale, rotation and horizontal flip worked separately in order to achieve a wider set, instead of the proposed overlapping method. Skew parameters were randomly selected from

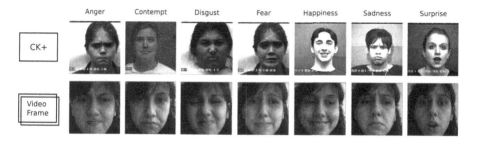

Fig. 2. CKP static images vs our frame sequence, both cropped to 224 × 224 pixels

$\{-0.1, 0, 0.1\}$, translation parameters were sampled from $\{0, \delta\}$, where δ is a random sample from a $[0, 4]$ set, scaling uses a δ value to define a random parameter $c = 47/(47 - \delta)$ and the rotation is dependent on the angle sampled randomly from $\{-\pi/18, 0, \pi/18\}$. The final augmentation version has 978, 288 and 132 images per class in training, validation and test phases, respectively. We included a set of images to the test phase, populated with frames (Fig. 2) from a real-time video framework composed by a Viola-Jones $OpenCV^3$ face tracker [16] along with our classifier. Images were resized to the classifier input shape (224 × 224 px), captured within 35 frames per second, and classified in 0.250s (average including cropping process) into an expression displayed in the command line. The experiments started from the determent of the network, questioning the relevant inner parameters comparing with an appropriate solver. Fixing the network, involved testing some prominent networks from the state of the art: the AlexNet from classifications of the ILSVRC2012 challenge [10], the recent GoogLeNet and the classic LeNet-5, to report a fair judgment. The non-augmented dataset was considered small enough for a CNN input and therefore a candidate to make a sanity check on the hyperparameters. In this context, we used GoogleNet and LeNet-5 and both passed the test, overfitting with an high accuracy between [0.9,0.92] in training, whereas the validation loss computed by summing the total weighted loss over the network (for the outputs with non-zero loss), reached a value of 0.8. In order to follow the right network in place, we conducted some preliminary classification experiments over a short amount of training time and the best gains in the test set came from the LeNet-5 as depicted in Fig. 3. Since the results over the training (Table 1) and validation set had a retarded loss decay, no further experiments on GoogleNet and Alexnet were developed because training memory and processing time would become an issue. Considering these marks, a new version of LeNet-5 (LeNet - Ov) was explored. Moreover, LeNet-5 baseline parameters were addressed from the *Caffe* standards.

The network was prepared taking into consideration the dependency on the convolutional neural networks to the feature extraction process, and how first layers and their high level of information are determinant to the success of

---

[3] http://opencv.org/.

the classification. Therefore we tested different ways of adjusting the weights involved. Our network was tested with three types of fillers, namely, *gaussian, positive unitball* and *xavier*. The *gaussian* filler only chooses values according a gaussian distribution, limiting non-zero inputs up to 3, and the standard deviation assume a 0.01 value (increased from the default 0.005). The *positive unitball*, fills a blob with values between [0, 1] such that $\forall i \sum_j xij = 1$. Finally, the *xavier* type (weight filler), initializes the incoming matrix with values from an uniform distribution within $[-\sqrt{\frac{3}{n}}, \sqrt{\frac{3}{n}}]$, where the n is the number of the input neurons. This *Caffe* version of Xavier differs from what was initially introduced by Glorot [7], removing the output information. Our version of the network is optimized with a stochastic gradient descent solver, since the Alexnet trained with Adagrad (see Fig. 3) was not expressive. The solver hyperparameters were highly influenced with the results [2] from an automatic hyperparameters optimization over the MNIST dataset. Our parameters fit their hyperspace, with 0.09 for the momentum, 0.0005 of weight decay and a initial learning rate of 0.001, dropping a factor of 10 in the last 10% of iterations. Our model produces a set of discrete class labels (0–6) or predicted classes which can be distinguished from the actual classes with the information provided by the confusion matrices, as shown in Fig. 3. It is possible then to extract from the test set classification (Table 2), three types of outcomes, as depicted in Table 3, and infer at least two scores, Precision (P) and Recall (R). From the framework used to test the model (Fig. 4) a confusion matrix with 30 frames per class is presented. Ideally we should gather with 132 different subjects to test our framework, equal to the number used to test CKP dataset, however this simple test is enough to infer the learning stage. During this test, we tried to mimic the CKP dataset then expecting same labels in return.

5   Results and Discussion

In this section we discuss the results from the preliminary experiments, as well as the final model achievements comparing to our video framework. Figure 3 shows the density of the first classification tests, trained just over 20000 iterations. This demonstration crossed with the Table 1 was used to select the best network shape for the problem, filtered by the most accurate results over a short period of training. From Table 1 we can also infer that the poor expressiveness between methods is representative of the lack of some augmentation diversity. As explained in Sect. 3, the most complex networks (AlexNet and Googlenet) were discarded due to their poor performance which reinforces the relation unbalance with the low dimension dataset (even augmented). The best results from the 3-type version of our assembled model used the *UnitBall* configuration trained over 100000 iterations and achieved a top recognition with F1-score of 0.906, representing an average value retrieved from Table 3. The version accuracy, also 90% on test set, is close to the current state of the art, with the CKP dataset represented by a 93% of accuracy. We could improve the accuracy with a threshold value to infer the classes with best confidence, however the purpose of this paper

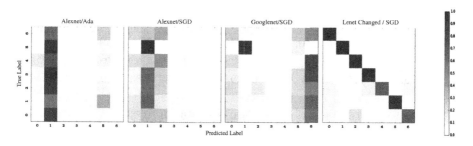

Fig. 3. Normalized confusion matrices - (a) Alexnet/Ada; (b) Alexnet/SGD; (c) Googlenet/SGD; (d) Lenet changed/SGD.

Table 1. Train loss results used to select the shallow network

Data	Solver	Net	Loss train
CKP + Augm	Ada	Alexnet	0.0082
CKP + Augm	SGD	Alexnet	0.0004
CKP + Augm	SGD	Googlenet	0.0001
CKP + Augm	SGD	**Lenet Ov**	0.0001

is to evaluate a sequence base system which is temporal dependent, due to this, we made an effort to reduce recognition time. According to Table 2 the model with the best results was the same used to perform the tests in the video framework that captured 30 frames per class. Figure 4 shows that in our framework, the expression of Contempt is the most expressive, misguiding the recognition of others. Moreover, the expression of Anger, Happiness and Sadness were not properly learned. Despite, the results not matching the desirable performance of the static test version, there are some significant achievements comparing to the state of the art, where some of the "emotions in the wild" [17] were also mis-classified or inexistent (such as Disgust in the test set). In this case, difficulty of video environment constraints were compared with the complexity of their dataset, which presents more candid images.

Table 2. Test accuracy with three versions of the CNN model (a), (b) and (c).

Weight/class	Anger	Contempt	Disgust	Fear	Happiness	Sadness	Surprise
Gaussian (a)	0.98	1	0.89	0.86	0.59	0.79	0.48
UnitBall (b)	0.98	1	1	0.91	0.78	0.95	0.70
Xavier (c)	0.98	1	1	0.92	0.75	0.92	0.70

Table 3. Precision, recall and F1 from three above versions (a) (b) (c).

	Gaussian (a)			UnitBall (b)			Xavier (c)		
	Prec.	Rec.	F1	Prec.	Rec.	F1	Prec.	Rec.	F1
Anger	0.747	0.985	0.850	0.949	0.985	0.967	0.833	0.985	0.903
Contempt	0.949	1	0.974	1	1	1	1	1	1
Disgust	0.652	0.894	0.754	0.820	1	0.901	0.830	1	0.907
Fear	0.820	0.864	0.41	0.924	0.917	0.920	0.897	0.924	0.910
Happiness	1	0.591	0.743	0.972	0.780	0.866	0.943	0.750	0.835
Sadness	0.840	0.795	0.817	0.881	0.955	0.916	0.884	0.924	0.904
Surprise	0.716	0.477	0.573	0.861	0.705	0.775	0.949	0.705	0.809

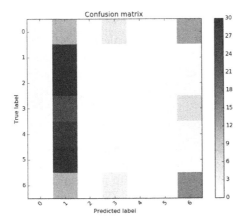

Fig. 4. Confusion matrix for the classification of the framework in the test set

6   Conclusion

In this paper, we present a deep inference model for emotion recognition from facial expressions. We developed a convolutional neural network tailored for the problem of emotion recognition from static facial expressions using the CKP dataset. After several enhancements, the results of our crafted network design are significant, attaining 90% of accuracy in the test set. Both the inclusion of a Dropout layer in architectural decisions and the augmentation performed on the dataset with several transformations, led to a superior performance with the seven facial expressions. Although the results are preliminary, the deep inference model is able to perform also in a sequence of video frames, in real-time, at least for some of the facial expressions (a problem that also exists in "wilder" datasets). This paper proved that we can start from static baselines to build a sequence model, thus answering in a positive way our initial question that we can rely on a deep emotion recognition model. Future work should focus on the improvement of the training set by increasing transformations and promoting different lighting and positions. Moreover, the real-time face tracker should also be improved by including active learning with human judgment.

References

1. Bettadapura, V.: Face Expression Recognition and Analysis: The State of the Art. CoRR abs/1203.6722 (2012)
2. Domhan, T., Springenberg, J.T., Hutter, F.: Speeding up automatic hyperparameter optimization of deep neural networks by extrapolation of learning curves. In: Proceedings of the 24th International Joint Conference on Artificial Intelligence (IJCAI) (2015)
3. Ebrahimi, S., Michalski, V., Konda, K., Memisevic, R., Pal, C.: Recurrent neural networks for emotion recognition in video. In: Proceedings of the ACM on International Conference on Multimodal Interaction, pp. 467–474. ACM (2015)
4. Ekman, P., Friesen, W.: Constants across cultures in the face and emotion. J. Pers. Soc. Psychol. **17**(2), 124–129 (1971)
5. Ekman, P., Heider, K.G.: The universality of a contempt expression: a replication. Motiv. Emot. **12**(3), 303–308 (1988)
6. Farfade, S., Saberian, M., Li, L.J.: Multi-view face detection using deep convolutional neural networks. In: Proceedings of the 5th ACM on International Conference on Multimedia Retrieval, pp. 643–650 (2015)
7. Glorot, X., Bengio, Y.: Understanding the difficulty of training deep feedforward neural networks. In: International Conference on Artificial Intelligence and Statistics (2010)
8. Goodfellow, I.J., et al.: Challenges in representation learning: a report on three machine learning contests. In: International Conference on Neural Information Processing (2013)
9. Jia, Y., Shelhamer, E., Donahue, J., Karayev, S., Long, J., Girshick, R., Guadarrama, S., Darrell, T.: Caffe: convolutional architecture for fast feature embedding. In: Proceedings of the ACM International Conference on Multimedia, pp. 675–678 (2014)
10. Krizhevsky, A., Sutskever, I., Hinton, G.: ImageNet classification with deep convolutional neural networks. In: Pereira, F., et al. (eds.) Advances in Neural Information Processing Systems 25, pp. 1097–1105. Curran Associates, Inc. (2012)
11. Lecun, Y., Bottou, L., Bengio, Y., Haffner, P.: Gradient-based learning applied to document recognition. Proc. of the IEEE **86**(11), 2278–2324 (1998)
12. Li, W., Li, M., Su, Z., Zhu, Z.: A deep-learning approach to facial expression recognition with candid images. In: 14th IAPR International Conference on Machine Vision Applications, pp. 279–282 (2015)
13. Lucey, P., Cohn, J.F., Kanade, T., Saragih, J., Ambadar, Z., Matthews, I.: The Extended Cohn-Kanade Dataset (CK+): a complete dataset for action unit and emotion-specified expression. In: IEEE Conference on Computer Vision and Pattern Recognition, pp. 94–101 (2010)
14. Srivastava, N., Hinton, G., Krizhevsky, A., Sutskever, I., Salakhutdinov, R.: Dropout: a simple way to prevent neural networks from overfitting. J. Mach. Learn. Res. **15**(1), 1929–1958 (2014)
15. Szegedy, C., Liu, W., Jia, Y., Sermanet, P., Reed, S., Anguelov, D., Erhan, D., Vanhoucke, V., Rabinovich, A.: Going deeper with convolutions. In: IEEE Conference on Computer Vision and Pattern Recognition, pp. 1–9. IEEE, June 2015

16. Viola, P., Jones, M.J.: Robust real-time face detection. Int. J. Comput. Vis. **57**(2), 137–154 (2004)
17. Yu, Z., Zhang, C.: Image based static facial expression recognition with multiple deep network learning. In: Proceedings of the ACM on International Conference on Multimodal Interaction (2015)

Sparse Linear Models Applied to Power Quality Disturbance Classification

Andrés F. López-Lopera$^{(\boxtimes)}$, Mauricio A. Álvarez, and Álvaro Á. Orozco

Electrical Engineering Program, Universidad Tecnológica de Pereira,
Pereira, Colombia
{anfelopera,malvarez,aaog}@utp.edu.co

Abstract. The automatic recognition of Power Quality (PQ) disturbances can be seen as a pattern recognition problem, in which different types of waveform distortion are differentiated based on their features. Similar to other quasi-stationary signals, PQ disturbances can be
decomposed into time-frequency dependent components by using time-
frequency or time-scale dictionaries. Short-time Fourier, Wavelets, and
Stockwell transforms are some of the most common dictionaries used
in the PQ community. Previous works about PQ disturbance classification have been restricted to the use of one of the above dictionaries.
Taking advantage of the theory behind sparse linear models (SLMs), we
introduce a sparse method for PQ representation, starting from overcomplete dictionaries. We apply Group Lasso. We employ different types of
time-frequency dictionaries to characterize PQ disturbances and evaluate their performance under different pattern recognition algorithms.
We show that SLMs promote the sparse basis selection improving the
classification accuracy.

Keywords: Disturbances classification · Overcomplete representation ·
Power quality · Sparse linear models · Statistical signal processing

1 Introduction

Power quality (PQ) evaluates the waveform of voltage and current signals with
respect to pure sinusoidal waveforms for a single frequency component (e.g.
60 Hz) [1]. Non-linear electronic devices such as alternators and current starters
introduce undesirable harmonic distortions to the network [2]. Other sources
that distort pure electric signals include high electrical loads, electrical faults,
and capacitors switching banks [2]. Depending on the source of PQ disturbances,
we can find different types of distortions such as harmonics, swell/sag events,
interruption events, voltage fluctuations, and transients [1].

The automatic classification of PQ disturbances can be seen as a pattern
recognition problem, in which the types of distortions are differentiated based on
their features [3]. For the feature extraction, PQ disturbances can be decomposed
into time-frequency dependent components by using time-frequency or time-scale
transforms, known as dictionaries [3,4]. Short-time Fourier transform (STFT),

© Springer International Publishing AG 2017
C. Beltrán-Castañón et al. (Eds.): CIARP 2016, LNCS 10125, pp. 521–529, 2017.
DOI: 10.1007/978-3-319-52277-7_63

Wavelets transform (WT), and S-transform (ST) are commonly used in PQ representation for the feature extraction [3,5]. For the pattern recognition of PQ disturbances, classifiers based on K nearest neighbours (K-NN), support vector machines (SVM), and artificial neural networks (ANN) have been employed [3].

PQ distortion classification researches have used different combinations of dictionaries and classifiers to improve the accuracy of PQ disturbance discrimination [3]. Previous works in PQ disturbance representation have been limited to the use of single dictionaries for the feature extraction step. This is, they only use either GT, WT or ST together with some specific classifiers [3,5]. Studies in signal processing have shown that combining several dictionaries for signal analysis improves the robustness for the feature extraction step [4,6]. Combinations of complete dictionaries for signal analysis are known as overcomplete representations (OR), and they are useful for increasing the richness of the representation by removing uncertainty when choosing the "proper" dictionary, and for increasing the flexibility for matching any structure from data [7]. However, OR increases the dimension of the coefficient of representation extracted from signals, including not relevant information [4].

In order to remove the redundant information before the classification step, research communities in machine learning have proposed different methods for automatically performing basis selection in linear models. These methods are known as sparse linear models (SLMs) [8]. SLMs tend to increase the PQ representation accuracy due to OR and reduce the PQ classification challenge due to the sparse tendency [6]. Because PQ disturbance representation requires the analysis of variables (coefficient of representations) that are grouped per each dictionary (STFT, WT or ST), we employ a SLM known as Group Lasso. The performance of Group Lasso is evaluated using different classifiers with either one dictionary at a time or with OR. We employ the discrete-time forms of the GT, WT, and ST as dictionaries, and we use different classifiers for the PQ classification, namely, K-NN, SVM, ANN, and two Bayesian approaches.

This paper is organized as follows. Materials and methods are given in Sect. 2. In Sect. 3, we discuss the results. Finally, Sect. 4 shows the conclusions.

2   Materials and Methods

2.1   Power Quality (PQ) Disturbances

PQ determines the fitness of the electric power to consumer devices, evaluating the quality of voltage and current signals. There are two main types of PQ disturbances: events and variations. Events describe sudden distortions which occur at specific time intervals. On the other hand, variations are steady-state or quasi-steady-state disturbances which require continuous measurements [1]. This paper is focussed on six types of common disturbances: sag/swell events, interruption events, voltage variations, harmonic distortions and oscillatory transient events [1–3]. Table 1 summarizes the characteristics of the PQ disturbances listed above. Figure 1 shows examples of simulated PQ distortions using the electrical power distribution model from Fig. 2.

Table 1. Summary of PQ disturbances.

Type of disturbance	Spectral content	Duration	Magnitude [p.u.]
1. Swell events		>10 ms	1.1–1.8
2. Sag events		>10 ms	0.1–0.9
3. Interruptions events		<1 min	0–0.1
4. Oscillatory transients			
4.1. Low frequency	<5 KHz	0.3–50 ms	0–4
4.2. High frequency	0.5–5 MHz	5 ms	0–4
5. Harmonic distortion	k = 2–40 (th)	Steady state	<0.2
6. Voltage fluctuation	<180 Hz	Intermittent	0.001–0.07

Spectral Content and Duration correspond to the typical frequencies and time lapses. Magnitude represents the variation in amplitude and it is given in per unit (p.u.).

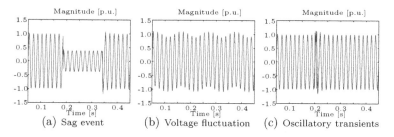

(a) Sag event   (b) Voltage fluctuation   (c) Oscillatory transients

Fig. 1. PQ disturbance examples for (a) a sag event, (b) a voltage fluctuations, and (c) an oscillatory transient, using the Simulink model from Fig. 2.

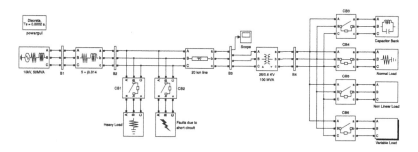

Fig. 2. Simulink diagram of the electrical power distribution model.

2.2 Overcomplete Representation

Let be $\mathbf{y} \in \mathbb{R}^N$ a vector which represents a discrete-time PQ disturbance of length N. In order to analyse the different types of waveform distortions, we want to completely represent the disturbance $\mathbf{y}$ given a linear superposition of atoms ϕ_γ, i.e. $\mathbf{y} = \Phi\boldsymbol{\beta}$, where $\Phi = (\phi_\gamma : \gamma \in \Gamma)$ contains the collection of atoms ϕ_γ, and it is known as a dictionary [4]. The vector $\boldsymbol{\beta} \in \mathbb{R}^M$ is the vector of

Table 2. Atom structures for the Gabor transform (GT), the Mexican hat Wavelet transform (MHWT), and the Stockwell transform (ST).

Dictionary	Interpretation for γ	Structure of the atoms $\phi_\gamma[n]$
GT	$\gamma = (f_k, m_l)$	$\frac{1}{\sqrt{2\pi\sigma_o^2}} \exp\left\{-(n - m_l)^2/2\sigma_o^2\right\} \cdot \cos\left(2\pi f_k(n - m_l)\right)$
MHWT	$\gamma = (m_l, \sigma_\nu)$	$\frac{2}{\sqrt{3\sigma_\nu}\,\pi^{1/4}}[1 - (n - m_l)^2/\sigma_\nu^2] \cdot \exp\{-(n - m_l)^2/2\sigma_\nu^2\}$
ST	$\gamma = (f_k, m_l, \sigma_\nu)$	$\frac{1}{\sqrt{2\pi\sigma_\nu^2}} \exp\left\{-(n - m_l)^2/2\sigma_\nu^2\right\} \cdot \cos\left(2\pi f_k(n - m_l)\right)$

Indexation: $f_k = kf_o$ $(k = 1, \cdots, K)$, $\sigma_\nu = \nu\sigma_o$ $(\nu = 1, \cdots, V)$, and $m_l = lm_o$ $(l = 1, \cdots, L)$. The constants K, V, and L represent the total amount of frequencies, scales, and locations, respectively, employed for the synthesis. The constants f_o, σ_o, and m_o are the principal components for frequency, scale, and location, respectively.

representation coefficients. According to the indexation of the variables of the dictionary Φ (e.g. time-frequency), the parameter γ has different interpretations.

Table 2 shows the structures of the atoms for the most common dictionaries applied in PQ disturbance analysis: Gabor transform (GT), Wavelet transform (WT), and Stockwell transform (ST). For the WT, we employ the Mexican Hat Wavelet Transform (MHWT) because their atoms can be easily implemented. However, it is possible to use other wavelet functions (e.g. Hamming, Hann).

An overcomplete representation (OR) is performed by the combination of several dictionaries, $\Psi = (\Phi_g : g \in G)$, where G represents the amount of dictionaries employed for the synthesis [4]. Note that each dictionary $\Phi_g = (\phi_{\gamma_g} : \gamma_g \in \Gamma_g)$ has different indexation parameter γ_g, allowing the combination of different types of atoms (e.g. frequency, time-frequency, time-scale). In this paper, we analyse the PQ disturbances in terms of the vector β (analysis). Then β is used for the feature extraction in order to feed the statistical classifiers. We also use the obtained vectors β to reconstruct the PQ disturbances (synthesis).

2.3   Sparse Linear Models for Grouped Variable Selection

Tibshirani in [9] proposed a method for the estimation in linear models known as Lasso. To solve the inverse problem, $\mathbf{y} = \Psi\beta$, he demonstrated that Lasso obtains a sparse representation by the minimization of the regularized cost function $\hat{\beta}_\lambda = \arg\min_\beta\{\|\mathbf{y} - \Psi\beta\|_2^2 + \lambda\|\beta\|_1\}$, where the vector $\hat{\beta}_\lambda \in \mathbb{R}^M$ depends of the regularization parameter λ. The norms $\|\cdot\|_2$ and $\|\cdot\|_1$ correspond to the ℓ_2-norm, and ℓ_1-norm, respectively. Lasso assumes that there is a unique correspondence between parameters and variables, performing the variable selection individually. However, the individual selection of variables produces a not satisfactory solution for grouped variables (e.g. when β_g is solely related to the group Φ_g from the dictionary $\Psi = (\Phi_g : g \in G)$) [8]. To deal with grouped variable estimation, a Group Lasso version was proposed in [10]. Group Lasso assumes that vector β is partitioned in G groups where the penalty is an intermediate between ℓ_1 and ℓ_2 regularizations [8]. Group Lasso has the attractive property that it performs the variable selection at the group level, promoting sparsity over $\hat{\beta}_\lambda$ for large values of λ [8]. For linear regression, the cost function for Group Lasso is given by

$$\hat{\boldsymbol{\beta}}_\lambda = \arg\min_{\boldsymbol{\beta}} \left\{ \|\mathbf{y} - \boldsymbol{\Psi}\boldsymbol{\beta}\|_2^2 + \lambda \sum_{g=1}^{G} \|\boldsymbol{\beta}_g\|_{\mathbf{K}_g} \right\}, \tag{1}$$

with $\|\boldsymbol{\beta}_g\|_{\mathbf{K}_g} = (\boldsymbol{\beta}_g^\top \mathbf{K}_g \boldsymbol{\beta}_g)^{1/2}$. The algorithm to solve the problem from Equation (1) is given in [10]. In this paper, each group g corresponds to a single dictionary.

2.4 Procedure

Dataset: we simulated the PQ disturbances mentioned in Sect. 2.1 from an electrical power distribution model based on [11]. We introduced a variable load module to simulate voltage fluctuations. Figure 2 shows the Simulink diagram of the electrical power distribution model. We generated 1200 PQ disturbances, 200 samples per each type. Each disturbance has 2000 discrete-time values for an interval between 0.05 and 0.45 s. The electrical parameters for the different types of PQ disturbance sources (e.g. RLC values, nominal voltage, and active/reactive power), were tuned manually to simulate disturbances according to the Table 1.

Group Lasso: we implement the algorithm proposed in [10], using a penalty factor $\lambda = 1 \times 10^{-2}$, and a tolerance equal to 1×10^{-10}. These values are chosen in order to obtain a signal reconstruction error lower than 1×10^{-3}. The percentage of sparsity for $\boldsymbol{\beta}$ is computed by the sum of the number of coefficients lower than 1×10^{-4}, and dividing the result by the length of the vector $\boldsymbol{\beta}$.

Dictionaries: we use a location factor $m_o = 0.005$ s. For the GT and the ST, we focus in the first 40 harmonics with $\sigma_o = 0.005$ s. For the MHWT, we use an scale $\sigma_o = 0.005$ s with $V = 10$. We fix six scale terms for the ST. To evaluate the SLM performance using OR, we combine GT, WT and ST dictionaries, and we denote it as the GWST dictionary. We also add a cosine/sine (Harmonics) dictionary for all the cases with the first 40 harmonics to improve the PQ synthesis.

Feature Extraction: we compute the following features over $\{\beta_p\}_{p=1}^{P}$ to create the feature vectors $\{\mathbf{x}_p\}_{p=1}^{P}$ for each signal p: mean of the absolute values (F_1), standard deviation (F_2), kurtosis (F_3), Shannon's energy (F_4) and RMS value (F_5). We also add the mean of the absolute values of the derivative β' (F_6). For the PQ disturbance classification step, we normalize the features by subtracting their mean values, and dividing the result using their standard deviation.

Classifiers: we use several classifiers from the state-of-the-art. The theory behind them can be found in deep in any pattern recognition textbook [3,8]. K-nearest neighbours (K-NN), Bayesian classifiers based on linear (LDC) and quadratic (QDC) discriminant functions, support vector machines (SVM), and neural networks (ANN) are employed. For the ANN-based classifier, we use the Neural Network Toolbox provided for Matlab R2013a. For the other classifiers, we use the toolbox for Pattern Recognition (PRTools Toolbox). We make experiments with a different number of neighbours for the K-NN classifiers. We chose the 1-NN and the 3-NN because they presented better behaviour. For the SVM, we use an RBF kernel, $k(\mathbf{x}, \mathbf{x}') = \exp(-\|\mathbf{x}-\mathbf{x}'\|^2/\sigma^2)$. The bandwidth parameter σ and the regularization parameter for the SVM are tuned by cross-validation. We design an ANN made of three hidden layers with 20, 15 and 10 neurons in each layer, respectively [12]. We use sigmoid transfer functions.

We test all the classifiers twenty times with different training sets. We select randomly the 70% of the total samples per each type of PQ disturbance for training, and then we use the other 30% for testing. The performance for each test experiment is computed by the sum of the successful cases, and dividing the result by the total number of test samples. Finally, we compute the mean μ and the standard deviation σ of the performance obtained for all the experiments. Figure 3 summarizes the procedure taken into account in this paper.

Fig. 3. Block diagram of PQ disturbance classification procedure.

3   Results and Discussions

To highlight the advantages of SLM for PQ representation, Fig. 4 shows the synthesis for (a) an example of a swell. The result is shown for two cases: (b) without sparsity, and (c) using Group Lasso. We used the GWST dictionary which we obtained by combining the GT, MHWT, and ST. Both methods can synthesize the PQ distortion in Figure (a), ensuring a low reconstruction error.

In order to quantify the level of the sparsity produced by Group Lasso over the GWST representation, the synthesis step is performed over all the PQ disturbance dataset. We obtain values lower than 1×10^{-2} for the sparsity percentages

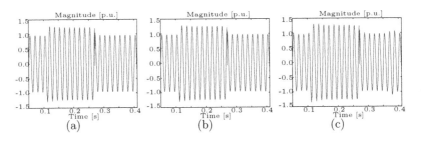

Fig. 4. In (a) we show the swell example. The synthesis results using GWST dictionary are showed for the approach (b) without SLM and (c) using Group Lasso.

produced by the method without SLMs, concluding that this approach tend to use all the coefficients from β. This makes difficult subsequent PQ studies (e.g. PQ disturbance classification). On the other hand, Group Lasso automatically selects representative coefficients required to synthesize the disturbances. We note that the Harmonics, MHWT, and ST dictionaries contain relevant components in terms of the disturbance representation when compared to GT. Table 3 shows the sparsity percentages produced by using Group Lasso according to Sect. 2.4. We compute the sparsity percentages per each type of PQ disturbance (rows), and each dictionary (columns).

To evaluate the performance of SLM for PQ classification, we perform the procedure described in Fig. 3 per each PQ disturbance through GT, MHWT, ST, and GWST. We repeat this procedure when Group Lasso is applied to obtain the feature set $\{x_p\}_{p=1}^P$. We train the different classifiers, and evaluate their performance on the test set as well as we describe in the procedure. Table 4 shows the PQ classification performance over the test set. When Group Lasso is not applied (first six rows), GT and ST show better results than WT and GWST, independently of the classifier employed. When Group Lasso is applied (last six rows), we notice that for any particular representation, and any classifier, the accuracy obtained by additionally applying Group Lasso is higher, when compared to the same representation, and the same classifier used without Group Lasso. For example, when the representation is ST and the classifier is ANN, applying Group Lasso increases the performance by almost 4%. The improvement is even higher when GWST is applied, and the classifier is almost LDC. In this case, the improvement is close to 28%. Due to the similarity of some accuracy results, we apply the Wilcoxon rank-sum test over the results per classifier [8], concluding that the differences are statistically significant. We note that the results from GWST without Group Lasso is lower than the other schemes for all classifiers. This is because the representation using GWST tends to introduce redundant information, producing misclassification results in each classifier. This redundancy is removed when we apply Group Lasso together with GWST, outperforming the classification accuracy for all the statistical classifiers.

Table 3. Sparsity percentages using GWST with Group Lasso.

PQ disturbance	Dictionary			
	Harmonics	GT	MHWT	ST
Harmonic distortion	6.0750	53.0061	21.9631	33.9341
Swell event	32.6000	47.8226	10.7996	38.1105
Sag event	13.4750	33.2939	6.6398	26.4847
Interruption event	9.4750	24.9780	4.8516	19.8153
Voltage fluctuation	10.9750	77.0085	14.1627	68.7561
Oscillatory transients	31.4750	67.1982	20.8467	46.0506

Table 4. Performance of the classifiers without SLM and using Group Lasso for GT, MHWT, ST and combining all the dictionaries (GWST).

Classifier	Dictionary			
	GT + Harmonics	MHWT + Harmonics	ST + Harmonics	GWST + Harmonics
	$\mu \pm \sigma$	$\mu \pm \sigma$	$\mu \pm \sigma$	$\mu \pm \sigma$
Without SLM				
1-NN	0.8710 ± 0.0149	0.6311 ± 0.0216	0.8322 ± 0.0175	0.7292 ± 0.0159
3-NN	0.8592 ± 0.0102	0.6360 ± 0.0197	0.8261 ± 0.0186	0.7246 ± 0.0166
LDC	0.5956 ± 0.0199	0.5363 ± 0.0203	0.6560 ± 0.0204	0.5121 ± 0.0296
QDC	0.8053 ± 0.0122	0.5808 ± 0.0327	0.7200 ± 0.0184	0.6550 ± 0.0180
SVM	0.8471 ± 0.0177	0.6546 ± 0.0170	0.8082 ± 0.0208	0.7144 ± 0.0179
ANN	0.8510 ± 0.0830	0.6399 ± 0.0336	0.8299 ± 0.0420	0.7285 ± 0.0459
With Group Lasso				
1-NN	0.8629 ± 0.0140	0.8908 ± 0.0117	0.8821 ± 0.0147	**0.9396 ± 0.0168**
3-NN	0.8628 ± 0.0121	0.8882 ± 0.0159	0.8763 ± 0.0152	**0.9382 ± 0.0179**
LDC	0.7565 ± 0.0198	0.7126 ± 0.0235	0.7101 ± 0.0220	**0.7925 ± 0.0191**
QDC	0.8179 ± 0.0170	0.7760 ± 0.0163	0.8157 ± 0.0102	**0.8996 ± 0.0177**
SVM	0.8649 ± 0.0143	0.8461 ± 0.0199	0.8454 ± 0.0226	**0.9128 ± 0.0133**
ANN	0.8457 ± 0.0447	0.8754 ± 0.0189	0.8681 ± 0.0375	**0.8978 ± 0.0675**

4   Conclusions

We introduced the concepts of overcomplete representations (OR) and sparse linear models (SLM) for PQ disturbance classification. We combined different time-frequency dictionaries, which are well known in the PQ literature. We introduced Group Lasso assuming that each dictionary is a group in the SLM.

As we showed experimentally, Group Lasso improves the performance of PQ disturbance classification for both linear and non-linear classifiers compared to methods without SLM. Due to SLMs can carry out OR, ensuring a high performance for PQ disturbance classification, this framework removes the uncertainty about which dictionary should be used for which type of distortion.

Acknowledgment. The authors would like to thank to the Maestría en Ingeniería Electrica at Universidad Tecnológica de Pereira for the support provided.

References

1. Bollen, M., Gu, I.H., Santoso, S., McGranaghan, M., Crossley, P., Ribeiro, M., Ribeiro, P.: Bridging the gap between signal and power. IEEE Signal Process. Mag. **26**, 12–31 (2009)
2. Benysek, G., Pasko, M.: Power Theories for Improved Power Quality. Power Systems. Springer, Heidelberg (2012)

3. Granados, D., Romero, R., Osornio, R., Garcia, A., Cabal, E.: Techniques and methodologies for power quality analysis and disturbances classification in power systems: a review. IET Gener. Transm. Distrib. **5**, 519–529 (2011)
4. Chen, S.S., Donoho, D.L., Saunders, M.A.: Atomic decomposition by basis pursuit. SIAM J. Sci. Comput. **20**, 33–61 (1998)
5. Eristi, H., Demir, Y.: Automatic classification of power quality events and disturbances using wavelet transform and support vector machines. IET Gener. Transm. Distrib. **6**, 968–976 (2012)
6. Manikandan, M., Samantaray, S., Kamwa, I.: Detection and classification of power quality disturbances using sparse signal decomposition on hybrid dictionaries. IEEE Trans. Instrum. Meas **64**, 27–38 (2015)
7. Ren, L., Lv, W., Jiang, S., Xiao, Y.: Fault diagnosis using a joint model based on sparse representation and SVM. IEEE Trans. Instrum. Meas. **PP**, 1–8 (2016)
8. Murphy, K.P.: Machine Learning: A Probabilistic Perspective (Adaptive Computation And Machine Learning Series). The MIT Press, Cambridge (2012)
9. Tibshirani, R.: Regression shrinkage and selection via the Lasso. J. R. Stat. Soc. **58**, 267–288 (1996)
10. Yuan, M., Yuan, M., Lin, Y., Lin, Y.: Model selection and estimation in regression with grouped variables. J. R. Stat. Soc. **68**, 49–67 (2006)
11. Khokhar, S., Mohd Zin, A., Mokhtar, A., Ismail, N.: MATLAB/Simulink based modeling and simulation of power quality disturbances, pp. 445–450. Institute of Electrical and Electronics Engineers Inc. (2014)
12. Monedero, I., Leon, C., Ropero, J., Garcia, A., Elena, J., Montano, J.C.: Classification of electrical disturbances in real time using neural networks. IEEE Trans. Power Deliv. **22**, 1288–1296 (2007)

Trading off Distance Metrics vs Accuracy in Incremental Learning Algorithms

Noel Lopes[1,2(✉)] and Bernardete Ribeiro[2]

[1] UDI, Polytechnic of Guarda, Guarda, Portugal
noel@ipg.pt
[2] CISUC, Department of Informatics Engineering,
University of Coimbra, Coimbra, Portugal
bribeiro@dei.uc.pt

Abstract. With the growth and development of data, the empirical evidence supporting a link between the distance metrics that are used in the instance-based algorithms and generalization has been mounting. In this paper, we look at distinct similarity measures to study its impact on the performance accuracy of incremental instance-based algorithms in pattern recognition problems. An in-depth analysis of the results of the proposed study for a variety of classification tasks (binary and multiway) from various different domains shines light on the trade off between the distance metrics and yielded accuracy.

Keywords: Distance metrics · Instance-based learning · Nearest neighbor · Incremental learning · Incremental Hypersphere Classifier (IHC)

1 Introduction

In recent years there has been much interest in incremental learning algorithms, mainly due to their potential to deal with large scale datasets and data streams. Contrasting with batch learning algorithms, commonly designed with the emphasis on effectiveness (e.g. classification performance) and under the assumptions that data is static and its volume manageable, incremental algorithms are typically designed with emphasis on efficiency (e.g. time required to produce a model) [11]. Rather than requiring access to the complete dataset, incremental algorithms are designed to rapidly update their models to incorporate new information on a sample-by-sample basis and therefore suitable for high-throughput.

In previous work we presented a novel incremental instance-based learning algorithm which presents good properties in terms of multi-class support, complexity, scalability and interpretability. The algorithm named Incremental Hypersphere Classifier (IHC) algorithm [6] is extremely versatile and highly-scalable, being able to accommodate memory and computational restrictions, while creating the best possible model with the amount of given resources. Moreover, since the algorithm's execution time grows linearly with the number of samples stored in the memory, creating adaptive models and extracting information in real-time from large-scale datasets and data streams is feasible.

© Springer International Publishing AG 2017
C. Beltrán-Castañón et al. (Eds.): CIARP 2016, LNCS 10125, pp. 530–538, 2017.
DOI: 10.1007/978-3-319-52277-7_64

Experimental results, using well-known datasets, demonstrated that the IHC is able to handle concept drifts scenarios, while maintaining superior classification performance. Additionally, the resulting models are interpretable, making this algorithm useful even in domains where interpretability is a key factor. Finally, since the IHC keeps samples that are at the odds of lying on the decision boundary while removing the noisy and less relevant ones, it represents a good choice for selecting a representative subset of the data for applying more sophisticated algorithms in a fraction of the time required for the complete dataset [7].

Despite these advantages, IHC is a distance based learning method and naturally sensitive to the choice of distance metrics. Therefore it is important to study their impact on IHC performance, in particular concerning incremental learning scenarios. Accordingly, in this paper we analyze the impact of distinct distance metrics in the IHC algorithm, which proved to be efficient in large-scale recognition problems and online learning. We provide a detailed empirical evaluation on fifteen datasets with several sizes and dimensionality.

The remainder of this paper is organized as follows. The next Section introduces the IHC algorithm. Section 3 presents and discusses the experimental results. Finally, in Sect. 4 the conclusions and future work are addressed.

2   Incremental Hypersphere Classifier (IHC) Algorithm

Let us consider a training dataset, $\{(\mathbf{x_i}, y_i) : i = 1, \ldots, N\}$, composed by N samples, each encompassing an input vector, $\mathbf{x_i} \in \mathbb{R}^D$, with D features, and the associated class label, $y_i \in \{1, \ldots, C\}$, where C is the number of classes. For each sample, i, IHC defines an hypersphere with center $\mathbf{x_i}$ and radius ρ_i:

$$\rho_i = \frac{\min(d(\mathbf{x_i}, \mathbf{x_j}))}{2}, \text{ for all } j \text{ where } y_j \neq y_i \tag{1}$$

where $d(\mathbf{x_i}, \mathbf{x_j})$ is the distance between $\mathbf{x_i}$ and $\mathbf{x_j}$ input vectors. Table 1 presents the distance metrics used in this study. For the Minkowsky metric, p was set to the number of features, D, in order to give more weight to the individual distance components as the space dimensionality increases [4].

The hypersphere's delineate the regions of influence of the associated samples and are used to classify new instances. Basically, given a new data point, $\mathbf{x_k}$, it is classified with the class associated to the nearest hypersphere (not the nearest sample). More precisely, $\mathbf{x_k}$ is associated to class y_i (i.e. $y_k = y_i$) provided that:

$$d(\mathbf{x_i}, \mathbf{x_k}) - ga_i\rho_i \leq d(\mathbf{x_j}, \mathbf{x_k}) - ga_j\rho_j, \text{ for all } j \neq i \tag{2}$$

where g (gravity) controls the extension of the zones of influence and a_i is the accuracy of sample i when classifying itself and the forgotten training samples for which i was the nearest sample in memory.

Note that for $g = 0$ the decision rule of the IHC is exactly the same as the one of the 1-Nearest Neighbor (NN) (see Eq. 2). Hence, by fine-tuning g, IHC will always yield better or equal performance than 1-NN. This is important because

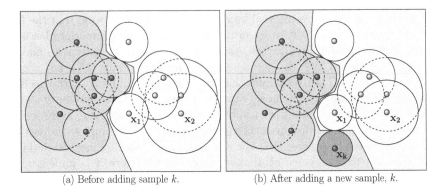

(a) Before adding sample k. (b) After adding a new sample, k.

Fig. 1. Hypersphere's and decision surface generated by IHC ($g = 1$) for a toy problem.

Table 1. Distance metrics' formulas. Note that Euclidean, Manhattan and Chebychev are special cases of Minkowsky, obtained respectively for $p = 2$, $p = 1$ and $p \to \infty$.

Metric	Formula
Euclidean	$d(\mathbf{x_i}, \mathbf{x_j}) = \left(\sum_{k=1}^{D} (x_{ik} - x_{jk})^2 \right)^{\frac{1}{2}}$
Manhattan	$d(\mathbf{x_i}, \mathbf{x_j}) = \sum_{k=1}^{D} |x_{ik} - x_{jk}|$
Canberra	$d(\mathbf{x_i}, \mathbf{x_j}) = \sum_{k=1}^{D} \frac{|x_{ik} - x_{jk}|}{|x_{ik}| + |x_{jk}|}$
Chebychev	$d(\mathbf{x_i}, \mathbf{x_j}) = \max(|x_{ik} - x_{jk}|)$
Minkowsky	$d(\mathbf{x_i}, \mathbf{x_j}) = \left(\sum_{k=1}^{D} |x_{ik} - x_{jk}|^p \right)^{\frac{1}{p}}$

Cover and Hart [3] demonstrated that for $N \to \infty$, the 1-NN error rate is never more than twice the minimum achievable error rate of an optimal classifier [2].

A major advantage of the IHC algorithm relies on the possibility of building models on a sample-by-sample basis. Figure 1 presents the hypersphere's generated by IHC and the resulting decision surface, (a) prior to and (b) after the addition of a new sample, for a toy problem. Note that adding a new sample might affect the radius of samples already in the model (in this case the ones with input vectors $\mathbf{x_1}$ and $\mathbf{x_2}$). Notice also that samples near the decision border have smaller radius than those far away (see Fig. 1). Hence, when the memory is full, samples with smaller radius – that play the most significant role in the construction of the decision surface – are kept, while those with bigger radius – that have less or no impact in the model – are discarded. Unfortunately, outliers will most likely have a small radius and end-up occupying the limited memory resources. Thus, although their impact is diminished by the accuracy variable in Eq. 2, it is still important to identify and remove them from memory. To address this problem IHC mimics the process used by the IB3 algorithm [1,9], which uses a significance test to remove all samples that are believed to be noisy.

Another advantage of IHC is that it can accommodate restrictions in terms of memory and computational power, creating the best model possible for the amount of resources given, instead of requiring systems to comply with its own requirements. Since we can control the amount of memory and computational power required by the algorithm and due to its scalability creating up-to-date models in real-time is feasible [7]. A more detailed description of the IHC can be found elsewhere [6,7] and a working version of the algorithm, including its source code, can be found at http://sourceforge.net/projects/ihclassifier/.

3 Experimental Results

Our goal consists of determining the impact of distance metrics in the IHC classification performance. Recently, we have analyzed the impact of distance metrics in batch scenarios, for both the NN and IHC algorithms [8]. Among the conclusions, we have found that the No-Free-Lunch theorem [10] still applies and the best distance metric is problem dependent. Accordingly, in batch learning configurations, it is desirable to perform a grid search both for the distance metric and g parameters in order to determine favorable parameter configurations [8]. Unfortunately, in incremental scenarios, performing a grid search is not feasible and knowing beforehand which distance metrics are likely to yield quality models becomes a fundamental aspect. Moreover, typically in incremental learning configurations, IHC must work with limited memory settings, being able to store only a small fraction of the samples. Therefore, adequate distance metrics play a vital role in choosing the core samples that delineate the decision borders.

In order to analyze the performance of distance metrics in incremental learning scenarios, we carried out extensive experiments in the same fifteen UCI databases [5] that were previously investigated in [8], comprehending distinct data distributions and characteristics (see Table 2). Altogether, five distinct memory configurations were considered, allowing IHC to store approximately 20%, 40%, 60%, 80% and 100% of the training samples. For statistical significance, each experiment was executed using repeated 5-fold stratified cross-validation. Altogether 30 different random cross-validation partitions were created, accounting for a total of 150 runs per benchmark and memory configuration. Overall, 2250 runs per benchmark (dataset) were performed. Given the large number of runs (33,750 in total) the experiments were performed only for $g = 1$. The results were compiled both for the unseen test data and for all the data (encompassing both training and test data). The latter, reflects the IHC performance on forgotten data and it is important because real-world databases often present a high-degree of redundancy with similar records being common [7]. Figure 2 presents the IHC results, obtained for the different memory settings. Note that, in general, higher memory configurations correspond to better results.

On average Euclidean and Manhattan metrics present the best performance results for most memory settings (except for the 100% memory configuration,

Table 2. Experimental dataset characteristics.

Database	Samples	Inputs	Classes
Balance	625	4	3
Breast cancer	569	30	2
Ecoli	336	7	8
German	1000	59	2
Glass	214	9	6
Haberman	306	3	2
Heart-statlog	270	20	2
Ionosphere	351	34	2
Iris	150	4	3
Pima	768	8	2
Sonar	208	60	2
Tic-tac-toe	958	9	2
Vehicle	946	18	4
Wine	178	13	3
Yeast	1484	8	10

in which case Canberra yields better results for the test data). In fact, there is strong statistical evidence compelling the choice of these two distance metrics (see Table 3). Overall, these two metrics attained competitive and in many cases top classification performance results for most benchmarks (Breast cancer, Ecoli, German, Heart-statlog, Ionosphere, Pima, Sonar, Vehicle, Wine and Yeast). Moreover, Manhattan also attained good results in the Glass dataset, achieving top results for the 20% and 100% memory configurations. In the remaining configurations, Canberra yielded the highest F-Scores. Additionally, Manhattan outperforms all other metrics for the Breast cancer and wine datasets.

Concerning performance in the individual datasets, for the Vehicle dataset both Manhattan and Euclidean yield superior classification performance, with Manhattan presenting better results when less memory is available. These two metrics also present good results in the Sonar dataset, with the Manhattan attaining the top performance for the 20%, 40% and 80% memory configurations and Euclidean and Canberra yielding the highest results respectively for the 60% and 40% configurations. In the German dataset, overall both Manhattan and Euclidean attained competitive results. Moreover, Manhattan achieved the top results on the test data for memory configurations of 20%, 40% and 60%, while the highest results for 80% and 100% were yielded by Canberra. Interestingly, despite Chebychev yielding the worst results for the test datasets, this metric attained some of the best results when considering all data, evidencing that the model is overfitting the training data. Concerning the Heart-statlog dataset, with the exception of the 100% memory configuration, once again both Manhattan

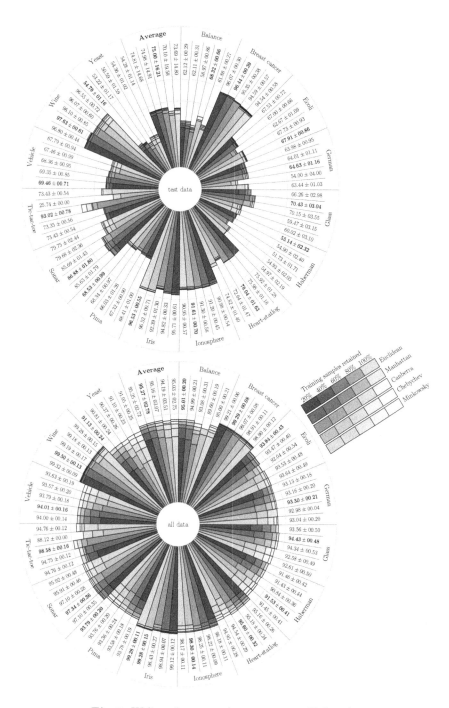

Fig. 2. IHC performance (macro-average F-Score).

Table 3. Null hypotheses (H_0 : F-Score$_X$ >= F-Score$_Y$) rejected using the Wilcoxon signed rank test, for the (a) test data and for (b) all data.

(a) Test data

Samples retained	Significance level	X distance metric	Y distance metric
20%	0.010	Canberra	
	0.025	Chebychev	
	0.025	Minkowsky	
40%	0.025	Canberra	
	0.025	Chebychev	
	0.010	Minkowsky	Manhattan
60%	0.025	Canberra	
	0.025	Chebychev	
	0.050	Minkowsky	
80%	0.025	Canberra	
	0.050	Chebychev	
100%	0.050	Canberra	
20%	0.025	Chebychev	
	0.010	Minkowsky	
40%	0.025	Chebychev	
	0.005	Minkowsky	
60%	0.005	Chebychev	Euclidean
	0.010	Minkowsky	
80%	0.025	Chebychev	
	0.010	Minkowsky	
100%	0.050	Chebychev	
	0.025	Minkowsky	
20%	0.010	Chebychev	Minkowsky

(b) All data

Samples retained	Significance level	X distance metric	Y distance metric
20%			
40%	0.050	Canberra	
60%			
80%	0.025	Canberra	Manhattan
	0.050	Chebychev	
100%	0.050	Canberra	
	0.050	Chebychev	
20%	0.025	Chebychev	
	0.025	Minkowsky	
40%	0.050	Chebychev	
	0.050	Minkowsky	Euclidean
80%	0.025	Chebychev	
	0.050	Minkowsky	
100%	0.005	Chebychev	
	0.005	Minkowsky	
20%	0.005	Chebychev	Minkowsky
100%	0.050	Minkowsky	

and Euclidean attained competitive results. Moreover, Manhattan yielded top results on the test dataset for the 20% and 40% memory configurations, while Canberra attained the top results for the remaining configurations. Manhattan and Euclidean also yielded good results for the Ecoli dataset, with the highest F-Score being obtained by Euclidean for the 20% memory setting, by Manhattan for the 40% memory configuration and by Minkowsky for the remaining configurations. With respect to the Haberman dataset, Manhattan yielded the highest results for the 60% and 80% memory settings, while Euclidean, Minkowsky and Chebychev attained the best F-Scores respectively for the 100%, 20% and 40% configurations. Regarding the Ionosphere, Manhattan yielded the dominant classification performance in the 40%, 60% and 80% memory settings, while the 20% and 100% were attained respectively by Canberra and Chebychev. The Euclidean distance metric excel all the others in the Yeast dataset. Moreover, concerning the Pima dataset, it also outperformed the other distance metrics, except in the 100% memory configuration for which Minkowsky yielded the best F-Score. The Canberra distance metric performed particularly well on the Tic-tac-toe problem, excelling by far the remaining distance metrics. Chebychev, on the other hand, yielded the worst results for this dataset and its performance significantly dropped with increase of available memory. As in the case of the German, the model overfits the training data, indicating that Chebychev-based IHC models are prone to overfitting the training data. In fact, on average Chebychev yielded the worst F-Scores for all memory settings. Nevertheless, this metric excelled all the others performance in the Balance dataset. Moreover, Chebychev also performed quite well on the Iris dataset attaining together with the Minkowsky distance metric the top classification performances.

4   Conclusions and Future Work

We are seeing a torrent of data coming in from sensors everywhere. This data is compounding daily, creating what is called "fast data". In this context, incremental algorithms are part of the solution for dealing with the data explosion that is happening at a massive scale. The big challenge is now speed and agility when building systems, in particular for dealing with anomaly detection and concept drifts that occur in many fields of science and society in general.

In this paper, we looked at the importance of distance metrics for assessment of similarity of patterns in incremental learning. To reinforce this idea, we interpreted the distance metric as a pivotal parameter for the success of many machine learning algorithms and models. We extended our previous research on the impact of distance metrics on batch learning to incremental scenarios. In the latter using grid-search like methods for determining favorable metrics is not feasible. Therefore, distance metrics play a vital role in the choice of core samples, which are expected to be representative of the whole dataset and will in practice shape the boundary decisions. To analyze the performance of distance metrics in incremental scenarios, we carried out extensive experiments using fifteen UCI databases, with distinct data distributions and characteristics. Altogether, five memory configurations were considered, allowing IHC to store approximately 20%, 40%, 60%, 80% and 100% of the samples and for statistical significance, each experiment was executed using repeated 5-fold stratified cross-validation.

This study demonstrates that the Euclidean and Manhattan, two of the most commonly used distance metrics, which consistently yield good results over a wide range of problems as shown in the experimental tests, are probably the best choices for distance based learning methods when performing a grid-search method is not a viable option. In this scenario, the Manhattan distance is preferred, in particular for large datasets, since it is computationally less demanding. Future work will focus as building ensembles using distinct distance metrics.

References

1. Aha, D., Kibler, D., Albert, M.: Instance-based learning algorithms. Mach. Learn. **6**(1), 37–66 (1991)
2. Bishop, C.M.: Pattern Recognition and Machine Learning. Springer, Heidelberg (2006)
3. Cover, T.M., Hart, P.E.: Nearest neighbor pattern classification. IEEE Trans. Inf. Theory **13**(1), 21–27 (1967)
4. de Geer, J.P.V.: Some Aspects of Minkowski Distance. Leiden University, Department of Data Theory (1995)
5. Lichman, M.: UCI machine learning repository (2013). http://archive.ics.uci.edu/ml

6. Lopes, N., Ribeiro, B.: An incremental class boundary preserving hypersphere classifier. In: Lu, B.-L., Zhang, L., Kwok, J. (eds.) ICONIP 2011. LNCS, vol. 7063, pp. 690–699. Springer, Heidelberg (2011). doi:10.1007/978-3-642-24958-7_80

7. Lopes, N., Ribeiro, B.: Machine Learning for Adaptive Many-Core Machines - A Practical Approach, vol. 7. Studies in Big Data. Springer, Heidelberg (2014)

8. Lopes, N., Ribeiro, B.: On the impact of distance metrics in instance-based learning algorithms. In: Paredes, R., Cardoso, J.S., Pardo, X.M. (eds.) IbPRIA 2015. LNCS, vol. 9117, pp. 48–56. Springer, Heidelberg (2015). doi:10.1007/978-3-319-19390-8_6

9. Wilson, D., Martinez, T.: Reduction techniques for instance-based learning algorithms. Mach. Learn. **38**(3), 257–286 (2000)

10. Wolpert, D.H.: The lack of a priori distinctions between learning algorithms. Neural Comput. **8**(7), 1341–1390 (1996)

11. Zhou, Z.: Three perspectives of data mining. Artif. Intell. **143**(1), 139–146 (2003)

Author Index

Acuña, Diego 318
Adán, A. 60
Agudelo-España, Diego 249
Aguilar, Pablo S. 257
Alcântara, Rafaela 451
Alexandrov, Mikhail 476
Aliquintuy, Marcelo 208
Allende, Héctor 318, 409
Allende-Cid, Héctor 318
Alvarado-Pérez, J.C. 334
Álvarez, A. 158
Álvarez, M. 125, 158, 443
Álvarez, Mauricio A. 249, 291, 493
    502, 521
Álvarez-Meza, A. 125, 443
Alvarez-Meza, A.M. 309
Alvarez-Meza, Andres 282
Anaya-Isaza, A.J. 334
Antensteiner, Doris 175
Arciniegas-Mejía, A.F. 343
Azpiroz, Fernando 326

Ballesteros L., Dora M. 27
Ballesteros, Cristian 401
Becerra, Miguel A. 426
Bobadilla, Julio Cesar Mendoza 117
Bolaños-Ledezma, M. 343
Bouwmans, Thierry 282
Bravo-Montenegro, M.J. 343
Busch, Christoph 385
Bustamante, Isneri Talavera 233
Bustos, Cristina 417

Caicedo, Juan C. 274, 393
Calvo, José Ramón 134
Camargo, Jorge E. 274
Campuzano-Alvarez, Mirlayne 184
Cappabianco, Fábio A.M. 459
Carbajal, Guillermo 352
Carvajal, Jacobo García 426
Carvajal, Juan A. 467
Castañeda, Benjamin 101
Castellanos-Dominguez, German 282, 309
Castelo-Fernández, César 192

Castro-Ospina, A.E. 343
Chaves, Deisy 401
Cornejo, Jadisha Yarif Ramírez 76
Cuellar, J. 158
Culquicondor, Aldo 192

de A. Araújo, Arnaldo 300
De La Pava, I. 125
de Marsico, Maria 233
Demertzi, Athena 434
Di Perri, Carol 434
Diaz, P. 334
dos Santos, Jefersson A. 300
Drozdzal, Michal 326
Duin, Robert P.W. 150

Ferri, Francesc J. 368
Fonseca, Pablo 101
Fonseca, Rainer Larin 200
Fonseca-Bruzón, Adrian 184
Formanová, Dominika 52
Frandi, Emanuele 208
Frazão, Xavier 511
Frucci, Maria 1, 19
Fuentes, Magdalena 352

Gago-Alonso, Andrés 109
García, Hernán F. 158, 291
García-Reyes, Edel 150
Garea Llano, Eduardo 167
Gil, Esteban 409
Gómez, Alvaro 352
Gomez, Augusto 68
Gómez, Francisco 434
Gómez, Juan C. 360
Gomez-Barrero, Marta 385
Gómez-Orozco, V. 158
Gonzalez, Hector 368
González, Julián Gil 502
Gonzalez-Vanegas, W. 443
Gousseau, Yann 484
Guasmayan-Guasmayan, F.A. 343
Gustin, I.D. 343

Haindl, Michal   44, 52, 84
Havlíček, Vojtěch   44
Heine, Lizette   434
Henao, O.   125, 158
Hernández-Durán, Mairelys   217
Hernandez-Muriel, J.   443
Hoyos, Jesus   401
Huber-Mörk, Reinhold   175
Hurtado, Julio   142

Ide, Jaime S.   459
Imbajoa-Ruiz, D.E.   343
Insuasti-Ceballos, David   282

Jourlin, Michel   36
Junior, Perfilino Ferreira   451

Laranjeira, Ana   511
Laureys, Steven   434
Lecumberry, Federico   257
Lellis, Lucas Santana   459
Lemus, Camilo   27
Li, Yan   150
Loog, Marco   150
Lopes, Noel   530
López, Erick   409
López-Lopera, Andrés F.   521
Luna, Jaime A. Guzmán   426

Machado, Alexei M.C.   300
Malagelada, Carolina   326
Martínez, Darwin   434
Martínez-Díaz, Yoanna   233
Mata, Francisco José Silva   233
Mazo, Claudia   401
Megrian, Daniela   257
Mejía, J.   125
Melo-Betancourt, L.G.   309
Méndez-Vázquez, Heydi   200, 217
Mendiola-Lau, Victor   233
Mendoza, Marcelo   142
Miranda, Paulo   459
Montalvo, Ana   134
Morell, Carlos   368
Mujica-Parodi, Lilianne R.   459

Ñanculef, Ricardo   142, 208
Negri, Pablo   241
Noyel, Guillaume   36

Ocampo-Blandon, Cristian   484
Orozco, A.   125, 158, 443
Orozco, Álvaro Á.   249, 291, 493, 502, 521
Orozco-Alzate, Mauricio   150
Osorio Roig, Dailé   167

Papa, João Paulo   192
Pascual, Guillem   326
Pedrini, Helio   76, 117, 225
Peluffo-Ordóñez, Diego H.   334, 343, 426
Peña-Unigarro, D.F.   334
Perea, Jhon   68
Pérez, Andrés   393
Pham, Tuan D.   93
Pimentel, André   511
Plasencia-Calaña, Yenisel   150, 200, 217
Prado-Romero, Mario Alfonso   109
Prieto, S.A.   60
Pujol, Oriol   377
Pulgarin-Giraldo, Juan D.   282, 309

Quevedo, Angélica   393
Quintana, B.   60

Radeva, Petia   326
Raja, Kiran B.   385
Ramachandra, Raghavendra   385
Ramos, Aline   451
Ramos-Bermudez, S.   309
Renza, Diego   27
Reyes, Angie K.   274
Ribeiro, Bernardete   511, 530
Riccio, Daniel   1, 19
Riera, Carles R.   377
Rojas, Oscar E.   266
Romero, Dennis G.   467
Rosero-Montalvo, P.   334
Rudas, Jorge   434

Sad, Gonzalo D.   360
Salazar-Castro, J.A.   334
Sánchez, Marcela Bedoya   426
Sanniti di Baja, Gabriella   1, 19
Santana, Tiago M.H.C.   300
Sappa, Angel D.   467
Sedláček, Matěj   52
Seguí, Santi   326
Serino, Luca   1, 19
Sigaard, Morten   385
Skitalinskaya, Gabriella   476
Soddu, Andrea   434

Sousa e Santos, Anderson Carlos   225
Souza, Vinicius M.A.   10
Stefanovskiy, Dmitry   476
Stokkenes, Martin   385
Štolc, Svorad   175
Suykens, Johan A.K.   208

Tejada, Javier   476
Terissi, Lucas D.   360
Therón, R.   334
Tobón, Catalina   426
Torres-Valencia, C.   443
Tozzi, Clesio Luis   266

Trujillo, Maria   68, 401, 417
Tshibanda, Luaba   434

Vácha, Pavel   84
Valenzuela, Ricardo   101
Valle, Carlos   409
Vargas Cardona, Hernán Darío   493
Vargas, Elizabeth   417
Vázquez, A.S.   60
Viñoles, Carolina   352
Vitrià, Jordi   326

Wainer, Jacques   101

Printed in the United States
By Bookmasters